The Engineering Science of Mineral Processing

The Engineering Science of Mineral Processing: A Fundamental and Practical Approach emphasizes the fundamentals of mineral processing to provide readers with a deep understanding of the science and phenomena that occur during the processing of ores. It also offers guidance on contemporary process implementation for improve sustainability through practical industry applications. It includes examples of dynamic simulations and practical execution of advanced software to guide operating plans to ensure optimal conditions that predict process constraints.

- Focuses on the science of mineral processing, including particulate systems, hydrodynamics, and physical chemistry.
- Discusses modeling, rheology, comminution, classification, flotation, and solid-liquid separation.
- Includes practical examples from real-world industrial applications.
- Provides information on dynamic process simulations and the application of digital twins in mineral processing plants to improve management, toward the eco efficient processing of ores.
- Details the future of mineral processing in the digital era.

Offering a balance between fundamentals and applications, this book will be of interest to researchers and industry professionals working to optimize mining, mineral, and chemical processing plants. It will also be of value to advanced students taking mineral processing and chemical engineering courses.

Fernando Concha A. was Professor Emeritus at the Water Research Center for Agriculture and Mining, CRHIAM Department of Metallurgical Engineering, University of Concepción, Chile.

Osvaldo A. Bascur is Principal and Consultant Fellow at OSB digital, LLC. He received his PhD in metallurgical engineering from University of Utah. He is author of *Digital Transformation in the Process Industries: A Road Map*, published by CRC Press/Taylor & Francis.

The Engineering Science of Mineral Processing

A Fundamental and Practical Approach

Fernando Concha A. and Osvaldo A. Bascur

CRC Press
Taylor & Francis Group
Boca Raton London New York

CRC Press is an imprint of the
Taylor & Francis Group, an **informa** business

Designed cover image: Osvaldo A. Bascur

First edition published 2024
by CRC Press
2385 NW Executive Center Drive, Suite 320, Boca Raton FL 33431

and by CRC Press
4 Park Square, Milton Park, Abingdon, Oxon, OX14 4RN

CRC Press is an imprint of Taylor & Francis Group, LLC

ISBN: 978-1-032-61419-9 (hbk)
ISBN: 978-1-032-61420-5 (pbk)
ISBN: 978-1-032-61421-2 (ebk)

DOI: 10.1201/9781032614212

Typeset in Times
by Apex CoVantage, LLC

To the memory of Professor Fernando Concha A.,
professor, colleague, friend, and mentor.

Contents

Preface

The objective of this book is to provide, in a unified form, the science and engineering science tools required to analyze the mineral processing field from a fundamental point of view. The main purpose is to use these tools to study several phenomena that occur during the processing of ores. The book emphasizes the fundamentals of one of the important fields of the mining industry, which is the processing of minerals to obtain valuable metals.

The book is intended to have a proper balance between fundamentals and mineral processing applications. This balance requires that fundamentals predominate, not only because they are needed, but because often they are not included in standard textbooks. We hope that the book will stimulate researchers to dig into the scientific basis of the many processes used in mineral processing.

We find that several books on mineral processing use science at an unnecessary elemental level, especially in the case of fluid dynamic applications. This work is planned as a graduate-level textbook, but because it analyzes in detail the fundamental phenomena and applications occurring during the processing of ores, it should be of great interest for undergraduate teaching and useful for practicing engineers.

Chapter 1, "Introduction to Mineral Processing," introduces the field of mineral processing and its development from unskilled labor to an industry governed by scientific disciplines and the concept of minerals and ore deposit. The stages of mineral deposit exploitation and the concentration process are briefly described, and several technologies are presented. The conceptual framework, methods, and language used to understand the fundamentals and to develop models of mineral processing operations are provided in later chapters. Here, the universality of hydrodynamics and physical chemistry phenomena occurring in the mineral processes are stressed.

Chapter 2 presents the fundamentals of modeling of operations in a mineral processing plant, establishing empirical and phenomenological models and clarifying their importance and the types and the quality of the mathematical models for physical phenomena.

Chapter 3 deals with the combination of hydrodynamics—so important for mineral processing and metallurgy. A detailed study of the Navier-Stokes equation and their limiting cases, potential flow, Stokes flow, and the theory of boundary layer, which forms the basis of the processes in a mineral processing plant or concentrator of the mining industry, are shown. Also included is motion of particles in fluids and sedimentation in diverse laminar flows. Convective diffusion in laminar and turbulent flows and models for the attachment of particles to bubbles are presented.

In Chapter 4, "Particulate Systems Technology," the size and size distribution of rocks play an important if not decisive role in the metallurgical efficiency of mineral processing of metallic and nonmetallic ores. Crushing and grinding equipment and their comminution circuits basically depend on the size of the ore fed to the circuit. In concentration processes, such as flotation and gravity separation, the size of the mineral particles establishes the boundary of applicability of each technique, and in solid-liquid separation processes, the particle size and shape influence their performance. In this chapter, we present the concept of a particle and collection of particles, their properties, and the ways to describe and measure them. Finally, methods to model the collection of particles are shown.

Chapter 5, "Physical Chemistry of Interfaces," discusses several important phenomena of particulate systems such as flotation, coagulation, flocculation, and thickening, which are studied with the objective of understanding the interactive forces between molecules and between molecules and particles. The DLVO, non-DLVO, and augmented DLVO theory are described, following with some applications to particle separation using froth flotation. DLVO theory, named after the four scientists Derjaguin, Landau, Verwey, and Overbeek, who developed the theory, has been the dominant paradigm for interpreting interparticle forces for more than half a century. DLVO theory is a theory of colloidal dispersion stability in which zeta potential is used to explain that as two particles approach one another, their ionic atmospheres begin to overlap, and a repulsion force is developed.

Chapter 6 covers suspension rheology of particulate systems, which is fundamental in all unit operations of a concentrator, where particles mixed with water at different concentrations behave in another way to their separate component. Applications of Bingham and Power law mathematical models lead to quantitative descriptions of the fluid mechanics variables of suspension velocity, percent solid by volume, particle size distribution, percent fines in the slurry, volume flow rates, average velocity, maximum velocity, wall shear stress, and shear.

In Chapter 7, "Particulate Comminution," several quantitative applications to grinding mills, such as ball charge motion, steel consumption, and ball charge optimization, are presented. The kinetics of grinding is discussed using population balances, and the empirical laws of continuous grinding are presented. The Bonds law is used to show the energy consumption in tumbling mills. Semiautogenous mills (SAG) and high-pressure grinding rolls (HPGRs) are briefly described using population balances. Finally, mass balances in closed-circuit grinding classification are described. Two examples of a steady state mass balance simulator (Optimill) and a dynamic closed-circuit grinding classification circuit (Dynamill) are discussed in detail. The importance of the dynamic simulation is presented to understand the effect of the sump/pump sizing with respect to the volume of the mills for design of operating control strategies in industrial operations.

Chapter 8, "Particulate Classification," gives a very detailed and profound analysis of the dynamics of the classification process with application to hydrocyclones, using the different hydrodynamics techniques, which were presented in Chapter 3, resulting in a theoretical phenomenological and mathematical model for hydrocyclone performance.

Chapter 9, "Particulate Flotation," introduces the sequential steps of conditioning of the solid particles, water, and air and the formation of air bubbles and froth. The attachment and detachment to bubbles and the transport of the mineralized bubbles to and from the froth is given. Each of these steps is based on hydrodynamics, surface chemistry of interfaces, and particle size distribution characterization of Chapters 3, 4, and 5. Flotation dynamics is presented next, including two phenomenological flotation models. The importance of the two-phase flow continuity equation as a restriction is presented because it has always been forgotten or neglected in the literature.

It is the most important strategy to control the hydrodynamics of a flotation vessel in an industrial plant.

Chapter 10, "Particulate Solid-Liquid Separation," shows the development of a phenomenological theory thickening, the sedimentation-consolidation process used, not only in the mineral industry but profusely in wastewater treatment, using hydrodynamics in the solid-liquid separation in diluted and concentrated suspensions. The processes of thickening and filtration are analyzed.

In Chapter 11, we present one of the important problems facing the mining industry today due to climate change—the scarcity of water and its management, which is essential for the processing of minerals. Copper mining is an excellent example of the challenges faced in industry today. The quantitative data on the type of water used, the amounts involved in each mining process, and the need for efficient technologies are provided from the Chilean experience.

Chapter 12, "Future of Mineral Processing in the Digital Era," is the final chapter. The mineral industry is following the trend of the Industrial Revolution. From (1) steam engine, (2) electricity, and (3) electronics emerges Industry 4.0 (also known as the *Digital Revolution*), which is an industry based on sciences, information technology, digitalization, computer power, and internet networks. This chapter discusses the effect of this tendency, which is starting to influence, and certainly will change, the future of mining. An example of a digital twin for application in mineral and water processing is presented showing an improved copper production rate with a large increase in water recovery necessary to feed grinding processes.

POST REVIEW AND EDITING

Early on, Fernando asked me to be his coauthor of the book, and after he passed away, his family asked me to assist with the final production of the book. This book was Fernando's dream from the time he finished his PhD thesis at the University of Minnesota. From the time that we started to work together while doing my engineering thesis, he always had this goal in his mind.

We worked together for many years in defining the foundation for providing a scientific approach for understanding the suspensions in sedimentation, which is the foundation of mineral processing. The rheological understanding of suspensions applies to all aspects of the wet processing of particulates systems in the process industries.

Today, with decreasing ore grades, more than ever, we need sustainable and efficient strategies to extract the valuable metals from minerals. Today, large amounts of energy and water are used in mineral processing, which creates large amounts of tailings.

The scientific approach to mineral processing is the foundation to build new processes and to operate plants more efficiently. These models will help in instructing the people and computers using artificial intelligence. A derivative of applying these models is the development of improved dynamic process simulators for training operators to run the industrial plants.

We hope that this book will provide a foundation for process analysis, design, and optimization of mineral processing plants in the Digital Age.

I am very honored to have been selected to be Fernando's coauthor of this scientific approach to mineral processing. I also wish to acknowledge the support of Fernando's widow, Margarita, who has been a sustaining presence during this endeavor—the publication of Fernando's opus.

Osvaldo A. Bascur
The Woodlands, Texas
September 2023

Acknowledgments

Several parts of this book are the result of research by Fernando Concha A. and his research group at the University of Concepción, Chile, with important contributions from M.C. Bustos and R. Bürger in Stuttgart, Germany, and the collaboration of Dr. Nickolaj Rulyov of the National Academy of Sciences of Ukraine; coauthors Drs. Arturo Barrientos Daniel Sbarbaro, Fernando Betancourt, Christian Goñi, Pablo Cornejo, and Marcelo Vergara; and many graduate and undergraduate students at the University of Concepción, who collaborated with their doctoral theses, Jorge Menacho, Oswaldo Bustamante, Elias Ovalle, Jaime Collantes, Julio Miranda; and master's theses, Bruno Chiné and Juan Daza; and engineering theses: Osvaldo A. Bascur, Arturo Barrientos, Arturo Christiansen, Jorge Montero, Elizabeth Alarcon, Jorge Etcheberry, Miguel Peña, Gonzalo Rojas, Guillermo Parga, Katherine Arellano, Samuel Vergara, Alonso Pereira, Hugo Muñoz, Diana Celi, Giovana García, and Daniel Pérez.

Thanks to Dr. Osvaldo A. Bascur from OSB Digital LLC, for his contribution in writing Chapter 12 and contributing to the chapters on comminution, classification, flotation, and solid-liquid separation, and in providing me with abundant literature in several areas of mineral processing such as comminution, classification, flotation, and water recovery. Thanks also to the faculty members Leopoldo Gutierrez, Fernando Betancourt, and former professor Christian Goñi at the Department of Metallurgical Engineering at the University of Concepción for taking time from their work to read chapters of the book and make useful suggestions.

FROM THE COAUTHOR

As the coauthor, for my part, I am very thankful for the collaboration with Dr. Arturo Barrientos, Mario Parada, Ivan Flores, Dr. John A. Herbst, Dr. Douglas W. Fuerstenau, Dr. J.D. Miller, Dr. Raj Rajamani, Dr. Edward Freeh, Dr. Karl Sastry, Dr. Heinrich Schubert, Dr. William Pate, Dr. Juan Luis Sepulveda, Dr. Jaime Sepulveda, Dr. H.Y. Sohn, Dr. Milton Wadsworth, Dr. J. Pat Kennedy, Dr. Daniel Hodouin, Dr. Alfredo Gordon, Eduardo Canales, Fernando Betancourt, Freddy Retamal, Simon Espinosa, Ricardo Montecinos, Nelver Benavides, Ted Gorry, Allison Buenemann and Allison Shatkin.

Special thanks are extended to my editor, Terese Platten, for her coaching and review of the manuscript, Prof. Douglas W. Fuerstenau for suggesting a modification and the post editing and indexing with Jayachandran Rajendiran.

1 Introduction to Mineral Processing

1.1 INTRODUCTION

The great importance of minerals in human civilization is recognized from the fact that no human activity is possible without their participation. In the prehistory, human progress has been attested by times marked by the conquest of metals and the techniques to obtain them from their minerals. The Ages of Stone, Bronze, and Iron are the consecutive steps man has taken in the technological conquest of Earth's crust. At present, any scientific, technological, and industrial activity necessarily uses minerals or their derivatives such as construction rocks, fertilizers, metals and their alloys, or complex salts produced directly from minerals, as well as cement, glass, ceramics, pigments, and many other salts, bases, or acids used in the industry.

Since mining is an important source of work and foreign source of income for a country, the impact of minerals on our civilization is not only of technological order but also influences the different branches of a country's social, political, and economic life. Minerals' growing influence on life and development stems from two fundamental facts, the steady population growth and economic expansion. Faced with the depletion of known mineral reserves, man has used its intelligence to make better use of them, to find new reserves of minerals or to replace those depleted materials, diversifying the possibilities of supply and use.

Mineral processing is not a modern undertaking. Whenever an ore was dressed to obtain a concentrate, two processes were used in an inseparable form—crushing and washing. There is evidence that in the IV Egyptian dynasty, some 2,500 years BC, the ancient Egyptians dug for and washed gold. There is also evidence of gold washing in Sudan in the XII dynasty. Nevertheless, the earliest written reference for crushing and washing in Egypt is that of Agatharchides, a Greek geographer who lived 200 years before Christ. Ardaillon, author of the book *Les mines du Laurion dans l'antiquité*, described in 1897 the process used in the extensive installation for crushing and washing ores in Greece between the fifth to the third century BC. A.J. Wilson (1994) describes in his book *The Living Rock*, mining of gold and copper in the Mediterranean from the fall of the Egyptian dynasties right to the Middle Ages and the Renaissance (Concha 2014; Concha and Bürger 2002).

The development of mineral processing from unskilled labor to craftsmanship and eventually to an industry governed by scientific discipline is largely due to the Saxons in "Germany" and Cornishmen in England, beginning in the sixteenth century. An international exchange of technology began between these two countries and continued for a long time. But it was in Saxony where Agricola wrote his "De Re Metallica," the first major contribution to the development and understanding of the mining industry, published in Latin in 1556, and shortly after translated to

DOI: 10.1201/9781032614212-1

German and Italian. Agricola's book had a tremendous impact not only in the mineral industry but also on society in general and continued to be the leading textbook for miners and metallurgist outside the English-speaking world for at least another 300 years. Apart from its immense practical value as a manual, the greatest influence of *De Re Metallica* was in preparing the ground for the introduction of a system of mining education, which, with various modifications to suit the local conditions, was later to be adopted internationally.

The first book on mining technology ever written, Georgius Agricola (1556), and published in the city of Freiberg in Germany shows that most procedures for recovering minerals from rocks use water (Figure 1.1). Reducing rock size and classifying its product, the separation of the useful material from the gangue and its concentration, the transport of concentrates and tailings are more efficient using water. In all these processes there is interaction between mineral particles and water, which allows the recovery of valuable mineralogical species. During the processes the particles move within the fluid, and due to the effect of the relative velocity between the solids and the fluid, the separation of the particles by various forces such as gravity, centrifugal force, oscillations, and adhesion forces is achieved (Wilsdorf 1987).

Agricola's book describes several methods of washing gold, silver, tin, and other metallic ores (Figure 1.2). He describes settling tanks used as classifiers, jigs, and thickeners and settling ponds used as thickeners or clarifiers (Figure 1.3). These devices operated in a batch or semicontinuous manner. A typical description is as follows (Agricola 1556):

> To concentrate copper at the Neusohol in the Carpathians, the ore is crushed and washed and passed through three consecutive washer-sifter. The fine particles are washed through a sieve in a tub full of water, where the undersize settles to the bottom of the

FIGURE 1.1 Landscape with mine and concentrator in Germany in the first half of the fourteenth century.

Source: Bartels 1994.

FIGURE 1.2 Washing and settling according to Agricola.

> tub. At certain stage of filling of the tub with sediment, the plug is drowned to let the water run away. Then, the mud is removed with a shovel and taken to a second tub and then to a third tub where the whole process is repeated. The copper concentrate which has settled in the last tub is taken out and smelted.

It is evident from these references that by using the washing and sifting processes, from the ancient Egyptians and Greeks to the mediaeval Germans and Cornishmen knew the practical effect of the difference in specific gravity of the various components of an ore and used sedimentation in operations that can now be identified as classification, clarification, and thickening. There is also evidence that in the early years no distinction was made between these three operations.

1.2 DEFINITION AND CLASSIFICATION OF MINERALS

Minerals are inorganic bodies with defined chemical composition and certain characteristic physical properties, originated by natural processes in the Earth's crust. The

FIGURE 1.3 Settling tanks described by Agricola.

inorganic composition of minerals is an indispensable condition because it immediately eliminates all flora, fauna, and bodies of inorganic composition but product of life such as animal bones, corals, pearls, shells, and others. The defined chemical composition, based on stochiometric relationships, provides its identity to minerals and individualizes them, which is complemented by physical properties such as its crystalline system, color, hardness, specific weight, electrical conductivity, and magnetic permeability among others. Finally, natural occurrence is a requirement that limits the origin of minerals to geological processes, definitively rejecting any artificial product such as metal alloys, synthetic stones, and other inorganic products, in which man has participated. Most minerals are in solid physical states, such as copper and iron, which also exist in a liquid state, such as water and mercury, and in gaseous state such as carbon dioxide and sulfur dioxide.

More than 1,500 different mineralogical species are known, and of these only 20 are chemical elements known as *native* minerals, with the rest consisting of more than one element. Mineralogical species are crystalline solids based on anionic and cationic chemical groups, which relates them to their geological origin and deposits. In mineralogy, the science that studies minerals and their properties, the minerals are classified according to anionic groups, such as oxides, carbonates, sulfates, and others, of which a total of 12 are distinguished. However, this classification does not fully satisfy those fields of application in which technological and economic interest predominates. In mining and extractive metallurgy, minerals are classified according to their practical utility, method of exploitation, and their specific applications. Minerals with economic value are called ores, and those without economic value are rocks. Minerals are subdivided into metallic minerals as sources of metals and industrial minerals that are used to obtain nonmetallic materials.

1.3 ORE DEPOSITS AND MINERALS

Ores are found in deposits that are isolated and very scarce points in Earth's crust. They are complex formations, both in their physical state and in their chemical composition. The physical complexity is due to the dissemination or intimate mixture between the minerals and the matrix rock, while the chemical complexity comes from the numerous mineralogical species that make up the deposit. Mineral deposits originate during geological processes, by the movement of mineral masses in a fluid or solid state, compact or dispersed, caused by changes in temperature or pressure in Earth's crust.

The deposits are classified according to the main geological process that participated in their formation. If the deposit was formed simultaneously with its matrix, they are called primary, and if the deposit remains in its original formation or if it has undergone some subsequent changes, those are called secondary. Knowledge of the geometric shape of the deposit and its dissemination is of great importance in mining extraction and subsequent processing as they indicate the correct method of exploitation and concentration. There are two types of minerals with respect to dissemination, those with high and almost exclusive concentration of useful minerals and those dispersed with different sizes in the rock.

The *grade* of a deposit, that is, the percentage of useful species contained in the rock, can be *low*, *medium*, or *high*. Low-grade minerals are those that need concentration before their metallurgical or chemical treatment and are characterized. They must be exploited on a large scale. Medium-sized-grade minerals should be concentrated before their metallurgical treatment, but economic profits are much higher. High-grade minerals lend themselves to direct metallurgical treatments.

1.4 STAGES IN MINERAL DEPOSIT EXPLOITATION

The exploitation of a mineral deposit has two stages, the exploitation itself of the ore and the *processing of the extracted ore*. The first is the search and discovery of the site, scientific and technological area that corresponds to the field of geology, which for its purposes uses methods such as geophysics, geochemistry, geobotany, and mineralogy. Essentially, this stage consists of two types of works: prospection with the purpose of the discovery of the deposit and exploration, which include the definition of the site, its valuation and exploitation. If the ore is close to the earth surface, the deposit is called *open pit mine* (Figure 1.4), and if on the contrary they lay deep in the ground, the deposit is called *underground mine* (Figure 1.5).

FIGURE 1.4 Escondida (BHP), the largest open pit copper mine in the world.

FIGURE 1.5 El Teniente (Codelco), the longest underground copper mine with 4,500 kilometers of tunnels.

1.5 MINERAL PROCESSING TECHNOLOGY

From the reception of the ore from the mine to the delivery of the metal concentrate to chemical or metallurgical for treatment or to the world market, the process to recover the valuable part of the ore includes technological and scientific activities. *Mineral processing technology* is the field of engineering dedicated to process the ore using their many physical and physicochemical properties such as mineralogic composition, particle size and distribution, mass density, surface charge, hydrophobicity, concentration, and several others that make it possible to obtain from them a concentrated material in terms of valuable species. The place in which these processes are implemented is called a *concentrator*.

The concentrator receives ores containing rocks that can measure from a few centimeters to blocks of one meter. The first step in treating the ore is to reduce it in size to that required to the subsequent processes. Reducing larger block sizes to a size of approximately 5 mm is called *crushing*. This operation is usually performed dry and in two or three stages. The reduction in particle size less than 5 mm receives the name of *grinding* and can be usually achieved in a dry or wet environment.

Size reduction is expensive because it consumes machinery and grinding media requiring a great amount of energy, around 80% of the total energy used in the concentrator. Consequently, grinding beyond the particle size strictly necessary for the next operation should be avoided. This requirement introduces the next step called *classification* that is an integral part of *size reduction circuits*. Classification removes from size reduction step any material that has already acquired the necessary size to send it to the next step of *concentration*. Classification like size reduction can be performed in wet conditions in hydro-classifiers or in dry equipment such as screens, depending on the place of the circuit where is located.

The objective of concentration is to separate from the ore the minerals from the barren rock, called *tailings*. To perform this task, the process makes use of the differences in physical or physicochemical properties between the several minerals and rocks contained in the ore. Some of these properties are the specific weight, magnetic permeability, electrical conductivity, hardness and color of minerals, and hydrophobicity, this last in the process of *flotation*. In general, these operations do not change the chemical identity of mineralogical species. In this sense, mineral processing differs from the field of *extractive metallurgy*, which with their hydro-, pyro-, and electrometallurgical processes inevitably change the chemical identity of the mineral.

Concentrates and tailings are suspensions in water called *pulps* of solid concentration from 20% to 60% solids. Concentrates, before being sent to smelting or transported to the world market, must be further dewatered by *filtration* to obtain a product of around 8% to 10% humidity. Water from tailings must be recovered and recycled by *thickening*. This solid-liquid separation step is crucial since, due to climate change, water is scarce in most parts of the world in the areas where mines are located. Tailing flowing out of the thickener are transported to *tailing dumps*, large reservoirs where the barren material are deposited. These dams contain around 45% water, which should also be recovered.

The transport of material, ores, minerals, pulps, and water must be between several processes, crushing, grinding, classification, concentration, dewatering, and

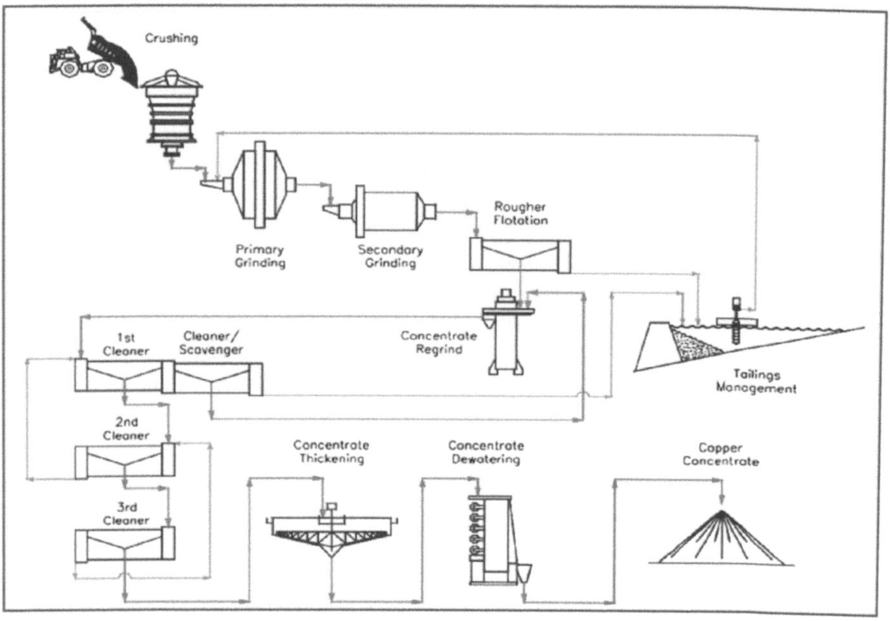

FIGURE 1.6 Mineral processing plant integrated process flow diagram.

tailing ponds are essential and form part of the knowledge needed to process the ores (Figure 1.6).

1.6 THE SCIENCE IN MINERAL PROCESSING

When engineers study macroscopic problems occurring in mineral processing, it is convenient to ignore the atomic and molecular nature of matter and postulate that solid or fluid bodies are continuous. This theory of *continuum mechanics* is equivalent to arguing that certain properties of matter can be continuously allocated in space. The reason why the mechanics of the continuous medium can be successfully applied to practical problems, in circumstances where we all know that matter is composed of atoms and molecules, is that the scales of these problems are much larger than the molecular scale and the time involved much longer than the vibration period of molecules. As a result, continuum mechanics constitute a fundamental part of the science needed to describe the fundamentals of mineral processing.

Continuum mechanics is the basis to describe the deformation and flow of solids and fluids; therefore, to understand the stresses during deformation and breakage of ores and the reduction of size by grinding, the science of *mechanics* is essential. In the case of fluids, solid-fluid, part of *fluid dynamics* is essential. Since in general particles in fluids are well mixed, the concept of *diffusion* and *turbulent flow* are important. These particles usually are electrically charged, and their values depend on pH; therefore *surface-chemistry* is part of the required knowledge. This same knowledge is applied in the phenomena of *hydrophobicity*, the basis of flotation, the process of

Aerial view of a Mineral Processing Plant	
Grinding	
Classification	

FIGURE 1.7 Global and individual process areas in a mineral processing plant.

Froth Flotation	
Thickening	
Filtration	

FIGURE 1.7 Continued

Tailings Pond	

FIGURE 1.7 Continued

recovery of the mineral species in a three-phase flow of solid water and air bubbles and of *coagulation* and *flocculation* and finally the transport of pulps between processes and from processes to deposits require understanding of *rheology*, the flow of high viscosity suspensions.

1.7 MINERAL TECHNOLOGY SUMMARY

The objective of this book is to present the scientific foundations that forms the basis for understanding the current technology applied to mineral processing and to help industrial operators, designers, and researchers to develop new technological tools to improve production processes. Alongside the development of these scientific bases, examples of how to use them in special cases will be provided. It will also show the current state of development of the main stages of processing a mineral, such as size reduction, classification, flotation, solid-liquid separation, rheology, and transport of pulps. It will provide a view of the use of the process models in the Digital Era with a view of the future of mineral processing operations assisted by predictive analysis and artificial intelligence. Competence centers will manage the mine, production, energy, water, and shipping in an integrated manner towards a sustainable eco-efficient beneficiation of low-grade ores.

REFERENCES

Agricola, G. 1556. *De Re Metallica*. Translated by H.C. Hoover and L.H. Hoover, 1950. New York: Dover.

Ardaillon, E. 1897. *Les mines du Laurion dans l'antiquité*. Paris: A. Fontemoing.

Bartels, C. 1994. Werk und wirkung: Bergbau un hüttenwessen. In *Georgius Agricola Bergwelten, 1494–1994, Bergbau der Agricola-Zeit, Europäische Montanwirtshaft zu Beginn der frühen Neuzet*. Essen, Germany: Glöchauf Verlag. p. 175.

Concha, F.A. 2014. *Solid Liquid Separation in the Mining Industry*. Cham: Springer.

Concha, F.A., and Bürger, R. 2002. A century of research in sedimentation and thickening. *KONA Powder and Particle* 20:38–70.

Wilsdorf, H. 1987. *Kulturgeshichte des Bergbaus: Ein Illustrierter Streifzug durch Zeten un Kontinente*. Essen: Glöchauf Verlag.

Wilson, A.J. 1994. *The Living Rock: The Story of Metals Since Earliest Times and Their Impact on Civilization*. Cambridge: Woodhead.

2 Modeling of Mineral Processing

2.1 INTRODUCTION

In trying to establish the origin, the types and the quality of mathematical models of physical phenomena, it is inevitable to become involved in a philosophical and a linguistic discussion. Empiricism, rationalism, and phenomenology are terms used to describe philosophical lines of thought that were finally adopted in science.

2.1.1 Empiricism, Rationalism, and Phenomenology as Basis for Modeling

In philosophy *empiricism* is the attitude of mind that emphasizes the part played by experience in acquiring knowledge against that part played by reasoning. An empirical doctrine states (1) that we have no ideas other than those derived from sense experience and (2) that statements, other than those of logic, can be known to be true only from experience. This attitude has been opposed to *rationalism*, the belief that some ideas are independent of experience and that some nonlogical truths can be known in a manner which prevents their being discarded by experience. All philosophers have regarded the truths of logic as necessary; that is, as capable of being known with a certainty beyond the possibility of refutation by experience. Rationalist have claimed this status for some metaphysical, moral, and even scientific truths. An empiricist is one who only trusts experience and who is skeptical of methods which seem theoretically sound but have not been proved in practice. If the word experience is understood in its usual sense, empiricism is almost trivially true as a theory of knowledge underlying both common sense and scientific beliefs for, however great the part allowed to theory construction in science, science is distinguished from metaphysics by the fact that it admits no statement which cannot, directly or indirectly, be tested in experience.

Phenomenology is the name for a philosophical movement whose primary objective is the direct investigation and description of phenomena as consciously experienced, without theories about their origin and as free as possible from preconceptions and presuppositions. Followers of phenomenology will hold that it is possible to obtain insight into the essential structures and relationships of phenomena based on careful study of concrete examples supplied by experience or imagination. In contrast to empiricism, with which it shares the unconditional respect for experience, phenomenology does not restrict these data to the range of sense experience but admits on equal terms non-sensuous data such as relations, that is, logical bonds between things, and values, embracing not only morally good but also such concepts as the beautiful, as long as they present themselves intuitively. Consequently, phenomenology does not reject generalizations and recognizes, in addition to the analytic a

DOI: 10.1201/9781032614212-2

priori and synthetic a posteriori, knowledge of synthetic a priori relationships based on insight, within the essential information of experience. In contrast to rationalism, which stresses conceptual reasoning at the expense of experience, phenomenology insists on the intuitive foundation and verification in experience for all formal concepts, and especially for all a priori claim. In contrast to an analytical philosophy, which uses the immediately given in all its complexity, phenomenology resists all reinterpretations of the given, analyzing it for what it is and on its own terms.

2.1.2 A Priori and Empirical Science

One important question is the distinction between empirical science and the a priori science. *Empirical science* can treat the physical world solely on the level of phenomena which can appear to our senses or by the extension of the sense-experience by sophisticated instruments that have proved, for example, the discontinuity of matter, the movements of atoms, or the bending of light, none of which could be observed directly with our senses. Empirical science uses the most recent discoveries in all their complexity to describe nature.

In contrast to empirical science, it has become increasingly apparent that a necessary condition for scientific observation is the postulation of certain basic data, things given or taken for granted. Science based on these beliefs are called *a priori science*. The result of theories such as the theory of relativity or the principles of uncertainty in atomic physics, which express the belief that observation affects the phenomena observed, run against the whole view of scientific empiricism and are of great importance for the consideration of the ultimate validity of science demonstration. It should not be assumed that, because exact predictions of physical phenomena may be impossible, physical action is ultimately undetermined. The cause of indeterminacy is in the methodology of research rather than in nature. An a priori scientific theory can generally be safely developed regardless of them. It should, however, be remembered that scientific conclusions are essentially partial or temporary and that they can at best describe only part of the physical phenomena.

2.1.3 Phenomenological Theory

The physical phenomena in nature cannot be described directly, and any attempt to do so begets a model. There are two of these useful models, the corpuscular and the phenomenological model. We all know that matter is composed of molecules and that, although they are not captured by our senses, they can be defined as the smallest proportions of a matter that exhibit characteristic properties. Molecules are made up of atoms and these of nucleus and electrons. The nucleus, in turn, is composed of other elementary particles, the behavior of which has only been partly elucidated. The science that studies and relates these particles is quantum mechanics.

The fact that many physical phenomena are independent of detail in the behavior of elementary particles, which is evidenced by the very similar attitude of very different corpuscular structure materials, has led to the formulation of a theory of the continuous medium. This is a medium that does not lose its properties when it is indefinitely divided. The continuous medium, or field, can be a seat of matter, movement, strength, and energy. Those theories that are expressed by fields are called phenomenological theories because they represent the phenomena of physical reality without attempting a description in corpuscular terms or other elementary magnitudes.

The concept of continuous medium can be illustrated by a scheme such as Figure 2.1, where the abscissa represents the volume of the medium considered and then ordinate any property on that volume, for example, density. For regions with volumes smaller than V^* it is very likely to find places in space without containing matter. This is a consequence of the movement of the molecules in this region that, for this reason, is called the *domain of molecular effects*. If the scale of the phenomena of our interest falls to the right of the pint V^* in the plot of figure, a point called *limit of the continuum*, the *mechanic of the continuous medium* gives results that can be used safely.

Although corpuscular and phenomenological theories are mutually contradictory as a model of physical phenomena, they can be agreed by statistical mechanics. From this point of view, it is possible to treat particulate systems of any kind, including those of point mass, by obtaining statistical averages which, it has been shown, satisfy the same equations as a continuous medium. As these results are accurate and not approximations, predictions of the general theory of the continuous medium can be considered accurate from the point of view of statistical averages. The results will be accurate only if the general theory is considered since any constraint will necessarily produce an approximation.

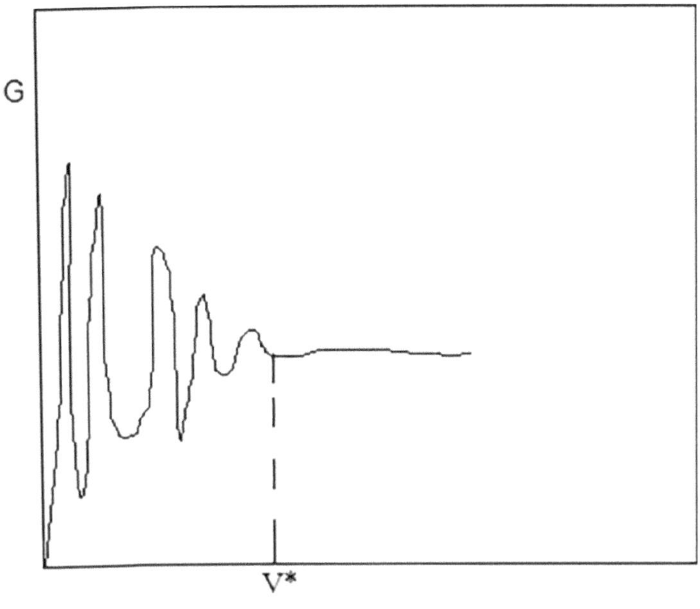

FIGURE 2.1 Concept of a continuous medium showing the change of property G versus the volume V^* being in consideration.

Continuum mechanics, like classical mechanics from which it is part, is based on the concepts of *body*, *Euclidean space-time*, and a *force system*. These three elements are linked through what is called a *dynamic process*, that is, a movement linked to a system of forces that must comply with two *axioms*, called *laws of mechanics*. These axioms are the laws of conservation of *linear momentum* and *angular momentum*. They are valid for all kinds of movements of all kinds of bodies. Above those laws the principle of *material indifference*, also known as *principle of material objectivity*, must be added. This law implies that the properties of a material are the same for all observers, that is, they are independent of the methods used to measure them. This property must be especially recognized because the laws of mechanics change when changing the observer called *frame of reference*.

The objective of formulating a field theory is to unify the study of dynamic processes by establishing conservation laws or balance which relate the variables involved. For regions of space where variables are continuous, the basic principles appear in an *integral* form but can be transformed into an equivalent *differential*. These relationships are called *field differential equations*. In regions where discontinuities exist, they give rise to *discontinuity conditions*. To complete the description of a continuous medium, axioms that define ideal materials must be added. These relationships called *constitutive assumptions* restrict the type of dynamic process a body can undergo. Constitutive assumptions which are expressed in the form of a functional relationship between *kinematic variables* and *dynamic variables* are called *constitutive equations*. Field equations together with constitutive equations, discontinuity conditions, and initial and boundary conditions define a theory that predicts the behavior of a particular material in specific problems.

2.2 MATHEMATICAL MODELS

While it is common that mathematics is needed in solving specific problems, many scientists regard the basic principles of the theory originating in intuition as the first step leaving mathematical treatment as belonging only to a later stage in the development of the theory. When finally, the mathematics begin, it may be entirely precise, but setting up the theory is for them an extra-mathematical operation. Truesdell (1984) indicates that the characteristics of a good theory are that the physical concepts themselves are made mathematical at the outset, and mathematics is used to formulate the theory. Although it is true that a priori science frequently defers the sense-experience, uses mathematical models, and then returns to the prediction or demonstration of phenomena.

> A theory should not collapse if some of its applications fail. It is possible that the description of some details is not sufficiently precise, but the general theory should remain valid. Experimental facts should not be disregarded, but the *danger is in the disrupting and disorganizing effect of masses of experimental data*. The scientist must order. The role of a theory is not to adjust fudge factors to data, but to draw much from little.
>
> (Truesdell and Noll 1965; Truesdell 1966a, 1966b)

It is well-known that all fields of engineering have a common background in the physical sciences. However, engineering shares not only the basic sciences such as chemistry, physics, and mathematics, but also disciplines which span subdivisions of science, the engineering sciences, such as thermodynamics, kinetics, and transport processes. With the current trend towards fundamental approaches to engineering, it is not surprising that more and more such disciplines have become of interest to engineers in common. The study of all phenomena involving the transport of mass momentum and energy has simplified and unified the study of processes involving fluid mechanics and of mass and heat transfer. Thus, the field of *transport phenomena* as a distinct engineering subject was introduced in the sixties and was adopted by chemical, mechanical, and metallurgical engineering. Mineral processing has profited by the application of transport phenomena to the study of the dressing of ores. In more recent years, thermodynamics and transport phenomena, jointly, are regarded as one important field of engineering science, the thermo-mechanics of continuous media.

When macroscopic phenomena are studied in engineering, it is convenient to use phenomenological theories. They unify the study of dynamic processes, establishing conservation balances that relate variables describing the phenomena. The basic principles, which are stated in integral form and are called *macroscopic balances*, may be transformed into a local form for regions of space where the variables are continuous. These differential equations are called *local field equations*. At discontinuities or singular surfaces, the macroscopic balances yield the *jump conditions*. To complete the picture, relationships describing the different materials must be incorporated into these relations; they are called *constitutive equations*. The local field equations with the jump conditions, constitutive equations, and appropriate initial and boundary conditions constitute a *phenomenological theory* that can describe the behavior of a material in a specific problem.

Rarely can the local field equations be applied directly in engineering problems to obtain a mathematical model of a process. Usually there is a lack of knowledge that prevents the development of detailed models. In other occasions there is no need to have a model containing detailed information.

2.3 TYPES OF PHENOMENOLOGICAL MODELS

Depending on the detail needed, phenomenological models can be classified into six categories as described in the following sections. It is our belief that all models developed with the claim of being based on first principles are phenomenological models or mechanistic models and that it is a sufficient classification of models.

2.3.1 Dynamic Models in Laminar Flows

Dynamic models in laminar flows are based on the direct application of the local conservation equations to specific problems. It is very rare when this category can be applied to a whole process or to an equipment. One such application is the dynamic sedimentation process. More frequently, they are applied to subprocesses within an operation. Examples of these are the models for the motion of a single particle in a fluid with a given unperturbed velocity field, the diffusion to, or from, a particle with chemical reaction at the surface, and the collision and attachment of a particle to a bubble in the flotation process.

2.3.2 Dynamic Models in Turbulent Flows

In most phenomena of interest to mineral processing we are confronted with turbulent flows. These flows are characterized by the fact that the transport of their properties is due not only to molecular diffusion and laminar convection, but by eddy diffusion, that is, by macroscopic mixing of portion of fluids. In this case, the local transport equations are similar to those of laminar flow, but the variables must be considered as averages in a certain time period, and the constitutive equations must include effective coefficients. This is the case, for example, of the recent phenomenological models of hydrocyclones. The main problems in these models is to describe properly the turbulence. Several empirical models have been proposed for the turbulent fluctuation of the fluid in a vessel, for example, the mixing length of Prandtl, the k- and k-RNG methods, and the Reynold stress models DSM and ASM. The recent development of accurate methods for velocity measurements, such as the laser-Doppler velocimeter, had brought a better insight to turbulence modeling.

2.3.3 Kinematic Models

Kinematic models are widely used when limited information exist on momentum transport and when this information is replaced by a constitutive equation based on experiments. Frequently these models are one-dimensional with a constant velocity profile. Examples are the batch and continuous Kynch sedimentation processes and the distributed parameter model of grinding in a rotary mill.

2.3.4 Macroscopic Models

In macroscopic models, the original integral balances are evaluated with the theory of mean value. They contain the least information and are suitable when we are satisfied with an inlet-outlet relationship. Two of the most popular such models are the so-called lumped parameter kinetic model of grinding in a rotary mill and the Coe and Clevenger's method of thickener design.

It is important to spend some time to establish the position of *population balance* models in relationship to phenomenological models (Himmelblau and Bishoff 1968). Population balance models are based on the application of statistical mechanics to a system of particles of any kind. If the particle number is high, the average properties of groups of particles obtained from statistical theory form a continuum and can be treated with the same tools as phenomenological models. Strictly speaking, population balances are phenomenological models if the particles they treat are not the elementary particles constituting an atom. However, they usually contain information on the particles, such as size, shape, rate of growth, or rate of wear, not available in the usual continuum mechanical models. It is common in mineral processing to use the name population balance model for some macroscopic model of grinding in a rotary mill. It is true that a macroscopic model for comminution in a ball mill can be derived directly from a population balance model, but this is not the way it is usually done. Real population balance models for comminution have been used recently in connection with semiautogenous grinding (SAG) mills and ball wear in tumbling mills. In addition, population balance models have been used to describe the pelletization, flotation, and flocculation processes. All models require experimental data and estimation of coefficients or parameters using least squares methods (Himmelblau 1970; Kelleher et al. 2015). In addition, process data requires to be reconciled prior to the estimation of process parameters.

2.3.5 Mechanistic Models

The term *mechanistic model* has never been carefully defined but can be used for those models based on an a priori science other than field theory. Examples of mechanistic models would be the recent work based on discrete elements for the motion of balls in a tumbling mill.

2.3.6 Hybrid Process Dynamic Models

Data-driven discovery is revolutionizing the modeling, prediction, and control of complex systems. With modern mathematical methods, enabled by unprecedented availability of data and computational resources, we are now able to tackle previously unattainable challenge problems. The combinational of process dynamic models with new data science is enabling to parameterize models for online prediction to avoid violating process constraints in the real time in the process industries (Brunton and Kutz 2019; Kelleher et al. 2015; Raschka 2016). Early work in mineral processing industry was documented in Control in the 1980s and 1990s by Herbst et al. (1984) and by Rajamani and Herbst (1990). Recently, many novel approaches can be found in the proceedings of the International Federation of Automatic Control (IFAC) on mining, mineral, and metal processing.

Chapter 12 shows practical examples in the mineral processing industry using a process historian with advanced data analytics using hybrid process models (Bascur et al. 2022; Bascur and Retamal 2023).

The use of scientific methods in process analysis is not new, but we certainly can observe a tremendous need for improving the process efficiency and optimization of industrial process plants. By scientific methods we mean the gathering of information, the analysis of this information by suitable techniques, synthesis and decision-making, all utilizing a mathematical framework. The increased use of these methods in the rapid evolution of computation power of computers, networks, and access to large amount of data collected by process historians.

The systems approach brings together the techniques of engineers of varied backgrounds and operating personnel who are the system's users. The goal must be one of an overall approach based on the functional requirements the system must satisfy.

The general strategy of analysis of complex processes follows a typical set of steps.

1. Formulation of the problem, objectives, and value criteria. Performance requirements.
2. Preliminary inspection and classification of the process to break it down into subsystems.
3. Preliminary determination of the relationships among the subsystems.
4. Analysis of the variables and relationships to provide as simple and consistent set as possible.
5. Mathematical modeling of the relationships in terms of variables and parameters.
6. Evaluation of how well the model represents the real process, using judgement to integrate nonmathematical representations (classification of operating conditions [Bascur 1988, 2019 and 2020]).
7. Application of the model for either process design and optimization or design of process control strategies.

Figure 2.2 shows the current strategy to use process data in the development of process models. Using an object-oriented approach, an algorithm is designed to evaluate the process operating conditions for all process units involved in the system. Characterization of the process operating conditions and estimation of the parameters to create the predictive models. Test of the current model against the new data available. The most important aspect of the design of these process models is to have a human interface to present the results of the online augmented data for decision-making, continuous improvement, cause and effect analysis. Deliver the process system to the users for the training and use for process analysis and optimization.

The five steps that are required to achieve an industrial process model are shown in Figure 2.2. The first step is to build the necessary data access based on business knowledge. This first step is critical, and it requires a current process flow diagram (PFD) to build the database for creation of the data model (Turton et al. 2018). The PFD should include the critical measurements to analyze the data and to build the process events generate the gross error detections and transformation of time series

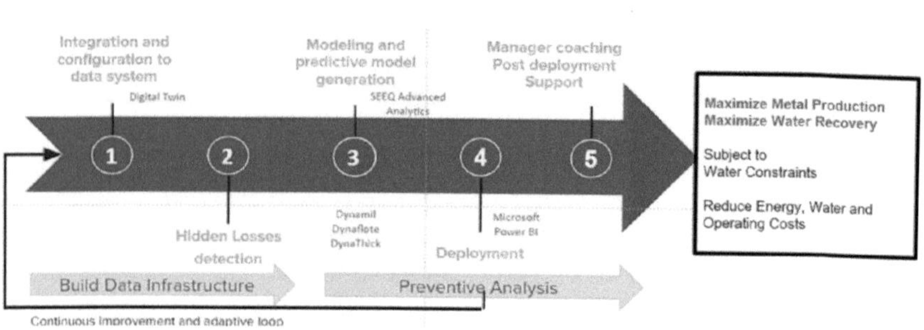

FIGURE 2.2 The five steps in providing a process model for process improvement.

data into information at the desired degree of detail. The second step is to classify the operational data to eliminate the data that will be useful to obtain the parameters for the hybrid models. The third step is using the advanced analytics tools to parameterize the process models required to understand the process outputs, disturbances, and observed variables. The fourth step is the design of the human machine interface to convey the results to the people, to the system for further analysis using Big Data tools such as Microsoft Power BI. And finally, the training, coaching, and continuous improvements of the models for constant adaptation of the changes that occur in the processes of continuous improvements and innovations (Bascur 2020).

Real problems are usually ill-defined and do not lend themselves easily to precise specification. One major premise involved in all such process analysis is that the entire process cab be broken down into distinguishable subsystems (elements) and that relationships exists among the subsystems, which, when assembled together, can simulate the process. Examples are provided in Chapters 7 and 9 through 12.

The other major problem is that the data is not perfect and requires pre-treatment prior to use in the estimation of coefficients. This is reason to use gross-error detection algorithm with real time data to produce robust online process models.

2.4 STATE OF THE ART OF MODELING IN MINERAL PROCESSING

2.4.1 Comminution

Comminution is one of the fields in which empirical, phenomenological, and mechanistic models have been developed and are used independently or are integrated in comminution systems.

In the last part of the nineteenth and in the first part of the twentieth centuries, several empirical models of comminution were developed. Their aim was to relate the reduction ratio with the energy consumed and, in this way, to select the proper size-reduction machines. Kick, Rittinger, and Bond "theories" are such models that are used until today. Beginning in the fifties, acquiring its full form in the sixties, and explored and applied since then to crushing and grinding in different comminution machines, the phenomenological models of comminution, known as "population balance models," developed into the successful "lumped parameter model," a macroscopic version and the not-so-used "distributed parameter model," a kinematical version. Until today, no complete dynamical model has been proposed for comminution.

The challenges today are, on the one hand, to transform the population balance model for SAG mills, which has been already developed into a working tool for automatic control and, on the other hand, to develop mechanistic models for describing the comminution parameters and develop adequate experimental techniques for parameter determination from single particle tests, from batch testing and by online measurements. Since no momentum or mechanical energy balance have been made for the comminution process, the energy consumption must be obtained by a different method. Here enters the recent work on a mechanistic model for the ball movement in a rotary ball mill. Even though these are the more relevant tasks today, it must be recognized that relations between the variables in the phenomenological models of comminution are still not well understood, and more work is needed in these areas.

2.4.2 Classification

Classification is another good example of the use of a combination of empirical and phenomenological models. The old Hazen (1904) sedimentation tank, which has served as a basis for modeling crossflow classifiers, is a phenomenological model developed in the early 1900s. Turbulence was added to that model to obtain more realistic results. However, the substitution of that type of classifier by hydrocyclones has diminished the interest in developing more sophisticated models for rake or spiral classifiers. The complicated flow in a hydrocyclone has retarded the development of phenomenological models. Empirical models, such as those developed by Lynch and Rao (1975) and Lynch (1977) and by Plitt (1976) and Plitt et al. (1987) are used extensively to date and have not yet been substituted by any phenomenological model. Himmelblau (1970) provides the basic foundations of empirical models using statistical methods. Currently, this has evolved into using machine learning tools (Raschka 2016; Kelleher et al. 2015; Brunton and Kutz 2019; Bascur et al. 2022).

Two types of phenomenological models of hydrocyclones are being presently developed. Both are based on the description of the flow pattern in the equipment and the trajectories that particle would follow subjected to the forces developed in the flow fields. Both are dynamic models in turbulent flow. The first assumes that the vortex flow in a hydrocyclone is well described by the Navier-Stokes equation, implying that the suspension within the equipment behaves as a Newtonian fluid. The numerical solution to the local turbulent conservation equations are made by either finite difference, finite element, or finite volume methods. Once the flow patterns are obtained, the linear momentum balances are solved for particles of several sizes, and the trajectories are followed to the overflow or the underflow. In that way, the partition curve is obtained. The problem of particle interaction is solved in several ways, for example using an effective fluid viscosity, corresponding to the local viscosity of the suspension affected by the local particle concentration.

Since the solution to the differential equations depends on the boundary conditions, and these are a consequence of the geometry of the walls, this method is useful for design purposes. For the time being the use of this model in commercial simulation packages is not possible due to the time consumed in the numerical calculations. The present challenge in this model is the three-dimensional solution of the flow pattern as a free boundary problem, recognizing the importance of the presence of the air core in the operation of a hydrocyclone. It would also be desirable to use the theory of mixture of continuum mechanics and solve the differential equations for the solid and the fluid components instead of using the Navier-Stokes equation. The result would be the local fluxdensity and the local concentration of particles at each point within the equipment.

The second phenomenological model of a hydrocyclone uses the boundary layer approach of fluid mechanics. Since the velocities within the equipment are very high, in the order of several meters per second, the flow field is divided into a boundary layer close to the walls and an inviscid external flow. The boundary layers are responsible for the short circuit and the leakage to the underflow and overflow respectively, and the external flow patterns permit the calculation of trajectories and separation by sizes. To be able to use this approach, the flow patterns in the equipment should be known, either by experimental determination or by prediction with the model previously discussed. Since the flow patterns in this second type of model are expressed in closed form, this phenomenological model can be implemented in a PC and can be incorporated into commercial simulation packages. The vortex motion in a hydrocyclone draws air into the equipment, which is responsible for the roping phenomenon. Roping has been the object of recent research, and the results are clarifying hydrocyclone operation. It is interesting to note that these two phenomenological models are not exclusive and that in the future they could be combined to obtain models for design and simulation purposes.

2.4.3 Flotation

The best example of the uneven progress of an applied discipline is the development of the theory of flotation. Flotation has greatly profited from the application of the theories of thermodynamics and physical chemistry. The result is that the chemistry and surface chemistry of flotation is at present time a well-known subject. This development of the theory of flotation has required from the workers in mineral processing a strong background in the physical chemistry of interfaces, and therefore it is not surprising to find surface chemistry a subject of advanced study for the extractive metallurgist.

In contrast the subject of flotation kinetics, or more properly, flotation dynamics, has seen less advance since studies were initiated in the early 1940s. There is no doubt in the writer's mind that this lack of development is due to the lower emphasis that subjects such as *continuum mechanics*, and its specialized branch of *physico-chemical hydrodynamics*, is given in the curricula of mineral processing and extractive metallurgy. Research into the dynamics of flotation began in the former Soviet Union, and it is interesting to note that these investigations were performed by people who were not metallurgists but physical chemists. These subjects became available to mineral processing researchers through the works of Deryagin and Dukhin (1961) in the 1950s, Levich (1962) in the 1960s, Shulze (1984) in the 1980s, and lately by Ruylov (2020). Phenomenological models incorporating the fundamental aspects of adsorption dynamics of surfactants in bubbles and its electrical interaction, particle-bubble collisions, particle adherence, particle dispersion in turbulent stirred tanks, and other hydrodynamic effects, are scarce. A good example is the work of Bascur (1982), which has been followed by Finnish researchers.

2.4.4 Gravity Concentration

The development of a theory that could explain the behavior of a suspension passing through any gravity equipment has still not been achieved. The application of the conservation equations for the mass and momentum of a two-phase suspension in a gravity concentration device remains an unsolved problem, that is, no dynamic phenomenological model exists. However, there are some isolated models for the sorting mechanism, which can be related to the behavior of certain individual gravity concentration devices. Here the equations can be purely empirical or based on the hydrodynamics of the specific phenomena with empirically determined parameters. Therefore, we can state that the modeling of gravity concentration is in the state of defining and describing phenomenological models of the subprocesses and that no integration of these sub-models have been made.

2.4.5 Magnetic Concentration

The development of models for magnetic concentration is described in detail in the book *Magnetic Methods for the Treatment of Minerals* by Svoboda (1987), and I will make no further comment on it.

2.4.6 Dewatering

The two most common processes of dewatering in the mineral industry, thickening and filtration, are good examples of mature phenomenological models. Both models were developed using mass and momentum balances and classify as dynamic models. In thickening, the starting point was a macroscopic model by Coe and Clevenger in 1916, and the development of a rational theory had to wait for more than 50 years, when the suspension, flocculated or not, began to be considered a continuous medium with two components, then local mass and momentum equations were applied, and a robust phenomenological theory was developed that includes conventional and high-capacity thickening. In the case of filtration, Tiller (1977) and coworkers in Houston established the fundamentals and experimental foundation, while Massarani and his group in COPPE, Rio de Janeiro, Brazil (1985), developed the theory, and Stahl and Mittarbeiter in Karlsruhe developed new types of filters.

REFERENCES

Bascur, O.A. 1982. Modeling and computer simulation of a flotation cell. PhD dissertation, University of Utah, Salt Lake City, Utah.

Bascur, O.A. 1988. A control data framework with distributed intelligence. *Advances in Instrumentation* 88:1553–1169.

Bascur, O.A. 2019. Chapter 2.6: Process control and operational intelligence. In *SME Mineral Processing and Extractive Metallurgy Handbook*. Edited by R.C. Dunne, S.K. Kawatra, and C.A. Young. Englewood, CO: SME.

Bascur, O.A. 2020. *Digital Transformation in the Process Industries: A Road Map*. Boca Raton, FL: CRC Press.

Bascur, O.A., Retamal, F., and Espinosa, S. 2022. Treating low grades ores using a digital in mining and mineral processing plants. In *Proceedings of the Copper 2022 International Conference*, IIMChile, Santiago. www.copper2022.cl

Bascur, O.A., and Retamal, S. 2023. Increasing water recovery to maximize copper production with low grade ores. In *Proceedings of COM2023*, Toronto, August 21–24.

Brunton, S.L., and Kutz, J.N. 2019. *Data-Driven Science and Engineering: Machine Learning, Dynamical Systems and Control*. Cambridge: Cambridge University Press.

Coe, H.S., and Clevenger, G.H. 1916. Methods for determining the capacity of slime settling tanks. *AIME Transactions* 55:356–385.

Deryagin, B.V., and Dukhin, S.S. 1961. Theory of flotation of small and medium size particles. *Bulletin of the Institution of Mining and Metallurgy (London)* 70:221.

Hazen, A. 1904. On sedimentation. Paper 980. *ASCE Transactions* (June 1):45–88.

Herbst, J.A., George, D.B., and Sastry, K.V.S., eds. 1984. *Control '84: Mineral and Metallurgy Processing*. New York: SME-AIME.

Himmelblau, D.M. 1970. *Process Analysis by Statistical Methods*. New York: Wiley.

Himmelblau, D.M., and Bishoff, K.B. 1968. *Process Analysis and Simulation: Deterministic Systems*. New York: Wiley.

Kelleher, J.D., MacNamee, B., and D'Arcy, A. 2015. *Fundamentals of Machine Learning for Predictive Analytics: Algorithms, Worked Examples and Case Studies*. Cambridge, MA: MIT Press.

Levich, G.V. 1962. *Physicochemical Hydrodynamics*. Englewood Cliffs, NJ: Prentice Hall.

Lynch, A.J. 1977. *Mineral Crushing and Grinding Circuits, Their Simulation, Optimization, Design and Control*. Amsterdam: Elsevier Scientific.

Lynch A.J., and Rao, T.C. 1975. Modeling and scale-up of hydrocyclone classifiers. In *Proceedings of the XI International Mineral Processing Congress*, Cagliari. p. 9.

Massarani, G. 1985. *Filtraçao*. Rio de Janeiro: Revista Brasileira de Engenharia. pp. 1–69.

Plitt, L.R. 1976. A mathematical model of the hydrocyclone classifier. *CIM Bulletin* (December):114.

Plitt, L.R., Flintoff, B.C., and Stuffco, T.J. 1987. Roping in hydrocyclones. In *Proceedings of the 3rd Conference on Hydrocyclones at Oxford*. Paper A-3. Edited by P. Wood. Amsterdam: Elsevier. pp. 21–34.

Rajamani, R.K., and Herbst, J.A., eds. 1990. *Control 90: Mineral and Metallurgical Processing*. Littleton, CO: SME.

Raschka, S. 2016. *Python Machine Learning*. Birmingham: Packt.

Ruylov, N. 2020. *Flotation: Physical Theory and Applications*. Kyiv: National Academy of Sciences of Ukraine. ch. 1–7.

Shulze, H.J. 1984. Physicochemical elementary process in flotation. In *Developments in Mineral Processing*, vol. 4. Edited by D.W. Fuerstenau. New York: Elsevier.

Svoboda, J. 1987. Magnetic methods for the treatment of mineral. In *Developments in Mineral Processing*, vol. 8. Edited by D.W. Fuerstenau. New York: Elsevier.

Tiller, F. 1977. Solid–liquid separation: An overview. *Chemical Engineering Progress* (October):65–75.

Truesdell, C. 1966a. *The Elements of Continuum Mechanics*. New York: Springer Verlag.

Truesdell, C. 1966b. Method and taste in natural philosophy. In *Six Lectures in Natural Philosophy*. New York: Springer Verlag. pp. 83–117.

Truesdell, C. 1984. *Rational Thermodynamics*, 2nd ed. New York: Springer.

Truesdell, C., and Noll, W. 1965. The non-linear theory of mechanics. In *Encyclopedia of Physics*, vol. 111/3. Edited by S. Flogge. New York: Springer Verlag.

Turton, R., Shaeiwitz, J.A., Bhattacharyya, D., and Whiting, W.B. 2018. *Analysis, Synthesis, and Design of Chemical Processes*, 5th ed. Boston, MA: Prentice-Hall. ch. 1.

FURTHER READING

Anlauf, H. 1986. *Entfeuchtung von Filterkuchen bei der Vaskuum, Druck/Vakuum Filtration, Fortschritte Berichte VOi*. Reihe 3: Verfahrenstechnik Nr. 114, VOi Verlag.

Austin, L.G., Barahona, C.A., and Menacho, J.M. 1987a. Investigations on autogenous and semi-autogenous grinding in tumbling mills. *Powder Technology* 51:283–294.

Austin, L.G., Klimpel, R.R., and Luckie, P.T. 1984. *Process Engineering of Size Reduction: Ball Milling*. New York: SME-AIME.

Austin, L.G., Menacho, J.M., and Pearcy, F. 1987b. A general model for semi-autogenous and autogenous milling. In *APCOM 87: Proceedings of the Twentieth International Symposium on the Application of Computers and Mathematics in the Mineral Industries. Volume 2: Metallurgy*. Johannesburg: Southern African Institute of Mining and Metallurgy. pp. 107–126.

Barrientos, A., and Concha, F. 1992. Chapter 21: Phenomenological model of classification in conventional hydrocyclones. In *Comminution: Theory and Practice*. Edited by K. Kawatra. New York: SME AIME. pp. 288–305.

Barrientos, A., Sampaio, R., and Concha, F. 1993. Effect of the air core on the performance of a hydrocyclone. In *Proceedings of the XVIII International Mineral Processing Congress*, vol. 1. Melbourne, Victoria: Australasian Institute of Mining and Metallurgy. pp. 267–269.

Bascur, O.A. 2021. Digital transformation in mining and metallurgical complexes. Technical Paper and Plenary Lecture. In *Proceedings of the 60th Annual Conference of Metallurgists (COM 2021)*. Montreal, Quebec: Canadian Institute of Mining, Metallurgy and Petroleum.

Bascur, O.A., and Herbst, J.A. 1985. Dynamic simulators for training personnel in the control of grinding/flotation systems. *Presented at the International Federation of Automatic Control (IFAC) Symposium on Automation in Mining, Mineral and Metal Processing*, Brisbane, Australia.

Bloor, M.I.G., and Ingham, D.B. 1983. Theoretical aspects of hydrocyclone flow. In *Progress in Filtration and Separation*, vol. 3. Edited by R.J. Wakeman. New York: Elsevier. pp. 57–148.

Bond, F. 1960. Crushing and grinding calculations. *British Chemical Engineering* 6:378–391, 543–548.

Bott, A.A. 1985. Zur Kontinuirlichen Druckfiltration. dissertation, Technische Hochschule Karlsruhe.

Bunge, R.C., and Fuerstenau, D.W. 1992. *The Modeling of Cross Flow Separations*. SME Preprint No. 92–121. Littleton, CO: SME.

Burt, A.O. 1985. Gravity concentration technology. In *Developments in Mineral Processing*, vol. 5. New York: Elsevier.

Bustos, M.C., and Concha, F. 1995. Chapter 2: Kynch theory of sedimentation. In *Sedimentation of Small Particles in a Viscous Fluid*. Edited by E. Tory. Southampton: Computational Mechanics Publications.

Camp, Th.R. 1943. The effect of turbulence in retarding settling. In *Proceedings of the 2nd Hydraulic Conference: Studies of Engineering*. Bulletin 27. Iowa City, IA: University of Iowa. pp. 307–317.

Chine, B., Concha, F., and Barrientos, A. 1995. A finite difference solution of swirling flow in the hydrocyclone. Presented at the IX International Conference on Finite Elements in Fluids, Venice, October.

Clarke, D.S., and Wilkes, N.S. 1989. *The Calculation of Turbulent Flows in Complex Geometries Using a Differential Stress Model*. Carlsbad, CA: AEA Technology.

Clarke, D.S., and Wilkes, N.S. 1991. *The Calculation of Turbulent Flows in Complex Geometries Using a Differential Stress Model*. Report No. 0216. Carlsbad, CA: AEA Technology.

Concha, F., and Barrientos, A. 1993. A critical review of thickener design methods. *Kona* 13(11):79–104.

Concha, F., Barrientos, A., Montero, J., and Sampaio, R. 1994. Air core and roping in hydrocyclones. In *Proceedings of the VIII European Symposium on Comminution*, Stockholm, May 17–19. pp. 814–823.

Concha, F., and Bustos, M.C. 1991. Settling velocities of particulate systems, 6. Kynch sedimentation processes: Batch settling. *International Journal of Mineral Processing* 32(3–4):193–212.

Concha, F., and Bustos, M.C. 1992. Settling velocities of particulate systems, 7. Kynch sedimentation processes: Continuous thickening. *International Journal of Mineral Processing* 34(1–2):33–51.

Concha, F., Bustos, M.C., and Barrientos, A. 1995. Chapter 3: Phenomenological theory of sedimentation. In *Sedimentation of Small Particles in a Viscous Fluid*. Edited by E. Tory. Southampton: Computational Mechanics Publications.

Concha, F., Lee, Ch.H., and Austin, L.G. 1992. Settling velocities of particulate systems. Part 8. Batch sedimentation of polydispersed suspensions of spheres. *International Journal of Mineral Processing* 35(3–4):159–175.

Davidson, M.R. 1991. Numerical prediction of the flow and particle separation in a gas cyclone with and without vortex finder. Presented at the Computational Techniques Conference, CTAC '91, Adelaide, Australia, July.

Dobbins, W.E. 1944. Effect of turbulence on sedimentation. *ASCE Transactions* 109:629–678.

Gaudin, A.M. 1939. *Textbook of Mineral Dressing*. New York: McGraw-Hill.

Gaudin, A.M. 1957. *Flotation*. New York: McGraw-Hill.

Herbst, J.A., and Fuerstenau, D.W. 1980. Scale-up procedure for continuous grinding mill design using population balance models. *International Journal of Mineral Processing* 7:1–31.

Holland-Batt, A.B. 1978. Design of gravity concentration circuits by use of empirical mathematical models. Paper 21. In *Proceedings of the Eleventh Commonwealth Mining and Metallurgical Congress*, Hong Kong, May. pp. 1–11.

Hsieh, K.T. 1988. Phenomenological model of the hydrocyclone. PhD dissertation, University of Utah, Salt Lake City, Utah.

Hsieh, K.T., and Rajamani, K. 1988a. Phenomenological model of the hydrocyclone: Model development and verification for single phase flow. *International Journal of Mineral Processing* 22:223.

Hsieh, K.T., and Rajamani, K. 1988b. Phenomenological model of the hydrocyclone: Numerical solution and verification of particle movement. In *Proceedings of the XVI International Mineral Processing Congress*, Stockholm. p. 377.

King, R.P. 2012. *Modeling and Simulation of Mineral Processing Systems*. Englewood, CO: SME.

Meloy, T.P., and Bergstrom, B.H. 1964. Matrix simulation of ball mill circuits considering impact and attrition grinding. In *Proceedings of the VII International Mineral Processing Congress*. Edited by N. Arbiter. New York: Gordon and Breach. pp. 19–31.

Menacho, J. 1986. Some solutions for the kinetics of combined fracture and abrasion breakage. *Powder Technology* 49:87–96.

Menacho, J., and Concha, F. 1986. Mathematical model of ball wear in grinding mills, I. Zero order wear rate. *Powder Technology* 47:87–96.

Menacho, J., and Concha, F. 1987. Mathematical model of ball wear in grinding mills, II. General solution. *Powder Technology* 52:267–277.

Mika, T.S., and Fuerstenau, D.W. 1971. The transient behavior of a distributed parameter mill model. Part 1: Shredding. *Dechema Monogrphien* 69(1292–1326):389–433.

Mishra, B.K., and Rajamani, R.K. 1992. *Numerical Simulation of Charge Motion in Ball Mills: Lifter Bar Effect*. SME Preprint No. 92–184. Littleton, CO: SME.

Mohammadi, B., and Pironneau, O. 1994. *Analysis of the k Turbulence Model*. New York: Wiley.

Rajamani, R.K., Agrawala, S., and Mishra, B.K. 1993. Mill scale-up: Ball collision frequency and collision energy density in laboratory and plant-scale mills. In *Proceedings of the XVIII International Mineral Processing Congress*. Series No. 3/93, vol. 1: Comminution and Classification. Melbourne, Victoria: Australasian Institute of Mining and Metallurgy. pp. 103–107.

Schubert, H. 1978. *Aufbereitung fester mineralisher Rohstoffe, Band II*. Leipzig: VEB Deutscher Verlag fur Grundstoffindustrie.

Schubert, H., and Nesse, T. 1973. The role of turbulence in wet classification. In *Proceedings of the X International Mineral Processing Congress*, IMM, London. pp. 213–239.

Sivamohan, R., and Forssberg, E. 1985. A review of the literature on intermediate and fine particle gravity concentration. *International Journal of Mineral Processing* 15:155–181.

Wilkes, N.S., and Clarke, D.S. 1987. *Turbulent Flow Prediction Using Algebraic Stress Models*. Carlsbad, CA: AEA Technology.

3 Hydrodynamics

3.1 INTRODUCTION

Cenegel and Cimbala (2018) in their book on fluid mechanics suggest that one of the first engineering solution using fluid mechanics was the supply of water for domestic use, and irrigation of crops was the construction of a series of pressurized lead and clay pipelines up to 45 km during 283 to 133 BC in the city of Pergamon (Turkey), and one of the well-known engineering works are the Roman aqueducts, some of which are still in use.

The first book on mining technology ever written, by Georgius Agricola (1556), and published in the city of Freiberg in Germany shows that most procedures for recovering minerals from rocks use water. Reducing rock size and classifying its product, the separation of the useful material from the gangue and its concentration, the transport of concentrates and tailings are more efficient using water. In all these processes there is interaction between mineral particles and water which allows the recovery of valuable mineralogical species. During the processes the particles move within the fluid and, due to the effect of the relative velocity between the solids and the fluid, the separation of the particles by various forces such as gravity, centrifugal force, oscillations, and adhesion forces is achieved.

3.2 FLUID MECHANICS

In Chapter 1, we indicated that a fluid is a material that, in the face of the slightest applied shear stress, is continuously deformed. This property sets it apart from a solid that can withstand great shear stresses without deformation. On the other hand, both solid and fluid can withstand normal stresses without deforming. In our study of the interaction between these two materials, we will use certain solid properties, such as size, particle size and shape distribution, and others, that we will treat in other chapters.

In many engineering problems, fluids behave as if they were Newtonian, that is, incompressible viscous linear, and the constitutive equation for the stress T is a linear function of the shear rate ∇v.

$$\boldsymbol{T} = p\boldsymbol{I} + \mu\left(\nabla \boldsymbol{v} + \nabla \boldsymbol{v}^T\right) \tag{3.1}$$

where p is fluid pressure, I is the dyadic unit tensor, $v(r,t)$ is the velocity field, and r and t are the position vector and time. Substituting this expression into the balance of linear momentum and adding the mass balance yields

DOI: 10.1201/9781032614212-3

$$\nabla \cdot \boldsymbol{v} = 0$$
$$\rho \frac{\partial \boldsymbol{v}}{\partial t} + \rho \nabla \cdot \boldsymbol{v} = -\nabla p + \mu \nabla^2 \boldsymbol{v} + \boldsymbol{b} \tag{3.2}$$

These are the Navier-Stokes equations, a powerful tool to solve problems related to mineral processing. To study specific problems, it is necessary to stipulate boundary conditions, the physical principle of which are as follows:

1. There is no penetration or slip between solid-fluid interfaces, so that both moves at the same velocity.
2. Normal component of the velocity for immiscible fluids is continuous.
3. Negligible shear stresses at fluid-gas interfaces.
4. The velocity should remain finite at the axis or plane of symmetry, for fluids with certain degree of symmetry, that is, the first derivative of the velocity should be zero with respect to the coordinates.

In this chapter, we present the basic fundamental concepts. Fundamental research in the physicochemical hydrodynamics in mineral processing systems has been examined by Levich (1962). Mineral processing systems are affected by many physicochemical factors on the motion of particles, bubbles, and drops (Clift et al. 1978). They describe the formation of bubbles and drops, which are very much important in the development of new flotation machines and in solvent extraction in hydrometallurgical processes.

3.2.1 Fluids at Equilibrium

Fluids at equilibrium are subjected to inertia, pressure, and body forces. Then from the Navier-Stokes:

$$\begin{aligned} \nabla \cdot \boldsymbol{v} &= 0 \\ \rho \dot{\boldsymbol{v}} &= \nabla p + \boldsymbol{b} \end{aligned} \tag{3.3}$$

If we consider gravity g as the only body force, integrating this equation yields the following:

$$\begin{aligned} \nabla p &= \rho(\boldsymbol{g} - \dot{\boldsymbol{v}}) \\ \int \nabla p \cdot d\boldsymbol{r} &= \int \rho(\boldsymbol{g} - \dot{\boldsymbol{v}}) \cdot d\boldsymbol{r} \\ \int \nabla p \cdot d\boldsymbol{r} &= -\int \rho(g\boldsymbol{k} - \dot{\boldsymbol{v}}) \cdot d\boldsymbol{r}, \quad \boldsymbol{k} \cdot d\boldsymbol{r} = dz, \end{aligned} \tag{3.4}$$

For incompressible fluids without acceleration:

$$\begin{aligned} \nabla p &= \rho g \\ p &= -\rho g z + C \\ p_L &= -\rho g L + C, \Rightarrow C = p_L + \rho g L \\ p &= p_L + \rho g (L - z) \end{aligned} \tag{3.5}$$

3.2.1.1 Forces on Submerged Bodies: Buoyancy

Buoyancy $\boldsymbol{f}_b$ is the force exerted by a fluid at equilibrium on a submerge body of surface S, then

$$\boldsymbol{f}_b = \oint_{S_p} \boldsymbol{T}\cdot\boldsymbol{n}dS = -\oint_{S_p} p\boldsymbol{n}dS \tag{3.6}$$

Using the Green Gauss Theorem (GGO) theorem,

$$\boldsymbol{f}_b = -\int_{V_p} \nabla\, pdV;$$

since $\nabla\, p = -\rho g k$

$$\boldsymbol{f}_b = -(\rho g V)\boldsymbol{k} \tag{3.7}$$

3.2.2 General Solutions to the Navier-Stokes Equation

The Navier-Stokes equations are known as very difficult solutions; however, there are special cases in which it is possible to obtain exact solutions and others in which one can take advantage in dividing the flow in categories depending on the relative value of the several terms of the equations. In the first case, the nonlinear terms are eliminated due to the form of the shear rates and in the second, when small terms may be neglected.

3.2.2.1 Exact Solutions

There are several simple, but important, cases of exact solution that appear in the field of mineral processing, such as the transport of water or diluted suspensions in a circular pipe, rheology measurement of suspensions of particles, and the flow in a viscometer or the low runoff of water in an inclined table. In what follows, we analyze these cases.

3.2.2.1.1 Flow in a Horizontal Circular Tube

For the horizonal flow of a liquid in a cylindrical tube, cylindrical coordinates are used (r,θ,z).

Continuity equation:

$$\frac{\partial\rho}{\partial t}+\frac{1}{r}\frac{\partial}{\partial r}(\rho r v_r)+\frac{1}{r}\frac{\partial}{\partial\theta}(\rho v_\theta)+\frac{\partial}{\partial z}(\rho v_z)=0 \tag{3.8}$$

Momentum equation:

$$\rho\left(\frac{\partial v_r}{\partial t}+v_r\frac{\partial v_r}{\partial r}+\frac{v_\theta}{r}\frac{\partial v_r}{\partial\theta}-\frac{v_\theta^2}{r}+v_z\frac{\partial v_r}{\partial z}\right)=-\frac{\partial p}{\partial r}+\frac{1}{r}\frac{\partial}{\partial r}\left(\frac{\partial}{\partial r}\left(rT_{rr}^E\right)\right) \tag{3.9}$$

$$+\frac{1}{r}\frac{\partial T_{r\theta}^E}{\partial\theta}-\frac{T_{\theta\theta}^E}{r^2}+\frac{\partial T_{rz}^E}{\partial z}+\rho g_r$$

$$\rho\left(\frac{\partial v_\theta}{\partial t}+v_r\frac{\partial v_\theta}{\partial r}+\frac{v_\theta}{r}\frac{\partial v_\theta}{\partial \theta}+\frac{v_r v_\theta}{r}+v_z\frac{\partial v_\theta}{\partial z}\right)=-\frac{1}{r}\frac{\partial p}{\partial \theta}+\frac{1}{r^2}\frac{\partial}{\partial r}\left(r^2T^E_{\theta r}\right)$$
$$+\frac{1}{r}\frac{\partial T^E_{\theta\theta}}{\partial \theta^2}+\frac{\partial T^E_{\theta z}}{\partial \theta}+\rho g_\theta$$

$$\rho\left(\frac{\partial v_z}{\partial t}+v_r\frac{\partial v_z}{\partial r}+\frac{v_\theta}{r}\frac{\partial v_z}{\partial \theta}+v_z\frac{\partial v_z}{\partial z}\right)=-\frac{\partial p}{\partial z}+\frac{1}{r}\frac{\partial}{\partial r}\left(rT^E_{zr}\right)$$
$$+\frac{1}{r}\frac{\partial T^E_{z\theta}}{\partial \theta}+\frac{\partial T^E_{zz}}{\partial z}+\rho g_z$$

Since the flow is one-dimensional in the z direction, all terms in (r,θ) are zero, and only the blue terms remain then,

$$\text{Continuity equation:}\frac{\partial}{\partial z}(v_z)=0$$

Momentum equation:

$$r \text{ component}: 0=-\frac{\partial p}{\partial r}+\rho g \tag{3.10}$$
$$\theta \text{ component}: 0=-\frac{\partial p}{\partial \theta}$$
$$z \text{ component}: 0=-\frac{\partial p}{\partial z}+\frac{1}{r}\frac{\partial}{\partial r}\left(rT_{rz}{}^E\right)$$

The solutions to Equation 3.10 are obtained by separation of variables,

$$T^E_{rz}(r)=\frac{1}{2}Kr,\quad K=-\frac{\Delta p}{L}$$
$$T^E_{rz}(r)=-\frac{1}{2}\frac{\Delta p}{L}r \tag{3.11}$$

Equation 3.11 is valid for any type of fluids since we still have not defined the type of fluid we are using. To describe a specific fluid, we have to formulate a *constitutive equation.*

3.2.2.1.2 Newtonian Fluid

The constitutive equation for the stress tensor $T_{r_z}^{E^2}(r)$ of a Newtonian fluid is a linear function of the shear rate $\partial v_z / \partial r$,

$$T^E_{rz}(r)=-\mu\frac{\partial v_z}{\partial r} \tag{3.12}$$

Then, the momentum equation becomes

$$\mu\frac{\partial v_z}{\partial r}=-\frac{1}{2}\frac{\Delta p}{L}r \tag{3.12a}$$

Integration leads to

$$v_z(r) = -\frac{1}{4}\frac{\Delta p}{\mu L} r^2 + C \tag{3.13}$$

If the tube is stationary, boundary condition 1 as mentioned earlier, the velocity vanishes at the tube wall; $C = \frac{1}{4}\frac{\Delta p}{\mu L} R^2$ and the velocity has a parabolic velocity distribution

$$v_z(r) = \frac{1}{4}\frac{\Delta p}{\mu L}\left(R^2 - r^2\right) \tag{3.14}$$

The pressure can be obtained from the r component of the momentum equation (Equation 3.10):

$$\begin{aligned} 0 &= -\frac{\partial p_0}{\partial r} + \rho g \\ p(r) - p_0(r) &= -\rho g(R - r)) \end{aligned} \tag{3.15}$$

It is usual to write $\tau(r) = T_{rz}^E(r)$ for the shear stress and $\dot{\gamma} = \partial v_z / \partial r$ for the shear rate. Figure 3.1 shows the shear stress and velocity distributions.

With velocity defined by Equation 3.14, the volume flow rate Q becomes

$$Q = \int_S v_z dS = \int_0^{2\pi}\left(\int_0^R v_z \times r dr\right) d\theta = \frac{\Delta p}{8\mu L}\pi R^2 \tag{3.16}$$

The average velocity is defined by $\overline{v}_z = Q/\pi r^2$, then

$$\overline{v}_z = \frac{\Delta p}{8\mu L} \tag{3.17}$$

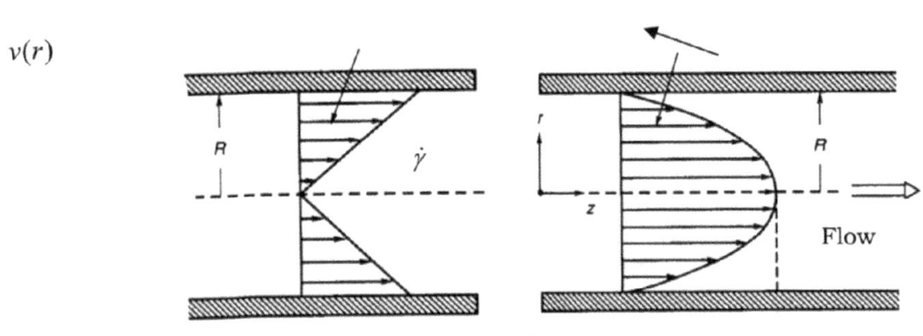

FIGURE 3.1 Shear stress, shear rate, and velocity distribution for the flow in a circular tube.

From Equation 3.11, the shear rate $\hat{\gamma}_w$ and shear stress τ_w at the tube wall are given by

$$\dot{\gamma}_w = -\frac{1}{2}\frac{\Delta pR}{\mu L} \text{ and } \tau_w = -\mu\frac{1}{2}\frac{\Delta pR}{\mu L} \tag{3.18}$$

3.2.2.1.3 *Flow in a Viscometer*

The flow between two concentric cylinders with radius R_1 and R_2 produced by the rotation of the cylinders with angular velocity of Ω_1 and Ω_1 respectively, the Navier-Stokes equation in cylindrical coordinates are (r,θ,z) as shown in Figure 3.2.

Continuity:

$$\frac{\partial \rho}{\partial t}+\frac{1}{r}\frac{\partial}{\partial r}(\rho r v_r)+\frac{1}{r}\frac{\partial}{\partial \theta}(\rho v_\theta)+\frac{\partial}{\partial z}(\rho v_z)=0 \tag{3.19}$$

Momentum:

$$\rho\left(\frac{\partial v_r}{\partial t}+v_r\frac{\partial v_r}{\partial r}+\frac{v_\theta}{r}\frac{\partial v_r}{\partial \theta}-\frac{v_\theta^2}{r}+v_z\frac{\partial v_r}{\partial z}\right)=-\frac{\partial p}{\partial r} + \mu\left(\frac{\partial}{\partial r}\left(\frac{1}{r}\frac{\partial}{\partial r}(r v_r)\right)+\frac{1}{r^2}\frac{\partial^2 v_r}{\partial \theta^2} + \frac{2}{r^2}\frac{\partial v_\theta}{\partial \theta}+\frac{\partial^2 v_r}{\partial z^2}\right)+\rho g_r \tag{3.20}$$

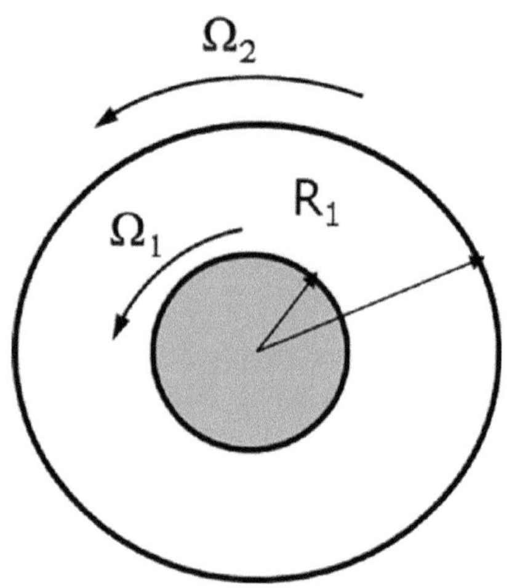

FIGURE 3.2 Flow between two rotating cylinders.

$$\rho\left(\frac{\partial v_z}{\partial t}+v_r\frac{\partial v_z}{\partial r}+\frac{v_\theta}{r}\frac{\partial v_z}{\partial \theta}+v_z\frac{\partial v_z}{\partial z}\right)=-\frac{\partial p}{\partial z}+\mu\left(\frac{1}{r}\frac{\partial}{\partial r}\left(r\frac{\partial v_z}{\partial r}\right)+\frac{1}{r^2}\frac{\partial^2 v_z}{\partial \theta^2}+\frac{\partial^2 v_z}{\partial z^2}\right)+\rho g_z \tag{3.21}$$

If only the blue terms are retained,

$$\frac{1}{r}\frac{\partial}{\partial \theta}(\rho v_\theta)=0 \tag{3.22}$$

$$\mu\frac{\partial}{\partial r}\left(\frac{1}{r}\frac{\partial}{\partial r}(rv_\theta)\right)=0 \tag{3.23}$$

$$\frac{\partial p}{\partial r}=\frac{v_\theta^2}{r},\qquad \frac{\partial p}{\partial z}=0 \tag{3.24}$$

In these equations, v_θ and p are the tangential velocity and the pressure, respectively.

3.2.2.1.4 *Pressure Distribution*

From Equation 3.24, the pressure distribution is

$$p(r)=v_\theta^2\ln r+C$$

$$p(R_2)=v_\theta^2\ln R_2+C=p_0$$

$$p(r)=v_\theta^2\ln r+p_0-v_\theta^2\ln R_2$$

$$p(r)=p_0+v_\theta^2\ln\left(\frac{r}{R_2}\right) \tag{3.25}$$

3.2.2.1.5 *Velocity Distribution*

Solving Equation 3.23 leads to

$$\frac{\partial}{\partial r}(rv_\theta)=C_1 r,\quad rv_\theta=\frac{C_1}{2}r^2+C_2$$

$$R_1 v_\theta(R_1)=\frac{C_1}{2}R_1^2+C_2$$

$$R_2 v_\theta(R_2)=\frac{C_1}{2}R_2^2+C_2$$

$$C_2=\frac{(\Omega_2-\Omega_1)R_2^2R_1^2}{R_2^2-R_1^2}$$

$$C_1 = 2\frac{\Omega_2 R_2^2 - \Omega_1 R_1^2}{\left(R_2^2 - R_1^2\right)} \qquad C_2 = \frac{\left(\Omega_2 - \Omega_1\right) R_2^2 R_1^2}{R_2^2 - R_1^2} \tag{3.26}$$

And the velocity distribution is

$$v_\theta(r) = \frac{\Omega_2 R_2^2 - \Omega_1 R_1^2}{R_2^2 - R_1^2} r + \frac{\left(\Omega_2 - \Omega_1\right) R_2^2 R_1^2}{R_2^2 - R_1^2} \frac{1}{r} \tag{3.27}$$

3.2.2.1.6 *Torque*

Torque is the moment of the force $\boldsymbol{T}_o = \int\limits_{S_{cilindro}} \boldsymbol{r} \times \left(-p\boldsymbol{n} + \boldsymbol{T}^E \bullet \boldsymbol{n}\right) dS$, and since $\boldsymbol{r}$ is perpendicular to n, $\boldsymbol{r} \times \boldsymbol{n} = \boldsymbol{0}$ and from Equation 3.1,

$$\boldsymbol{T}_o = \int\limits_{S_{cilindro}} \boldsymbol{r} \times \mu r \frac{\partial}{\partial r}\left(\frac{v_\varphi}{r}\right) \boldsymbol{e}_\theta \underbrace{(\boldsymbol{n} \bullet \boldsymbol{n})}_{1} dS$$

From Equation 3.27, and integrating

$$\boldsymbol{T}_o = \pm 4\pi L\left(\Omega_2 - \Omega_1\right) \frac{R_1^2 R_2^2}{\left(R_2^2 - R_1^2\right)} \boldsymbol{e}_\theta$$

where the sign depends on which cylinder the torque is measured.

In a Searle viscometer, only the interior cylinder rotates, so $\Omega_1 = 0$ and the torque is

$$\boldsymbol{T}_o = -4\pi L \Omega_2 \frac{R_1^2 R_2^2}{\left(R_2^2 - R_1^2\right)} \boldsymbol{e}_\theta \tag{3.28}$$

3.2.3 Examples

3.2.3.1 Flow in an Inclined Table

For a Newtonian fluid draining over an inclined table, consider the following assumptions: the fluid properties are constants. Choosing Cartesian coordinates (x, z), $v_z = v_z(x, z)$, $v_x = v_y = 0$ as shown in Figure 3.3.

Boundary conditions are

$$v_z(0, z) = 0, \quad \left.\frac{\partial v_z}{\partial x}\right|_L = 0$$
$$p(L) = 0 \tag{3.29}$$

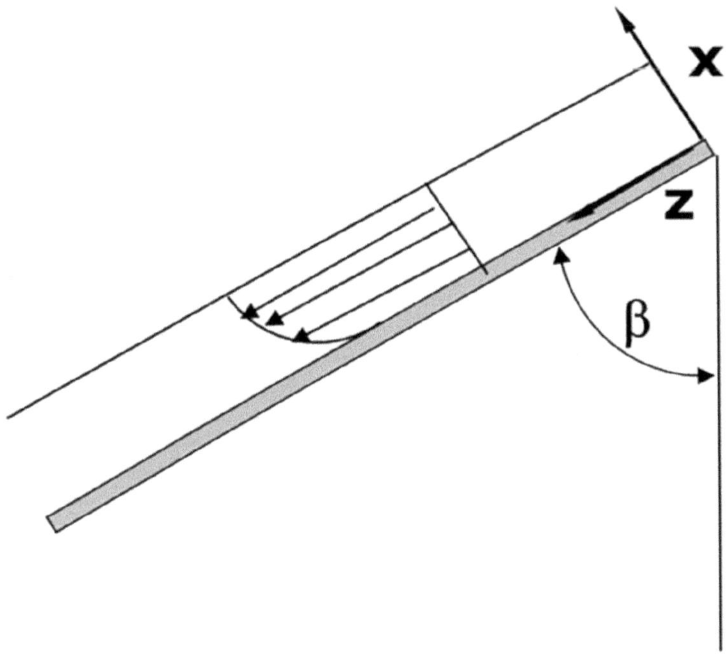

FIGURE 3.3 Fluid falling over an inclined table.

The Navier-Stokes equation in Cartesian coordinates are (x, y, z).

Continuity:

$$\frac{\partial \rho}{\partial t}+\frac{\partial}{\partial x}(\rho v_x)+\frac{\partial}{\partial y}(\rho v_y)+\frac{\partial}{\partial z}(\rho v_z)=0 \tag{3.30}$$

Momentum:

$$\rho\left(\frac{\partial v_x}{\partial t}+v_x\frac{\partial v_x}{\partial x}+v_y\frac{\partial v_x}{\partial y}+v_z\frac{\partial v_x}{\partial z}\right)=-\frac{\partial p}{\partial x}+\mu\left(\frac{\partial^2 v_x}{\partial x^2}+\frac{\partial^2 v_x}{\partial y^2}+\frac{\partial^2 v_x}{\partial z^2}\right)+\rho g_x \tag{3.31}$$

$$\rho\left(\frac{\partial v_y}{\partial t}+v_x\frac{\partial v_y}{\partial x}+v_y\frac{\partial v_y}{\partial y}+v_z\frac{\partial v_y}{\partial z}\right)=-\frac{\partial p}{\partial y}+\mu\left(\frac{\partial^2 v_y}{\partial x^2}+\frac{\partial^2 v_y}{\partial y^2}+\frac{\partial^2 v_y}{\partial z^2}\right)+\rho g_y \tag{3.32}$$

$$\rho\left(\frac{\partial v_z}{\partial t}+v_x\frac{\partial v_z}{\partial x}+v_y\frac{\partial v_z}{\partial y}+v_z\frac{\partial v_z}{\partial z}\right)=-\frac{\partial p}{\partial z}+\mu\left(\frac{\partial^2 v_z}{\partial x^2}+\frac{\partial^2 v_z}{\partial y^2}+\frac{\partial^2 v_z}{\partial z^2}\right)+\rho g_z \tag{3.33}$$

The reduced equation becomes
Continuity:

$$\frac{\partial}{\partial z}(\rho v_z) = 0; \rightarrow \quad v_z = v_z(x) \tag{3.34}$$

Momentum pressure:

$$\frac{\partial p}{\partial x} = -\rho g \mathrm{sen}\theta, \rightarrow p(x) = -\rho g x \mathrm{sen}\theta + C$$
$$p(L) = -\rho g L \mathrm{sen}\theta + C = 0, \rightarrow C = \rho g L \mathrm{sen}\theta$$

$$p(x) = \rho g (L - x)\mathrm{sen}\theta \tag{3.35}$$

Momentum velocity:

$$\mu \frac{\partial^2 v_z(x)}{\partial x^2} = \rho g \cos\theta$$
$$\frac{\partial v_z(x)}{\partial x} = \frac{\rho g \cos\theta}{\mu} x + C_1$$
$$v_z(x) = \frac{\rho g \cos\theta}{2\mu} x^2 + C_1 x + C_2$$
$$\left.\frac{\partial v_z}{\partial x}\right|_{x=L} = \frac{\rho g \cos\theta}{\mu} L + C_1 = 0 \rightarrow C_1 = -\frac{\rho g \cos\theta}{\mu} L$$
$$v_z(x) = \frac{\rho g \cos\theta}{2\mu} x^2 - \frac{\rho g \cos\theta}{\mu} Lx + C_2$$
$$v_z(0) = 0$$

$$v_z(x) = \frac{\rho g L^2 \cos\theta}{2\mu}\left(\left(\frac{x}{L}\right)^2 - 2\left(\frac{x}{L}\right)\right) \tag{3.36}$$

3.2.3.2 Flowrate and Average Velocity

The flowrate is given by

$$Q = \int_0^L v_z(x) a dx \text{ and } \overline{v}_z = \frac{Q}{aL} \tag{3.36a}$$

where a and L are the width and length of the table.

$$Q = \int_0^L \frac{\rho g L^3 \cos\theta}{2\mu}\left(\left(\frac{x}{L}\right)^2 - 2\left(\frac{x}{L}\right)\right) d\left(\frac{x}{L}\right)$$

$$Q = \frac{\rho g a L^3 \cos\theta}{2\mu}\left(\frac{1}{3}\xi^3 - \xi^2\right)_0^1$$

$$Q = \frac{\rho g a L^3 \cos\theta}{3\mu} \quad \text{and} \quad \bar{v}_z = \frac{\rho g L^2 \cos\theta}{3\mu} \tag{3.37}$$

3.2.4 Dimensional Analysis of the Navier-Stokes Equation

In studying the behavior of fluid, we use the *similarity theory* and *dimensional analysis* theory, which form the scientific basis for modeling of fluid phenomena (Matz 1954; Ibsen 1960). Similar processes are those having a constant relationship between corresponding characteristic magnitudes, such as the following:

- Geometry: $L, l;\ l^* = l_1/L_1 = l_2/L_2 = \cdots l_n/L_n$
- Time: $t, T;\ \ t^* = t_1/T_1 = t_2/T_2 \cdots t_n/T_n$
- Physical magnitudes: $u, U;\ u^* = u_1/U_1 = u_2/U_2 = \cdots u_n/U_n$
- Initial and boundary conditions: valid if the three characteristics, geometry, time, and physical magnitudes, are similar

Terms like l^*, t^* and u^* are known as *similitude invariants*, and terms with composite magnitudes, such as $\mathrm{Re} = du\rho/\mu$, are called *similitude* or *dimensionless numbers*; both constitute the *similitude criteria*.

With this definition, we can express experimental results as function of similitude criteria. Consider n similitude criteria $\pi_1, \pi_2\ \pi_3 \cdots \pi_n$, then we can write

$$f(\pi_1, \pi_2\pi_3 \cdots \pi_n) = 0 \tag{3.38}$$

In those cases, in which it is not possible to obtain an exact solution, we can make an order of magnitude study of the equations to find out if there are terms in the equation that can be deleted. First, we need to set the Navier-Stokes equations in dimensional form. Choose the following characteristic parameters for the flow. Length L, speed u_o, time t_o, pressure p_o, and body force g, and define the dimensional variables.

$$\begin{aligned} &x^* = x/L, \quad \boldsymbol{v}^* = \boldsymbol{v}/u_o, \quad t^* = t/t_o \quad \boldsymbol{b}^* = \boldsymbol{b}/g \\ &p^* = p/p_o, \ \nabla^* = \nabla/L, \ \partial/\partial t^* = t_o\,\partial/\partial t \end{aligned} \tag{3.39}$$

Substituting in the Navier-Stokes equations yields

$$\nabla^* \bullet \boldsymbol{v}^* = 0$$

$$\left[\frac{\rho u_o}{t_o}\right]\frac{\partial v^*}{\partial t^*} + \underbrace{\left[\frac{\rho u_o^2}{L}\right]\nabla^* v^* \bullet v^*}_{\text{Convective term}} = -\left[\frac{p_o}{L}\right]\nabla^* p^* + \underbrace{\left[\frac{\mu u_o}{L^2}\right]\nabla^{*2}\boldsymbol{v}^*}_{\text{Difusive term}} + [\rho g]\boldsymbol{b}^* \tag{3.40}$$

Dividing by the convective coefficient term all terms in the second equation of Equation 3.40, yields

$$\nabla^* \bullet \boldsymbol{v}^* = 0$$

$$\left[\frac{L}{t_o u_o}\right]\frac{\partial v^*}{\partial t^*} + \underbrace{\nabla^* v^* \bullet v^*}_{\text{Convective term}} = -\left[\frac{p_o}{\rho u_o^2}\right]\nabla^* p^* + \underbrace{\left[\frac{\mu}{\rho u_o L}\right]\nabla^{*2}\boldsymbol{v}^*}_{\text{Difusive term}} + \left[\frac{gL}{u_o^2}\right]\boldsymbol{b}^* \quad (3.41)$$

Defining the dimensionless numbers

$$\text{Stouhal } St = \frac{t_o u_o}{L}, \quad \text{Rouark } Ru = \frac{\rho u_o}{p_o}, \quad \text{Reynolds } \mathrm{Re} = \frac{\rho u_o L}{\mu} \text{ and}$$

$$\text{Froude } Fr = \frac{u_o^2}{Lg} \quad (3.42)$$

Equation 3.41 can be written in the dimensionless form

$$\nabla^* \bullet \boldsymbol{v}^* = 0$$

$$\frac{1}{St}\frac{\partial v^*}{\partial t^*} + \underbrace{\nabla^* v^* \bullet v^*}_{\text{Convective term}} = -\frac{1}{Ru}\nabla^* p^* + \underbrace{\frac{1}{\mathrm{Re}}\nabla^{*2}\boldsymbol{v}^*}_{\text{Difusive term}} + \frac{1}{Fr}\boldsymbol{b}^* \quad (3.43)$$

The Reynolds number, as defined in Equation 3.42, is the ratio of the convective to the diffusive coefficients in the flow.

3.2.4.1 Euler's Flow

In all those phenomena where the convective forces are much greater than the diffusive forces, the Reynolds number is very high, making the diffusive term negligible. Reducing the Navier-Stokes equation to Euler's dimensionless inviscid equation:

$$\nabla^* \bullet \boldsymbol{v}^* = 0$$

$$\frac{1}{St}\frac{\partial v^*}{\partial t^*} + \underbrace{\nabla^* v^* \bullet v^*}_{\text{Convective term}} = -\frac{1}{Ru}\nabla^* p^* + \frac{1}{Fr}\boldsymbol{b}^* \quad (3.44)$$

And in dimensional form, Euler's equation for inviscid flow is as follows:

$$\nabla \bullet \boldsymbol{v} = 0$$

$$\rho\frac{\partial \boldsymbol{v}}{\partial t} + \rho\nabla\boldsymbol{v} \bullet \boldsymbol{v} = -\nabla p + \boldsymbol{b} \quad (3.45)$$

3.2.4.2 Stokes Flow

If the convective forces are much smaller than the diffusive forces, the Reynolds number would be very small, and the diffusive terms become negligible. In those cases, we have Stokes flow.

$$\nabla^* \bullet \boldsymbol{v}^* = 0$$

$$\frac{1}{St}\frac{\partial v^*}{\partial t^*} = -\frac{1}{Ru}\nabla^* p^* + \underbrace{\frac{1}{\text{Re}}\nabla^{*2}\boldsymbol{v}^*}_{\text{Difusive term}} + \frac{1}{Fr}\boldsymbol{b}^* \tag{3.46}$$

And in dimensional form

$$\nabla \bullet \boldsymbol{v} = 0$$
$$\rho\frac{\partial \boldsymbol{v}}{\partial t} = -\nabla p + \mu\nabla^2\boldsymbol{v} + \boldsymbol{b} \tag{3.47}$$

3.2.4.3 Prandtl's Flow

In between those extremes, when both the convective and the diffusive terms are significant, no term can be neglected, and the equations become difficult to solve. Fortunately, for our purpose a simplification result, if the flow is the relative flow over a solid surface. Consider a high-velocity flow over a flat surface, as shown in Figure 3.4, where body forces are neglected, represented by the following equations.

$$\frac{\partial v_x}{\partial x} + \frac{\partial v_y}{\partial y} = 0$$
$$v_x\frac{\partial v_x}{\partial x} + v_y\frac{\partial v_x}{\partial y} = -\frac{1}{\rho}\frac{\partial p}{\partial x} + \nu\left[\frac{\partial^2 v_x}{\partial x^2} + \frac{\partial^2 v_x}{\partial y^2}\right] \tag{3.48}$$
$$v_x\frac{\partial v_y}{\partial x} + v_y\frac{\partial v_y}{\partial y} = -\frac{1}{\rho}\frac{\partial p}{\partial y} + \nu\left[\frac{\partial^2 v_y}{\partial x^2} + \frac{\partial^2 v_y}{\partial y^2}\right]$$

where $\nu = \mu/\rho$ is the kinematical viscosity. Boundary conditions are

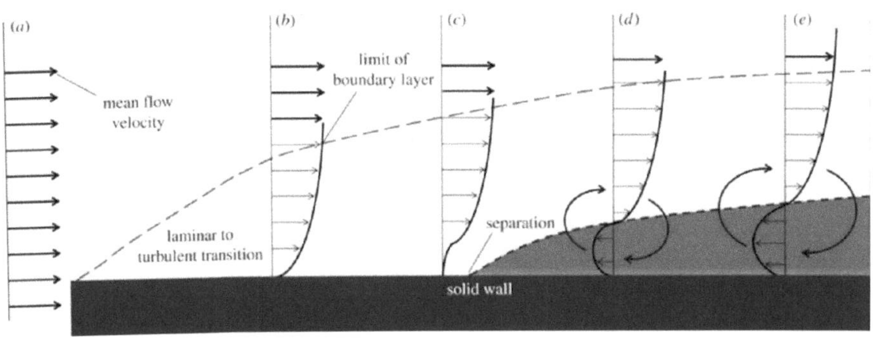

FIGURE 3.4 Flow past a flat plate at high Re.

Source: Fletcher, T., Altringham, J., Peakall, J., Wignall, P., and Dorrell, R. 2014. Hydrodynamics of fossil fishes. ***Proceedings of The Royal Society B*** **281:20140703**

$$\text{for } v_x = v_y = 0 \text{ and for } y \to \infty,\ v_x = u_0$$

$$\text{for } y = 0; v_x = v_y = 0 \text{ and for } y \to \infty; v_x = u_o \tag{3.49}$$

Due to the nature of the flow, we must define different characteristic parameters for the x and for the y directions, L_y and u_y respectively, and P_o for the pressure. With this choice, the following dimensionless variables result:

$$\begin{aligned} x^* &= x/L_x, & v_x^* &= v_x/u_x, & \frac{\partial}{\partial x^*} &= L_x\frac{\partial}{\partial x}, & p^* &= \frac{p}{p_o} \\ y^* &= x/L_y, & v_y^* &= v_y/u_y, & \frac{\partial}{\partial y^*} &= L_y\frac{\partial}{\partial y} \end{aligned} \tag{3.50}$$

With these dimensionless variables, and the definitions of the Ruark number, $Ru = \rho u_x^2 / p_c$ and $\text{Re} = \rho u_x L_x / \mu$, Equation 3.48 for high Reynolds number becomes

$$\begin{aligned}
&\underbrace{\frac{\partial v_x^*}{\partial x^*}}_{O(1)} + \underbrace{\left[\frac{L_x}{L_y}\frac{u_y}{u_x}\right]}_{O(1)}\underbrace{\frac{\partial v_y^*}{\partial y^*}}_{O(1)} = 0 \\
&\underbrace{v_x\frac{\partial v_x^*}{\partial x^*}}_{O(1)} + \underbrace{\left[\frac{L_x}{L_y}\frac{u_y}{u_x}\right]v_y^*}_{O(1)}\underbrace{\frac{\partial v_x^*}{\partial y^*}}_{O(1)} = -\underbrace{\frac{1}{Ru}\frac{\partial p^*}{\partial x^*}}_{O(1/Ru)} + \underbrace{\frac{1}{Re}\frac{\partial^2 v_x^*}{\partial x^{*2}}}_{O(1/Re)} + \underbrace{\left[\frac{L_x}{L_y}\right]^2}_{O(Re)}\underbrace{\frac{1}{Re}\frac{\partial^2 v_x^*}{\partial y^{*2}}}_{O(1/Re)} \\
&\underbrace{v_x\frac{\partial v_y^*}{\partial x^*}}_{O(1)} + \underbrace{\left[\frac{L_x}{L_y}\frac{u_y}{u_x}\right]v_y^*}_{O(1)}\underbrace{\frac{\partial v_y^*}{\partial y^*}}_{O(1)} = -\underbrace{\left[\frac{L_x}{L_y}\frac{u_x}{u_y}\right]}_{O(Re)}\underbrace{\frac{1}{Ru}\frac{\partial p^*}{\partial y^*}}_{O(1/Ru)} + \underbrace{\frac{1}{Re}\frac{\partial^2 v_y^*}{\partial x^{*2}}}_{O(1/Re)} + \underbrace{\left[\frac{L_x}{L_y}\right]^2}_{O(Re)}\underbrace{\frac{1}{Re}\frac{\partial^2 v_y^*}{\partial y^{*2}}}_{O(1/Re)}
\end{aligned} \tag{3.51}$$

Considering the order of magnitude shown, these equations reduce to dimensionless Prandtl equations.

$$\begin{aligned}
&\frac{\partial v_x^*}{\partial x^*} + \frac{\partial v_y^*}{\partial y^*} = 0 \\
&v_x\frac{\partial v_x^*}{\partial x^*} + v_y^*\frac{\partial v_x^*}{\partial y^*} = -\frac{1}{Ru}\frac{\partial p^*}{\partial x^*} + \frac{\partial^2 v_x^*}{\partial x^{*2}} \\
&0 = \frac{\partial p^*}{\partial y^*}
\end{aligned} \tag{3.52}$$

The corresponding equations in dimensional form are known as the *Prandtl boundary layer equations*.

$$\begin{gathered}\frac{\partial v_x}{\partial x}+\frac{\partial v_y}{\partial y}=0\\ v_x\frac{\partial v_x}{\partial x}+v_y\frac{\partial v_x}{\partial y}=-\frac{1}{\rho}\frac{\partial p}{\partial x}+\frac{\partial^2 v_x}{\partial x^2}\\ 0=\frac{\partial p}{\partial y}\end{gathered} \tag{3.53}$$

An important result for these equations is that there is no pressure gradient in the y direction, which means that the pressure is the same for all values of y.

3.2.4.4 Euler's Flow Harmonic Function

From Equation 3.45, Euler's flow can be written in the form

$$\begin{gathered}\nabla\bullet\boldsymbol{v}=0,\ \text{with}\quad \boldsymbol{v}=-\nabla\phi\\ \rho\frac{\partial\boldsymbol{v}}{\partial t}+\rho\left(\frac{1}{2}v^2\right)=-\nabla p-\nabla\widehat{\Phi}\\ \nabla\phi^2=0\end{gathered} \tag{3.54}$$

The solutions to Laplace's equation are called *harmonic function* ϕ.

3.2.4.5 Bernoulli's Theorem for an Ideal Fluid

The term in brackets is known as the Bernoulli, then Equation 3.54 can be written in the form

$$\begin{gathered}\frac{\partial\boldsymbol{v}}{\partial t}+\nabla\left(\underbrace{\frac{1}{2}v^2+\frac{p}{\rho}+\Phi}_{B}\right)=0\\ \nabla\left(-\frac{\partial\phi}{\partial t}+B\right)=0\\ \int_{L(t)}\nabla\left(-\frac{\partial\phi}{\partial t}+B\right)\cdot d\boldsymbol{r}=\int_{L(t)}d\left(-\frac{\partial\phi}{\partial t}+B\right)=0\\ -\frac{\partial\phi}{\partial t}+B=f(t)\end{gathered} \tag{3.55}$$

In steady flows, the Bernoulli becomes

$$B\equiv\frac{1}{2}v^2+\frac{p}{\rho}+\phi=0 \tag{3.56}$$

If a point 0 in the flow is known, this equation may be written in the form

$$p - p_0 = \frac{1}{2}\rho\left(v_0^2 - v^2\right) + \rho\left(\widehat{\Phi}_0 - \widehat{\Phi}\right) = 0 \tag{3.57}$$

3.3 POTENTIAL FLOW

The potential flow is Eulerian, isochoric, and irrotational, therefore $\nabla x v = 0$, the flow equations are (Essentials of Fluid Mechanics 2014, 2018)

$$\begin{aligned} &\nabla \bullet v = 0; \quad \nabla \times v = 0 \\ &\nabla^2 \phi = 0 \\ &p = p_0 + \frac{1}{2}\left(v_0^2 - v^2\right) + \rho\left(\widehat{\Phi}_0 - \widehat{\Phi}\right) \end{aligned} \tag{3.58}$$

In a potential flow, the velocity distribution depends exclusively on the geometry of the problem through the boundary conditions of the first equation dynamics appear only when the pressure calculated. In potential nonstationary flows, the velocity distribution is time independent, and time will only enter as a boundary condition for pressure.

3.3.1 Point Source and Sink

Consider a potential stationary flow of magnitude Q that arises (or disappears in) at a point of a plane, then potential, velocity, and stream function are as follows:

$$\begin{aligned} &v_\theta = -\frac{\partial \phi}{\partial r} = -\frac{1}{r}\frac{\partial \psi}{\partial \theta} \\ &\frac{\partial^2 \phi}{\partial r^2} + \frac{1}{r}\frac{\partial \phi}{\partial r} = 0, \\ &p = p_0 + \frac{1}{2}\left(v_0^2 - v^2\right) + \rho_0\left(\widehat{\Phi}_0 - \widehat{\Phi}\right) \end{aligned} \tag{3.59}$$

Integrating the Laplace equation yields, as shown in Figure 3.5:

$$\phi(r) = K \ln r$$

$$\boldsymbol{v}(r,\theta) = -\nabla \phi = -\frac{K_1}{r}\boldsymbol{e_r}$$

The volume flow is $Q = \int_S \boldsymbol{v} \bullet \boldsymbol{n} dS$

$$= \int_{2\pi} v_r r d\theta = -\int_{2\pi} K d\theta$$

$$= -2\pi K, \rightarrow K_1 = -\frac{Q}{2\pi}$$

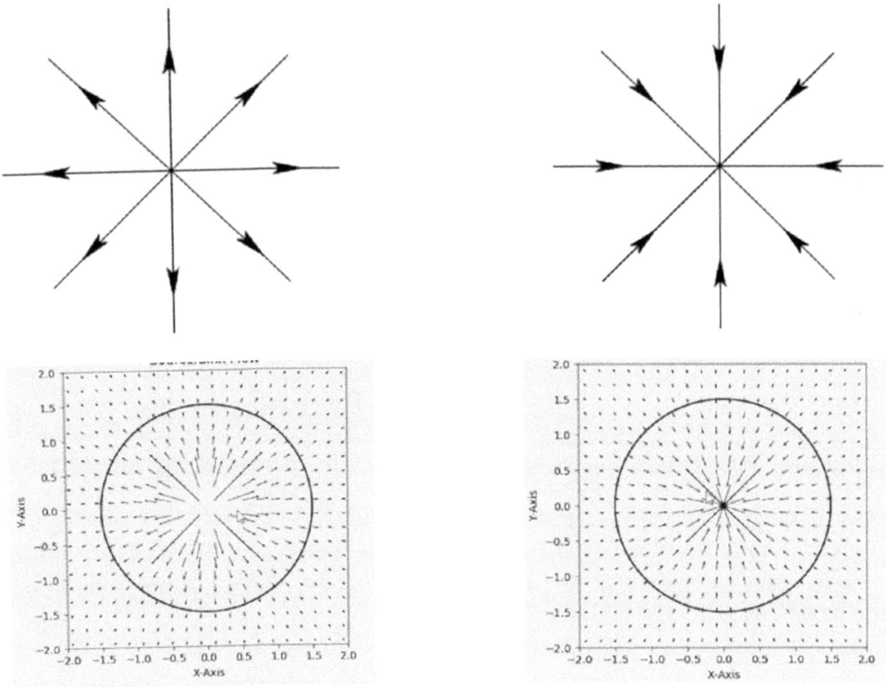

Source. Sink.

FIGURE 3.5 Velocity vectors.

$$v(r,\theta) = \frac{Q}{2\pi r}\boldsymbol{e}_r, \quad v_r(\theta) = \boldsymbol{v} \bullet \boldsymbol{n} = \frac{Q}{2\pi r}\boldsymbol{e}_r \bullet \boldsymbol{n} \tag{3.60}$$

$$\text{For a source}: \boldsymbol{n} = \boldsymbol{e}_r \rightarrow v_r(\theta) = \frac{Q}{2\pi r};$$

$$\text{For a sink}: \boldsymbol{n} = -\boldsymbol{e}_r \rightarrow v_r(\theta) = -\frac{Q}{2\pi r} \tag{3.61}$$

From Equations 3.57 and 3.61, the pressure distribution is as follows:

$$p - p_0 = -\frac{1}{2}\rho\left(\frac{Q}{2\pi r}\right)^2 = -\frac{1}{8}\rho\frac{Q^2}{\pi^2 r^2} \tag{3.62}$$

3.3.2 Examples

3.3.2.1 Free and Forced Vortex: Simple Model of a Hydrocyclone

A forced vortex is a combination of a free vortex and a sink (rigid motion). Consider a potential stationary flow with circulation Γ due to a concentrated vortex filament

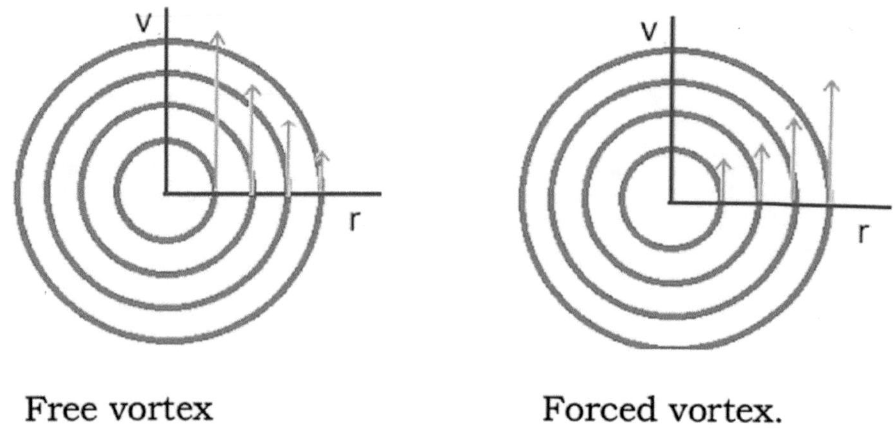

FIGURE 3.6 Velocity vectors.

at a given point on a plane. The combination of a free vortex, for a radius r from $b \le r < \infty$, and a forced vortex, for a radius r from $0 < r \le b$, is the simplest hydrocyclone model, as shown in Figure 3.6.

$$\begin{aligned} &\nabla^2\phi = 0 \\ &\frac{1}{r^2}\frac{\partial^2\phi}{\partial\theta^2} = 0, \quad \rightarrow \quad \frac{\partial\phi}{\partial\theta} = C_2\,, \rightarrow \phi(\theta) = C_2\theta \\ &v_\theta = -\frac{1}{r}\frac{\partial\phi}{\partial\theta} \quad = \frac{\partial\psi}{\partial\theta} \\ &\quad = -\frac{C_2}{r} \quad = \frac{\partial\psi}{\partial\theta};\ \Gamma = -\oint_{2\pi} C_2 d\theta;\ \Gamma = -2\pi C_2, \ \rightarrow \ C_2 = -\frac{\Gamma}{2\pi} \end{aligned} \tag{3.63}$$

$$\underbrace{v_\theta = \frac{\Gamma}{2\pi r},}_{\text{Free vortex}} \qquad \underbrace{v_\theta = \frac{\Gamma}{2\pi b}}_{\text{Forced vortex}} \tag{3.64}$$

If the highest point of the tangential velocity is $r = b$, as shown in Figure 3.7, the forced vortex can be written as follows:

$$v_\theta = \frac{\Gamma}{2\pi b^2} r \tag{3.65}$$

Combining Equations 3.64 and 3.65, we obtain Figure 3.8.

3.3.2.2 Pressure Distribution

The pressure distribution can be written in terms of the circulation. Consider Equation 3.57 neglecting the body force.

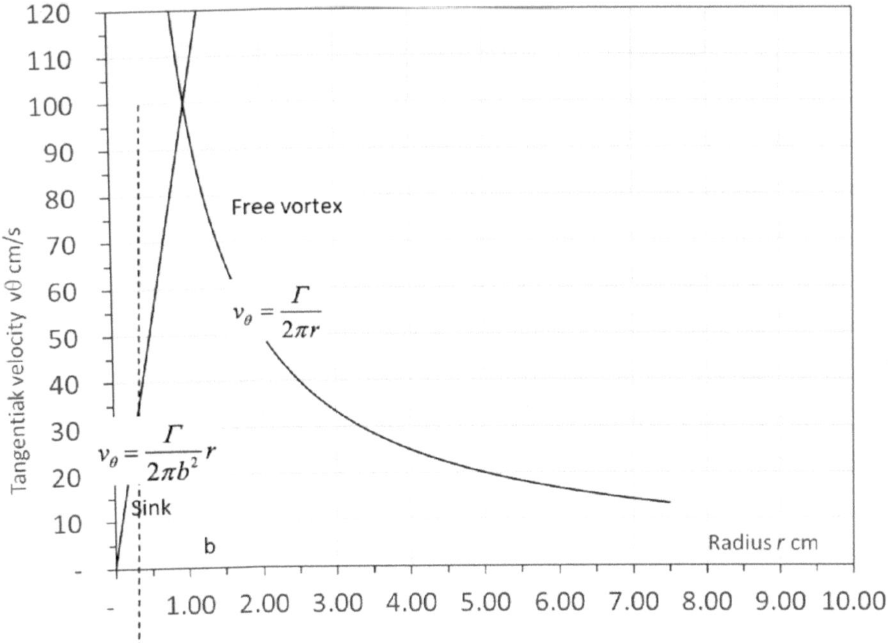

FIGURE 3.7 Free vortex and a sink.

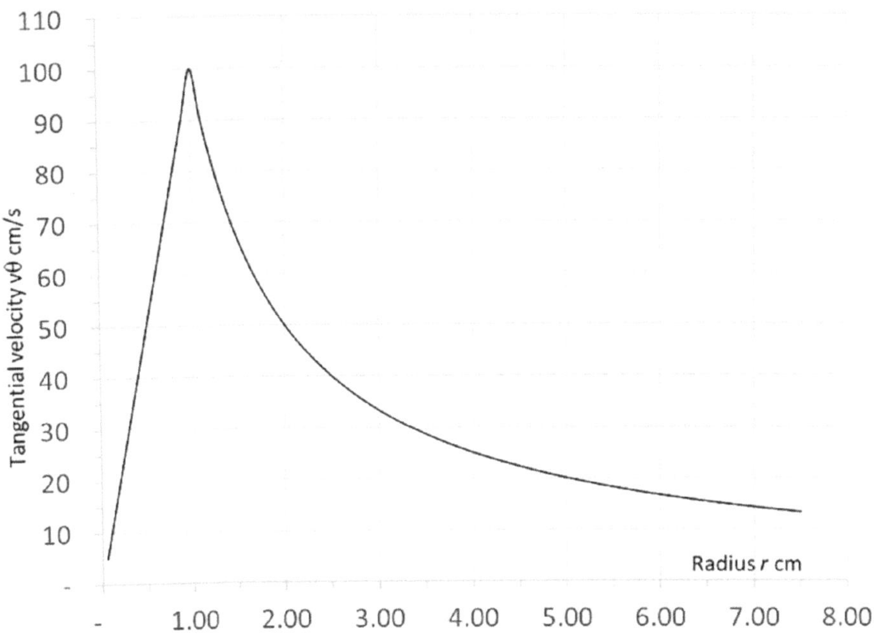

FIGURE 3.8 Simple hydrocyclone model.

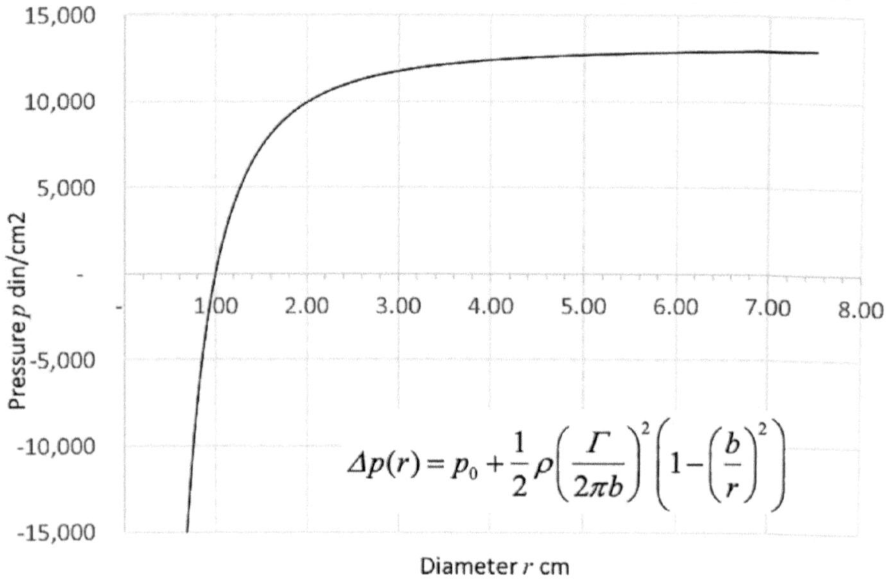

FIGURE 3.9 Pressure distribution in a forced vortex.

$$p(r) = p(a) + \frac{1}{2}\rho\left(v_{\theta a}^2 - v_\theta^2\right), \quad p(a) = p_0$$

$$v_\theta(r) = \frac{\Gamma}{2\pi}\frac{1}{r}; \quad v_\theta(a) = \frac{\Gamma}{2\pi}\frac{1}{a}$$

$$p(r) = p_0 + \frac{1}{2}\rho\left(\left(\frac{\Gamma}{2\pi a}\right)^2 - \left(\frac{\Gamma}{2\pi r}\right)^2\right)$$

$$p(r) = p_0 + \frac{1}{2}\rho\left(\frac{\Gamma}{2\pi b}\right)^2\left(1-\left(\frac{b}{r}\right)^2\right), \quad \text{for } 0 < r \le \infty$$

$$p(r) = p_0 + \frac{1}{2}\rho\left(\frac{\Gamma}{2\pi b}\right)^2\left(1-\left(\frac{b}{r}\right)^2\right), \quad \text{for } 0 < r \le \infty$$

The result is shown in Figure 3.9.

3.3.3 Potential Flow around a Particle

The potential flow around a particle is of particular importance in the overall formulation of a phenomenological theory for particle sedimentation in fluids. When the Reynolds number tends to infinity ($\mathrm{Re} \to \infty$) and the flow becomes isochoric and irrotational, meeting the following equations:

$$\nabla \bullet \boldsymbol{v} = 0, \ \nabla \times \boldsymbol{v} = 0 \ \ \nabla^2 \phi = 0, \ B = B_0 \tag{3.66}$$

A harmonic solution for the Laplace equation exists and is unique if boundary conditions are stipulated at ϕ or ϕ'. Then, if $\phi\left(|r| \rightarrow \infty\right) = \text{constant}$, the solutions depend on the functional form of the boundary geometry, that is, on the geometry of the surface of the particle only.

The flow of an inviscid isochoric fluid over a stationary spherical particle, with velocity $\boldsymbol{v}(\boldsymbol{r})$ with a constant value of $\boldsymbol{u}$ far from the particle, is given by the following:

$$\begin{aligned} &\boldsymbol{v} = -\nabla \phi, \quad \boldsymbol{v}(\boldsymbol{r} \rightarrow \infty) = \boldsymbol{u} \\ &v_\theta = \frac{1}{r}\frac{\partial \phi(r,\theta)}{\partial \theta}, \quad v_r = -\frac{\partial \phi(r,\theta)}{\partial r} \\ &\text{at } S_p: \ \boldsymbol{v} \bullet \boldsymbol{n} = 0; \ v_\theta = -usen\,\theta, \ v_r = u\cos\theta \end{aligned} \tag{3.67}$$

From Equation 3.63:

$$\begin{aligned} &\nabla^2 \phi = 0 \\ &\frac{1}{r^2}\frac{\partial^2 \phi}{\partial \theta^2} = 0, \quad \rightarrow \frac{\partial \phi}{\partial \theta} = C, \end{aligned}$$

$$v_\theta = -\frac{1}{r}\frac{\partial \phi}{\partial \theta} = -\frac{1}{r}C \tag{3.67a}$$

The tangential component of the velocity at the particle surface is a linear function of the relative solid-fluid velocity, and the radial component is equal to zero:

$$v_\theta(\Theta) = \left(\frac{3}{2}\operatorname{sen}\theta\right)u, \quad \text{and} \quad v_r = 0 \tag{3.68}$$

The pressure is given by a nonlinear function called the Bernoulli equation:

$$\begin{aligned} &\mathrm{p}(\Theta) + 1/2\,\rho_f v_\theta^2 = p + 1/2\,\rho_f u^2 = \text{constants} \\ &\mathrm{p}(\Theta) - p = \frac{1}{2}\rho_f u^2\left(1 - \left(\frac{v_\theta}{u}\right)^2\right) \end{aligned} \tag{3.69}$$

Defining pressure coefficient by $C_p = \left(\mathrm{p}(\Theta) - p\right)/(1/2)\,\rho_f u^2$, Bernoulli's equation may be expressed as follows:

$$C_p = 1 - \frac{9}{2}\operatorname{sen}^2\theta \tag{3.70}$$

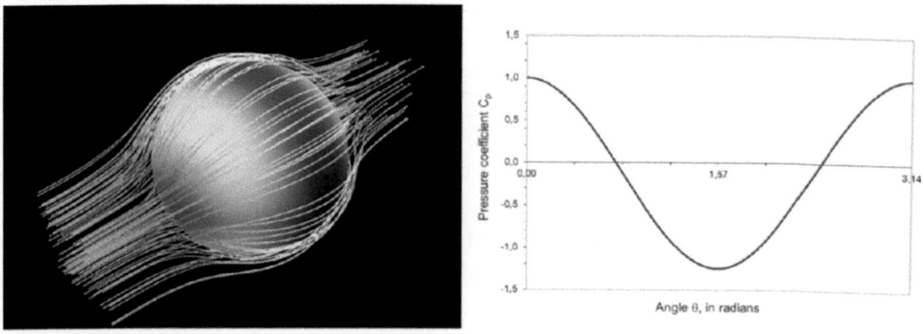

Streamlines. Pressure coefficient.

FIGURE 3.10 Inviscid flow around a sphere.

Source: Nathan, P. 2011. Potential flow over a sphere. www.youtube.com/watch?v=Bndj6eX7-_I. Accessed August 2023.

Figure 3.10 shows Equation 3.70 in graphic form, where $p(\theta)$ and $v(\theta)$ are, respectively, the pressure and tangential velocity at an angle θ on the surface of the sphere, and p and u are the values of the same variables at the bulk of the flow.

Since an inviscid flow has no viscosity, the hydrodynamic force is zero. This result is because the friction drag is zero in the absence of viscosity and the form drag depends on the pressure distribution over the surface of the sphere, which is symmetric leading to a zero-net force.

3.4 STOKES FLOW

If the convective forces in each flow are smaller than the viscous forces, the Reynolds number tends to zero, and the convective term in Navier-Stokes equation is neglected, yielding *Stokes flow* (see Section 3.1). Stokes flow, where gravity is the only body force, is represented by the following equation:

$$\nabla \cdot v = 0$$

$$\rho_0 \frac{\partial \boldsymbol{v}}{\partial t} = -\nabla p + \mu \nabla^2 \boldsymbol{v} + \rho_0 g \tag{3.71}$$

One of the most important example of Stokes flow is the motion of a particle in a viscous fluid. The principal characteristics of Equation 3.1 are their linearity, which allows the study of the motion of a particle in a fluid as the sum of solutions from simple motions. On the other hand, this linearity suggests certain general conclusions, valid for particles of arbitrary shape, although explicit solutions are only feasible for particles having a high degree of symmetry. In fact, in those cases where no numerical calculations are possible, the sole knowledge of the existence of a solution, and of its form, is of great importance because it provides a guide for the development of semi-empirical relationships.

In this section, we will study rectilinear, rotary, and oscillatory motions of particles of arbitrary shape through a viscous fluid at low Reynolds numbers. For each case, we will calculate the velocity and pressure fields and the forces and torques induced on the particle. These studies will constitute the fundamentals of several of the mineral processing operations.

3.4.1 Rectilinear Uniform Translation of a Particle

Consider the slow motion of a particle of arbitrary shape translating with constant velocity u through a fluid otherwise moving at constant velocity $\boldsymbol{u}$, at the point X, measured with respect to a fixed frame of reference $\{\boldsymbol{B},t\}$ at point 0 (Figure 3.11), where Q^* is the center of a reference frame fixed at the particle $\{\boldsymbol{B}^*,t\}$ and the vectors marked (*) refer to this reference frame.

The following set of equations represent the problem with respect to the frame of reference fixed to a particle at point Q^*:

$$\begin{gathered}\nabla p = \mu \nabla^2 \boldsymbol{v}^* + \rho_0 \boldsymbol{g} \\ \nabla \cdot \boldsymbol{v}^* = 0\end{gathered} \tag{3.72}$$

With boundary conditions:

$$\begin{gathered}\text{At } Sp \quad \boldsymbol{v}^*\left(\boldsymbol{r}^*\right) = 0 \\ \text{At} \left|\boldsymbol{r}^*\right| \to \boldsymbol{v}^*\left(\boldsymbol{r}^*\right) = \boldsymbol{u}_0 - \boldsymbol{u}, \quad p\left(\boldsymbol{r}^*\right) = p_0, z^* = 0\end{gathered} \tag{3.73}$$

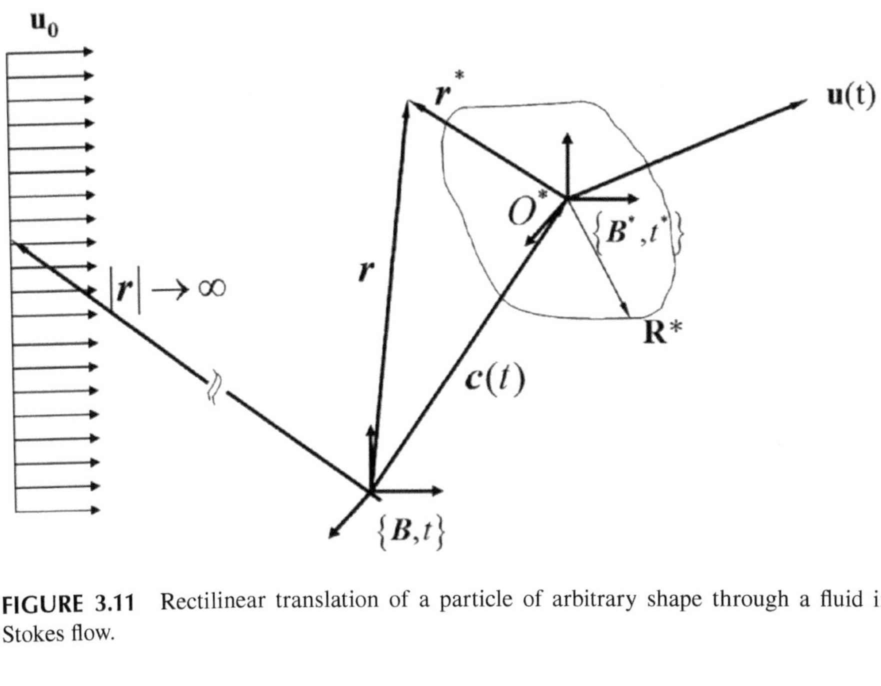

FIGURE 3.11 Rectilinear translation of a particle of arbitrary shape through a fluid in Stokes flow.

Since Equation 3.72 is linear, it is convenient to separate this boundary problem in two parts, a static (h) part and a dynamic part (d), with $v^* = v_h^* + v_d^*$ and $p = p_h + p_d$, according to the following scheme:

1. Static problem:

$$\begin{gathered} \nabla p_h = \rho_0 \boldsymbol{g}, v_h^* = 0 \\ \nabla \cdot \boldsymbol{v}_h^* = 0 \end{gathered} \tag{3.74}$$

With boundary conditions:

$$At \left| \boldsymbol{r}^* \right| \to p_h \left(\boldsymbol{r}^* \right) = 0, z_0 = z_0^* \tag{3.75}$$

Since the static problem is independent of the motion, we can integrate Equation 3.74 to give the *hydrostatic pressure field* $p_h\left(z^*\right)$,

$$\begin{aligned} \int \nabla p_h &= \int \rho_0 g \cdot d\boldsymbol{r}^* \\ \int dp_h &= -\int \rho_0 g dz^* \\ p_h &= -\rho_0 g \left(z^* - z_0^* \right) \end{aligned} \tag{3.76}$$

2. Dynamic problem:

$$\begin{gathered} \nabla p_d = \mu \nabla^2 \boldsymbol{v}_d^* \\ \nabla \cdot \boldsymbol{v}_d^* = 0 \end{gathered} \tag{3.77}$$

With boundary conditions:

$$\begin{gathered} At\ Sp \quad v_d^*\left(r^*\right) = 0 \\ At \left| r^* \right| \to \infty, \quad v_d^*\left(r^*\right) = u_0 - u, \quad p_d\left(r^*\right) = 0, \quad z^* = z_0^* \end{gathered} \tag{3.78}$$

Where the dynamic pressure field $p_d\left(r^*\right)$ is measured with respect to p_0, the pressure at $\left|\boldsymbol{r}^*\right| \to \infty$.

3.4.1.1 Velocity Field

Because Equation 3.78 and its boundary conditions are linear in $\boldsymbol{u} - \boldsymbol{u}_0$, the solution to the differential equation should also be a linear function of $\boldsymbol{u} - \boldsymbol{u}_0$, then,

$$v\left(r^*\right) = V\left(r^*, f\left(r^*\right)\right) \cdot \left(u - u_0\right) \tag{3.79}$$

Where the tensor V depends on the particle position $\boldsymbol{r}^*$ and its surface shape $f\left(r^*\right)$ and is independent of the fluid properties.

$$\boldsymbol{v}_d\left(\boldsymbol{r}^*\right) = \boldsymbol{V}\cdot\left(\boldsymbol{u}-\boldsymbol{u}_0\right) \tag{3.80}$$

Equation 3.80 shows that, when a particle of arbitrary shape moves slowly at a constant velocity $\boldsymbol{u}$ through a viscous fluid with unperturbed velocity field $\boldsymbol{u}_0$, the velocity field induced by the particle is a linear function of the unperturbed velocity field, and the linear transformation *dyadic velocity* $\boldsymbol{V}\left(\boldsymbol{r}^*, f\left(\boldsymbol{r}^*\right)\right)$ is a function of the position vector and the shape of the particle only.

3.4.1.2 Dynamic Pressure Field

Substituting Equation 3.80 in the Stokes equation (Equation 3.77) and integrating yields the following:

$$\begin{aligned}\nabla p_d &= \mu\nabla^2\left(\boldsymbol{V}\cdot\left(\boldsymbol{u}-\boldsymbol{u}_0\right)\right)\\ &= \mu\nabla^2\boldsymbol{V}^T\cdot\left(\boldsymbol{u}-\boldsymbol{u}_0\right)\end{aligned} \tag{3.81}$$

$$\begin{aligned}\int\nabla p_d\cdot \mathrm{d}\boldsymbol{r}^* &= \mu\left[\int\nabla^2\boldsymbol{V}^T\cdot \mathrm{d}\boldsymbol{r}^*\right]\cdot\left(\boldsymbol{u}-\boldsymbol{u}_0\right)\\ \int_0^{r^*}\mathrm{d}_p &= \mu\left[\int_\infty^{r^*}\nabla^2\boldsymbol{V}^T\cdot \mathrm{d}\tfrac{3}{4}^*\right]\cdot\left(\boldsymbol{u}-\boldsymbol{u}_0\right)\end{aligned} \tag{3.82}$$

$$p_d\left(\boldsymbol{r}^*\right) = \left[\int_\infty^{r^*}\nabla^2\boldsymbol{V}^T\cdot \mathrm{d}\xi^*\right] \tag{3.83}$$

Defining the vectorial pressure field by

$$\boldsymbol{p}\left(\boldsymbol{r}^*\right) = \left[\int_\infty^{r^*}\nabla^2\boldsymbol{V}^T\cdot \mathrm{d}\xi^*\right] \tag{3.84}$$

$$\text{yields}\quad p_d\left(\boldsymbol{r}^*\right) = \mu p\left(\boldsymbol{r}^*\right)\left(\boldsymbol{u}-\boldsymbol{u}_0\right) \tag{3.85}$$

Equation 3.85 shows that, in Stokes flow, the dynamic pressure field p_d is also a linear function of the unperturbed velocity field, where the proportionately function is the *vector pressure field* $p\left(r^*\right)$. This result is different to potential flow, where the velocity field is a linear function of the unperturbed velocity field, but the pressure is not, which happens because the Stokes equation is linear whereas Euler's equation is not.

3.4.1.3 Total Pressure Field

The total pressure field is the sum of the hydrostatic and the dynamic pressure fields:

$$p\left(\boldsymbol{r}^*\right) = \mu\boldsymbol{p}\left(\boldsymbol{r}^*\right)\left(\boldsymbol{u}-\boldsymbol{u}_0\right) + \rho_0 g\left(z_0^* - z^*\right) \tag{3.86}$$

3.4.1.4 Stresses

The constitutive equation of a Newtonian fluid is

$$T = -p\boldsymbol{I} + \mu\left(\nabla \boldsymbol{v} + \nabla \boldsymbol{v}^T\right)$$

Substituting Equations 3.80 and 3.85 yields

$$\begin{aligned} \boldsymbol{T} &= -\mu \boldsymbol{I}\boldsymbol{p}\cdot\left(\boldsymbol{u}-\boldsymbol{u}_0\right)+\mu\left(\nabla \boldsymbol{V}\cdot\left(\boldsymbol{u}-\boldsymbol{u}_0\right)+\nabla^T\boldsymbol{V}\cdot\left(\boldsymbol{u}-\boldsymbol{u}_0\right)\right) \\ &= \mu\left(-p\boldsymbol{I}+\nabla \boldsymbol{V}+\nabla^T\boldsymbol{V}\right)\cdot\left(\boldsymbol{u}-\boldsymbol{u}_0\right) \end{aligned} \tag{3.87}$$

Defining the *triadic* stress tensor $P\left(r^*\right)$

$$\boldsymbol{P}\left(\boldsymbol{r}^*\right) = -p\boldsymbol{I} + \nabla \boldsymbol{V} + \nabla^T \boldsymbol{V} \tag{3.88}$$

Then, Equation 3.87 can be written in the form

$$\boldsymbol{T}\left(\boldsymbol{r}^*\right) = \mu \boldsymbol{P}\left(\boldsymbol{r}^*\right)\left(\boldsymbol{u}-\boldsymbol{u}_0\right) \tag{3.89}$$

3.4.1.5 Hydrodynamic Force

The force $\boldsymbol{F}\left(\boldsymbol{r}^*\right)$ exerted by a fluid on a particle is expressed by

$$\boldsymbol{F}\left(\boldsymbol{r}^*\right) = \oint_{S_p} \boldsymbol{T}\left(\boldsymbol{r}^*\right)\cdot \boldsymbol{n}\mathrm{d}S \tag{3.90}$$

Substituting Equation 3.89 gives

$$\begin{aligned} \boldsymbol{F} &= \oint_{S_p} \mu \boldsymbol{P}\cdot\left(\boldsymbol{u}-\boldsymbol{u}_0\right)\cdot \boldsymbol{n}\mathrm{d}S \\ &= \mu\left(\oint_{S_p} \boldsymbol{P}^T\cdot \boldsymbol{n}\mathrm{d}S\right)\cdot\left(\boldsymbol{u}-\boldsymbol{u}_0\right) \end{aligned} \tag{3.91}$$

Defining the translation tensor $\boldsymbol{K}$, which depends on the shape of the particle

$$\boldsymbol{K}\left(f\left(\boldsymbol{r}^*\right)\right) = \oint_{S_p} \boldsymbol{P}^T\left(\boldsymbol{r}^*\right)\cdot \boldsymbol{n}\mathrm{d}S \tag{3.92}$$

Then, Equation 3.91 can be written by

$$\boldsymbol{F} = \mu \boldsymbol{K}\cdot\left(\boldsymbol{u}-\boldsymbol{u}_0\right) \tag{3.93}$$

Equation 3.93 is known as *Stokes equation.* It shows that the hydrodynamic force exerted by a fluid on a solid translating with constant velocity through the fluid is a linear function of the relative particle fluid velocities. The linear transformation $\boldsymbol{K}\left(f\left(\boldsymbol{r}^*\right)\right)$ is called the *translation tensor*, which is a function of the shape of the particle and is independent of the fluid properties. Brenner (1966) demonstrate that the translation tensor is symmetric $\boldsymbol{K} = \boldsymbol{K}^T$.

A fixed particle has velocity $\boldsymbol{u} = \boldsymbol{0}$, in that case

$$\boldsymbol{F} = \mu \boldsymbol{K} \cdot \boldsymbol{u}_0 \tag{3.94}$$

For the special case of a spherical particle, the translation tensor has the value (Concha 1968)

$$\boldsymbol{K} = 6\pi R\boldsymbol{I} \tag{3.95}$$

3.4.2 Sedimentation of a Spherical Particle in a Quiescent Fluid

Consider a small solid sphere with density ρ_p submerged in a viscous fluid of density and viscosity ρ_f and μ and suspended with by a string. The forces acting on the particles are gravity F_g that pulls the sphere down, the buoyancy $\boldsymbol{F}_b$, pressure forces of the fluid on the particle that pushes the sphere upwards and the string resistance F_{string}, that supports the particle from falling. The force balance gives

$$\boldsymbol{O} = \boldsymbol{F}_{\text{string}} + \underbrace{\boldsymbol{F}_g}_{-\rho_p V_p \boldsymbol{g}} + \underbrace{\boldsymbol{F}_b}_{+\rho_f V_p \boldsymbol{g}} \tag{3.96}$$

$$\boldsymbol{F}_{\text{string}} = \underbrace{\Delta\rho V_p \boldsymbol{g}}_{\text{Netweight}} \tag{3.97}$$

If the string is cut, the particle will accelerate, and three terms appear: acceleration, drag force, and added mass. For a sphere, with volume $V_p = (\pi/6)d^3$,

$$\underbrace{\rho_p V_p \dot{u}(t)}_{\text{Accelaration}} + \frac{1}{2}\underbrace{\rho_p V_p \dot{u}(t)}_{\text{Added mass}} = \underbrace{\Delta\rho V_p g}_{\text{Net weight}} - \underbrace{6\pi\mu R u(t)}_{\text{Drag force}}$$

$$\dot{u}(t) + \frac{2}{3}\left(\frac{18\mu}{\rho_p d^2} u(t)\right) - \frac{2}{3}\frac{\Delta\rho g}{\rho_p} = 0 \tag{3.98}$$

The solution to this differential equation is

$$u(t) = \frac{1}{18}\frac{\Delta\rho d^2 g}{\mu}\left(1 - \exp\left(-\frac{2}{3}\frac{\mu}{18\rho_p d^2} t\right)\right) \tag{3.99}$$

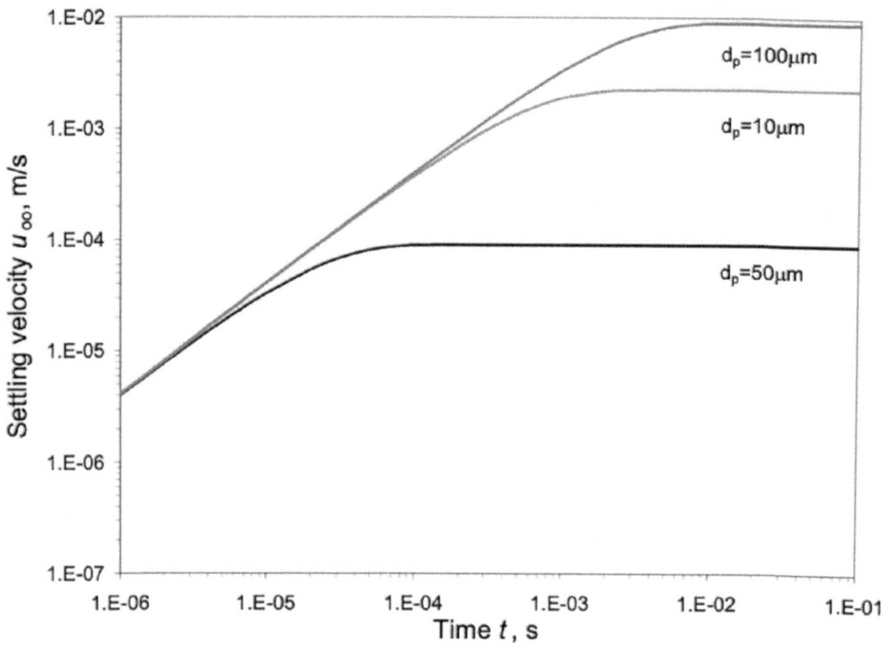

FIGURE 3.12 Evolution of the settling velocity and terminal velocity.

Figure 3.12 gives the function of sedimentation and time for spheres of 100, 50, and 10 μm and 2.64 g/cm^3 density in water at 20°C.

For time $t = \infty$, the exponential terms vanish, and Equation 3.99 yields the terminal settling velocity of a sphere in a viscous fluid u_∞, known as *Stokes law*.

$$u(t) = \frac{1}{18}\frac{\Delta\rho d^2 g}{\mu} \tag{3.100}$$

3.4.3 Sedimentation of a Particle in Oscillating Fluids

The motion of solid particles in oscillating fluids forms part of several mineral processing operations. Before the advent of flotation, gravity methods were the most important of the concentration processes. Agricola (1556) shows equipment such as trough, strakes, canvas, and jigs for "washing" tin ores.

Equipment for gravity concentration such as shaking tables and jigs are still used today for concentration and pre-concentration operations of heavy minerals.

A shaking table is a concentration device, consisting of a deck inclined slightly from the horizontal, on which a film of water flows continuously, and particles are fed. The combined effect of oscillation and water flow segregate the particles according to specific gravity and particle size.

In the case of jigging (Gaudin 1939; Taggart 1951; Richard and Locke 1940; Jong et al. 1996; Srinivasa et al. 1999; Tavares 1999), the oscillations are vertical and produce the motion of a consolidated bed of solid and fluid particles of different sizes and densities. This phenomenon is known as *consolidation trickling*, in which the coarse particles touch each other at some point during the jigging cycle, forming bridges that immobilize them, leaving the small some freedom to settle through the bed. In this form, layers of similar specific gravity have wide range of particle sizes.

In both of this processes, tabling and jigging, the characteristic feature is the use of an unsteady fluid field, which give rise to extra hydrodynamic forces and drag between the solid and the fluid.

Several works have been published on the settling of particles in oscillating flow field studying the effect of vertical oscillation (Baird et al. 1967), particle trajectories (Houghton 1966, 1968), settling particles of arbitrary shape (Concha 1968), and particle motion (Hwang 1983; Ikeda 1989).

In this section we direct our attention in solving the equation of motion of particles of arbitrary shape and spheres as an example, settling under the influence of gravity and oscillatory velocity fields.

3.4.3.1 Oscillating Fluids

Consider a viscous oscillating fluid, the following dimensionless variables:

$$r^* = r/d, t^* = \omega t, v^* = v/u_0, p^* = d/\mu u_0, \nabla^* = d\nabla, b^* = b/g \tag{3.101}$$

where ω is the frequency of oscillation, u_0 is the characteristic velocity, and d is the characteristic particle dimension. With these new variables, the dimensionless momentum equation

$$\rho\frac{\partial \boldsymbol{v}}{\partial t} + \rho\nabla\boldsymbol{v}\bullet\boldsymbol{v} = \nabla p + \mu\nabla^2\boldsymbol{v} + \rho\boldsymbol{b} \tag{3.101a}$$

is expressed in the form

$$\left(\rho\omega u_0\right)\frac{\partial \boldsymbol{v}^*}{\partial t^*} + \left(\frac{\rho d u_0}{\mu}\right)\nabla^*\boldsymbol{v}^*\bullet\boldsymbol{v}^* = \left(\frac{\mu u_0}{d^2}\right)\nabla^* p^* + \left(\frac{\mu u_o}{d^2}\right)\nabla^{*2}\boldsymbol{v}^* + \rho g\boldsymbol{b}^* \tag{3.101b}$$

Dividing by the coefficient of the viscous term:

$$\left(\frac{d^2\omega}{\nu}\right)\frac{\partial \boldsymbol{v}}{\partial t} + \left(\frac{d\rho}{\nu}\right)\nabla^*\boldsymbol{v}^*\cdot\boldsymbol{v}^* = \nabla^* p^* + \nabla^{*2}\boldsymbol{v}^* + \left(\frac{\rho d^2 g}{\mu u_0}\right)\boldsymbol{b}^* \tag{3.102}$$

Defining the following dimensionless numbers:

Vibrational Reynolds number: $\mathrm{Re}_v = \dfrac{\rho d^2\omega}{\mu}$

Translational Reynolds number: $\mathrm{Re}_t = \dfrac{\rho d u_0}{\mu}$

Froude number: $Fr = \dfrac{u_0^2}{dg}$

$$\frac{\mathrm{Re}_t}{Fr} = \frac{\rho d^2 g}{\mu u_0}$$

Using these dimensionless numbers in Equation 3.102 results in the final form of the dimensionless *Navier-Stokes equation for oscillatory motions*:

$$\mathrm{Re_v}\frac{\partial v^*}{\partial t^*} + \mathrm{Re_t}\underbrace{\nabla^* v^* \cdot v^*}_{\text{Convective tem}} = -\nabla^* p^* + \underbrace{\nabla^{*2} v^*}_{\text{Difisise tem}} + \frac{\mathrm{Re}_t}{Fr}\boldsymbol{b}^* \tag{3.103}$$

We can expect those simple solutions of the dimensionless Navier-Stokes equation are possible for extreme values of one of the Reynolds numbers. For oscillation in Stokes regimes $\mathrm{Re}_{t_t} = 0$ so that if we use for this case $\dot{v} \equiv \partial \boldsymbol{v} / \partial t$ the dimensional Navier-Stokes equation can be written in the form,

$$\begin{aligned} \nabla \cdot \boldsymbol{v} &= 0 \\ \dot{\boldsymbol{v}} &= -\frac{1}{\rho}\nabla p + \nu \nabla^2 + \rho \boldsymbol{g} \end{aligned} \tag{3.104}$$

with boundary conditions for a reference frame fixed on the particle,

$$\boldsymbol{v}(\boldsymbol{R},t) = 0, \boldsymbol{v}(\boldsymbol{r} \to \infty, t) = \boldsymbol{u}(t) \text{ and } p(\boldsymbol{r} \to \infty, t) = 0 \tag{3.105}$$

3.4.3.2 Velocity and Pressure Fields

Because Equation 3.104 is linear, the velocity field, the pressure field, and boundary conditions are linear functions of $\boldsymbol{u}$ and $\dot{\boldsymbol{u}}$ at any point in the fluid (Concha 1968).

$$\boldsymbol{v}(t) = \boldsymbol{V} \cdot \boldsymbol{u}(t) \text{ and } p(\boldsymbol{r},t) = \mu \boldsymbol{P}_1 \cdot \boldsymbol{u}(t) + \rho \cdot \boldsymbol{P}_2 \cdot \dot{\boldsymbol{u}}(t) \tag{3.106}$$

Where $\boldsymbol{V}$, $\boldsymbol{P}_1$, and $\boldsymbol{P}_2$ are the dyadic velocity field and the vector pressure fields (Concha 1968).

$$\boldsymbol{V} = \nabla\nabla f(\boldsymbol{R}) - \nabla^2 f(\boldsymbol{R})\boldsymbol{I} \text{ and } \boldsymbol{P}_1 = \nabla\nabla^2 f(\boldsymbol{R}) \text{ and } \boldsymbol{P}_2 = -\nabla f(\boldsymbol{R}) \tag{3.107}$$

were the function $f(\boldsymbol{R})$ describes the geometry of the particle surface shape.

3.4.3.3 Hydrodynamic Force

The hydrodynamic force is given by $\boldsymbol{F}_h = \oint_{S_p} \boldsymbol{T} \cdot \boldsymbol{n} dS$, with the stress tensor given by

$$\boldsymbol{T} = \mu\left(-\boldsymbol{P}_1\boldsymbol{I} + \nabla\boldsymbol{V} + \nabla\boldsymbol{V}^T\right) \cdot \boldsymbol{u} - \rho \boldsymbol{P}_2 \boldsymbol{I} \cdot \dot{\boldsymbol{u}} \tag{3.108}$$

$$F_h = -\mu \oint_{S_P} \underbrace{\left\{ \left(P_1^T + \nabla V + \nabla V^T \right) \cdot n dS \right\}}_{X(r, f(R))} \cdot u - \rho \oint_{S_P} \underbrace{\left\{ P_2^T I \cdot n dS \right\}}_{Y(r, f(R))} \cdot u \quad (3.109)$$

$$F_h = -\mu X(r, f(R)) \cdot u - \rho Y(r, f(R)) \cdot \dot{u} \quad (3.110)$$

X and Y are second order tensor depending on the position r in the fluid and the surface shape of the particle $f(R)$ and R represents the surface of the particle.

3.4.3.4 Particle Settling in Vertically Viscous Oscillating Fluids

Consider a particle of arbitrary shape settling without rotation in a vertically oscillating fluid. Assume the following conditions (Concha 1968):

1. The oscillatory motion of the fluid is caused by some quasi-static exterior force independent of the solid particle.
2. The fluid velocity varies slightly over the order of the particle dimension.
3. The unperturbed fluid velocity field is given by $v(t) = v_0 e^{-i\omega t}$ and $\dot{v}(t) = -i\omega v_0 e^{-i\omega t}$.
4. Gravity is the only body force.

The linear momentum balance is given by

$$\underbrace{\frac{dm_p}{dt}}_{\substack{\text{Particle}\\ \text{momentum}}} + \underbrace{\frac{dm_f}{dt}}_{\substack{\text{Fluid}\\ \text{momentum}}} = \underbrace{F_g}_{\substack{\text{Gravity}\\ \text{force}}} + \underbrace{F_b}_{\substack{\text{Buoyancy}\\ \text{force}}} + \underbrace{F_d}_{\substack{\text{Drag}\\ \text{force}}} \quad (3.111)$$

From Equation 3.110,

$$\underbrace{\rho_p V_p \dot{u}}_{\substack{\text{Momentum}\\ \text{rate of change}}} = \underbrace{\rho_p V_p g}_{\text{weight}} - \underbrace{\rho_f V_p g + \rho_f V_p \dot{v}}_{\text{buoyancy}} - \underbrace{\mu X \cdot (u - v) - \rho_f Y \cdot (\dot{u} - \dot{v})}_{\text{Drag}} \quad (3.112)$$

Defining $\lambda = \rho_g / \rho_f$ and rearranging

$$\begin{aligned} \dot{u} + \frac{\mu}{\rho_f}\left(\lambda V_p I + Y\right)^{-1} X \cdot u = (\lambda - 1) V_p \left(\lambda V_p I + Y\right)^{-1} \cdot g \\ + \left(\lambda V_p I + Y\right)^{-1} \left(V_p I + Y\right) \cdot \dot{v} + \\ + \frac{\mu}{\rho_f}\left(\lambda V_p I + Y\right)^{-1} X \cdot v \end{aligned} \quad (3.113)$$

Defining the second order tensors A, B, and C by

$$A = \frac{\mu}{\rho_f}\left(\lambda V_p \boldsymbol{I} + \boldsymbol{Y}\right)^{-1} \cdot \boldsymbol{X}$$

$$\boldsymbol{B} = (\lambda - 1)V_p\left(\lambda V_p \boldsymbol{I} + \boldsymbol{Y}\right)^{-1} \cdot \boldsymbol{g} \tag{3.114}$$

$$\boldsymbol{C} = \left(\lambda V_p \boldsymbol{I} + \boldsymbol{Y}\right)^{-1} \cdot \left(V_p \boldsymbol{I} + \boldsymbol{Y}\right)$$

The sedimentation velocity of the particle may be written as

$$\dot{u} + A\bullet\boldsymbol{u} = \boldsymbol{B} + \boldsymbol{C}\cdot\dot{v} + A\bullet\boldsymbol{v} \tag{3.115}$$

Substituting the value of $v(t) = v_0 e^{-i\omega t}$ and $v(t) = -i\omega v_0 e^{-i\omega t}$ in Equation 3.115 yields the differential equation for the sedimentation of the particle in the oscillatory field

$$\dot{\boldsymbol{u}} + \boldsymbol{A}\cdot\boldsymbol{u} = \boldsymbol{B} - \left(\boldsymbol{C}\cdot\boldsymbol{v}_0\right)i\omega e^{-i\omega t} + \left(\boldsymbol{A}\cdot\boldsymbol{v}_0\right)e^{-i\omega t} \tag{3.116}$$

This is an inhomogeneous first order differential equation, the solution of which is the sum of a homogeneous $\boldsymbol{u}_h$ and a particular solution $\boldsymbol{u}_p$.

$$\boldsymbol{u}_h = \boldsymbol{D}\cdot \mathrm{e}^{-\boldsymbol{A}t}, \boldsymbol{u}_p = \boldsymbol{E} + \boldsymbol{H}e^{-i\omega t} \text{ and } \dot{\boldsymbol{u}}_p = -i\omega\boldsymbol{H}e^{-i\omega t} \tag{3.117}$$

Substituting $\boldsymbol{u}_n$, $\boldsymbol{u}_p$, and $\dot{u}_p$ from Equation 3.117 in Equation 3.116 gives

$$\boldsymbol{H}\cdot(\boldsymbol{A} - i\omega\boldsymbol{I})e^{-i\omega t} + \boldsymbol{A}\cdot\boldsymbol{D}\cdot\mathrm{e}^{-\boldsymbol{A}t} + \boldsymbol{A}\cdot(\boldsymbol{E}+\boldsymbol{H})e^{-i\omega t} = \boldsymbol{B} + (\boldsymbol{A}\cdot - i\omega\boldsymbol{C}\cdot)\boldsymbol{v}_0 e^{-i\omega t} \tag{3.118}$$

Equation 3.118 is the general solution to Equation 3.116. Separating the real and imaginary part leads to

$$\boldsymbol{A}\cdot\boldsymbol{D}\cdot\mathrm{e}^{-\boldsymbol{A}t} + \boldsymbol{A}\cdot\boldsymbol{E} = \boldsymbol{B}$$

$$\boldsymbol{H}\cdot(\boldsymbol{A} - i\omega\boldsymbol{I})e^{-i\omega t} = (\boldsymbol{A} - i\omega\boldsymbol{C}\cdot)\boldsymbol{v}_0 e^{-i\omega t}$$

The solution of which is

$$\boldsymbol{E} = \boldsymbol{A}^{-1}\cdot\boldsymbol{B}$$

$$\boldsymbol{H} = (\boldsymbol{A} - i\omega\boldsymbol{I})^{-1}\cdot(\boldsymbol{A} - i\omega\boldsymbol{C})\cdot\boldsymbol{v}_0 \tag{3.119}$$

Substituting Equation 3.119 in Equation 3.115, the value of the velocity field is

$$\boldsymbol{u} = \boldsymbol{D}\cdot e^{-\boldsymbol{A}t} + \boldsymbol{A}^{-1}\cdot\boldsymbol{B} + (\boldsymbol{A} - i\omega)^{-1}\cdot(\boldsymbol{A} - i\omega\boldsymbol{C})\cdot\boldsymbol{v}_0 e^{-i\omega t} \tag{3.120}$$

If initially the particle is at rest, the initial conditions is $u(0) = 0$, then

$$\boldsymbol{D} = -\boldsymbol{A}^{-1} \cdot \boldsymbol{B} - (\boldsymbol{A} - i\omega \boldsymbol{I})^{-1} \cdot (\boldsymbol{A} - i\omega \boldsymbol{C}) \cdot \boldsymbol{v}_0 \cdot \boldsymbol{e}^{-i\omega t} \tag{3.120a}$$

Substituting in Equation 3.120 yields

$$\boldsymbol{u}(t) = \left(\boldsymbol{A}^{-1} \cdot \boldsymbol{B}\right) \cdot \left(\boldsymbol{I} - \boldsymbol{e}^{-\boldsymbol{A}t}\right) + (\boldsymbol{A} - i\omega)^{-1} \cdot (\boldsymbol{A} - i\omega \boldsymbol{C}) \cdot \left(\boldsymbol{I}\boldsymbol{e}^{-i\omega t} - \boldsymbol{e}^{-\boldsymbol{A}t}\right) \cdot \boldsymbol{v}_0(t) \tag{3.121}$$

At quasi-steady state

$$\boldsymbol{u}(t) = \left(\boldsymbol{A}^{-1} \cdot \boldsymbol{B}\right) + (\boldsymbol{A} - i\omega)^{-1} \cdot (\boldsymbol{A} - i\omega \boldsymbol{C}) \cdot \boldsymbol{v}_0(t) \tag{3.122}$$

Equation 3.122 represents the settling of a particle of arbitrary shape in a harmonic oscillatory fluid field at steady state.

3.4.3.5 Sedimentation of Spherical Particles in Sinusoidal Oscillations

For spherically isotropic particles settling in a vertically oscillating fluid, with their axis of symmetry parallel to the gravitational field direction, all second order tensors are isotropic and can be expressed in the generic form $\boldsymbol{Z}_j = Z_j \boldsymbol{k}$, then Equation 3.122 becomes

$$\boldsymbol{u} = \frac{B}{A}\boldsymbol{k} + \frac{(A - i\omega C)}{(A - i\omega)} v_0 \boldsymbol{k} e^{-i\omega t} \tag{3.122a}$$

Writing $e^{-i\omega t} = \cos \omega t - i \operatorname{sen} \omega t$

$$\boldsymbol{u} = \frac{B}{A}\boldsymbol{k} + \frac{(A - i\omega C)(A - i\omega)}{A^2 - \omega^2} (\cos \omega t - i \operatorname{sen} \omega t) v_0 \boldsymbol{k} \tag{3.122b}$$

Conserving the real part only results in

$$\boldsymbol{u} = \frac{B}{A}\boldsymbol{k} + \left(\frac{A^2 + \omega^2 C}{A^2 + \omega^2} \cos \omega t + \frac{A\omega(1 - C)}{A^2 + \omega^2} \operatorname{sen} \omega t \right) v_0 \boldsymbol{k} \tag{3.123}$$

Since the amplitude of oscillation is v_0 and the phase lag by θ, defining

$$\frac{A^2 + \omega^2 C}{A^2 + \omega^2} v_0 = u_0 \cos\theta, \frac{A\omega(1 - C)}{A^2 + \omega^2 C} v_0 = u_0 \operatorname{sen} \theta, \text{ then } \tan\theta = \frac{A\omega(1 - C)}{A^2 + \omega^2} \tag{3.124}$$

Equation 3.123 becomes

$$\boldsymbol{u} = \frac{B}{A}\boldsymbol{k} + u_0 \cos(\cos \omega t - \theta)\boldsymbol{k} \tag{3.125}$$

Combining both terms in Equation 3.124 yields

$$\frac{u_0}{v_0}=\frac{\sqrt{A^2\omega^2(1-C)^2+\left(A^2+\omega^2C\right)^2}}{\left(A^2+\omega^2\right)} \tag{3.126}$$

Defining $\alpha = A/\omega$, the third term of Equation 3.124 may be written

$$\tan\theta=\frac{\alpha(1-C)}{\alpha^2+C} \tag{3.127}$$

For a spherical particle the following relationship holds the following (Concha 1968):

$$\begin{aligned}&K=6\pi R,\quad \mathrm{Re}_v=\frac{(2R)^2\omega}{v},\quad X=6\pi R\phi_1,\quad Y=\frac{2}{3}\pi R^3\phi_2\\&\phi_1=1+\frac{\sqrt{2}}{4}\mathrm{Re}_v^{1/2},\quad \phi_2=1+9\sqrt{2}\mathrm{Re}_v^{-1/2},\quad \phi_3=36\mathrm{Re}_v^{-1}\end{aligned} \tag{3.128}$$

The parameters $\boldsymbol{A}$, $\boldsymbol{B}$, and $\boldsymbol{C}$ become (Concha 1968):

$$\begin{aligned}&A=\frac{9u\phi_1}{2\lambda+\phi_2}\frac{1}{R^2},\quad B=\frac{2(\lambda-1)g}{2\lambda+\phi_2},\quad C=\frac{2+\phi_2}{2\lambda+\phi_2},\quad \alpha=\frac{\phi_1\phi_3}{2\lambda+\phi_2},\quad \omega=\frac{v\mathrm{Re}_v}{4R^2}\\&u=\frac{2(\lambda-1)gR^2}{9v\phi_1}+\cos(\omega t-\theta)u_0\\&\theta=\arctan\left(\frac{2\phi_1\phi_3(\lambda-1)}{\phi_1^2\phi_3^2+(2\lambda+\phi_2)(2+\phi_2)}\right)\end{aligned} \quad 3.129)$$

The terminal settling velocity u_t is the first term of Equation 3.129.

$$u_t=\frac{2\left(\rho_p-\rho_f\right)R^2g}{9\mu\left(1+\frac{\sqrt{2}}{4}\mathrm{Re}_v^{1/2}\right)} \tag{3.130}$$

Plotting u_i and u_i/v_0 versus the vibrational Reynolds number is shown in Figure 3.13.

Figure 3.13 shows that the terminal velocity of the particles diminishes with the vibrational Reynolds number, which indicates that at a high value of this parameter the particle would remain suspended in the fluid. It also shows the magnitude of the settling velocity may be regulated with the frequency of oscillations.

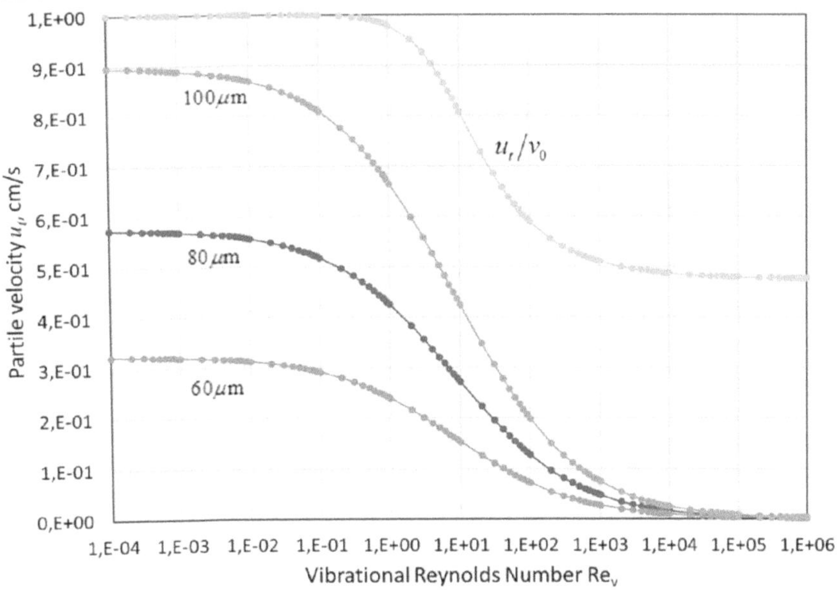

FIGURE 3.13 Terminal settling velocities u_t in cm/s as function of the vibrational Reynolds number for spherical particles of 100, 80, and 60 μm and u_t/v_0, where v_0 is the velocity of the fluid.

3.5 PRANDTL'S FLOW

For intermediate values of the Reynolds number, inertial and viscous forces in the fluid are of the same order of magnitude (see Section 3.2.2). In this case, the flow may be divided into two parts, an external inviscid flow far from the particle and an internal flow near the particle, where the viscosity plays an important role. This picture forms the basis of the *boundary-layer theory* (Schlichting 1955; Bird et al. 1964).

In the external inviscid flow, Euler's equations are applicable, and for the potential flow around a particle, the velocity and pressure distributions are obtained from Equations 3.138 and 3.139. The region of viscous flow near the particle is known as the boundary layer, and it is there where a steep velocity gradient permits the nonslip condition at the solid surface to be satisfied. The energy dissipation produced by the viscous flow within the boundary layer retards the flow, and at a certain point, aided by the adverse pressure gradient, the flow reverses its direction. These phenomena force the fluid particles outward and away from the solid particle producing the phenomena called *boundary layer separation*, which occurs at an angle of separation given by Lee and Barrow (1968). The boundary layer created on the surface of particles changes according to the roughness of the surface. Figure 3.14 shows the analysis and experiments conducted by Scobie et al. (2014) with golf ball and a cricket ball using an infrared camera to capture thermal images using

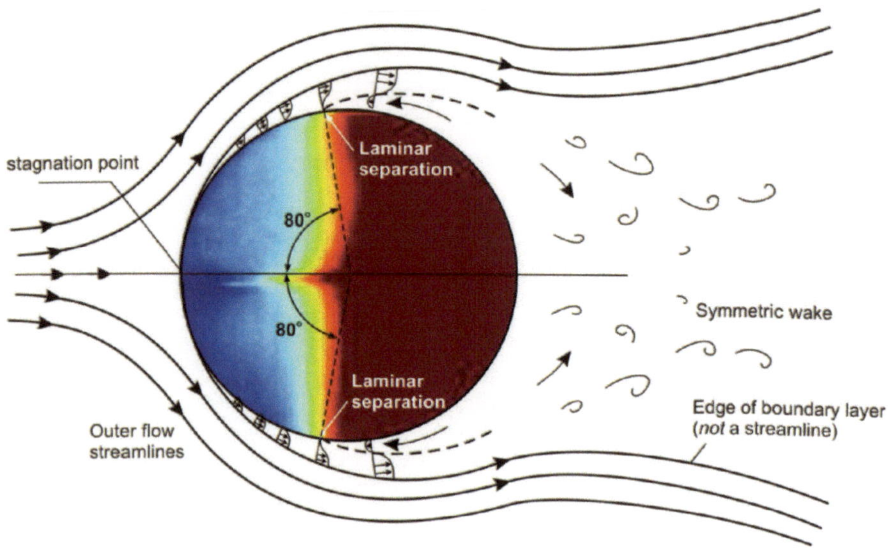

FIGURE 3.14 Schematic view of the boundary layer for the flow over a smooth sphere.

Source: Scobie, J.A., Sangan, C.M., and Lock, G.D. 2014. Flow visualization experiments on sports balls. *Procedia Engineering* 72:738–743.

vanadium oxide Micro-Bolometer Array (MBA) type. As such the different shapes of the boundary layer are detected based on the type of roughness. This effect allows a particle to travel a longer distance such as a golf ball compared to a cricket ball for example.

The thickness δ of the boundary layer, defined as the distance from the solid surface to the region where the tangential velocity reaches 99% of the value of the external inviscid flow, is proportional to $Re^{-0.5}$ and, at the point of separation, may be written in the form:

$$\frac{\delta}{R} = \frac{\delta_0}{Re^{1/2}} \tag{3.131}$$

giving a value of $\delta_0 = 9.06$.

The separation of the boundary layer prevents the recovery of the pressure at the rear of the sphere, resulting in an asymmetrical pressure distribution with a higher pressure at the front of the sphere. Figure 3.15 shows a plot of the pressure coefficient C_p of a sphere in terms of the distance from the front stagnation point over the surface of the sphere in an inviscid flow and in boundary layer flow.

The asymmetry of the pressure distribution explains the origin of the form-drag in an inviscid fluid, the magnitude of which is closely related with the position of the point of separation. The farther the separation points from the front stagnation

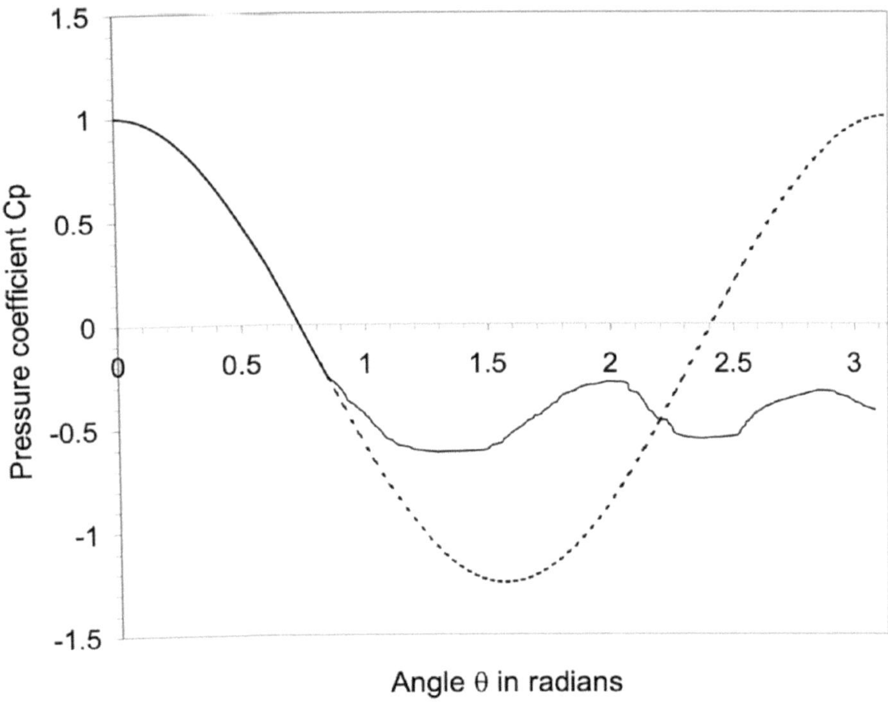

FIGURE 3.15 Pressure coefficient versus angle θ at the surface of the sphere.

Source: Concha, F.A. 2014. *Solid–Liquid Separation in the Mining Industry.* Cham: Springer. p. 52, Fig. 4.8.

point, the smaller the form-drag. For a sphere at high Reynolds number between $10^4 < \mathrm{Re} < 1.5 \times 10^5$, the position of the separation point does not change except with the change of flow from laminar to turbulent. Therefore, the form-drag will remain approximately constant between those Reynolds numbers. At the same time, the friction-drag, also called skin friction, falls proportional to $\mathrm{Re}^{-1/2}$. From these observations, we can conclude that, for Reynolds numbers of about $\mathrm{Re} = 1.000$, the viscous interaction force has diminished sufficiently for its contribution to the total interaction force to be negligible. Therefore, between $10^3 < \mathrm{Re} < 10^4$, the drag coefficient is approximately constant at $C_D \approx 0.44$.

For Reynolds numbers greater than $\mathrm{Re} \approx 150.000$, the flow changes in character and the boundary layer becomes turbulent. The increase in kinetic energy of the external region permits the flow in the boundary layer to reach further to the back of the sphere, shifting the separation point to values close to $\theta_s \approx 110°$ and also permitting the base pressure to rise. The effect of these changes on the drag coefficient is a sudden drop and, after that, a sharp increase with the Reynolds number. Figure 3.16 shows a plot of standard experimental values of the drag coefficient versus the Reynolds number where this effect is shown.

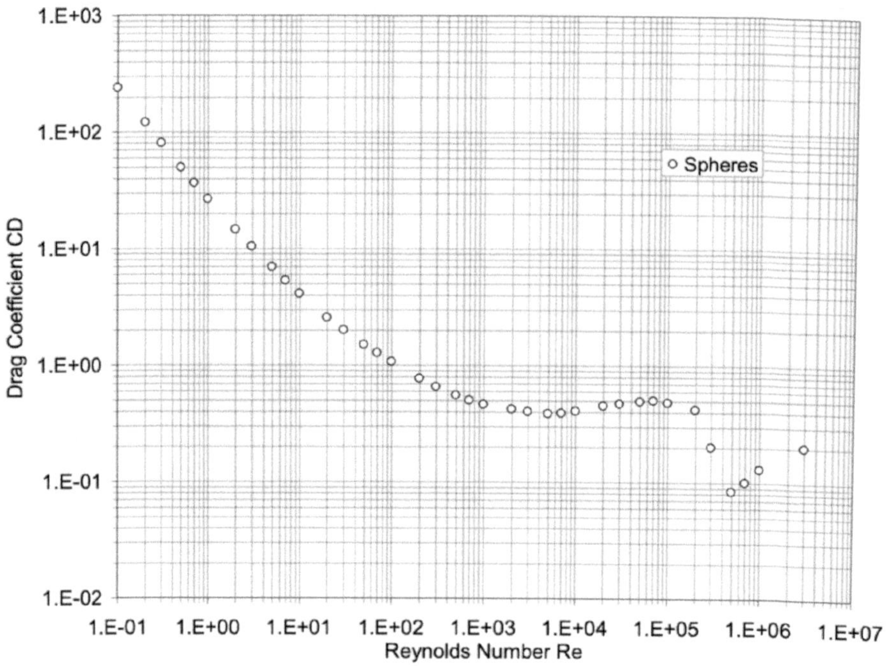

FIGURE 3.16 Drag coefficient C_D versus Reynolds number Re according to standard data from Lapple and Shepherd (1940). See also Perry and Chilton (1973, p. 5.62).

3.5.1 Drag Coefficient for a Sphere for the Range 0 < Re < 150.000

Figure 3.17 confirms that for small Reynolds number, the drag coefficient varies as $\mathrm{Re}^{-1/2}$. To obtain a general equation relating the drag coefficient to the Reynolds number, we use the concept of boundary layer and the knowledge that, for a given position at the surface of the sphere, the pressure inside the boundary layer is equal to the pressure in the inviscid region just outside the boundary layer before the separation point, and it is a constant beyond. We should also remember that the point of separation and the base pressure are constant for Reynolds numbers greater than $\mathrm{Re} \approx 10^3$.

Consider a solid sphere of radius R with an attached boundary layer of thickness δ submerged in a flow at high Reynolds number. Assume that the system sphere and boundary layer have a spherical shape with a radius equal to $a \approx R + \delta$, as shown in Figure 3.17 (Concha and Almendra 1979).

Outside the spherical shell of radius a and up to the point of separation θ_s, the flow is inviscid and therefore the fluid velocity and the pressure distributions are given by the following:

$$u_\theta(\theta) = \frac{3}{2} u \sin\theta \text{ and } p(\theta) = \frac{1}{2}\rho_f u^2 \left(1 - \left(\frac{u_\theta}{u}\right)^2\right) \text{from } 0 \le \theta \le \theta_s \quad (3.132)$$

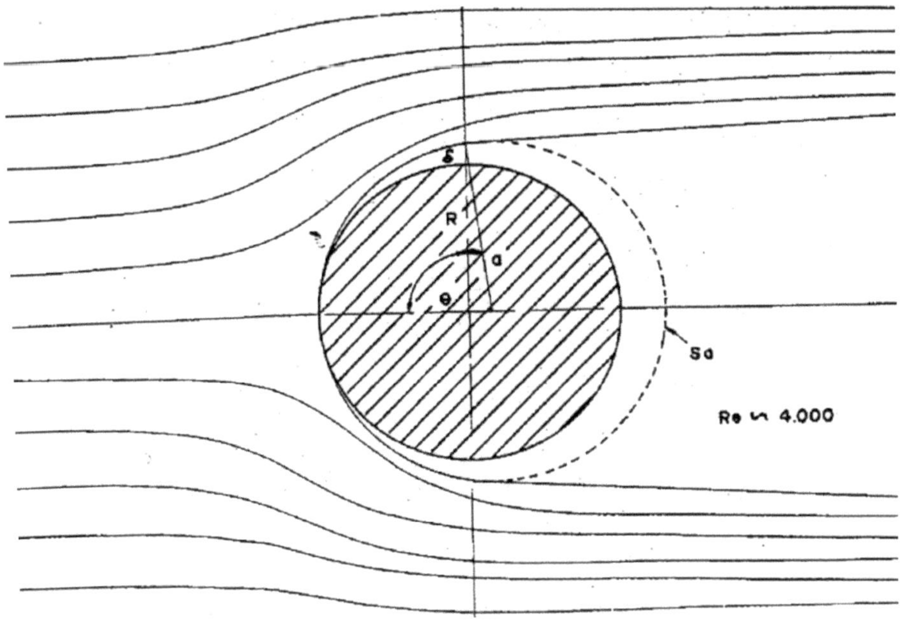

FIGURE 3.17 Physical model for the flow in boundary layer around a sphere.

Source: Concha, F., and Almendra, E.R. 1979. Settling velocities of particulate systems, 1. Settling velocities of individual spherical particles. *International Journal of Mineral Processing* 5:349–367.

Beyond the separation point, a region exists where the pressure is constant and equal to the base pressure p_b

$$p(\theta) = \frac{3}{2} usen\theta \text{ and } p(\theta) = p_b \text{ from } \theta_s \leq \theta \leq \pi \tag{3.133}$$

Since the effect of viscosity is confined to the interior of the sphere of radius a, the drag excerpted by the fluid on a, consists of the form drag only, then

$$F_D = \oint_{S_a} p(\theta) \cos\theta dS \tag{3.134}$$

The element of surface of the sphere of radius a is as follows:

$$dS = a^2 sen\theta \; d\theta \, d\phi \tag{3.135}$$

where ϕ is the azimuthal coordinate. Replacing into Equation 3.134 results as follows:

$$F_D = \int_0^{2\pi} \int_0^{\pi} p(\theta) sen\theta \cos\theta d\theta d\phi = 2\pi a^2 \int_0^{\pi} p(\theta) sen\theta \cos\theta d\theta$$

Since the values of $p(\theta)$ is different before and after the separation point, the integrals may be separate in two parts, from 0 to θ_s and from θ_s to π:

$$F_D = 2\pi a^2 \left(\int_0^{\theta_s} p(\theta) sen\theta \, d(sen\theta) + \int_{\theta_s}^{\pi} p(\theta) sen\theta \, d(sen\theta) \right) \quad (3.135a)$$

Substituting the values of $p(\theta)$ from Equation 3.133 and integrating we obtain:

$$F_D = \pi a^2 \rho_f u^2 \left(\frac{1}{2} sen^2\theta_s - \frac{9}{16} sen^4\theta_s - \frac{1}{2} p_b^* sen^2\theta_s \right) \quad (3.136)$$

Substituting $a = k + \delta$ and defining the function $f(\theta_s, p_b^*)$ in the form:

$$f(\theta_s, p_b^*) = \frac{1}{2} sen^2\theta_s - \frac{9}{16} sen^4\theta_s - \frac{1}{2} p_b^* sen^2\theta_s \quad (3.137)$$

Equation 3.136 may be written

$$F_D = \rho_f \pi R^2 u^2 \left(1 + \frac{\delta}{R} \right)^2 f(\theta_s, p_b^*). \quad (3.138)$$

In terms of the drag coefficient C_L :

$$C_D = 2f(\theta_s, p_b^*) \left(1 + \frac{\delta}{R} \right)^2 \quad (3.139)$$

and defining the new parameter C_0 in the form:

$$C_0 = 2f(\theta_s, p_b^*) \quad (3.140)$$

and using Equation 3.139 yields:

$$C_D = C_0(\theta_s, p_b^*) \left(1 + \frac{\delta_0}{Re^{1/2}} \right)^2 \quad (3.141)$$

The value of $f(84^\circ, -0.4) = 0.142$ so that from Equation 3.140, $C_0 = 0.284$. The value of, $\delta_0 = 9.06$ therefore

$$C_D = 0.284 \left(1 + \frac{9.06}{Re^{1/2}} \right)^2 \quad (3.142)$$

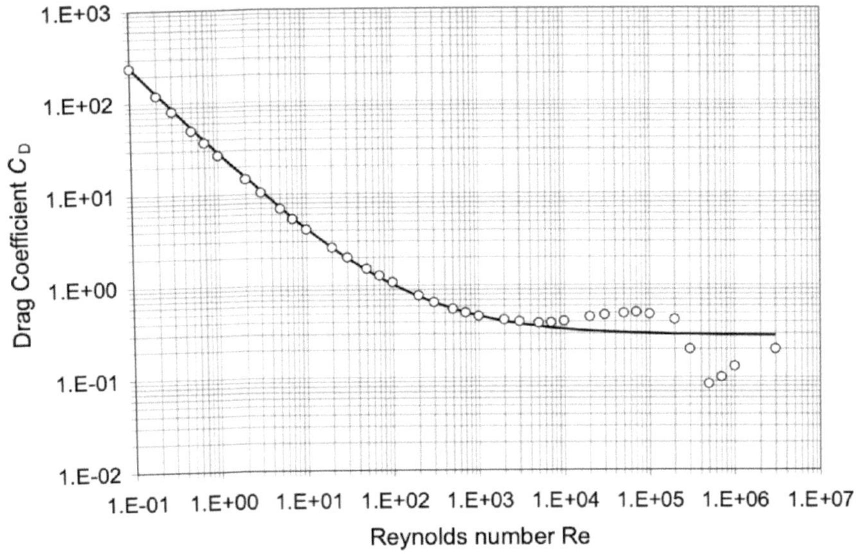

FIGURE 3.18 Drag coefficient versus Reynolds number for a sphere.

Source: Concha, F., and Almendra, E.R. 1979. Settling velocities of particulate systems, 1. Settling velocities of individual spherical particles. *International Journal of Mineral Processing* 5:349–367.

Equation 3.142 represents the drag coefficient for a sphere in boundary layer flow (Concha and Almendra 1979). A comparison with the experimental data of Lapple and Shepherd (1940) is shown in Figure 3.18.

Dallavalle (1954) presented a similar equation in the form $C_D = (0.63 + 4.8/\mathrm{Re}^{1/2})^2$ that, if we put it in the same form of Equation 3.142, yields $C_D = 0.63\left(1 + 7.62/\mathrm{Re}^{1/2}\right)^2$.

3.5.2 Sedimentation of a Sphere for Higher Reynolds Numbers

We have seen that when a particle settles at terminal velocity u_∞ the drag force is the net weight of the particle

$$F_D = -\left(-\rho_p V_p g + \rho_f V_p g\right) \equiv \Delta\rho V_p g \tag{3.143}$$

This equation shows that, once the volume and the solid-fluid density are known, the drag force for a particle in sedimentation is known beforehand. For a spherical particle, and $V_p = 4/3\pi R^3$.

The drag coefficient and the Reynolds number are

$$\#C_D = \frac{4}{3}\frac{{}''\rho dg}{\rho_f u_\infty^2} \quad \text{and} \quad \mathrm{Re}_\infty = \frac{du_\infty \rho_f}{\mu_f} \tag{3.144}$$

where the sphere diameter is $d = 2R$ and u_∞ is the terminal settling velocity of a sphere in an infinite fluid.

Combining both terms yields two dimensionless numbers (Heywood 1962):

$$C_D \,\mathrm{Re}_\infty^2 = \left(\frac{4}{3}\frac{\Delta\rho\rho_f g}{\mu_f^2}\right) d^3 \qquad C_D \frac{Re_\infty}{C_D} = \left(\frac{3}{4}\frac{\rho_f^2}{\Delta\rho\mu_f g}\right) u_\infty^3 \tag{3.145}$$

Concha and Almendra (1979) defined the characteristic parameters P and Q of the solid-liquid system by

$$P = \left(\frac{3}{4}\frac{\mu_f^2}{\Delta\rho\rho_f g}\right)^{1/3} \qquad Q = \left(\frac{4}{3}\frac{\Delta\rho\mu_f g}{\rho_f^2}\right)^{1/3} \tag{3.146}$$

so that Equation 3.145 may be written in the form:

$$C_D Re_\infty^2 = \left(\frac{d}{P}\right)^3 = d^{*3} \qquad \frac{Re_\infty}{C_D} = \left(\frac{u_\infty}{Q}\right)^3 \equiv u_\infty^{*3} \tag{3.147}$$

Equation 3.147 defines a dimensionless size d^* and a dimensionless velocity u^*, which are characteristics of a solid-liquid system:

$$d^* = \left(\frac{d}{P}\right) \qquad u_\infty^* = \left(\frac{u_\infty}{Q}\right) \tag{3.148}$$

Since there is a direct relationship between the drag coefficient and the Reynolds number, see for example Equation 3.142, there must be a similar relationship between the dimensionless groups $C_D Re_\infty^2$ and Re_∞ / C_D. Multiplying the two terms in Equation 3.147, we can observe that the Reynolds number may be written in terms of the dimensionless size and velocity:

$$\mathrm{Re}_\infty = d^* u^* \tag{3.149}$$

Replacing Equations 3.141 and 3.149 with Equation 3.147, we obtain:

$$d^{*3} = C_0 \left(1 + \frac{\delta_0}{\left(u_\infty^* d^*\right)^{1/2}}\right)^2 \left(u_\infty^* d^*\right)^2 \tag{3.150}$$

$$u_\infty^* d^* + \delta_0 \left(u_\infty^* d^*\right)^{1/2} - \frac{d^{*3/2}}{C_0^{1/2}} = 0 \tag{3.151}$$

Solving these algebraic equation, explicit expressions for the dimensionless settling velocity u_∞^* of a sphere of dimensionless size d^* are obtained (Concha and Almendra 1979):

$$u_\infty^* = \frac{1}{4}\frac{\delta_0^2}{d^*}\left(\left(1+\frac{4}{C_0^{1/2}\delta_0^2}d^{*3/2}\right)^{1/2}-1\right)^2 \tag{3.152}$$

$$d^* = \frac{1}{4}C_0 u_\infty^{*2}\left(1+\left(1+\frac{4\delta_0}{C_0^{1/2}}u_\infty^{*-3/2}\right)^{1/2}\right)^2 \tag{3.153}$$

Equations 3.152 and 3.153 are known as *Concha and Almendra's equations* for a sphere. These two equations are general for spheres settling in a fluid at any Reynolds number. Introducing the values of $C_0 = 0.28, \delta_0 = 9.06$, the following final equations are obtained:

$$u^* = \frac{20.52}{d^*}\left(\left(1+0.0921d^{*3/2}\right)^{1/2}-1\right)^2 \tag{3.154}$$

$$d^* = 0.07u^{*2}\left(1+\left(1+68.49u^{*-3/2}\right)^{1/2}\right)^2 \tag{3.155}$$

Table 3.1 shows values for the parameters P and Q for several solid densities.

3.5.2.1 Corrections to the Field Equations

The preceding sections involved hydrodynamic forces on particles moving through unbounded fluids.

As a particle settles in a suspension, each particle is surrounded by other particles hindering its motion. This hindrance is due to several effects, for example when the particle changes its position, it can find the new site occupied by another particle and will collide with it, changing its trajectory. The more concentrated the suspension is, the greater the chance of collision. The result is that hindrance is a function of concentration. On the other hand, the settling of each particle of the suspension produces a back flow of the fluid decreasing the particle velocity and retarding its sedimentation.

The problems encountered in the field of mineral processing always involve the presence of boundaries, the walls of a container, for example. The presence of walls increases the total hydrodynamic force on the particle that depends on the particle properties and to the relative value of its distance to the wall.

3.5.2.2 Correction for Particle-Particle Interaction

Several theoretical works have studied the interaction of particles in a suspension during sedimentation. Tory (1996) discusses these types of interactions.

TABLE 3.1
***P* and *Q* Parameters**

ρ_s (g/cm³)	ρ_f (g/cm³)	μ_f (g/cms)	*P* (cm)	*Q* (cm/s)	ρ_s (g/cm³)	ρ_f (g/cm³)	μ_f (g/cms)	*P* (cm)	*Q* (cm/s)
1.10	1	0.0100	9.1470E-03	1.0933E+00	3.35	1	0.0100	3.1934E-03	3.1314E+00
1.15	1	0.0100	7.9906E-03	1.2515E+00	3.40	1	0.0100	3.1711E-03	3.1535E+00
1.20	1	0.0100	7.2600E-03	1.3774E+00	3.45	1	0.0100	3.1494E-03	3.1752E+00
1.25	1	0.0100	6.7396E-03	1.4838E+00	3.50	1	0.0100	3.1282E-03	3.1967E+00
1.30	1	0.0100	6.3422E-03	1.5767E+00	3.55	1	0.0100	3.1076E-03	3.2179E+00
1.35	1	0.0100	6.0245E-03	1.6599E+00	3.60	1	0.0100	3.0876E-03	3.2388E+00
1.40	1	0.0100	5.7622E-03	1.7354E+00	3.65	1	0.0100	3.0681E-03	3.2594E+00
1.45	1	0.0100	5.5404E-03	1.8049E+00	3.70	1	0.0100	3.0490E-03	3.2798E+00
1.50	1	0.0100	5.3492E-03	1.8694E+00	3.75	1	0.0100	3.0304E-03	3.2999E+00
1.55	1	0.0100	5.1819E-03	1.9298E+00	3.80	1	0.0100	3.0123E-03	3.3198E+00
1.60	1	0.0100	5.0338E-03	1.9866E+00	3.85	1	0.0100	2.9945E-03	3.3394E+00
1.65	1	0.0100	4.9013E-03	2.0403E+00	3.90	1	0.0100	2.9772E-03	3.3588E+00
1.70	1	0.0100	4.7817E-03	2.0913E+00	3.95	1	0.0100	2.9603E-03	3.3780E+00
1.75	1	0.0100	4.6730E-03	2.1400E+00	4.00	1	0.0100	2.9438E-03	3.3970E+00
1.80	1	0.0100	4.5735E-03	2.1865E+00	4.05	1	0.0100	2.9276E-03	3.4158E+00
1.85	1	0.0100	4.4820E-03	2.2311E+00	4.10	1	0.0100	2.9118E-03	3.4343E+00
1.90	1	0.0100	4.3974E-03	2.2741E+00	4.15	1	0.0100	2.8963E-03	3.4527E+00
1.95	1	0.0100	4.3189E-03	2.3154E+00	4.20	1	0.0100	2.8811E-03	3.4709E+00
2.00	1	0.0100	4.2457E-03	2.3553E+00	4.25	1	0.0100	2.8663E-03	3.4889E+00
2.05	1	0.0100	4.1772E-03	2.3940E+00	4.30	1	0.0100	2.8517E-03	3.5067E+00
2.10	1	0.0100	4.1129E-03	2.4314E+00	4.35	1	0.0100	2.8375E-03	3.5243E+00
2.15	1	0.0100	4.0524E-03	2.4677E+00	4.40	1	0.0100	2.8235E-03	3.5417E+00
2.20	1	0.0100	3.9953E-03	2.5029E+00	4.45	1	0.0100	2.8098E-03	3.5590E+00
2.25	1	0.0100	3.9413E-03	2.5372E+00	4.50	1	0.0100	2.7963E-03	3.5761E+00
2.30	1	0.0100	3.8901E-03	2.5706E+00	4.55	1	0.0100	2.7831E-03	3.5931E+00

2.35	1	0.0100	3.8415E-03	2.6032E+00	4.60	1	0.0100	2.7702E-03	3.6098E+00
2.40	1	0.0100	3.7952E-03	2.6349E+00	4.65	1	0.0100	2.7575E-03	3.6265E+00
2.45	1	0.0100	3.7511E-03	2.6659E+00	4.70	1	0.0100	2.7450E-03	3.6430E+00
2.50	1	0.0100	3.7089E-03	2.6962E+00	4.75	1	0.0100	2.7328E-03	3.6593E+00
2.55	1	0.0100	3.6686E-03	2.7258E+00	4.80	1	0.0100	2.7207E-03	3.6755E+00
2.60	1	0.0100	3.6300E-03	2.7548E+00	4.85	1	0.0100	2.7089E-03	3.6915E+00
2.65	1	0.0100	3.5929E-03	2.7832E+00	4.90	1	0.0100	2.6973E-03	3.7075E+00
2.70	1	0.0100	3.5574E-03	2.8111E+00	4.95	1	0.0100	2.6858E-03	3.7232E+00
2.75	1	0.0100	3.5232E-03	2.8384E+00	5.00	1	0.0100	2.6746E-03	3.7389E+00
2.80	1	0.0100	3.4902E-03	2.8651E+00	5.05	1	0.0100	2.6635E-03	3.7544E+00
2.85	1	0.0100	3.4585E-03	2.8914E+00	5.10	1	0.0100	2.6527E-03	3.7698E+00
2.90	1	0.0100	3.4279E-03	2.9172E+00	5.15	1	0.0100	2.6420E-03	3.7850E+00
2.95	1	0.0100	3.3983E-03	2.9426E+00	5.20	1	0.0100	2.6315E-03	3.8002E+00
3.00	1	0.0100	3.3698E-03	2.9676E+00	5.25	1	0.0100	2.6211E-03	3.8152E+00
3.05	1	0.0100	3.3422E-03	2.9921E+00	5.30	1	0.0100	2.6109E-03	3.8301E+00
3.10	1	0.0100	3.3154E-03	3.0162E+00	5.35	1	0.0100	2.6009E-03	3.8449E+00
3.15	1	0.0100	3.2895E-03	3.0400E+00	5.40	1	0.0100	2.5910E-03	3.8596E+00
3.20	1	0.0100	3.2644E-03	3.0633E+00	5.45	1	0.0100	2.5812E-03	3.8741E+00
3.25	1	0.0100	3.2400E-03	3.0864E+00	5.50	1	0.0100	2.5716E-03	3.8886E+00
3.30	1	0.0100	3.2164E-03	3.1091E+00	5.55	1	0.0100	2.5622E-03	3.9029E+00

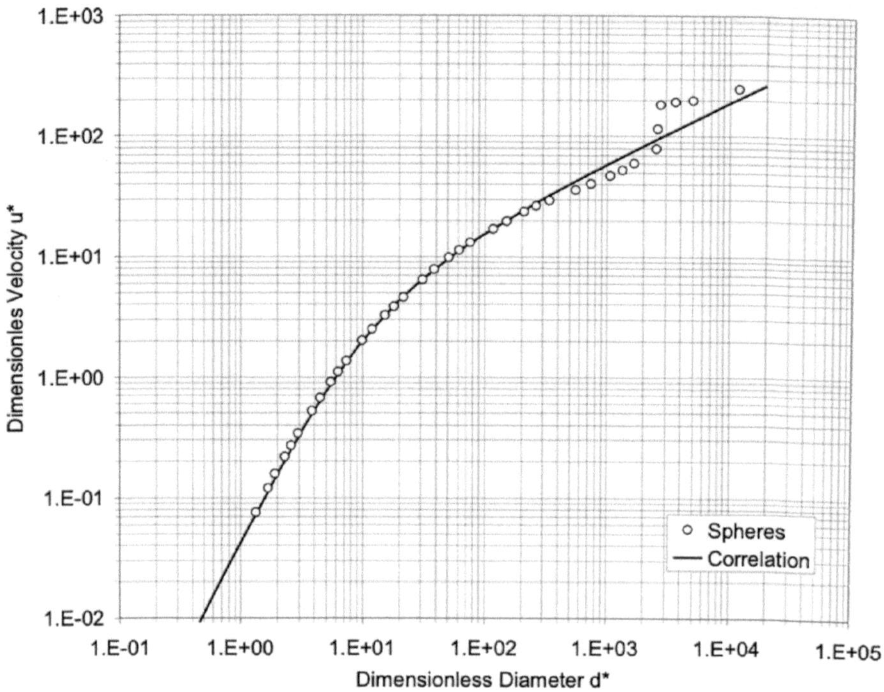

FIGURE 3.19 Dimensionless velocity versus dimensionless diameter for the sedimentation of spheres.

Source: Concha, F., and Almendra, E.R. 1979. Settling velocities of particulate systems, 1. Settling velocities of individual spherical particles. *International Journal of Mineral Processing* 5:349–367.

3.6 SETTLING VELOCITIES

For suspensions of spheres, Concha and Almendra (1979) assumed that the same equation used for individual spherical particles is valid for a suspension of particles with a correction for the presence of other particles. Then, the sedimentation of a suspension can be described by Equation 3.154 replacing the parameters P and Q with parameters depending on the volume fraction of solids φ, $P(\varphi)$ and $Q(\varphi)$. For suspensions of spheres Equation 3.154 can now be written in the form:

$$U^* = \frac{20.52}{D^*}\left(\left(1+0.0921D^{*3/2}\right)^{1/2}-1\right)^2 \tag{3.156}$$

$$\text{where } \; D^* = \frac{d}{P(\varphi)} \; \text{ and } \; U^* = \frac{u}{Q(\varphi)} \tag{3.157}$$

It is convenient to express the properties of a suspension affected though its concentration, such as the viscosity, as the product of the same property of the pure fluid with a function of concentration. Assume that $P(\varphi)$ and $Q(\varphi)$ can be related with P and Q in that form, then:

$$P(\varphi) = Pf_p(\varphi) \text{ and } Q(\varphi) = Qf_q(\varphi) \tag{3.158}$$

Replacing Equation 3.158 into Equation 3.157, and using the definitions of d^* and u^* from Equation 3.147, results in the following:

$$D^* = \frac{d^*}{f_p(\varphi)} \text{ and } U^* = \frac{u^*}{f_q(\varphi)} \tag{3.159}$$

With these definitions Equation 3.156 may be written in the form:

$$u^* = \frac{20.52}{d^*} f_p(\varphi) f_q(\varphi) \left(\left(1 + 0.0921 f_p^{-3/2} d^{*3/2}\right)^{1/2} - 1 \right)^2 \tag{3.160}$$

This expression, known as *Concha and Almendra's equation for a suspension of spheres* (Concha and Almendra 1979), permits the calculation of the settling velocity of a sphere of any size and density when it forms part of a suspension of volume fraction φ. To calculate the values of parameters $P(\varphi) = Pf_p(\varphi)$ and $Q(\varphi) = Qf_q(\varphi)$, consider small values of the Reynolds number, $Re \to 0$. The following expressions may be derived from Equations 3.154 and 3.160:

$$\begin{aligned} &\text{If } 0.0921 d^{*3/2} \ll 1 \Rightarrow u_\infty^* = 20.52 \left(\frac{0.0921}{2} \right)^2 d^{*2} \\ &\text{If } 0.0921 d^{*3/2} f_p^{-3/2} \ll 1 \Rightarrow u^* = 20.52 \left(\frac{0.0921}{2} \right)^2 d^{*2} f_p^{-2}(\varphi) f_q(\varphi) \end{aligned} \tag{3.161}$$

In these expressions, the symbols u_∞^* and u_∞ indicate the settling velocity of a particle in an infinite medium and the velocity of the same particle in a suspension. The quotient between these two terms is as follows:

$$\frac{u^*}{u_\infty^*} = f_p^{-2}(\varphi) f_q(\varphi) \text{ for Re} \to 0 \tag{3.162}$$

With a similar deduction than for Equation 3.161, but for high Reynolds numbers, Equations 3.154 and 3.160 reduce to

$$\begin{aligned} 0.0921 d^{*3/2} \gg 1 &\Rightarrow u_\infty^* = 20.52 \times 0.0921 \times d^{*1/2} \\ 0.0921 d^{*3/2} f_p^{-3/2} \gg 1 &\Rightarrow u^* = 20.52 \times 0.0921 \times d^{*1/2} f_p^{-1/2}(\varphi) f_q(\varphi) \end{aligned}$$

The quotients between these two equations are as follows:

$$\frac{u^*}{u^*_\infty} = f_p^{-1/2}(\varphi)f_q(\varphi) \quad \text{for} \quad \text{Re} \to \infty \tag{3.163}$$

To find functional forms for the functions $f_p(\varphi)$ and $f_q(\varphi)$, experimental values for the settling velocities u_∞ and u are needed.

Richarson and Zaki (1954) presented the most comprehensive and most cited work for u/u_∞. They performed careful sedimentation and liquid fluidization test with mono-sized spherical particles in a wide range of particles sizes and fluid densities and viscosities. They expressed their result in the form:

$$u/u_\infty = (1-\varphi)^n \quad \text{for} \quad \text{Re} \to 0 \quad \text{and} \quad u/u_\infty = (1-\varphi)^m \quad \text{for} \quad \text{Re} \tag{3.164}$$

From Equations 3.162, 3.163, and 3.164, we can write

$$\begin{aligned} f_p^{-2}(\varphi)f_q(\varphi) &= (1-\varphi)^n \\ f_p^{-1/2}(\varphi)f_q(\varphi) &= (1-\varphi)^m \end{aligned} \tag{3.165}$$

by solving this algebraic sets, we obtain the following:

$$f_p = (1-\varphi)^{(2/3)(m-n)} \quad \text{and} \quad f_q = (1-\varphi)^{(1/3)(4m-n)} \tag{3.166}$$

Using Equation 3.160, the best values for m and n were

$$n = 3.90 \quad \text{and} \quad m = 0.85 \tag{3.167}$$

$$\text{Then,} \quad f_p(\varphi) = (1-\varphi)^{-2.033}, \quad f_q(\varphi) = (1-\varphi)^{-0.167} \tag{3.168}$$

A plot of Equation 3.160 is given in the Figures 3.20 and 3.21, and Figure 3.22 shows the settling velocity U^* versus D^* for suspensions of spheres in water at 20°C for different values of the concentration φ. Figure 3.20 can be used to visualize the state of flow of particulate system.

3.6.1 Drag Coefficient

From Equations 3.152 and 3.160, we deduce that

$$\tilde{\delta}_0^2 = \delta_0^{\ 2} f_p(\varphi)f_q(\varphi) \text{ and } \tilde{C}_0^{\ 1/2}\,\tilde{\delta}_0^2 = C_0^{1/2}\delta_0^{\ 2} f_p^{3/2}(\varphi)$$

therefore, the parameters of the drag coefficient are

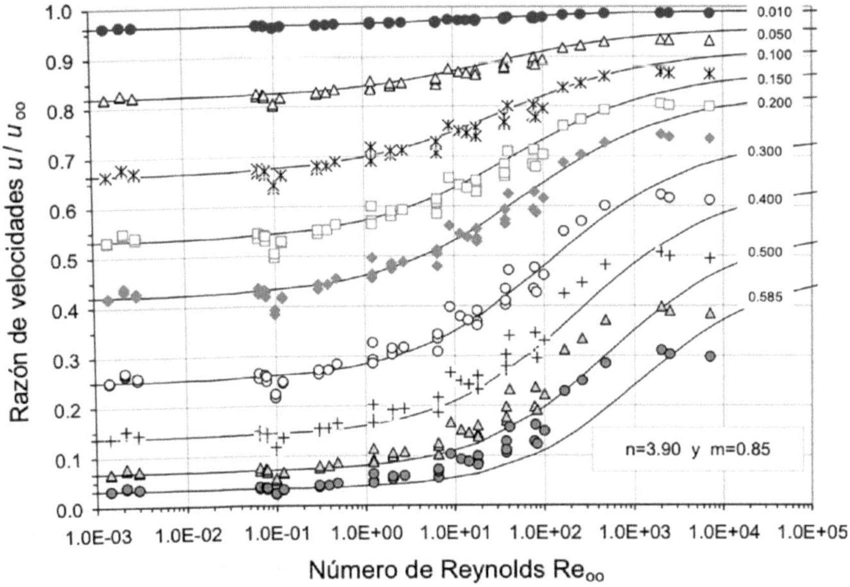

FIGURE 3.20 Dimensionless settling velocities versus Reynolds number, together with data obtained from Richarson and Zaki (1954) experimental results.

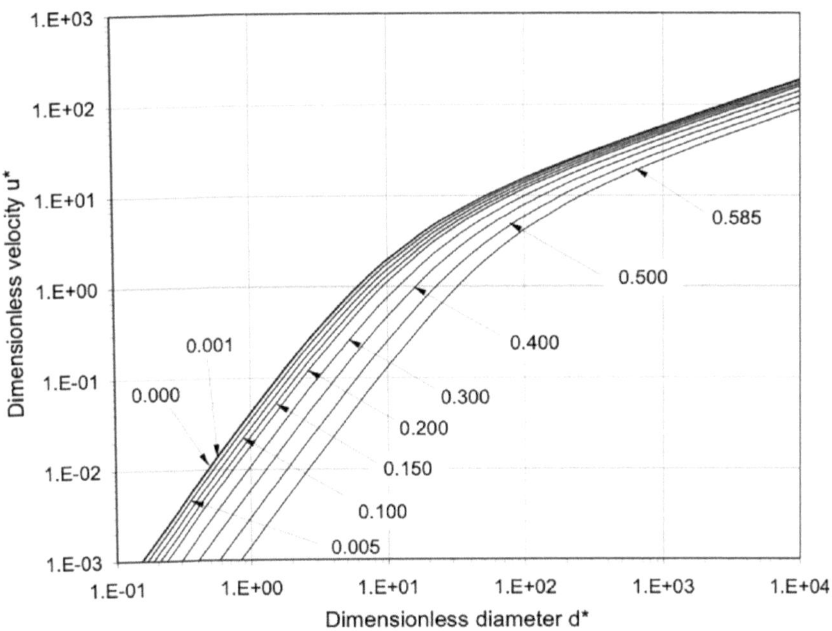

FIGURE 3.21 Simulation of the dimensionless velocity u^* versus the dimensionless size d^* for spherical particles settling in a suspension in water at 20°C with the volume fraction of solids a parameter.

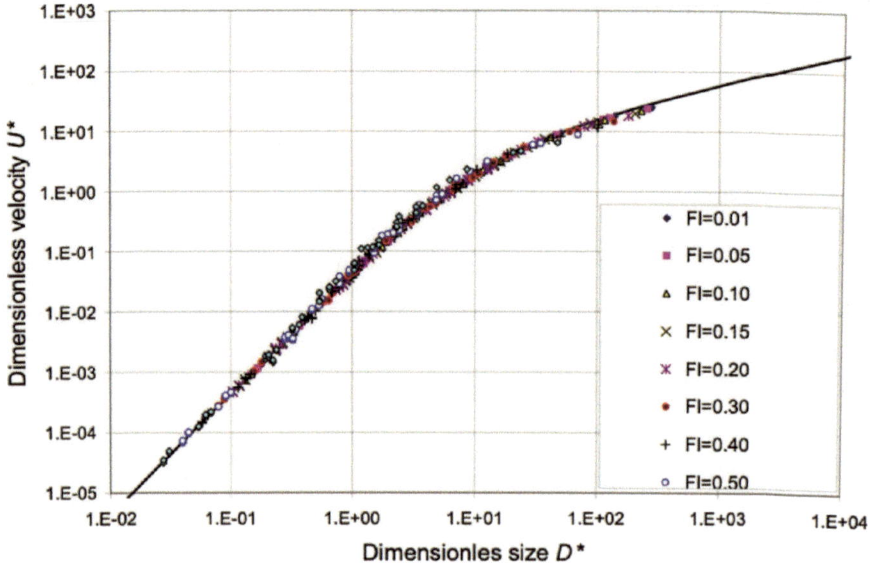

FIGURE 3.22 Dimensionless velocity U^* for suspensions of spheres of any size and density versus the dimensionless diameter D^*.

$$\tilde{C}_0 = C_0 f_p(\varphi) f_q^{-2}(\varphi) \text{ and } \tilde{\delta}_0 = \delta_0 f_p^{1/2}(\varphi) f_q^{1/2}(\varphi) \tag{3.169}$$

and the drag coefficient of the sphere can be written in the form

$$C_D = C_0 f_p(\varphi) f_q^{-2}(\varphi)\left(1 + \frac{\delta_0 f_p^{1/2}(\varphi) f_q^{1/2}(\varphi)}{\mathrm{Re}^{1/2}}\right)^2 \tag{3.170}$$

Using the values for $f_p(\varphi)$ and $f_q(\varphi)$ from Equation 3.168, we finally obtain

$$C_D = 0.28(1-\varphi)^{-2.01}\left(1 + \frac{9.08(1-\varphi)^{-1.83}}{\mathrm{Re}^{1/2}}\right)^2 \tag{3.171}$$

3.6.2 Corrections to the Field Equations Due to the Presence of Walls

The preceding sections involved hydrodynamic forces on particles moving through unbounded fluids. Now we will develop the interactions between particles and between particles and the walls of the container. The problems encountered in the field of mineral processing always involve the presence of boundaries, the walls of a container, for example.

The problems encountered in the field of mineral processing always involve the presence of boundaries, the walls of a container, for example. For particles having a given drag force, the size of the particle, presence of walls, and its distance to the particle affects its settling velocity. Therefore, the wall-effect is important in laboratory experiments such as in settling of particles in tubes. For example, in a tube 100 times the size of the particle, the settling velocity is only 75% of that in an unbounded media; therefore in industrial equipment, there is not necessary to make any correction.

In the presence of walls, Stokes' solution to the problem of the settling of particles should be corrected incorporating a new boundary condition at the wall surface. Exact determination of the new resistance dyadic is impossible; however the method of reflection (Brenner 1964) furnishes a useful approach when the ratio of particle size to wall distance is small. If we write the drag force for particle of arbitrary shape $\boldsymbol{F}_\infty = -\Delta\rho V_p \boldsymbol{g}$ moving with constant velocity u_∞ through an unbounded fluid by

$$F_\infty = -\mu \boldsymbol{K}_\infty \cdot \boldsymbol{u}_\infty \tag{3.172}$$

$\boldsymbol{K}_\infty$ is the resistance dyadic which depends on the properties of the particle only.

Define a new parameter of the dimensionless translation dyadic $\boldsymbol{\Phi}_\infty$ for particles in an unbounded fluid

$$\boldsymbol{\Phi}_\infty = \frac{\boldsymbol{K}_\infty}{6\pi R} \tag{3.173}$$

where R is a characteristic particle dimension. Now Equation 3.172 can be written in the form

$$\boldsymbol{F}_\infty = -6\pi\mu R \boldsymbol{\Phi}_\infty \cdot \boldsymbol{u}_\infty \tag{3.174}$$

If the particle is now moving at a smaller velocity $\boldsymbol{u}$ in a bounded fluid, the drag force $\boldsymbol{F}$ can be written as

$$\boldsymbol{F} = -6\pi\mu R \boldsymbol{\Phi} \cdot \boldsymbol{u} \tag{3.175}$$

Since the drag force on the particle is not affected by the walls (its velocity is), $F = F_\infty$ and

$$\boldsymbol{\Phi} \cdot \boldsymbol{u} = \boldsymbol{\Phi}_\infty \cdot \boldsymbol{u}_\infty \rightarrow \boldsymbol{u} = \boldsymbol{\Phi}^{-1} \cdot \boldsymbol{\Phi}_\infty \cdot \boldsymbol{u}_\infty \tag{3.176}$$

For spherical particles the characteristic dimension R is its radius and $\boldsymbol{\Phi}_\infty = \boldsymbol{I}$. From Equations 3.174 and 3.176, we finally have the correction for a spherical particle in the presence of a vertical wall.

$$\frac{u}{u_\infty} = 1 - k\frac{R}{l} + O(c/l) \tag{3.177}$$

TABLE 3.2
Values of *k* for Different Boundaries

Boundary shape	Location of the particle	Value of *k*
Circular cylinder	Cylinder axis $R = L$	2.10444
	Eccentric $R_0 = L$	$K = f(R/R_0)$
	R_0 is the cylinder radius	
Single plane wall	ℓ is the distance to the wall	*k*
Parallel to plane walls	Midway between walls	1.004
	¼ distance between walls ℓ is the distance to the nearest wall	0.6526
One flat wall	ℓ = perpendicular distance to the wall	9=8
	ℓ = parallel distance to the wall	9=16

TABLE 3.3
Values of $f(R/R_0)$

R/R_0	$f(R/R_0)$	R/R_0	$f(R/R_0)$
0,00	2,1044	0,45	2,0483
0,01	2,1436	0,50	2,0657
0,05	2,0103	0,60	2,1697
0,10	2,0973	0,70	2,4596
0,20	2,0794	0,80	3,2316
0,30	2,0569	0,90	5,9050
0,35	2,0481	1,00	∞
0,40	2,0440		

The values of *k* are given by Happel and Brenner (1965) in Tables 3.2 and 3.3.

$$R/R_0$$
$$f\left(R/R_0\right)$$

3.7 CONVECTIVE DIFFUSION

Most processes of the chemical-metallurgical industry are determined by hydrodynamic factors, with special importance of heterogeneous transformation into liquids and gases. In this category of phenomena, we can include the fundamental heterogeneous catalytic and non-catalytic chemical reactions, adsorption, desorption, dissolution and crystallization of a solution, electrochemical reactions, and many other phenomena performed in the interface of solids, liquids, and gases.

All heterogeneous reactions take place in stages, most of which relate to the transport of chemical reagent and products and to the heterogeneous chemical reaction itself. The first step in this sequence is the transport of the reagents to the surface, the second the reaction itself or the desorption, dissolution, or precipitation on the surface and the last the transport of the products outside the reaction zone. The overall or net speed of the process is determined by the combined effect of these three stages. However, if one of these stages is significantly slower than the others, it will govern the final reaction speed.

Many heterogeneous reactions are controlled by diffusion of the reagents in a fluid medium, which becomes the reason for the importance of the study of phenomenon of "convective diffusion."

The transport of a solute though a fluid is governed by two mechanisms, convection and diffusion. Convection is the transport associated with fluid movement, therefore its driving force is the fluid velocity. Molecular diffusion is the consequence of concentration gradients which produces a flow tending to establish the balance in concentration.

For a multi-component system, the convective diffusion is expressed by

$$\frac{dc}{dt} + \nabla c \cdot \boldsymbol{v} = \mathcal{D}\nabla^2 c \tag{3.178}$$

The boundary conditions for this equation are usually a constant concentration c_0 away from the reaction surface and a flow of mass in the fluid in a normal direction to the surface equal to the heterogeneous reaction rate.

$$\begin{gathered} c(|r| \to \infty) = c_0 \\ c(R) = c_s; \quad \mathcal{D}(\nabla c \cdot n) = kc_s^n \end{gathered} \tag{3.179}$$

The mass flow is given by Fick's law, the constitutive equation $\boldsymbol{j} = -\mathbb{D}\nabla c$, then from Equation 3.179

$$j_n = \mathcal{D}(\nabla c \cdot n) \text{ for } y = 0 \tag{3.180}$$

where n is the unit vector normal to the reaction surface and c_s is the concentration of the solute at the surface.

It is advisable to write the convective diffusion in a dimensionless form with the aim of studying its solution in cases where any of the transport mechanisms is dominant. Consider the dimensionless time, $t^* = t/t_0$, dimensionless velocity, $v^* = v/u_0$, dimensionless concentration $c^* = c/c_0$, and dimensionless gradient $\nabla^* = L\nabla(\cdot)$ and $j^* = (L/\mathcal{D}c_0)\,j$, then

$$\left(\frac{c_0}{t_0}\right)\frac{dc^*}{dt^*} + \left(\frac{c_0 u_0}{L}\right)\nabla^* c^* \cdot \boldsymbol{v}^* = \left(\frac{\mathcal{D}c}{L^2}\right)\nabla^{*2} c^* \tag{3.181}$$

$$\left(\frac{L/u_0}{t_0}\right)\frac{dc^*}{dt^*} + \nabla^* c^* \cdot \boldsymbol{v}^* = \left(\frac{\mathcal{D}}{Lu_0}\right)\nabla^{*2} c^*$$

Defining the dimensionless numbers Stanton $St = t_0 / (L / u_0)$ and Péclet $Pe = Lu_0 / \mathcal{D}$ the dimensionless convective diffusion equation becomes

$$\frac{1}{St}\frac{dc^*}{dt^*} + \nabla^* c^* \cdot v^* = \frac{1}{Pe}\nabla^{*2} c^* \tag{3.182}$$

The boundary conditions are

far from the particle $|r| \to \infty$ and $c = c_0$; and $|r^*| \to \infty$ and $c = 1$

At the particle surface

$$\begin{gathered} r = L, c = c_s, \quad j = \frac{\mathcal{D}c_o}{L}\underbrace{(\nabla c \cdot \boldsymbol{n})}_{j^*} = \left(c_o^n k\right) c_s^n \\ |r^*| \to \infty, c^* = 1, \underbrace{\left(\nabla^* c^* \cdot \boldsymbol{n}\right)}_{j^*} = \underbrace{\left(\frac{c_o^n k}{\mathcal{D}c_o / L}\right)}_{Da_{II}} c_s^{*n}, \end{gathered} \tag{3.183}$$

where Da_{II} is the Damköhler number. The boundary conditions then are

$$\begin{gathered} |r^*| \to \infty, c^* = 1 \\ r = 1, \quad c^* = c_s^*, \frac{j^*}{\mathbb{D}a_{II}} = c_s^{*n} \end{gathered} \tag{3.184}$$

Many times, there is no chemical reaction at the surface of a solid, but only mass transfer, such as the dissolution of a solid. In this case $j = k_m (c_0 - c_s)$ and

$$j^* = \underbrace{\left(\frac{k_m c_o}{\mathcal{D}c_o / L}\right)}_{Sh}\left(1 - c_s^*\right) \tag{3.185}$$

So that the boundary conditions becomes in this case

$$\begin{gathered} |r^*| \to \infty, c^* = 1 \\ r = 1, \quad c^* = c_s^*, \quad j^*(1) = Sh\left(1 - c_s\right) \end{gathered} \tag{3.186}$$

In general, the boundary condition (Equation 3.184) is nonlinear and the solution of the convective diffusion equation becomes difficult and one has to resort to use approximate methods of processing to calculate the concentration c_s^*. However, if one of the stages of the processes, the diffusion or reaction, is slower, the boundary condition can be simplified.

From Equation 3.183, two cases may be observed for the boundary condition (Equation 3.184) $\left|\boldsymbol{r}^*\right| \to \infty$ one if the chemical reaction $c_0^n k$ is much greater than the diffusion rate $\mathcal{D}c_o/L$, or otherwise.

3.7.1 Reaction Rate Much Greater Than Diffusion Rate

If the diffusion rate in boundary condition (Equation 3.184) is $\mathbb{D}a_{II} \to \infty$, the concentration at the surface is approximate zero, $c_s \approx 0$, which means that the reaction rate is very large so that all material transported to the surface is instantly transformed, consumed there, and both the concentration gradient in the solution and the flow rate to the surface are at maximum. This rate is called *flow limit*, and the process is controlled by diffusion.

3.7.2 Reaction Rate Much Lower Than Diffusion Rate

If the diffusion coefficient $\mathbb{D}_{II} \to 0$, the reaction rate is much smaller than the gradient of concentration in Equation 3.183, and the concentration is constant in all the reaction field, $c_s = c_o$ and the process will be controlled by the reaction.

3.7.3 Solution of the Convective Diffusion Equation for Extreme Values of the Péclet Number

In Equation 3.182, the Péclet number represents the ratio of convective to diffusive flows $Pe = Lu_0/\mathcal{D}$. If one of these predominates, that is, if the Péclet number is very large or very small, one of the terms of this equation can be neglected. Defining the Schmidt number as the ratio of the kinematic viscosity v to the diffusion coefficient $\mathcal{D}$ of the fluid $Sc = v/\mathcal{D}$

$$Pe = \frac{Lu_0}{v}\frac{v}{\mathcal{D}} \to Pe = \text{Re} \times Sc \tag{3.187}$$

In the previous section, hydrodynamic problems of interest in mineral processing are in the range of Reynolds numbers of 10^{-3} and 10^{-5}. Moreover, the Schmidt number can take the value of $Sc \sim 1$ in gases and $Sc \sim 10^3$ in liquids.

When the Reynolds number is small (Stokes flow), Péclet number may also be small or of an intermediate value depending on Schmidt's number. Moreover, for intermediate Reynolds numbers, the Péclet number can be intermediate or high, and at high Reynolds number, Péclet number will also be high. In Table 3.4, the different possible values of the Péclet number are shown.

TABLE 3.4
Values of the Péclet Numbers for Liquid and Gases

Material	Regime	Re ≈	Sc ≈	Pe	Observation
Gases	Stokes	10^{-1}	1	10^{-1}	$Pe \ll 1$
	Prandtl	10^{3}	1	10^{3}	$Pe \gg 1$
	Euler	∞	1	∞	$Pe = \infty$
Liquid	Stokes	$10^{-3} - 10^{-1}$	10^{3}	$10^{-1} - 10^{-2}$	$10^{-3} < Pe < 10^{3}$
	Prandtl	10^{3}	10^{3}	10^{6}	$1 < Pe < 10^{8}$
	Euler	∞	10^{3}	∞	$Pe \to \infty$

3.7.4 Convective Diffusion to a Sphere at Small Values of the Péclet Number: Diffusional Boundary Layer

From Table 3.4, we can see that small numbers of Péclet can only be obtained, for both liquids and gases, only at flows with small Reynolds numbers. Consider a fluid flow at low Reynolds number around a stationary sphere at steady state with the dimensionless velocity, $v^* = v / u_0$, dimensionless concentration $c^* = c / c_0$, and dimensionless gradient $\nabla^*(\cdot) = R\nabla(\cdot)$, and where R is the sphere radius, then from Equations 3.182 and 3.184, the dimensionless convective diffusion equation at low Péclet number becomes

$$\nabla^* c^* \cdot \mathbf{v}^* = \frac{1}{Pe} \nabla^{*2} \mathbf{v}^* \tag{3.188}$$

with boundary condition from Equation 3.184

$$\begin{aligned} \text{For } Pe \ll 1, \quad & \nabla^{*2} \cdot v^* = 0 \\ & c^*(1) = c_s^*, j^*(1) = \mathbb{D}a_{II} c_s^{*n} \\ & c(\infty) = 1 \end{aligned} \tag{3.189}$$

Using spherical coordinates this equation becomes

$$\frac{1}{r^{*2}} \frac{d}{dr^*} \left(r^{*2} \frac{dc^*}{dr^*} \right) = 0 \tag{3.190}$$

$$\begin{aligned} & \frac{dc^*}{dr^*} = \frac{a}{r^{*2}}, \quad c^* = -\frac{a}{r^*} + b \\ & \text{for } r^* = \infty \qquad c^*(\infty) = 1 = b \\ & \text{for } r^* = 1 \qquad \left. \frac{dc^*}{dr^*} \right|_1 = \mathbb{D}a_{II} c_s^{*n} = a \end{aligned} \tag{3.190a}$$

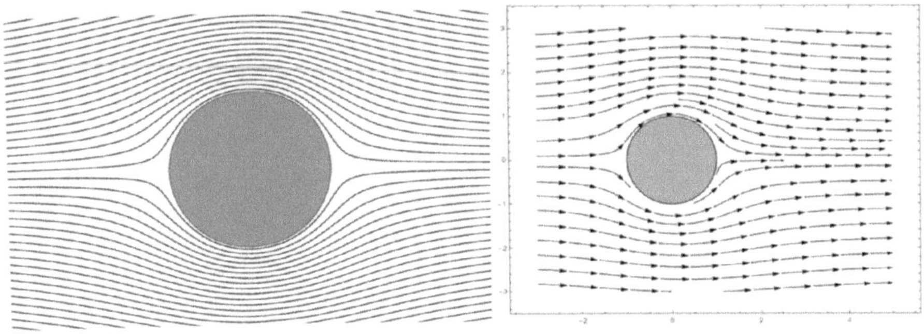

Streamlines Velocity vectors

FIGURE 3.23 Sphere in a Stokes flow.

then

$$j^* = \frac{\mathbb{D}a_{II} c_s^{*n}}{r^{*2}} \text{ and } c^*(1) = 1 - \mathbb{D}a_{IIs} c_s^{*n} \tag{3.191}$$

At the surface, $r^* = 1$, $c_s^* = 1 - \mathbb{D}a_{II} c_s^*$ and

$$j^* = \left(1 - c_s^*\right) \tag{3.192}$$

For the special case of a first order reaction ($n = 1$) and since $j = \left(Dc_0 / L\right) j^*$

$$\left.\begin{aligned} c_s^* &= \frac{1}{1 + \mathbb{D}a_{II}} \\ j &= \frac{\mathcal{D}c_o}{R}\left(1 - c_s^*\right) \end{aligned}\right\} \rightarrow j = \frac{\mathcal{D}c_o}{R}\left(\frac{1}{1 + \mathbb{D}a_{II}}\right) \tag{3.193}$$

Since $\mathbb{D}a_{II} = Rk / \mathcal{D}$ and $n = 1$,

$$j = \frac{\mathcal{D}c_o}{\left(R + \frac{\mathcal{D}}{k}\right)} \tag{3.194}$$

Defining the term $\delta = R + \mathcal{D} / k$, Equation 3.194 may be written finally as

$$j = \frac{\mathcal{D}c_o}{\delta_c} \tag{3.195}$$

For $Pe \ll 1$, the flow to a sphere of radius R may be interpreted as a diffusion through a diffusional film of thickness δ_c. The results are presented in Figure 3.23.

3.7.5 Convective Diffusion to a Sphere at High Values of the Péclet Number

As we have already seen, a flow with large Péclet number may be obtained under Stokes, Prandtl, or Euler's regime. In this section we will look at the problem of diffusion-convective to large Péclet number toward a sphere with first-order heterogeneous chemical reaction on its surface in these three regimes.

3.7.5.1 Convective Diffusion

Consider a stationary flow over a sphere with first order heterogenous reaction at the surface.

From Equations 3.188 and 3.179, the mass and momentum equations, at steady state with boundary conditions, in dimensionless variables are

$$\begin{aligned} \nabla^* c^* \cdot v^* &= \frac{1}{Pe} \nabla^{*2} c^* \\ |r| \rightarrow \infty, c &= c_0 \\ r = R, \quad c = c_s, \quad j_0^* &= \mathbb{D}a_{II} c_s^* \end{aligned} \tag{3.196}$$

Where $c^* = c / c_0$, $\mathbf{v}^* = \mathbf{v} / u_0$ and $Pe = Ru_0 / \mathcal{D}$

In spherical coordinates we have

Continuity
$$\frac{1}{r^*}\frac{\partial}{\partial r^*}\left(r^{*2} v_r^*\right) + \frac{1}{sen\theta}\frac{\partial}{\partial \theta}\left(v_\theta^* sen\theta\right) = 0 \tag{3.197}$$

Momentum
$$\begin{aligned} v_r^* &= \left[\frac{3}{2}\frac{1}{r^*} - \frac{1}{2}\frac{1}{r^{*3}} - 1\right]\cos\theta \\ v_\theta^* &= \left[-\frac{3}{4}\frac{1}{r^*} - \frac{1}{4}\frac{1}{r^{*3}} + 1\right] sen\theta \end{aligned} \tag{3.198}$$

Convective diffusion
$$v_r^* \frac{\partial c^*}{\partial r^*} + \frac{v_\theta^*}{r^*}\frac{\partial c^*}{\partial \theta} = \frac{1}{Pe}\left(\frac{\partial^2 c^*}{\partial r^{*2}} + \frac{1}{r^{*2}}\frac{\partial^2 c^*}{\partial \theta^2}\right) \tag{3.199}$$

Applying a transformation: $\overline{x} = \int_0^{x^*} s^{*2}$, $\overline{y} = s^* y^*$, $\overline{c} = c^*$, Equation 3.199 becomes

$$\overline{v}_x\left(s^{*2}\frac{d\overline{c}}{d\overline{x}} + \frac{\overline{y}}{s^*}\frac{ds^*}{dx^*}\frac{\partial \overline{c}}{\partial \overline{y}}\right) + \left(s^*\overline{v}_y - \frac{\overline{y}}{s^*}\frac{ds^*}{dx^*}\overline{v}_x\right)s^*\frac{\partial \overline{c}}{\partial \overline{y}} = \frac{1}{Pe}s^{*2}\frac{\partial^2 \overline{c}}{\partial \overline{y}}$$

$$\bar{v}_x s^{*2}\frac{d\bar{c}}{d\bar{x}}+\underbrace{\frac{\bar{y}}{s^*}\frac{ds^*}{dx^*}\bar{v}_x\frac{\partial\bar{c}}{\partial\bar{y}}-\frac{\bar{y}}{s^*}\frac{ds^*}{dx^*}\bar{v}_x\frac{\partial\bar{c}}{\partial\bar{y}}}_{0}+s^*\bar{v}_y\frac{\partial\bar{c}}{\partial\bar{y}}=\frac{1}{Pe}s^{*2}\frac{\partial^2\bar{c}}{\partial\bar{y}}$$

$$\bar{v}_x\frac{d\bar{c}}{d\bar{x}}++\bar{v}_y\frac{\partial\bar{c}}{\partial\bar{y}}=\frac{1}{Pe}\frac{\partial^2\bar{c}}{\partial\bar{y}} \tag{3.200}$$

Collecting momentum and convective diffusion equations, we can write for the diffusional boundary layer at high Péclet number, which are the same as those for a flat plate

$$\begin{gathered}\bar{v}_x\frac{\partial\bar{v}_x}{\partial\bar{x}}+\bar{v}_y\frac{\partial\bar{v}_y}{\partial\bar{y}}=0\\ \bar{v}_x\frac{\partial\bar{v}_x}{\partial\bar{x}}+\bar{v}_y\frac{\partial\bar{v}_x}{\partial\bar{y}}=-\frac{\bar{p}}{\partial\bar{x}}+\frac{\partial^2\bar{v}_y}{\partial\bar{y}^2}\\ \bar{v}_x\frac{d\bar{c}}{d\bar{x}}++\bar{v}_y\frac{\partial\bar{c}}{\partial\bar{y}}=\frac{1}{Pe}\frac{\partial^2\bar{c}}{\partial\bar{y}}\end{gathered} \tag{3.201}$$

With boundary conditions

$$\begin{aligned}&\bar{c}(\bar{x},0)=\bar{c}_s,\bar{j}=\left.\frac{\partial\bar{c}}{\partial\bar{y}}\right|_{\bar{y}=0}=\frac{1}{s^*}\mathbb{D}a_{II}\bar{c}_s^n\\ &\bar{c}(\bar{x},\infty)=1\end{aligned} \tag{3.202}$$

3.8 TURBULENT FLOW

A laminar flow regimen is found only in cases where the fluid velocity is relatively small. In most cases of interest to the industry in which fluid movement in equipment and pipes is carried out in a turbulent regime, that is, in a disorderly manner, and particles follow paths without apparent law. The transport of mass, momentum, and energy is mainly produced by the mixture of small macroscopic portions of the fluid called eddies in a process called *eddy diffusion* or *eddy dispersion*.

It can be demonstrated that all field equations defined previously are valid for a turbulent flow as long as the relevant properties are considered averages and constituent equations for turbulent flow densities are defined or, equivalently, effective transport coefficients are defined.

3.8.1 Origin and Description of the Turbulence

When tangential forces are applied to a viscous fluid, movements originate in it according to the magnitude of the applied stress and the shape of the walls. Within the established velocity field, secondary flows can originate; some of which are ordered due to pressure gradients and irregular ones called turbulence. The appearance of turbulence does not depend only on the walls but requires mechanisms that generate movements in directions other than those of the applied efforts. These mechanisms should be sought in the flow itself. In many cases a flow is initially laminar and then transforms into turbulent. According to experimental evidence the initiation of turbulence is sudden and happens in certain limited zones of the laminar regime, for example near walls and in sudden pressure gradients.

When our interest lies in the main flow, we can ignore the fluctuating terms and treat the mass and linear momentum flow in the same way as it is done for a laminar flow, that is,

$$\begin{aligned} \frac{\partial \rho \boldsymbol{v}}{\partial t} + \rho \nabla \cdot \boldsymbol{v}\boldsymbol{v} &= \nabla \cdot \boldsymbol{T} + \boldsymbol{b} \\ \nabla \cdot \rho \boldsymbol{v} &= 0 \end{aligned} \tag{3.203}$$

However, new average properties must be defined for the variables involved, $\boldsymbol{v} = \bar{\boldsymbol{v}} + \boldsymbol{v}'$, $\boldsymbol{T} = \bar{\boldsymbol{T}} + \boldsymbol{T}'$, and $\boldsymbol{b} = \bar{\boldsymbol{b}} + \boldsymbol{b}'$, where the tilde represents the property fluctuations. Integrating the resulting equations for short times generates the Reynolds equations:

$$\begin{aligned} \frac{\partial \rho \bar{\boldsymbol{v}}}{\partial t} + \rho \nabla \cdot \overline{\boldsymbol{v}\boldsymbol{v}} &= \nabla \cdot \left(\bar{\boldsymbol{T}} + \boldsymbol{T}'\right) + \bar{\boldsymbol{b}} \\ \nabla \cdot \rho \bar{\boldsymbol{v}} &= 0 \\ \boldsymbol{T}' &= -\rho \boldsymbol{v}'\boldsymbol{v}' \end{aligned} \tag{3.204}$$

Where $\bar{T}$ and $\bar{b}$ are the mean velocity, mean stress tensor, and mean body force fields, while $\boldsymbol{v}'$ is the fluctuating velocity field, ρ is the constant fluid density, and $\boldsymbol{T}_t = -\rho \boldsymbol{v}'\boldsymbol{v}$ is the Reynolds stresses.

For the turbulent transport of mass, the time-average diffusion equation is

$$\frac{\partial \bar{c}_i}{\partial t} + \nabla \cdot \overline{c_i \bar{\boldsymbol{v}}} = -\nabla \cdot \left(\bar{\boldsymbol{J}}_i + \boldsymbol{J}_i^t\right) \quad with \quad \boldsymbol{J}_i^t = c_i \boldsymbol{v} \tag{3.205}$$

Where $\bar{\boldsymbol{J}}_i$ is the mean flux-density function, $c_i^{'}$ the fluctuating concentration, and $\boldsymbol{J}_i^t$ the turbulent flux-density.

3.8.2 Eddy Diffusivity Model of the Turbulence

To solve the previous equations, it is necessary to postulate constitutive equations for the turbulent flows. Therefore, the structure of the turbulence should be analyzed to obtain relationships between the transport properties and the characteristics of the main flow. The resulting motion in a turbulent flow consists of the overlapping eddies of different size and directions, which deform most strongly in the direction of the main flow and produce rotation within the chaotic movements performed by the fluid.

The quantification of turbulence can be evaluated by the spatial correlation between the flow variables between two points in space. The movement between these two points can be considered random, or isotropic, if this correlation is very small. The limiting size for which isotropy is known as the microscale λ of the turbulence. The magnitude of turbulence is measured by the *macro-scale* defined as the average of the distances for which there is correlation between the properties of the flow

$$L = \int_0^{\infty} R(\xi) d\xi \tag{3.206}$$

where ξ is the distance between two points and $R(\xi)$ the coefficient of correlation of properties between these two points.

For large eddies of the order of the macro-scale of turbulence, the fluctuating Reynolds number is very large so that the flow can be considered inviscid.

$$\text{Re}_L \equiv Lu / v \gg 1 \tag{3.207}$$

where $u' = \sqrt{v_i v_i'}$ is the fluctuating velocity, or root mean square velocity (RMS), a turbulent velocity scale.

Meanwhile for small eddies, the inertial and viscous forces have the same order of magnitude, and the viscosity becomes important

$$\text{Re}_\lambda \equiv Lu' / v \ll 1 \tag{3.208}$$

This indicates that energy dissipation occurs in the small eddies λ.

These two cases demonstrate that in a turbulent flow, energy enters through the swirls of the order of the macro-scale l and dissipate by eddies of the order of the micro-scale λ. Even if the energy dissipates in the small eddies, the amount of energy dissipated depends on the big eddies, as they capture it. This means that the function is micro-scale independent and depends only on the macro-scale.

3.8.3 Turbulent Flow Near Walls

There are several theories to describe the behavior of turbulent flows near solid walls. One of the two most used empirical theories is the *power law* and the second is the *Prandtl mixing length.*

3.8.3.1 Power Law

The power law comes from the Blasius empirical formula for the drag coefficient friction in smooth circular section pipes $C_D \approx 0.06651/\text{Re}^{1/4}$, and the Reynolds number is defined by $\text{Re} = v_m \delta / \upsilon$, where v_m is the maximum velocity, δ is the thickness of the boundary layer, and $\upsilon = \mu_f / \rho_f$ are the fluid viscosity-density ratio as shown in Figure 3.24. The shear rate τ_0 at the wall and shear velocity $\dot{\gamma}_0$ are defined by

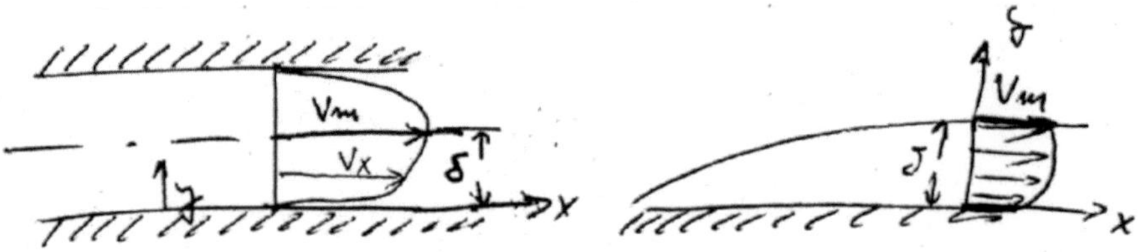

FIGURE 3.24 Turbulent boundary layer.

$$C_D(\text{Re}) = \frac{\tau_0}{(1/2)\rho_f v_m^2} \quad \text{and} \quad \dot{\gamma}_0 = \sqrt{\frac{\tau_0}{\rho_f}} \tag{3.209}$$

From these two relations, the shear rate at the wall becomes

$$\dot{\gamma}_0 = 0.71 C_D^{1/2} v_m \tag{3.210}$$

and from in the boundary layer (Schlichting 1955)

$$\frac{v_x}{v_m} = \left(\frac{y}{\delta}\right)^{1/7} \quad \text{or} \quad \frac{v_r}{v_m} = \left(\frac{r}{\delta}\right)^{1} \tag{3.211}$$

3.8.3.2 Prandtl Mixing Length Theory

Prandtl described the mixing length as the diameter of masses of fluid, or blobs, moving with at an average velocity and other properties over a length scale l_m within a fluid whole, before it becomes mixed in with neighboring masses where $\overline{u}$ is the average velocity (Figure 3.25).

$$\text{If} \quad \begin{array}{l} u' \simeq \overline{u} + \dfrac{\partial \overline{u}}{\partial y} l_m - \overline{u} \text{ and } v' \simeq \dfrac{\partial \overline{u}}{\partial y} l_m \\ \left|\overline{u'v'}\right| \simeq -l_m^2 \dfrac{\partial \overline{u}}{\partial y} \left|\dfrac{\partial \overline{u}}{\partial y}\right| \end{array} \tag{3.212}$$

The absolute values in this equation is necessary for the Reynolds stresses to have the same sign as in the viscous shear (Cebeci 1982). For free shear flows, l_m depends on the nature of the flow and, in general, is space dependent. For different flows, the mixing lengths are given as

$$l_m = \begin{cases} 0.090l \text{ for plane jets} \\ 0.075l \text{ for circular jets} \\ 0.070l \text{ for mixing layers} \\ 0.050l \text{ for shear layers} \end{cases} \tag{3.213}$$

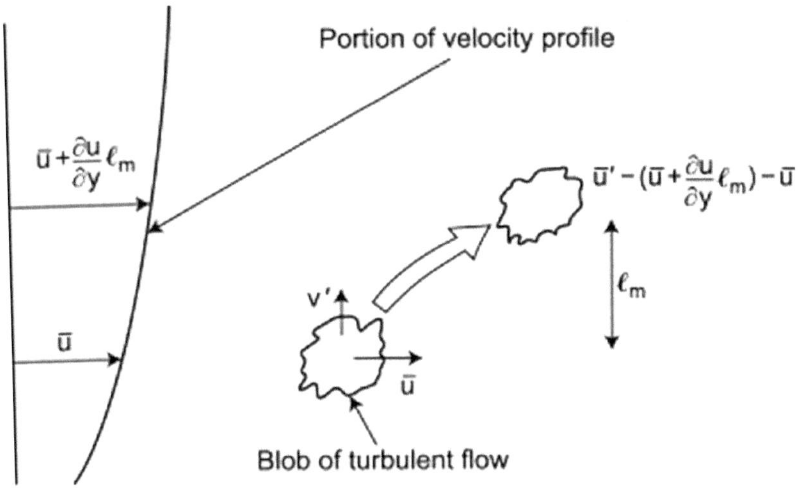

FIGURE 3.25 Concept of Prandtl mixing length.

Source: Houghton et al. 2016

where l is the turbulent scale or mixing length, then the constitutive equation for the shear rate τ can be written in the form

$$\tau = \alpha \rho l^2 \left(\frac{\partial \overline{u}}{\partial l} \right)^2 \tag{3.214}$$

where α is a numerical value.

Prandl assume that the shear stresses were constant in all the flow field except near the wall where he called them τ_0, then the shear rate at the wall is

$$\frac{\partial \overline{u}}{\partial l} = \frac{1}{\sqrt{\alpha}} \sqrt{\frac{\tau_0}{\rho}} \frac{1}{l} \rightarrow \frac{\partial \overline{u}}{\partial l} = \frac{1}{\sqrt{\alpha}} \frac{u_0}{l} \rightarrow u_0 = \sqrt{\alpha} l \frac{\partial \overline{u}}{\partial l} \tag{3.215}$$

To integrate the second expression of Equation 3.215, it is assumed that in the vicinity of the wall $l = \beta y$, so that $\partial \overline{u} / \partial l = \beta \partial \overline{u} / \partial y$, where l is numerical parameter, and with the characteristic turbulence velocity Equation 3.215 may be integrated to give

$$\frac{\overline{u}}{u_0} = \frac{1}{\beta \sqrt{\alpha}} \ln y + \kappa \tag{3.216}$$

This equation is known as the logarithmic velocity profile. The constant κ may be calculated at δ_0 where the velocity is $\overline{u}_0$ (Figure 3.26).

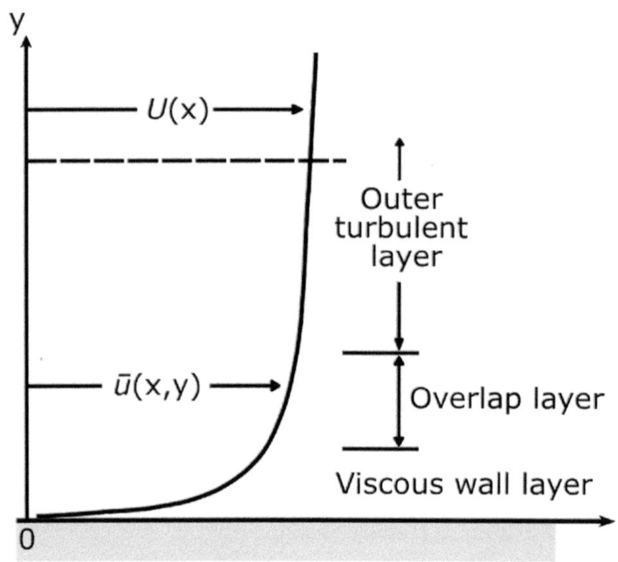

FIGURE 3.26 Boundary layer over a flat plate.

$$\kappa = \frac{\overline{u}_0}{u_0} - \frac{1}{\beta\sqrt{\alpha}} \ln \delta_0$$

$$\frac{\overline{u}}{u_0} = \frac{1}{\beta\sqrt{\alpha}} \ln y + \frac{\overline{u}_0}{u_0} - \frac{1}{\beta\sqrt{\alpha}} \ln \delta_0$$

$$\frac{\overline{u}}{u_0} = \frac{1}{\beta\sqrt{\alpha}} \ln\left(\frac{y}{\delta_0}\right) + \frac{\overline{u}_0}{u_0} \tag{3.217}$$

Since $l\left(\frac{\partial \overline{u}}{\partial l}\right) \sim u'$, we can see that $u_0 \sim u'$, that is u_0, is the scale of the turbulence.

3.8.4 Prandtl Boundary Layer $\delta \geq y \geq \delta_0$

As the fluid approaches the surface of the solid, it reaches a zone where $y = \delta_0$. Here the turbulence decreases to $y = \delta_0$ damping the turbulence so that the viscous effects begin to dominate. For $l \sim y \leq \delta_0$, $\beta = 0$, the Reynolds number is approximately 1.

$$\mathrm{Re} \sim \frac{u_0 \delta_0}{\nu} \sim 1 \quad \rightarrow \quad \delta_0 = a\frac{\nu}{u_0} \tag{3.218}$$

These expressions indicate that the Equation 3.217 is valid in the range $\delta \leq y \leq \delta_0$. Replacing the expression for δ_0 from Equation 3.218 yields

$$\frac{\bar{u}_0}{u_0} = \frac{1}{\sqrt{\alpha}} \ln\left(\frac{u_0 y}{\nu}\right) + \left\{\frac{\bar{u}_0}{u_0} - \frac{1}{\sqrt{\alpha}} \ln(a)\right\} \tag{3.219}$$

Calling $u^* = \bar{u}_0 / u_0$, $y^* = u_0 y / \nu$, and $A = 1/\sqrt{a}$, $B = \left\{\bar{u}_0 / u_0 - 1/\sqrt{\alpha}\right\}$ this equation can be written in the form

$$u^* = A \ln y^* + B \text{ for } \delta_0 \leq y \leq \delta \tag{3.220}$$

3.8.4.1 Zone with $y \leq \delta_0$

The region of values of y less than δ_0 is called *viscous wall layer* or *laminar sublayer*. In this region, shear stresses are predominantly viscous.

$$\bar{u} = \int \frac{\tau_0}{\rho \nu} dy + C, \quad \text{with } \bar{u}(0) = 0$$

Integrating $\bar{u} = \dfrac{\tau_0}{\rho \nu} y$, and since $\sqrt{\dfrac{\tau_0}{\rho}} = \bar{u}_0$, then $\dfrac{\bar{u}}{\bar{u}_0} = \dfrac{\bar{u}_0 y}{\nu}$ and

$$u^* = y^* \tag{3.221}$$

3.8.5 Turbulent Stirred Tank Reactors

One of the applications of the theory of turbulent flow is the hydrodynamic characteristics of the flow in an agitated reactor. These devices are widely used in the mining industry (leaching, flotation) as well as in other industries. In the flotation process, a finely ground mineral pulp is mixed with an airflow in the form of small bubbles in the presence of a reagent that produces a mineralized foam that overflows the pond with the valuable product. This process intents obtaining the best contact between the solid the liquid and the air bubbles which depends on the hydrodynamic conditions of the reactor.

In general, turbulent flows are not homogeneous. The energy incorporated into high Reynolds number flows produces large eddies that disintegrate into ever smaller eddies, causing most of the energy to dissipate through the small ones. When the microscale of the process is large compared to the scale of the eddies that dissipate the energy, there is a large spectrum of intermediate sizes that dissipate a small quantity of energy in different directions, so that information of the direction of large swirls is gradually lost. Therefore, any eddy of scale much smaller than the macro scale is statistically independent of it, and the only information it receives, and transfers, is the amount of kinetic energy.

For flows with high Reynolds numbers ($\text{Re} > 10^5$) and volumes under consideration of the order of $|\boldsymbol{r}|$, small compared to the scale of the main flow, which are of the order of the rotor of the tank, eddies of scale λ much smaller than the macro scale L of the swirls ($|\boldsymbol{r}| \sim \lambda \ll L$), the dissipation rate of properties of the system can be assumed to have *local isotropy* throughout the turbulent flow. For a rotor with size D with angular speed ω, $l \sim D$ and $\bar{u} \sim \omega \frac{D}{2}$. For example, an industrial reactor with a rotor of 40 [cm] to stir 1 [m^3] of water, the consumption of energy is about 2HP. If we are interested in a scale of $\lambda = 25\mu$ (the size of very small particles), $\frac{L}{\lambda} \gg 10^3$ and local isotropy is ensured.

In suspensions, the flow in the vicinity of a particle affects the transport of matter to and from it. If this mass transport is controlled by the thickness of the boundary layer around the particle, the influence of the main flow is minimal, and the phenomenon is controlled by what happens in a small volume surrounding the particle. The theory of local isotropy (Kolmogoroff 1991) is extremely useful in these cases as it allows a definition of the flow regimen in a small volume of the violently agitated fluid. It is then possible to conclude that the properties in the vicinity of a particle in the turbulent flow in a stirred tank are determined only by the local dissipation of energy $\hat{\in}(|\mathbf{r}|)$. For this reason, the dissipation of energy can describe the state of stirring in the reactor, rotor size and speed, pulp height, and other properties that affect the process through a change in the magnitude of the local dissipation energy only. Unfortunately, in general the value of the average of the local dissipation of the energy is known, implying that the representation of the properties will be valid only for similar system. As a conclusion, the size of an industrial reactor can then be obtained from pilot-scale tests using the local isotropy theory and similarity analysis.

3.8.5.1 Energy Dissipation in a Continuous Stirred Tank Reactor

A continuous stirred tank reactor (CSTR) is a mixing tank equipped with an impeller to provide highly effective mixing, in which different processes may achieve given tasks to reach a specified output. It is extensively used in the chemical, environmental, and mining industry. In mining engineering, the best examples are the leaching tanks and the flotation cells. Figure 3.27 shows a schematic drawing of a flotation cell.

3.8.5.1.1 Dimensional Analysis

The principal characteristic of a CSTR is that its content is perfectly mixed, without radial axial, or angular gradients in their properties, for example, in velocity, component concentrations, and energy dissipation.

The energy dissipated in a stirred tank reactor can be correlated with system variables through dimensional analysis. Consider the following variables that affect the system; rotor diameter D[L], rotor rotation speed $\omega[1/\mathrm{T}]$, power P in kW or Hp consumed $\left[\mathrm{ML}^2/\mathrm{T}^3\right]$, dissipation rate $\in = P/V$ in kW/ton $\left[\mathrm{L}^2/\mathrm{T}^3\right]$, fluid density $\rho\left[\mathrm{M}/\mathrm{L}^3\right]$, fluid viscosity $\mu[\mathrm{M}/\mathrm{LT}]$, the acceleration of gravity $g\left[\mathrm{M}/\mathrm{T}^2\right]$, and the geometry of the reactor (as shown in Figure 3.28).

$$f(D, N, \rho, \mu, g, P, \hat{\in}, E, H, C, B, b) = 0 \tag{3.222}$$

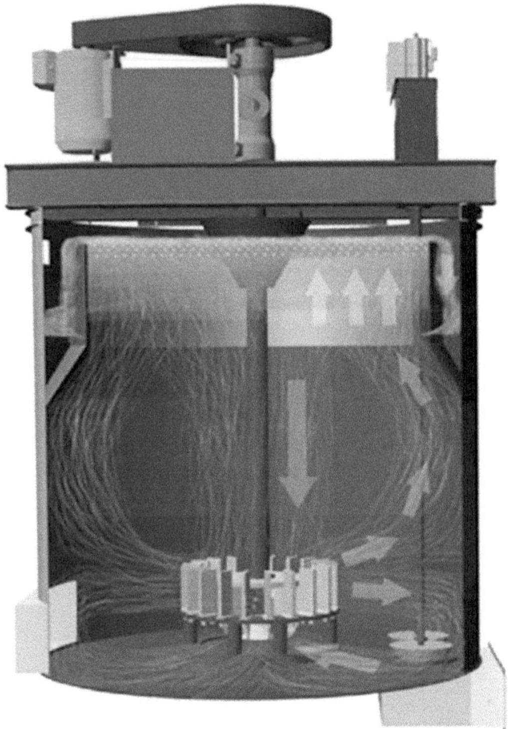

FIGURE 3.27 CFD drawing of a flotation cell.

Source: Michaud, L. 2016. Different types of flotation cells. *911 Metallurgist*, May 22. www.911metallurgist.com/blog/different-types-of-flotation-cells. Accessed August 2023.

The dimensional matrix is shown in Table 3.5.
From Table 3.5, we can write

$$f\left[\frac{\mu_t}{\rho_f D^2 N};\frac{g}{DN^2};\frac{P}{\rho_f D^5 N^3};\frac{\hat{\epsilon}}{D^2 N^3};\frac{L}{D};\frac{H}{D};\frac{C}{D};\frac{B}{D};\frac{b}{D}\right] \quad (3.223)$$

$$\text{Reynolds number: Re} = \frac{DN^2}{\nu} \text{ and Froud N° } Fr = \frac{DN^2}{g} \quad (3.224)$$

$$\text{Power number: } N_P = \frac{P}{\rho D^5 N^3} \text{ and Dissipation N° } N_{\hat{\epsilon}} = \frac{\hat{\epsilon}}{D^2 N^3} \quad (3.225)$$

$$\text{Air flow number: } N_{Qa} = \frac{Q_a}{D^3 N} \quad (3.226)$$

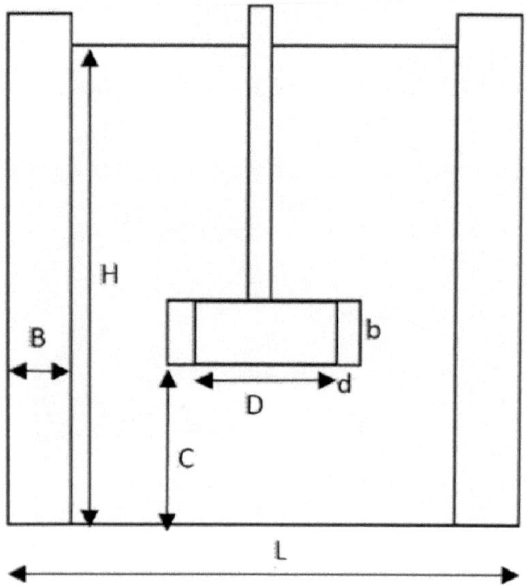

FIGURE 3.28 Dimensioned stirred tank reactor.

TABLE 3.5
Dimensional Matrix

	D	N	ρ_f	μ_t	g	P	L	H	C	B	b	Q_a	
L	1	0	3	–1	1	2	1	1	1	1	1	3	2
T	0	–1	0	–1	–2	–3	0	0	0	0	0	–1	–1
M	0	0	1	1	0	1	0	0	0	0	0	0	0
D	1	0	2	2	1	5	2	1	1	1	1	3	1
N	0	1	1	1	2	3	3	0	0	0	0	1	1
ρ_f	0	0	1	1	0	1	0	0	0	0	0	0	0

A flotation cell is a stirred vessel with Reynolds number determined by Equation 3.224,

$$\mathrm{Re} = \frac{l\bar{u}}{\nu} \tag{3.227}$$

For a cell with rotor equal to the macro scale of turbulence $D = 12.7[\mathrm{cm}]$, rotating at speed $N = 14[\mathrm{rpm}]$ and with a kinematic viscosity $\nu = 1\times10^{-4}[\mathrm{m}^2/\mathrm{s}]$, the value of $\mathrm{Re} = 0.127\times14/10^{-4} = 2.3\times10^3$. This value indicates that the flow in a flotation cell is highly turbulent and, considering that the intensity of the turbulence is given by Equation 3.227, leads to $\bar{u} \sim 1.78\left[\frac{m}{s}\right]$. The microscale of turbulence is $\lambda_0 = \frac{l}{\mathrm{Re}^{\frac{3}{4}}} \sim 7.33\times10^{-5}$, which ensures that the scale ratio is $\lambda_0 \ll l$.

In flotation we are interested in two scales, that of the size of the particles and that of the bubbles. As an average value, one can indicate particles of $100[\mu]$ and bubbles of $0.02[\mathrm{mm}]$, so that $\lambda \sim 2\times10^{-4}[\mathrm{m}]$ and $l \sim 1.27\times10^{-1}[\mathrm{m}]$, then $\frac{\lambda}{l} \sim 1.5\times10^{-3}$ and the local isotropy is satisfied. According to Kolmogoroff's theory the three important variables in the flows in stirred tanks with local isotropy are the micro-scale λ, the macro-scale l, and the turbulent energy dissipation $\hat{\in}[\mathrm{kW/ton}]$. Table 3.6 shows the scale, speed, and acceleration of turbulent flow.

Table 3.6 shows the relationships for the turbulence scale dependent on the λ values where v_λ and a_λ are the eddy velocity and acceleration, v' the intensity of the turbulence, $\mathrm{Re}_\lambda = \frac{\lambda v_\lambda}{\nu} \sim \left(\frac{\lambda}{l}\right)^{\frac{4}{3}}$ *Re* is the Reynolds number of the eddies and $\hat{\in}\left[\frac{L^2}{T^3}\right] \sim \left(\frac{\bar{u}^3}{l}\right)$, with $\bar{u}$ the average velocity of the principal flow.

3.8.5.1.2 *Aeration in Stirred Tanks*

For stirred tank reactor admitting a forced air flow Q_a, the power decreases and

$$N_p = K \times \mathrm{Re}^{\alpha} \times N_{Qa}^{\gamma} \tag{3.228}$$

TABLE 3.6
Properties of a Turbulent Flow with Local Isotropy

Turbulent scale	Turbulent velocity	Acceleration
$\lambda_0 \ll \lambda \ll l$	$v_\lambda \sim \left(\gamma\hat{\in}\right)^{\frac{1}{3}} \sim v'\left(\frac{\lambda}{l}\right)^{\frac{1}{3}}$	$a_\lambda \sim \left(\frac{\hat{\in}^2}{\lambda}\right)^{\frac{1}{3}}$
$\lambda \sim l$	$v_\lambda \sim \left(\nu\hat{\in}\right)^{\frac{1}{4}} \sim v'\left(\frac{1}{Re}()^{\frac{1}{4}}\right)$	$a_\lambda \sim \left(\frac{\hat{\in}^3}{\nu}\right)^{\frac{1}{4}}$
$\lambda \ll \lambda_0$	$v_\lambda \sim \left(\frac{\hat{\in}}{\nu}\right)^{\frac{1}{2}}\lambda \sim v'\mathrm{Re}^{\frac{1}{2}}\left(\frac{\lambda}{l}\right)$	$a_\lambda \sim \left(\frac{\hat{\in}}{\nu}\right)\lambda$

Geometrical similar stirred tank reactors Nagata (1975) may be written in terms of a Reynolds and a Froude numbers

$$N_p = K \times \mathrm{Re}^{\alpha} \times Fr^{\beta} \times N_Q^{\gamma} \tag{3.229}$$

and well-designed reactors, that is, without sedimentation, $\beta = 0$, then

$$N_p = K \times \mathrm{Re}^{\alpha} \times N_{\underline{Q}}^{\gamma} \text{ or } N_{\epsilon} = K_{\epsilon} \times \mathrm{Re}^{\alpha} \times N_{\underline{Q}}^{\gamma} \tag{3.230}$$

3.8.5.1.3 Liquid Systems

If a flotation cell is operated with water only, $N_Q = 0$ and

$$N_p = K \times \mathrm{Re}^{\alpha} \tag{3.231}$$

Figure 3.29 shows the correlations of power number N_P as a function of the Reynolds number for flotation cells (Arbiter et al. 1976).

3.8.5.1.4 Thee-Phase System Solid-Liquid-Air

Real flotation systems are three-phase systems solid-liquid-air for which the expression of last sections remain valid. If the system is kept at high Reynolds numbers, the

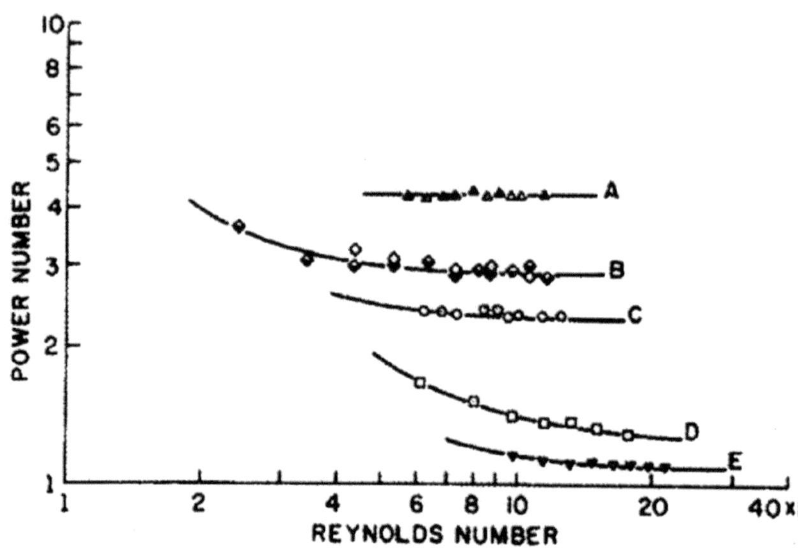

FIGURE 3.29 Power number for flotation cells. A = Agitair small wedge post impeller; B = Fagergreen; C = Agitair small round post impeller; D = Wedag medium; E = Denver D-2.

Source: Arbiter, N., Harris, C.C., and Yap, R.F. 1976. The air flow number in flotation machines scale-up. *International Journal of Mineral Processing* 3:257–280.

energy variable is eliminated since it becomes independent of the Reynolds number (Schubert 1999)

$$N_{\epsilon} = K_{\epsilon} \times N_Q^{\gamma} \tag{3.232}$$

3.8.5.1.5 Sedimentation and Suspension of Solids

By increasing the power in a stirred tank reactor from zero, the solid deposited in the bottom begins to be suspended, reaching the critical value when it is completely suspended. According to Nagata (1975) the increase in power is proportional to the ratio of the pulp and the fluid densities. If P_s and ρ are the power and the pulp density and P_o the power without solids, the power number at the critical point is

$$N_P = N_0 \times \frac{\rho}{\rho_f} \tag{3.233}$$

Hughmark (1974) demonstrated that the drag for a particle suspended in a stirred tank reactor is greater than that of the one in a quiescent fluid. Figure 3.30 and Equations 3.164 and 3.165 show the comparison.

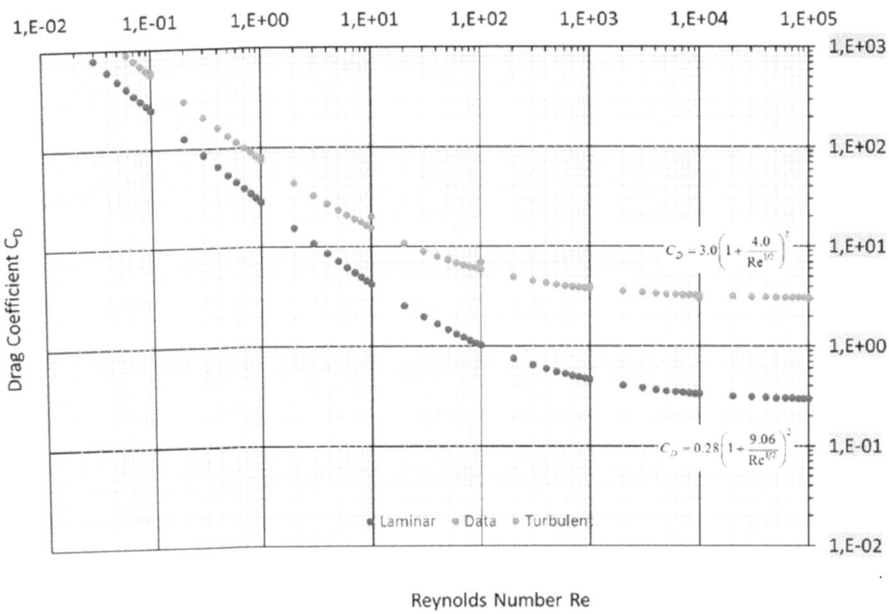

FIGURE 3.30 Drag coefficient versus Reynolds number for the settling of a sphere in laminar and turbulent flow.

These curves can be correlated with the drag coefficient (Concha and Almendra 1979)

$$C_D = C_0\left(1+\delta_0 / \mathrm{Re}^{1/2}\right)^2 \tag{3.234}$$

for laminar flow $C_0 = 0.28 \quad \delta_0 = 9.06$ and turbulent flow

$$C_0 = 3.0 \quad \delta_0 = 4.0 \tag{3.235}$$

3.8.5.2 Relative Gas Bubble-Fluid Velocity

A gas bubble in a stirred tank reactor, besides being affected by gravity, will also be dragged by the fluid. In this case the density of the air in the bubble is $\rho_b \sim 1.2$ kg/m^3 while for water is $\rho_f \sim 1100$ kg/m^3. The vertical component of the velocity is

$$u' \sim \left(\frac{\rho_f}{\rho_b}\right)^{1/2}\left[\frac{g-\left(\frac{\hat{\epsilon}^2}{\lambda}\right)^{1/3}}{\left(\frac{1}{\lambda}+\frac{3}{4}C_D\frac{\rho_f}{\rho_b}\frac{1}{d_b}\right)}\right]^{1/2} \quad \text{for } L \sim 12.7 \quad 10^{-3} < \lambda < 10 \tag{3.236}$$

Since $\rho_f / \rho_b \sim 10^3$ and $\frac{3}{4}C_D\frac{\rho_f}{\rho_b}\frac{1}{d_b} \sim 10^7$, the second term of Equation 3.236 is much greater than the first, so that

$$u' \sim \left(1.15\frac{1}{C_D^{1/2}}\right)\left[g-\left(\frac{\hat{\epsilon}^2}{\lambda}\right)^{1/3}\right]^{1/2} d_b^{1/2} \tag{3.237}$$

3.8.6 Turbulent Diffusion to a Particle in a Stirred Tank Reactor

In the same way as turbulence transports momentum in a fluid, it can also carry other properties such as mass. This transport can be characterized by an eddy diffusion coefficient $\mathbb{D}_t$ such that the constitutive equation of the turbulent mass flow will be

$$J_t = \mathbb{D}_t \nabla c_i \tag{3.238}$$

where J_t is the mass flux-density. The equation is similar to Fick's law of molecular diffusion, but its coefficient of turbulent diffusion is not a thermodynamic property but depends on the scale and other variables that characterize turbulence.

$$\mathbb{D}_t = \mathbb{D}_t\left(L, \rho, u'\right) \tag{3.239}$$

From the dimensional analysis of Table 3.1, we can find the relationship between $\mathbb{D}_t$ and the fluctuating velocity u.

$$\mathbb{D}_t \sim NL \tag{3.240}$$

The size of the particles in a flotation cell is between the macro and the microscale of turbulence, $\lambda_0 \le d \le l$; therefore, applying the theory of local isotropy, the turbulent diffusion coefficient is determined by

$$\mathbb{D}_t(\lambda) \sim v_\lambda \lambda \tag{3.241}$$

Since $\lambda \gg \lambda_0$, from Table 3.2, $v_\lambda \sim (\hat{\epsilon}\lambda)^{1/3}$, then

$$\begin{aligned}\mathbb{D}_t(\lambda) &= \alpha \hat{\epsilon}^{1/3} \lambda^{1/3} \\ &= \alpha l u' \left(\frac{\lambda}{l}\right)^{1/3}\end{aligned} \tag{3.242}$$

3.8.6.1 Relative Solid Particle-Fluid Velocity

Consider a particle moving in a turbulent fluid inside a stirred tank reactor. The linear momentum the particle acquires is at the expense of the momentum of the fluid, therefore the balance of the particle and fluid momentum are

$$\rho_p V_p \dot{\boldsymbol{v}}_p = \rho_p V_p \boldsymbol{g} - \rho_f V_p \left(\boldsymbol{g} - \dot{\boldsymbol{v}}_f\right) - \frac{1}{2} C_D \rho_f A_p \left|u'\right| \boldsymbol{u} \tag{3.243}$$

Subtracting term $\rho_p V_p \dot{v}_f$ from both members of Equation 3.243, and considering the relative solid-fluid velocity is $\boldsymbol{u}' = \boldsymbol{v}'_p - \boldsymbol{v}'_f$, Equation 3.243 becomes

$$\overline{u}' = \frac{\Delta\rho}{\rho_p} g - \frac{1}{2} C_D \frac{\rho_f}{\rho_p} \frac{A_p}{V_p} u'^2 \tag{3.244}$$

And the solid fluctuating velocity is

$$\dot{v}'_s = \dot{v}'_f + \frac{\Delta\rho}{\rho_p} g - \frac{1}{2} C_D \frac{\rho_f}{\rho_p} \frac{A_p}{V_p} u'^2 \tag{3.245}$$

Mineral particles of sizes contained in the reactor are $10^{-3} < d_p < 10^{-2}$, densities $2 < \rho_p < 7$, and bubble sizes of de order of $d_b \sim 0.02$ cm. For a particle in the range of $l \le d_p \ll L$, one can use the local isotropy theory so that, from section 3.4 we have

$$v_\lambda \sim (\lambda\hat{\epsilon})^{1/3}, \alpha_\lambda \sim (\lambda\hat{\epsilon})^{1/3}, u' = \sqrt{v_x^{12}} = \sqrt{v_y^{12}} = \sqrt{v_z^{12}} \quad (3.246)$$

At steady state $\dot{u} = u^2 / \lambda, \dot{v}_f = a_\lambda^2$, then from Equations 3.243 and 3.246

$$\rho_p V_p \frac{u'^2}{\lambda} + \frac{1}{2} C_D \rho_f A_p u'^2 = \Delta\rho V_p g - \rho_f V_p a_\lambda^2$$

$$\left(1 + \frac{1}{2} C_D \frac{\rho_f}{\rho_p} \frac{A_p}{V_p}\right) u'^2 = \frac{\Delta\rho}{\rho_p} \lambda \left(g - (\lambda\hat{\epsilon})^{2/3}\right)$$

$$u' = \left(\frac{\Delta\rho}{\rho_p}\right)^{1/2} \left[\frac{\left(\lambda g - (\lambda\hat{\epsilon})^{2/3}\right)}{\left(1 + \frac{1}{2} C_D \frac{\rho_f}{\rho_p} \frac{A_p}{V_p}\right)}\right]^{1/2} \quad \begin{array}{c} \text{for } L \sim 12.7 \\ 10^{-3} < d_p < 10^{-2} \\ 10^{-3} < \lambda < 10 \end{array} \quad (3.247)$$

3.8.6.1.1 Example

Consider the flotation cell shown in Figure 3.27 with $L \sim D$ and $l \sim \frac{L}{Re^{3/4}}$, $\mathrm{Re} = Lu / v \equiv D^2 N / v$, $D = 12\,\mathrm{cm}$, $N = 14\,\mathrm{rpm}$, $V = 10^{-2}\,\mathrm{cm}^2/\mathrm{s}$, $u = 180\,\mathrm{cm/s}$, $L \sim D = 12.7\,\mathrm{cm}$, $\mathrm{Re} = 2.3 \times 10^5$, $l = 1.21 \times 10^{-3}\,\mathrm{cm}$.

$$C_D = 4\left(1 + \frac{3.8}{\mathrm{Re}^{1/2}}\right)^2 = 4\left(1 + \frac{3.8}{4.8 \times 10^2}\right)^2 = 4.00$$

$$A_p = \frac{\pi d^2}{4}; \; V_p = \frac{\pi d^3}{6} \rightarrow \frac{A_p}{V_p} = \frac{3}{2} \frac{1}{d_p} \quad (3.248)$$

$$\hat{\epsilon} \sim \frac{u^3}{L} = \frac{180^3}{12.7} = 4.6 \times 10^5 \text{ and } 10^{-3} < \lambda < 10$$

$$5.4 \times 10 < (\lambda\hat{\epsilon})^{2/3} < 1.52 \times 10^4 \text{ and } 1 < g\lambda < 10$$

The terms in Equation 3.248 shows that for all cases $g\lambda \ll (\lambda\hat{\epsilon})^{2/3}$ in the square bracket of Equation 3.247, then Equation 3.247 becomes

$$u'_{\max} \sim \left(\frac{\Delta\rho}{\rho_p}\right)^{1/2} (\lambda\hat{\epsilon})^{1/3} \left(1 + \frac{3}{4} C_D \frac{\rho_f}{\rho_p} \frac{\lambda}{d_p}\right)^{-1/2} \quad (3.249)$$

In Equation 3.249, let $\left(\frac{\Delta\rho}{\rho_p}\right)^{1/2}(\lambda\hat{\epsilon})^{1/3} = \alpha$ and $\frac{3}{4}C_D\frac{\rho_f}{\rho_p}\frac{\lambda}{d_p} = \beta$ (3.250)

then Equation 3.249 can be expressed as

$$u' \sim \frac{\alpha\lambda^{1/3}}{(1+\beta\lambda)^{1/2}} \tag{3.251}$$

For the maximum velocity $u_{\max}$, $du/d\lambda = 0 \rightarrow \lambda_m = 2/\beta$, and replacing in Equation 3.251 gives

$$u'_{\max} \sim \frac{\alpha(2/\beta)^{1/3}}{(1+\beta\times 2/\beta)^{1/2}} = 0.73\frac{\alpha}{\beta^{1/3}} \tag{3.252}$$

$$u'_{\max} \sim 0.80\left(\frac{\Delta\rho}{\rho_p}\right)^{1/2}\left(\frac{\rho_p}{\rho_f}\right)^{1/3}(C_D)^{-1/3}\left(\hat{\epsilon}d_p\right)^{1/3} \tag{3.253}$$

Since $v_{\lambda=d_p} = \left(\hat{\in} d_p\right)^{1/3}$ we have

$$\frac{u'_{\max}}{v'_{\lambda=d_p}} \sim 0.80\left(\frac{\Delta\rho}{\rho_p}\right)^{1/2}\left(\frac{\rho_p}{\rho_f}\right)^{1/3}(C_D)^{-1/3} \tag{3.254}$$

For particles in the range $10^{-3} < d_p < 10^{-2}$,

$$0.2 < u'_{\max} / v'_{h-d_p} < 0.5 \tag{3.255}$$

For example, for $d_p = 50\,\mu\text{m}$ and $\rho_p = 4$, the ratio of the maximum fluctuating solid velocity to the fluid velocity

$$u'_{\max} / v'_{\lambda=d_p} \sim 0.34$$

REFERENCES

Agricola, G. 1556. *De Re Metallica*. Translated by H.C. Hoover and L.H. Hoover, 1950. New York: Dover. p. 347.

Arbiter, N., Harris, C.C., and Yap, R.F. 1976. The air flow number in flotation machines scale-up. *International Journal of Mineral Processing* 3:257–280.

Baird, M.H., Senior, M.G., and Thompson, R.J. 1967. Terminal velocities of spherical particles in a vertically oscillating liquid. *Chemical Engineering Science* 22:552–558.

Bird, R.B., Stewart, W.E., and Lightfoot, E.N. 1964. *Fenómenos de Transporte, Un estudio sistemático de los fundamentos del transporte de materia, energía y cantidad de movimiento*. Barcelona, S.A.: Editorial Reverte.

Brenner, H. 1964. Effect of finite boundaries on Stokes resistance of an arbitrary particle. *Journal of Fluid Mechanics* 18:146.

Brenner, H. 1966. Hydrodynamic resistance of small Reynolds numbers. In *Advances in Chemical Engineering*, vol. 6. Edited by T.B. Drew, J.W. Hoopes, and T. Vermeulen. New York: Academic Press. p. 324.

Cebeci, T., ed. 1982. *Numerical and Physical Aspects of Aerodynamic Flows*. Cham: Springer Verlag.

Cenegel, Y.A., and Cimbala, J.M. 2018. *Fluid Mechanics: Fundamentals and Applications*, 4th ed. New York: McGraw-Hill.

Clift, R., Grace, J.R., and Weber, M.E. 1978. *Bubbles, Drops and Particles*. New York: Academic Press.

Concha, F. 1968. Application of transport phenomena to mineral dressing. PhD dissertation, University of Minnesota. pp. 79, 182, 189.

Concha, F., and Almendra, E.R. 1979. Settling velocities of particulate systems, 1. Settling velocities of individual spherical particles. *International Journal of Mineral Processing* 5:349–367.

Dallavalle, J.M. 1954. *Micromeritics: The Technology of Particles*, 2nd ed. New York: Pitman.

Essentials of Fluid Mechanics. 2014. www.youtube.com/watch?v=x8_pfo1-MmI.

Essentials of Fluid Mechanics. 2018. www.youtube.com/watch?v=rB83DpBJQsE.

Gaudin, A.M. 1939. *Principles of Mineral Dressing*. New York: McGraw-Hill.

Happel, J., and Brenner, H. 1965. *Low Reynolds Hydrodynamics*. Englewood Cliffs, NJ: Prentice-Hall. p. 163.

Heywood, H. 1962. Uniform and non-uniform motion of particles in fluids. In *Proceeding of the Symposium on the Interaction Between Fluid and Particles*, Institute of Chemical Engineering, London. pp. 1–8.

Houghton, E.L., Carpenter, E.L., Collicott, S.H., and, Valentine, D.T. 2016. *Aerodynamics for Engineering Students*, 7th ed. Cambridge, MA: Butterworth-Heinemann is an imprint of Elsevier.

Houghton, G. 1966. Particle trajectories and terminal velocities in vertically oscillating fluids. *Canadian Journal of Chemical Engineering* 44:90–95.

Houghton, G. 1968. Particle retardation in vertically oscillating fluids. *Canadian Journal of Chemical Engineering* 46:79–81.

Hughmark, G.A. 1974. Hydrodynamics and mass transfer for suspended solid particles in a turbulent liquid. *AIChE Journal* 20:1.

Hwang, P.A. 1983. Fall velocity of particles in oscillating flow. *Journal of Hydraulic Engineering* 111(3):485–502.

Ibsen, D.C. 1960. *Units, Dimensionless Numbers*. New York: McGraw-Hill.

Ikeda, S. 1989. Fall velocity of single spheres in vertically oscillating fluids. *Fluid Dynamics Research* 5:203–216.

Jong, T.P.R., Witteveen, H.J., and Dalmijin, W.L. 1996. Penetration velocities in a homogenous jig bed. *International Journal of Mineral Processing* 46:277–291.

Kolmogoroff, A.N. 1991. The local structure of turbulence in incompressible viscous fluid for very large Reynolds numbers. *Proceedings of the Royal Society of London A* 434:9–13.

Lapple, C.E., and Shepherd, C.B. 1940. Calculation of particle trajectories. *Industrial & Engineering Chemistry* 32:605.

Lee, K., and Barrow, H. 1968. Transport process flow around a sphere with particular reference to the transfer of mass. *International Journal of Heat and Mass Transfer* 11:1020.

Levich, V.G. 1962. *Physicochemical Hydrodynamics*. Englewood Cliffs, NJ: Prentice-Hall.

Matz, W. 1954. *Anwendung des Änlichkeitsgrundsatzes n the der verfarens Technik*. Berlin: Springer Verlag.
Nagata, S. 1975. *Mixing: Principles and Applications*. Tokyo: Kodansha. pp. 24–58.
Perry, R.H., and Chilton, C.H. 1973. *Chemical Engineers Handbook*, 5th ed. Tokyo: McGraw-Hill Kogakusha, Ltd. p. 5.24.
Richard, R.H., and Locke, C.E. 1940. *Textbook of Ore Dressing*. New York: McGraw-Hill.
Richarson, J.F., and Zaki, W.N. 1954. Sedimentation and Fluidization: Part 1. *Transactions of the Institution of Chemical Engineers* 32:24–53.
Schlichting, H. 1955. *Boundary Layer Theory*. New York: McGraw-Hill.
Schubert, H. 1999. On the turbulence-controlled microprocesses in flotation machines. *International Journal of Mineral Processing* 56:257–276.
Scobie, J.A., Sangan, C.M., and Lock, G.D. 2014. Flow visualization experiments on sports balls. *Procedia Engineering* 72:738–743.
Srinivasa, B.K., Nishra, B.K., and Methrotra, S.P. 1999. Simulation of particle stratification in jigs. *Coal Preparation* 20:55–70.
Taggart, A.F. 1951. *Elements of Ore Dressing*. New York: Wiley.
Tavares, L.M. 1999. Monte Carlo simulation on the potential energy theory of jigging. *Coal Preparation* 20:71–83.
Tory, E.M. 1996. *Sedimentation of Small Particles in a Viscous Fluid*. Ontario, Canada: Computational Mechanics Publications, Inc.

4 Particulate Systems

4.1 SIZE REDUCTION EQUATION

The purposes of science in comminution are to understand the mechanism of the fracture of rocks and give the foundation for the process of *size reduction* in a mining operation and lead to a theory that would permit the development of models of the process with the objective of predicting, controlling, and improving its performance. Comminution is based on solids physics, mechanics, mineralogy, and fluid mechanics. Physics is primarily interested in the structure of solids, the behavior of materials in the face of the application of stresses, the physics of rupture and surfaces. The mechanics concern with pressure or collision forces, producing and expansion of cracks by the application of stresses, and the deformation and fracture of solids. The study of comminution involves (1) the development of a theory of materials, (2) the physics of deformation and fracture, and (3) the technology and development of models of size reduction (Schönert 1972, 1996).

Continuum mechanics is associated with the state of a given solids, for which deformations and stretches must be linked with applied stresses with mathematical formulations in laws for simple materials. Materials can be classified into elastic, inelastic, and elastoplastic. An elastic isotropic material shows a linear relationship between stress σ and deformation ε which can be expressed in form of Hooks law.

$$\sigma = Y\varepsilon \tag{4.1}$$

An inelastic material is continuously deformed when stress is applied. Most inelastic bodies have combined both relationships. The first part of the deformation is elastic followed by a ductile stage. If at his point the stresses are withdrawn, these materials do not return to their original state since some of the work used in the deformation dissipates. Further increase on the application of stress will break it. An example is shown in Figures 4.1 and 4.2.

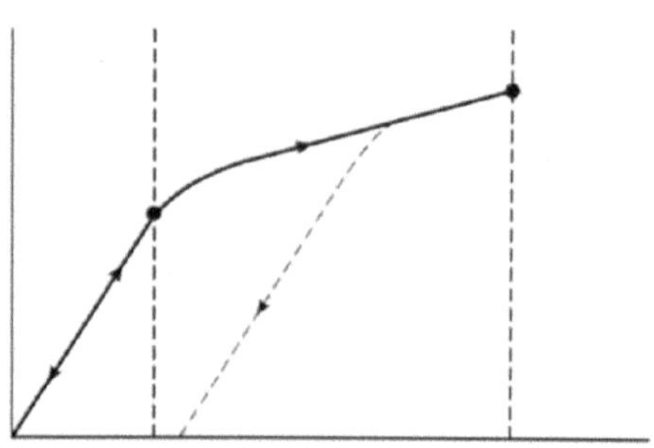

FIGURE 4.1 Inelastic material.

DOI: 10.1201/9781032614212-4

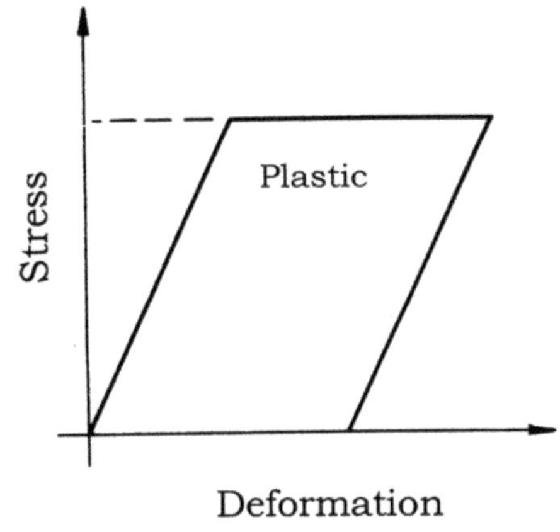

FIGURE 4.2 Elastoplastic material.

An *ideal* elastic material is broken when the breakage limit has been exceeded, that is when all atomic bonds on a certain plane are broken. Real elastic materials contain flaws on both a microscopic and macroscopic scale. In the field of material science, microscopic failures are called *dislocations*, and in rock mechanics they are known as *Griffith's cracks* (Griffith 1921). The existence of these failures on material explains their low mechanical resistance compared to that calculated for an elastic solid. Fracture theory studies the initiation of cracks from faults and their spread in the solid. When these are subjected to stresses, they activated the faults concentrating the forces on the tip of the faults multiplying the traction and leading to breakage. For example, the traction σ_0 is increased to σ_{max}

$$\frac{\sigma_{max}}{\sigma_0\sqrt{\frac{2l}{r}}}, \tag{4.2}$$

where σ_0 is the unperturbed stress perpendicular to the flaw, $2l$ is the fault length, and r the radius of the position vector at the fault tip.

Rocks and minerals are heterogeneous and anisotropic materials, the failure of which are called *Griffith cracks*. The tensile stress occurs along the micro crack edge inside soil particle leading to a crack inside particle as the stress increases. Fragile solids are broken with efforts in the order of one-tenth of Young's module, which represents the ideal conditions. Important factors in defining the resistance of such a material are the size and shape of the particle, the magnitude of the applied force, and the direction and speed in which it is applied. The resistance of rocks to pressure is much greater than the resistance to tension, and the rupture typically occurs along shear planes. Figure 4.3 shows the stress concentration at the tip of the failure (Schönert 1980). A rock salt particle in the form of a parallelepiped has a greater strength than another sample of the same material but in the form of a cube (Hiramatsu and Oka 1966). When stresses are applied slowly, the deformation rate

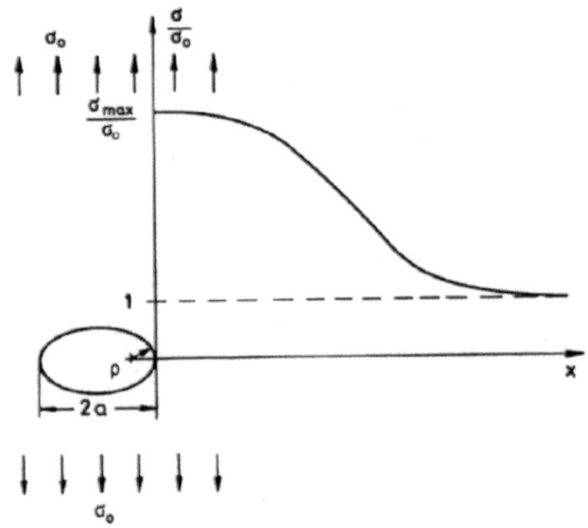

FIGURE 4.3 Stress concentration at the tip of the failure.

Source: Schönert, C. 1980. *Zerkleinern, Mechanishe Verfahrenstechnik*, 4th ed. Karlsruhe: Universität Karlsruhe.

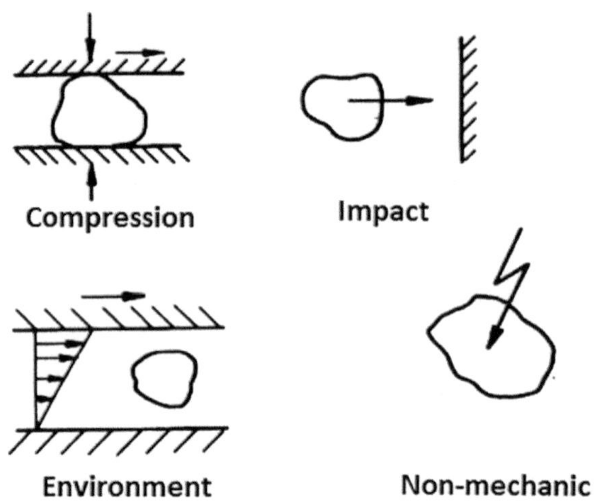

FIGURE 4.4 Forms in which the forces are applied.

Source: Schönert, C. 1980. *Zerkleinern, Mechanishe Verfahrenstechnik*, 4th ed. Karlsruhe: Universität Karlsruhe.

of the material is also slow, and less plastic deformation occurs before the breakage. This leads to the rupture at lower efforts and therefore with lower power consumption than when the speed of application of the efforts is higher. However, it has also been shown that when the speed of application of stress on a rock, such as high-speed impacts, is very large, the energy required for rupture is minimal. Figure 4.4 shows four types of forces for communication.

Since larger particles have the probability of having more Griffith cracks than smaller particles, they would break faster and with lower energy consumption. The size distribution resulting from the fracture of a rock is controlled by the initial fault distribution or by the distribution of the applied stress. The first case is obtained when the force is evenly applied in the sample (compression) and the second when the application of the force is concentrated (impact). Rocks contain many macroscopic failures in addition to Griffith's cracks, therefore the ideas developed in the preceding paragraphs are representative of those cases in which the big faults have disappeared, that is, in fine grinding.

Gilvarry (1961), assuming a Poisson distribution for Griffith's faults, developed a representation of the size distribution produced by the fracture of a particle given by the equation

$$F_3(x) = 1 - \exp(-\gamma x) \tag{4.3}$$

Where $F_3(x)$ is the mass fraction of particles smaller than size x, and γ is a measure of the Griffith faults activated. Figure 4.5 shows the distribution function of six different blasted rocks. It is interesting that Schumann's and Rosin-Rammler's equations are generalization of this function.

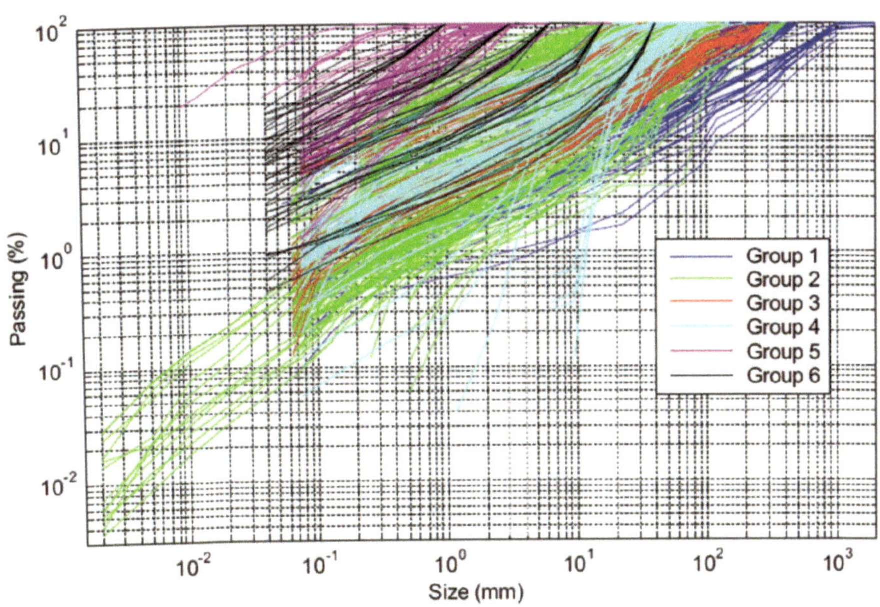

FIGURE 4.5 Particle size distribution of blasted rocks.

Source: Sanchidrián, J.A., Ouchterlony, F., Moser, P., Segarra, P., and López, L.M. 2012. Performance of some distributions to describe rock fragmentation data. *International Journal of Rock Mechanics & Mining Sciences* 53:18–31.

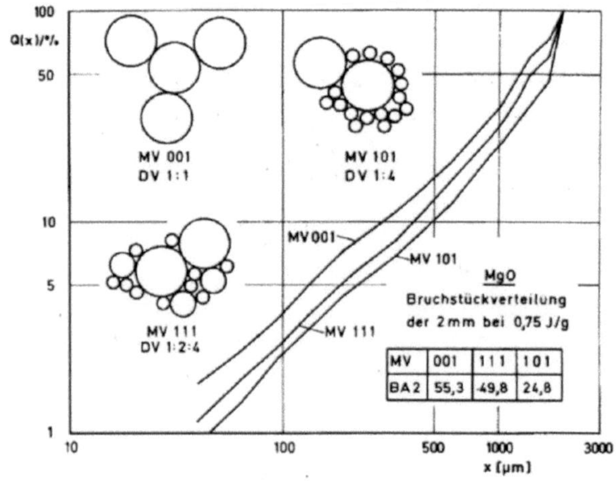

FIGURE 4.6 Particle size distribution of 2 [mm] MgO spheres in a bed of different particle sizes from 1:1, 1:4, and 1:2:4 with energy consumption of $W_M = 0.2085[\text{kWh/ton}]$.

Source: Hoffmann, N. 1976. The size distribution shapes during the fracture of binary and tertiary mixtures. *Chemie Ingenieur Technik* 48(4):329.

The result of compression of a particle is significantly influenced by the presence of other particles during the fracture, either because the material consists of more than one particle, it is a particle bed, or it contains the product of the original particle fracture. The result is that there is more than one fracture event present. It has been estimated that a significant part of the energy consumed in the fracture of the compression of a particle is transformed into kinetic energy of the progenies which, although not fractured, contain many failures. Figure 4.6 shows that the presence of particles of similar size produce a finer particle size distribution shape but consumes more energy (Hoffmann 1976).

When comminution events are applied successively to a set of particles, the problem becomes complex, and the energy cannot be used as a separate variable, but instead the total energy consumed is equal to the power of breakage events in the overall time of comminution, that is, the probability of fracture and time.

$$E_M = \int_0^t P d\xi \tag{4.4}$$

Bourgeois (1993) showed that particles of similar sizes have a wide variation of fracture energies due to their wide distribution of microscopic crack and the randomness of the fracture process. Nevertheless, a range of larger particles show a faster breakage but a smaller total consumption of energy than the breakage of a group of

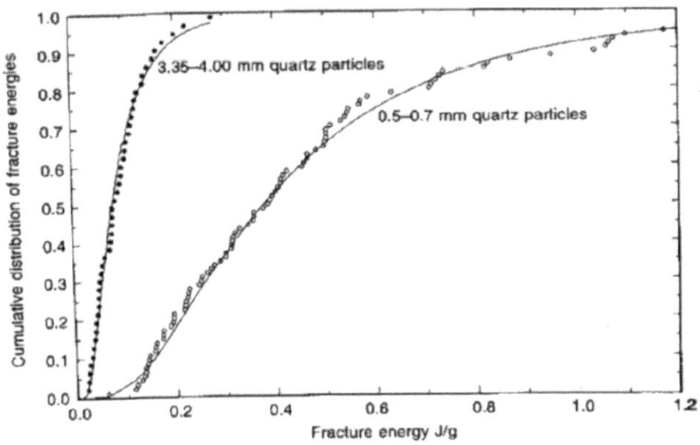

FIGURE 4.7 Measures distribution of particle fracture energy e_M for two range sizes of quartz particles.

Source: Bourgeois, F. 1993. Microscale modelling of comminution processes. PhD thesis, University of Utah, Salt Lake City.

smaller particles (Figure 4.7). This conclusion has great implication indicating that the early stages of size reduction are more energy efficient (Bourgeois et al. 1992).

4.1.1 Single Particle Breakage

Single-particle fracture studies have provided the basis for most of the fundamental particle comminution research with important contributions from researchers in the areas fracture phenomena [9–11], energy utilization (Schönert 1991; Rumpf 1973; Tavares 2000), particle properties and geometry (Yashima et al. 1987; Tavares and King 1998), energy size reduction relationship (Bergstrom and Sollenberger 1961; Rumpf 1973; Saeidi 2010) breakage characteristics (Napier-Munn et al. 1996), and material deformation response under applied stress (Schönert 1991; Sikong 1990).

A variety of testing methods have been used to measure the breakage characteristics of single particles subject to compression, each of which allows investigation over a restricted range of deformation rates. These tests can be classified as single impact, double impact, and slow compression. Figure 4.8 shows different types of single-particle breakage tests (Tavares 2007).

4.1.1.1 Breakage by Single Impact

There are two types of single impact procedures to break a particle, one using low-impact velocity represented by the freefall test known as drop weight test (Figure 4.8). The other uses high impact velocity, represented by the Rotary Impact Test.

The impact test produces a kinetic energy given by

$$E = \frac{1}{2} m_d v_I^2 \tag{4.5}$$

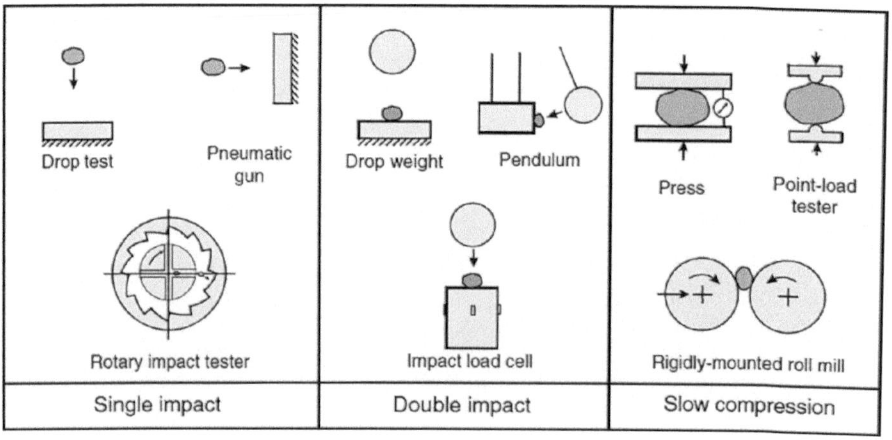

FIGURE 4.8 Different types of single-particle breakage tests.

Source: Tavares, L.M. 2007. Breakage of single particles: Quasi-static. In *Handbook of Powder Technology*, vol. 12: Particle Breakage. Edited by A.D. Salman, M. Ghadiri, and M.J. Hounslow. New York: Elsevier. pp. 3–66.

Where v_I is velocity of the drop weight at the impact instant and m_d the mass of the drop weight.

4.1.1.1.1 Low Impact Tester

For a free fall condition without air friction, that in this case is a good approximation, the velocity at the impact is $v_I = \sqrt{2gh}$, where h is the height from which the mass M_d is dropped, then, the energy involved is

$$E = M_d gh \tag{4.6}$$

The simplest type of single-impact testers is the drop weight test (DWT) where a body of mass is released on a particle lying on a metal plate. Figure 4.9a shows a schematic drawing of the tester and Figure 4.9b and 4.9c, two JKTech versions of the instrument producing an impact energy of $0.01\left[\frac{J}{g}\right], \left(1J/g = 0.278 \text{ kWh/ton}\right)$. In an instrument of about 100 [cm] on height (1) is the electromagnetic head, (2) a fixed lead head mass of 5.87 kg, (3) a steel anvil of 15.5 cm in diameter, (4) the steel base, (5) a ruler for drop height adjustment, (6) a mechanical arm, and (7) a drop-weight rail (Figure 4.9b).

The result of the DWT single particle tester is a single function t_{10}, which depends on a resulting normalized size distribution. The idea is that, if particles of different initial sizes break with the same t_{10} value, their size distributions will be similar when plotted in size-normalized x-axis (Powell et al. 2014)

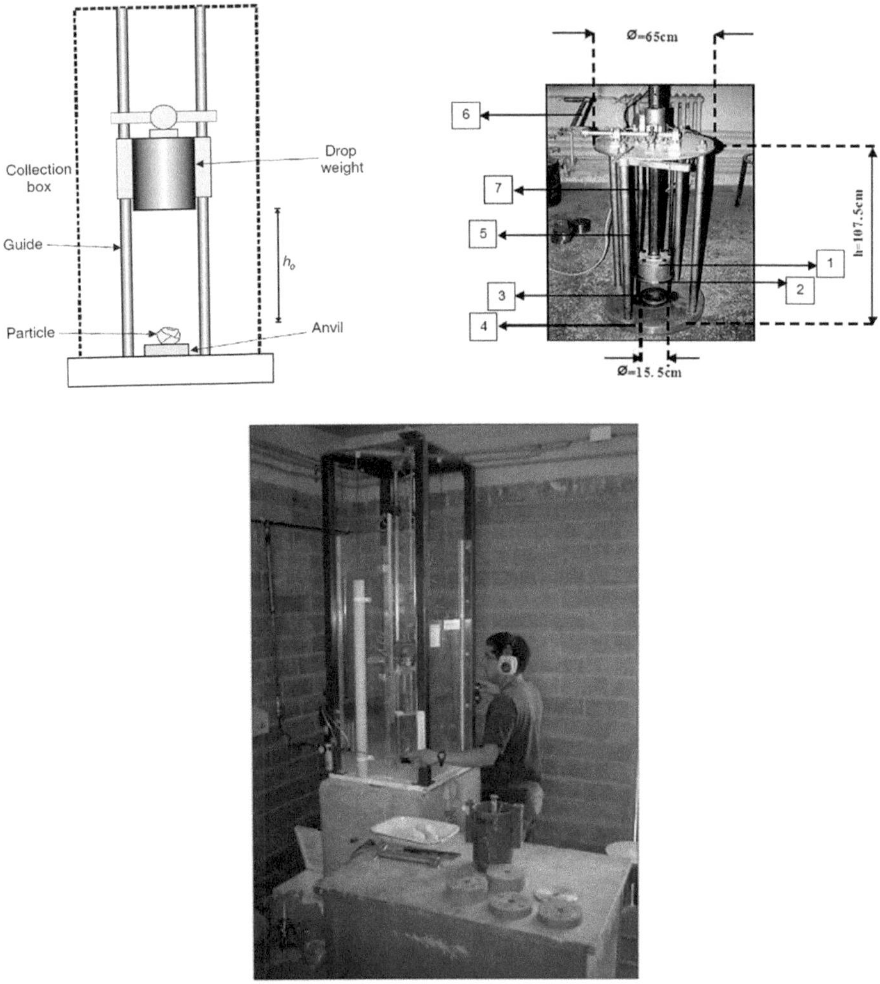

FIGURE 4.9 Single drop tester.

Source: SGS Minerals Services. 2013. JK drop weight test. T3 SGS 151. https://es.scribd.com/document/455245524/SGS-MIN-WA232-JK-Drop-Weight-Test-EN-11-pdf.

$$t_{10} = A\left(1 - \exp\left(bE_M\right)\right) \tag{4.7}$$

where A is the asymptotic limit to the t_{10} value with increasing impact energy and b is the curvature of the exponential decay and E_M is the applied specific energy in kWh/ton.

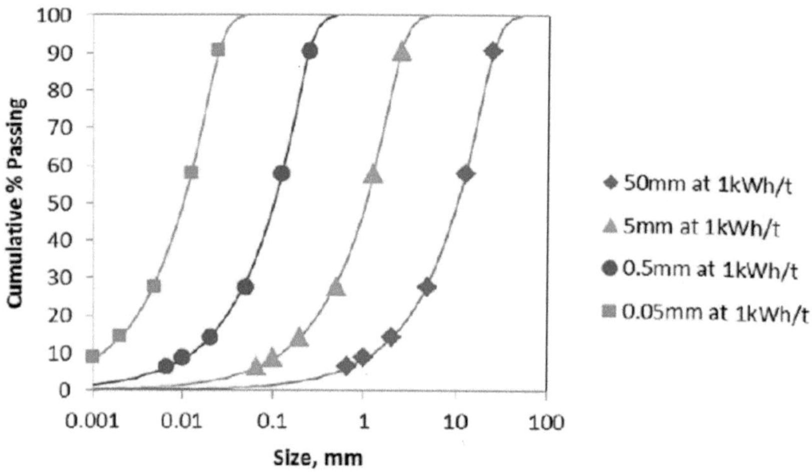

FIGURE 4.10 Size distribution at $1\left[\frac{\text{kWh}}{\text{ton}}\right]$.

Source: Powell, M., Hildenh, M., Ballantyne, G., Liu, J., and Tavares, L.M. 2014. The appropriate, and inappropriate, application of the JKMRC t_{10} relationship. Presented at IMPC XXVIII, Santiago, Chile.

For example, for an ore where $A = 50$ and $b = 0.69$, a plot of an applied specific energy $E_M = 1[\text{kWh/ton}]$ to four particle sizes gave the same $t_{10} = 24.92$, as shown in Figure 4.10.

The t_{10} parameter applies reasonably well for breakage of big particles in crushers, but for smaller sizes, the condition for successful characterization of an ore is that of *single impact breakage* to single particles between two metal plates applying the same energy. For low energy impact causing *incremental breakage* as opposed to single breakage, the applied energy should be related to a specific particle size. Powell et al. (2014) suggested E_{M75} defined as the energy to reduce the original particle to $75\mu m$.

4.1.1.1.2 High Impact Tester

A typical high impact tester is the Rotary Impact Test (RIT), where single particles enter the rotor centrally and are accelerated in one of the radial channels by the centrifugal force. Neglecting friction in the grinding chamber, the impact velocity of the particles on the test ring is given by the ejecting velocity from the rotor v_e as a function of the rotational speed N (Karpinski and Tervo 1964).

$$v_e = \sqrt{2}\pi D_R N \tag{4.8}$$

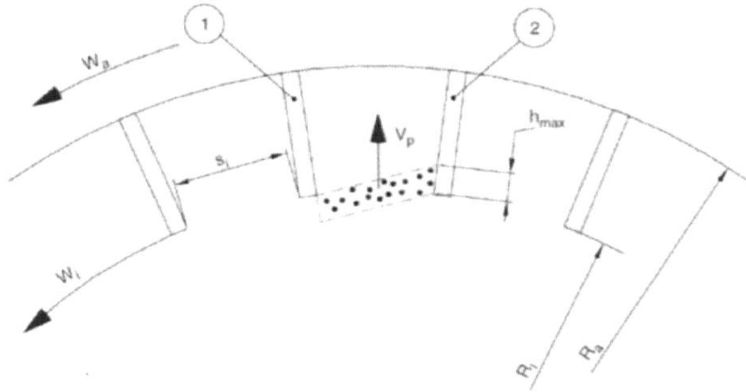

FIGURE 4.11 Geometry of the RIT and penetration of the particles between the impact beaters.

where D_R is the outer diameter and N the rotational speed of the rotor. Impact velocities of 60 [m/s] can give impact energies of up to 10 $\left[\frac{J}{g}\right]$ (0.5 kWh/ton).

According to Nied (2007), particles move radially in the breakage chamber before being ejected from the RIT at the particle ejection speed $v_p = 4.3[m/s]$ for a rotor diameter of 300[mm] and tangential velocity $v_I = 73[m/s]$ with a distance between the impact beaters s_i (Figure 4.11). The particle ejection speed can be described by

$$v_p = \sqrt{2}\left(1 - \frac{D_I^2}{D_A^2}\right)^{0.5} N \tag{4.9}$$

where D_I and D_A are the internal and external diameter of the RIT and N is the circumferential speed.

4.1.1.2 Slow Compression

Several types of particle testing machine have been used over the years based on the original Split Hopkinson Pressure Bar developed by Hopkinson. A good example is the Weichert and Herbst (1986) Ultrafast Load Cell (UFLC) consisting of a long steel rod equipped with strain gauges on which a single particle or a bed of particles is placed and impacted by a falling steel ball (Figure 4.12). The compressive wave resulting from the impact travels down the rod and is sensed by the solid-state strain gauges. This results in a voltage change in the Wheatstone bridge, which is then recorded as a function of time using a digital oscilloscope.

In Figure 4.12, u_b is calculated from a momentum balance of the falling ball and u_r from the measured compression of the rod (Tavares 2007).

Tavares and King (1998) present two important parameters related to the internal state of stress of a particle subjected to a slow point load compression stress

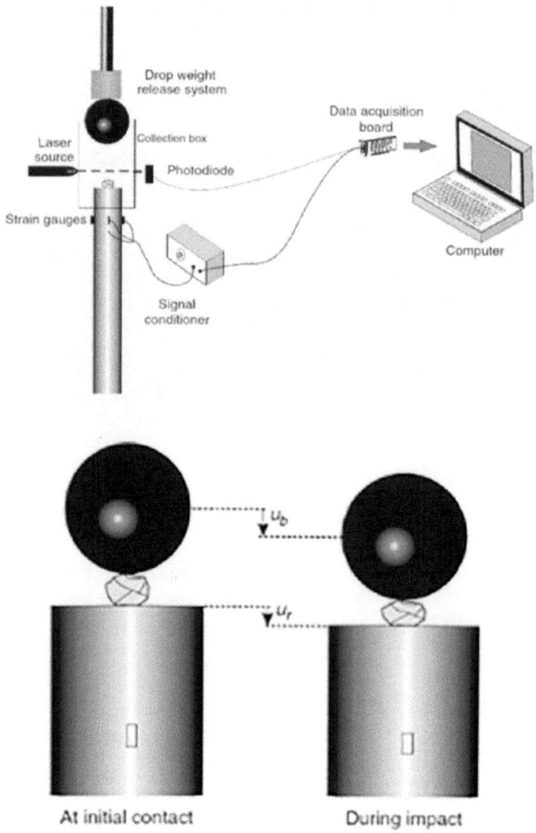

FIGURE 4.12 Ultrafast load cell: (a) Instrumentation. (b) Principle to determine the deformation.

Source: Tavares, L.M., and King, R.T. 1998. Single-particle fracture under impact loading. *International Journal of Mineral Processing* 54(1):1–28.

with fracture, (a) the fracture energy E and (b) the particle strength σ_p, which are uniquely related.

4.1.1.2.1 Fracture Energy

The fracture energy E_c is related to the applied force by

$$E_c = \int_0^{\alpha_c} F(\xi)\,d\xi \tag{4.10}$$

$$F(t) = \frac{1}{3} K d_p^{\frac{1}{2}} \alpha^{\frac{3}{2}} \tag{4.11}$$

Then the fracture energy is

$$E_c = \frac{2}{15} K d_p^{\frac{1}{2}} \alpha_c^{\frac{5}{2}} \tag{4.12}$$

Where d_p is the equivalent diameter of the particle and $\alpha_c = u_b - u_r$ is the deformation at primary fracture (Figure 4.12), and K is the local deformation coefficient.

4.1.1.2.2 *Particle Strength*

The particle tensile strength is

$$\sigma_p = \frac{2.8F_c}{\pi d_p^2} \tag{4.13}$$

From the force exerted at fracture and the energy equation we obtain a value of the local of the deformation constant K

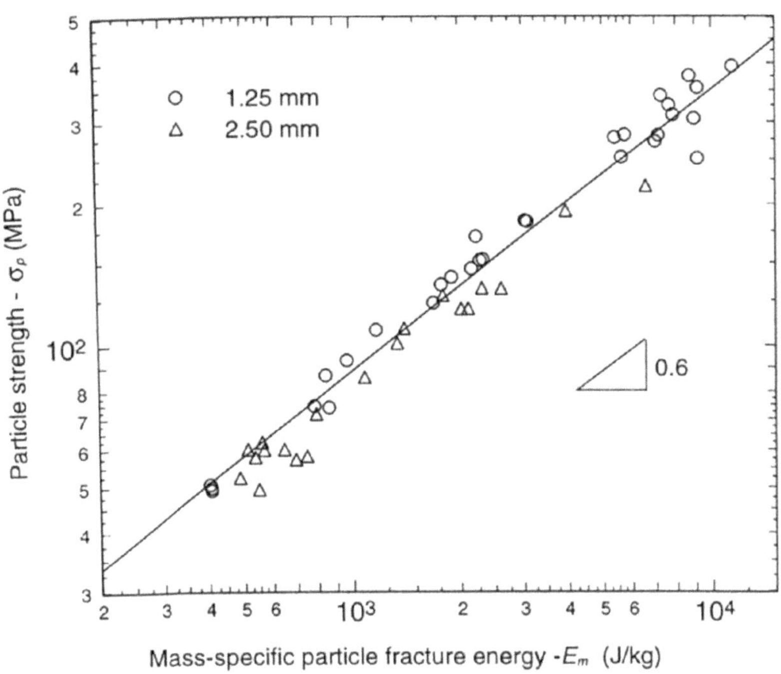

FIGURE 4.13 Particle strength versus mass specific fracture energy for 1.25 and 2.50 mm for glass spheres.

Source: Tavares, L.M., and King, R.T. 1998. Single-particle fracture under impact loading. *International Journal of Mineral Processing* 54(1):1–28.

$$K = \left(0.576 \frac{F_c^5}{d_p E^3} \right)^{\frac{1}{2}} \rightarrow F_c = 1.12 d_p^{\frac{1}{5}} E^{\frac{3}{5}} K^{\frac{2}{5}} \quad (4.14)$$

g the F_c from Equation 4.14 in 4.13 yields

$$\sigma_p = 1.0 \frac{K^{\frac{2}{5}}}{d_p^{\frac{9}{5}}} E^{\frac{3}{5}} \quad (4.15)$$

Expressing the fracture energy as the specific mass energy defined by

$$E_m = \frac{E}{\beta \rho_p d_V^3} \quad (4.16)$$

We obtain the final relationship between the particle strength and the mass specific breakage energy (Figure 4.13).

$$\sigma_p = K^{\frac{2}{5}} \left(\beta \rho_p E_m \right)^{\frac{3}{5}} \quad (4.17)$$

4.1.2 Particle Size Progenies

Several researchers (Gilvarry and Bergstrom 1961a, 1961b; Klotz and Shubert 1982; Gutsche et al. 1993; King 2001; Saho 2006) have shown that *primary single compressible breakage* produces particle size distributions represented by the following expression:

$$B(x, y) = \left(\frac{x}{y} \right)^n \quad (4.18)$$

Where $B(x, y)$ is the cumulative weight percentage of particles less than x and y is the original particle size.

In the case of successive *fracture after the primary breakage*, the product is made up of a mixture of two separate populations $\left(\frac{x}{y} \right)^{n_1}$ corresponding to the primary breakage and $\left(\frac{x}{y} \right)^{n_2}$ the results of further fracture. King (2001) presents the most useful three-parameter equation expression

$$B(x, y) = K(x / y)^{n_1} + (1 - K) \left(\frac{x}{y} \right)^{n_2} \quad (4.19)$$

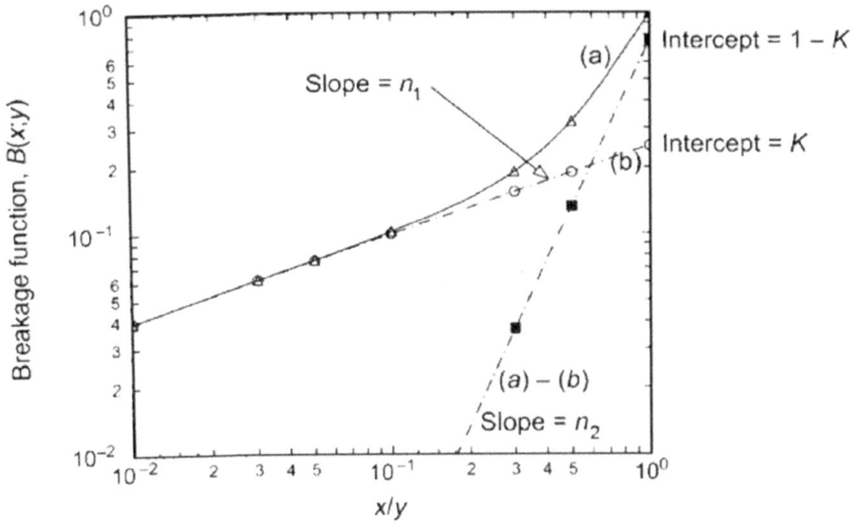

FIGURE 4.14 Progeny distribution for primary at low energy of 350 [J/kg] function. (b) Parameter determination.

Source: King, R.P. 2001. Modeling and Simulation of Mineral Processing Systems. Oxford: Butterworth-Heinemann. p. 140.

where with K is the intercept in Figure 4.14 of the line tangent to the curve for low values of $\frac{x}{y}$ while n_1, n_2 are slope of the tangent to the curve for small and high values of $\frac{x}{y}$ respectively. This form of representing the breakage of single particles has shown to be of great value in the case of practical size reduction in mineral processing for crushers and mills.

Typical breakage function $K = 12$, $n_1 = 5.1$, $n_2 = 1.12$. Parameter determination.

The product of single particle breakage of a material differs for distinct compressive load producing particle sizes with different distributions, which can be related to the volume specific breakage energy E_c. Figure 4.15 gives results from Baumbgardt et al. (1975).

4.1.3 Particle Beds

Data from testing single particles in rigidly mounted roll, along with data from bed breakage tests, have been used to describe comminution in High-Pressure Grinding Rolls (HPGR) (Kapur et al. 1993). However, establishing the link between single-particle breakage and mill performance is more difficult in high-pressure roll grinding, given the severe interaction effects that exist in the confined bed, which results in waste of as much as 50% of the mill energy in interparticle friction and

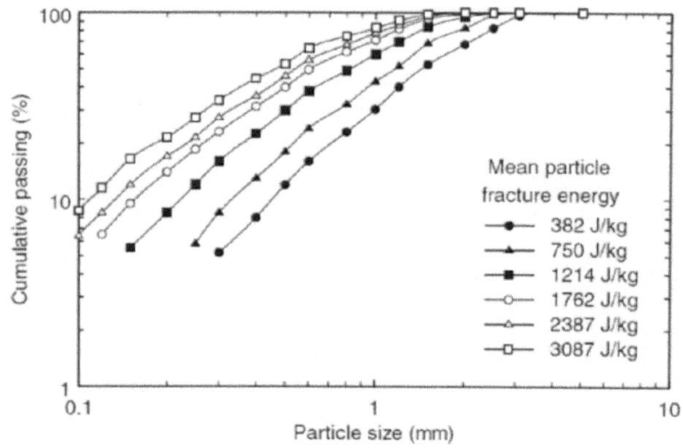

FIGURE 4.15 Particle size distribution product of slow breakage of 5 mm glass spheres at low compression at different fracture energy.

Source: Baumbgardt, S., Buss, B., May, P., and Schubert, H. 1975. On the comparison of results in single grain crushing under different load. In *Proceedings of the XI International Mineral Processing Congress*. Westmount, QC: Canadian Institute of Mining, Metallurgy and Petroleum. p. 3.

agglomeration of the product (Liu and Schönert 1996). Gutsche et al. (1993) present the following expression to calculate the compressive force F_α and the energy E_1 invested in the breakage of a single particle:

$$E_1 = D\int_0^{\alpha_1} \left(F(\alpha)\cos\alpha \sin\alpha\right) d\alpha \tag{4.20}$$

Morrell and Morrison (1989) demonstrated that single-particle impact-breakage data from the drop weight tester, combined with piston-press particle-bed breakage data, can be successfully used to model HPGR, as shown in Figure 4.16.

In a recent work, Barrios et al. (2020) presented the breakage of particles in beds under impact and slow compression using Discrete Element Method (DEM). In this model, every time a critical failure criterion is met, a particle is replaced by progeny fragments. This aspect of DEM is called particle replacement model (PRM). Tests were conducted on iron ore pellets as calibration of the breakage of single particles in two instruments shown in Figure 4.17a, an impact cell for high velocity tests, and 4.17b, a mechanical press with a velocity 10^{-4} (that of the impact load cell), and results of calibration experiments are shown in Figure 4.18.

Once calibration was completed, impact and slow compression test was performed on confined particle bed in the equipment shown in Figure 4.19, with DEM simulation shown in Figure 4.20 and the experiments in Figure 4.21.

The experiments and results in Figure 4.21 show that DEM simulations of particle bed compression of iron ore pellets, using the piston and die press system and particle replacement model represented, reasonably the compressive force versus

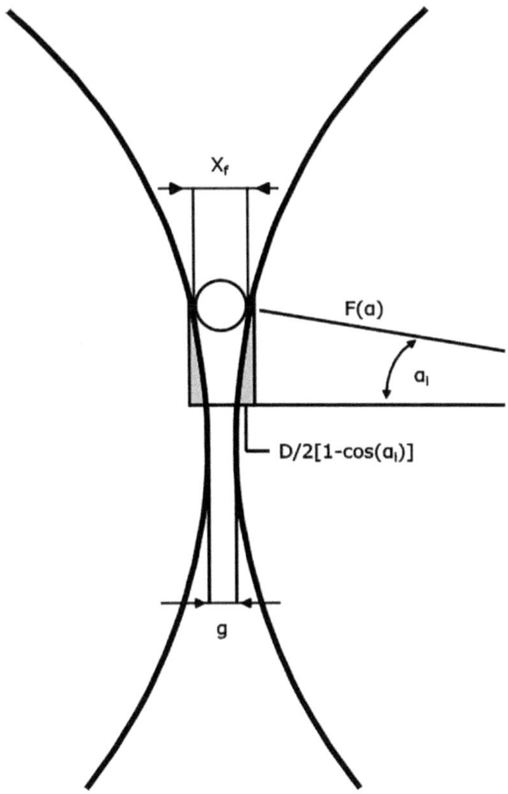

FIGURE 4.16 Schematic representation of nip size X_p nipped at angle α_l.

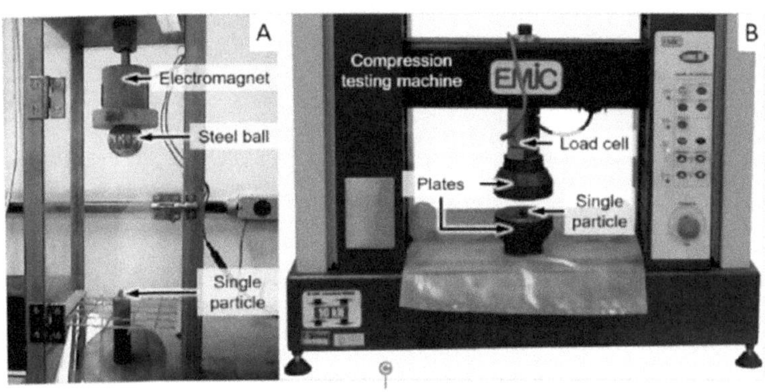

FIGURE 4.17 Experimental layout of the impact load cell (A), and the EMIC DL 1000 slow compression press (B) for single particle breakage tests.

Source: Barrios, G.K.P., Jiménez-Herrera, N., and Tavares, L.M. 2020. Simulation of particle bed breakage by slow compression and impact using a DEM particle replacement model. *Advanced Powder Technology* 31(7):2749–2758.

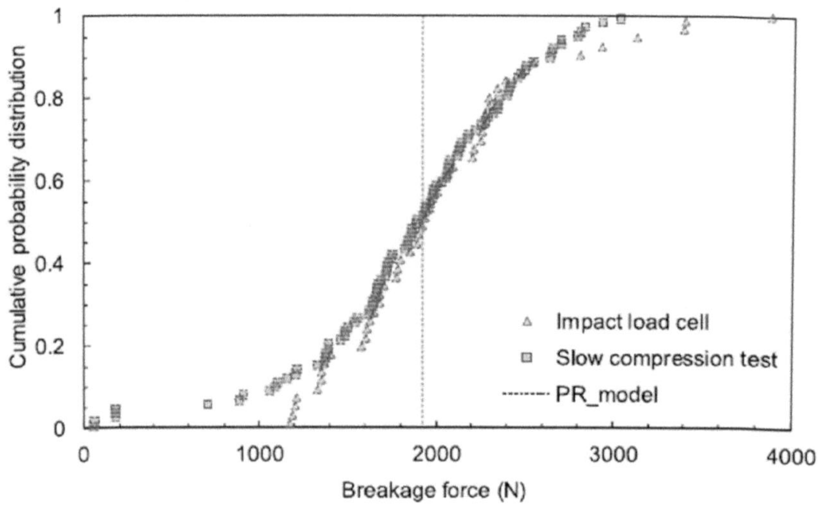

FIGURE 4.18 Probability distribution of the particle breakage forces measured using impact and compression loading tests for single iron ore pellets in the 12.5–9.0 mm size range.

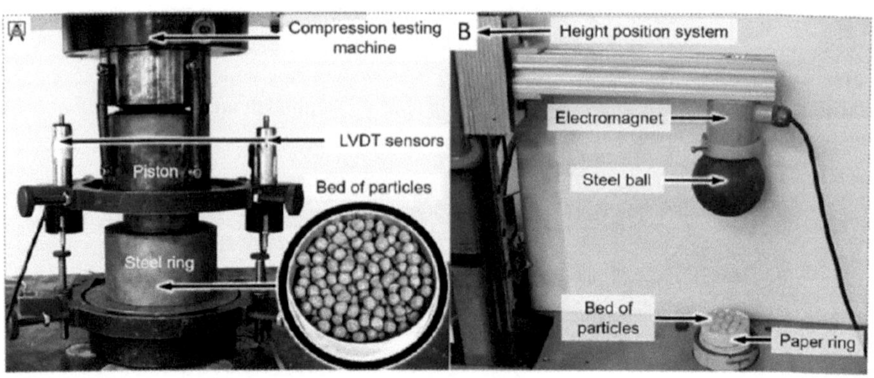

FIGURE 4.19 Experimental layout of the piston and die press test (A) and the impact load cell (B) for particle bed breakage tests.

Source: Barrios, G.K.P., Jiménez-Herrera, N., and Tavares, L.M. 2020. Simulation of particle bed breakage by slow compression and impact using a DEM particle replacement model. *Advanced Powder Technology* 31(7):2749–2758.

packing density and the amount of pellets broken as a function of compressive force from experiments. In addition, the particle size distributions from the tests were well described as a function of the maximum compressive force applied.

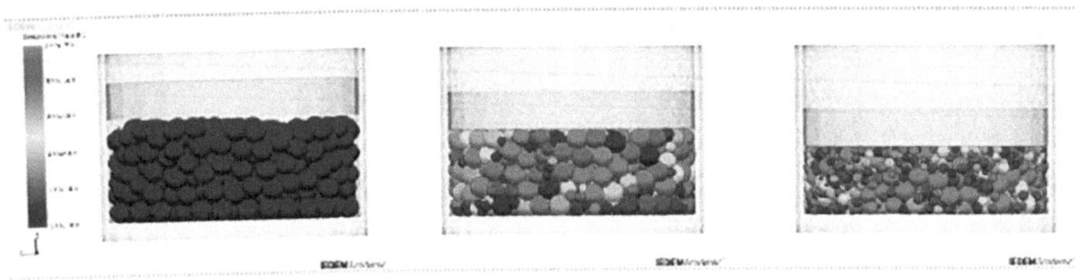

FIGURE 4.20 Illustrates a simulation showing the bed compressive force inside the piston and die press system along the simulation time.

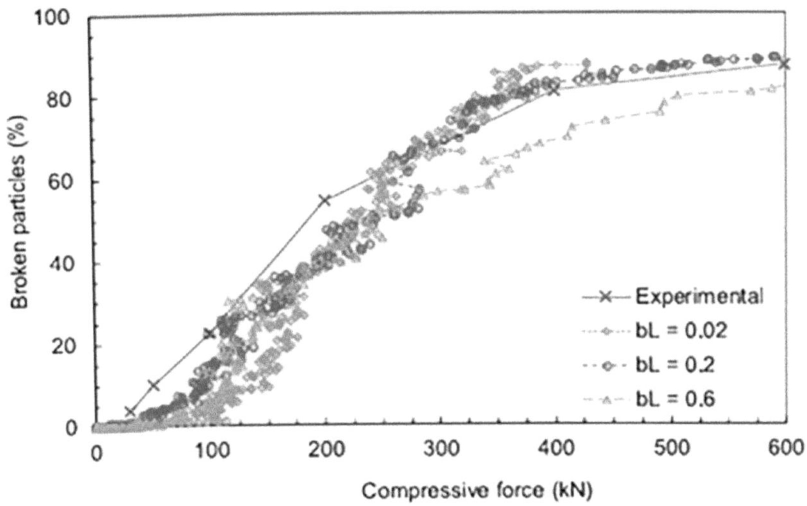

FIGURE 4.21 Comparison of the experimental results of the particle bed broken and the simulations using the PRM as a function of the compressive force using different stress relaxation factors bL.

4.2 LIBERATION

The detailed characterization of multiphase particulate systems is fundamental to guarantee an efficient separation of the valuable minerals mineral processing, allowing process engineers to achieve the best metallurgical recovery possible. One of the most important aspects of the ore is mineral liberation (Oliveira 2021). The importance of liberation in mineral processing was first recognized in the 1930s (Gaudin 1939), and it has become an active field in mineral processing since. Operational mineralogy developed as a branch of process mineralogy, especially by the developments sophisticated automated mineral analysis system instrumentation such as the Mineral Liberation Analyzer MLA and the Quantitative Evaluation of Minerals by Scanning Electron Microscopy (Brough et al. 2019; Fandrich et al. 2007; Cutting 1979; Cutting and Barber 1980; Herbst et al. 1988; Laslett et al. 1990).

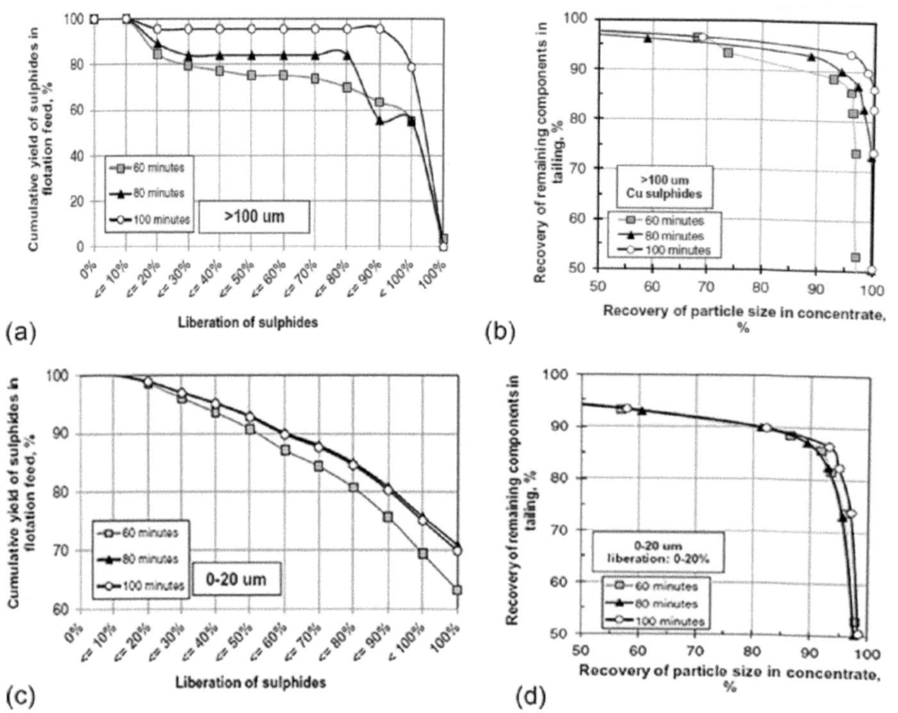

FIGURE 4.22 Liberation of sulfides particles (a) $> 100\mu m$ and (c) $< 20\mu m$ and flotation recovery of particles (b) $> 100\mu m$ and (d) $< 20\mu m$.

Bakalarz et al. (2017) presents an example and how to use liberation information to establish its influence and that of particle size in flotation of copper ore at laboratory scale. Three different milling time of flotation was used, 60, 80, and 100 min. They noticed that for particle size $> 100\mu m$ the liberation of sulfides increases the recovery with milling time, while for particles liberations up to $< 70\mu m$ they do not. See Figure 4.22.

One conclusion was that the differences in flotation requirements for fine and coarser minerals was justified to classify the feed into two or even more streams and processing in a narrow size distribution. The grinding of narrow size fraction separately and then float should result in better overall recovery and lower power consumption.

4.2.1 Modeling

Comminution models are used both for process design and process simulation. Yet few unit models operate at the liberation level texture is. Texture, defined as the distribution of minerals and pores in a defined volume, has been carefully studied at meso-scale and micro-scale. The behavior of an ore in processing circuit is influenced by the texture of the particles created during the comminution stages. Koch and Rosenkranz (2017) presented a table of liberation models that use different types of information grains morphology, grain size, phases, breakage model, dimensions, and particle shape (Table 4.1).

TABLE 4.1
Literature on Liberation Models

Reference	Grains morphology	Grain size	Phases	Mode of breakage	Dimensions	Shape of particles
Gaudin 1939	Cubes	Constant	2	Random	3	Cubes
Wiegel 1976	Cubes	Variable	2	Random	3	Cubes
Andrews and Mika 1975	Any	Variable	2	Random	1	NA
Steiner 1975	Any	Variable	3	Random	1	NA
King 1979	Any	Variable	2	Random	1	NA
Klimpel and Austin 1983	Any	Variable	2	Random	1	NA
Meloy and Gotoh 1985	Spheres	Variable	2	Random	2	Any
van Egmond and Meloy 1989	Rods, plates, diamonds	Constant	2	Non-random	2	Any
Barbery and Leroux 1988	Polyhedra	Variable	2	Random	2	Any
Ferrara et al. 1989	Bands, rods and spheres	Constant	2	Random	3	Any
Hsih and Wen 1994	Squares	Vanable	2	Non-random	1	NA
Bonifazi and Massacci 1995	Pixels	Variable	2-1	Random	2	Cubes
King and Schneider 1998	Any	Variable	2	Non-random	2	NA
Wei and Gay 1999	Any	Variable	2	Random	3	Cubes
Gay 2004, 1999	Disks	Variable	2	Random	2	Disks
Stamboliadis 2008	Spheres	Constant	2	Random	3	NA
Djordje vic 2013	Pixels	Variable	Any	Non-random	2	Any
Zhang and Subasinghe 2013	Polyhedra	Variable	2	Random	3	Convex
Evans et al. 2013	Cubes	Variable	Any	Random	3	Cubes
Ghazvinian et al. 2014	Voronoi polyhedra	Constant	1	Non-random	3	NA
Wang 2015	Pixels	Variable	Any	Non-random	3	Any
van der Wielen and Rollinson 2016	Voronoi polygons	Variable	Any	Random	2	Voronoi polygons
This study	Pixels	Variable	Any	Non-random	2	Any
King and Schneider 1998	Any	Variable	2	Non-random	2	NA
Wei and Gay 1999	Any	Variable	2	Random	3	Cubes
Gay 1999, 2004	Disks	Variable	2	Random	2	Disks
Stamboliadis 2008	Spheres	Constant	2	Random	3	NA
Djordjevic 2013	Pixels	Variable	Any	Non-random	2	Any
Zhang and Subasinghe 2013	Polyhedra	Variable	2	Random	3	Convex

(Continued)

TABLE 4.1 ***Continued***

Reference	Grains morphology	Grain size	Phases	Mode of breakage	Dimensions	Shape of particles
Evans et al. 2013	Cubes	Variable	Any	Random	3	Cubes
Ghazvinian et al. 2014	Voronoi polyhedra	Constant	1	Non-random	3	NA
Wang 2015	Pixels	Variable	Any	Non-random	3	Any
van der Wielen and Rollinson 2016	Voronoi polygons	Variable	Any	Random	2	Voronoi polygons
King and Rosenkranz	Pixels	Variable	Any	Non-random	2	Any

Source: Koch P., and Rosenkranz, J. 2017. Texture-based liberation models for comminution. Diva, Luleå University of Technology Division of Minerals and Metallurgical Engineering (MiMeR) Luleå. ORCID:0000-0003-4800-9533

A general approach based on particles and their mineralogy has been described by Lamberg and Vianna (2009). The idea is to use the information from the geological model (spatial model) to generate particles and simulate the comminution and concentration process to forecast concentrate and tailings. Both the texture simulation and breakage mechanisms are based on statistical physics models warranting realistic results (Koch and Rosenkranz 2017).

Lamberg and Vianna (2009) indicated the relevance and value of liberation analysis for plant design and optimization. They developed a systematic methodology for reconciliation of multiphase particles. The method consists of seven steps to minimize the error propagation of liberation data: (1) overall mass balance, (2) size-by-size mass balance, (3) conversions from elemental assays to mineral quantities, (4) liberation data reconciliation, (5) particle classification and binning, (6) smoothing of liberation data, (7) mass balance of particle classes (i.e., liberation mass balance of multiphase mineral particles).

In converting chemical analysis to mineral quantities, the residual R of converting chemical analysis to mineral quantities from the set of mass balance equations is minimized (Lamberg and Vianna 2009).

$$R = \sum_{e=1}^{i} \left(< G_e - \sum_{m=1}^{j} \frac{1}{100} G_{e,m} \times G_{ms} \right) \tag{4.21}$$

Where i is to number of elements and j the number of minerals, $G_{e,m}$ is the element mineral and $G_{m,s}$ is the mineral sample.

They indicate that particle data from the liberation analysis is not consistent with the mineral grades of the samples calculated in the previous step, nor is it consistent with the chemical composition of the samples because of the following:

- Normally liberation output has more minerals than estimated by chemical mass balancing.

- The chemical composition of minerals in the samples is not constant but varies, and certain composition has been selected in both the liberation analysis and in the chemical mass balancing to represent the average composition.
- The quality of chemical analysis is generally better than the modal composition based on liberation analysis (see previous section).

In processing the liberation data:

1. Minerals of the liberation analysis are grouped to match with the minerals estimated in step 3.
2. The weight proportion of particles is adjusted to match with the mineral grades estimated in step 3. Mineral composition of the particles is not changed.

The correction value $Q_{m,f,s}$ is calculated for each mineral m in a size fraction f of a stream s before each iteration round:

$$C_{m,f,s} = \frac{G_{m,f,s}}{\sum_{p=1}^{k} \frac{1}{100} G'_{p,f,s} \times G_{m.p}} \quad (4.22)$$

where $C_{m,f,s}$ is the correction of mineral, p refers to particle, $G_{m,f,s}$ is the mineral fraction size, and $G_{m,p}$ is the mineral particle. The mass proportion of particle p is recalculated on each iteration round using equation where $G_{p,f,s}$ is the particle fraction stream

$$G_{p,f,s} = G'_{p,f,s} \times \sum_{m=1}^{j} G_{m,p} \times C_{m,f,s} \quad (4.23)$$

$G_{m,p}$ refers to mineral particle and $C_{m,f,s}$ the mineral, fraction size correction.

4.3 SIZE ENLARGEMENTS

The reduction of size is necessary to achieve the appropriate fine particle sizes for their recovery by concentration from metalliferous ores due to the fine dissemination in which the minerals are found. As these fine sizes are not suitable for subsequent processes, such as thickening when there is a need to recover water from the tailings or when oxidized minerals are lixiviated, it is necessary to increase the size of these fines again by agglomeration. When agglomerating, particle of a suspension increases in size and acquire a greater sedimentation velocity essential to obtain a good separation by thickening. The agglomerates also form more permeable cakes accelerating filtration processes. Different methods are available to increase the size of solid particles. The most common are *coagulation*, by reduction of the interparticle electrostatic repulsion, *flocculation* by bridging particles with polymeric agents (flocculants) which are water-soluble, and *oil agglomeration*, in which an oily hydrocarbon is used to agglomerate hydrophobic particles suspended in water, hence producing hydrophobic agglomerate (Concha 2014).

4.3.1 Coagulation

In Chapter 7 we saw that the DLVO theory consider that suspension of particles in water possess electrical charges, mainly negative, that maintain them in suspension avoiding sedimentation. To make them settle, we must set the physicochemical condition in the ionic solution; electrostatic attraction, for particles to collide with another and form coarse entities that, subjected to gravity, will settle. This is the *critical coagulation concentration* (CCC), condition of neutralization of the surface charge by providing salts with ionic charge opposite to that of the particles. The aggregation efficiency follows Schulze-Hardy rule that propose that the critical coagulation concentration decreases as the inverse sixth power of the counterion valence (Trefalt et al. 2020). However, for diffuse potentials multivalent counterions, the CCC increases with the surface charge density, inducing the strong decrease of the CCC with valence leading to a scaling of CCC with z is stronger than z^2 depending on the nature of the counterion (Lyklema 2013; see Chapter 6).

Particle suspensions present in the several stages of mineral processing are much larger than those of a colloidal system. The particle range for the fines in these processes is from 1 to 10 microns, therefore the dominant forces are inertial and convective forces, which are produced by an agitator, a pump, or the action of gravity. These flow conditions may impose enormous energy to the particles, permitting them to overcome the energy barriers of electrostatic repulsive forces without the necessity of neutralizing the charge. An excessively violent agitation may re-disperse the particles and coagulation would not occur. The results of these two types of coagulation, by charge neutralization or by hydrodynamic surpassing the energy barrier, are different, and in the first case a neutral coagulum is obtained, while in the second the aggregate has an electric charge.

4.3.1.1 Coagulation Kinetics

Orthokinetic coagulation of colloidal systems is described by the Smoluchowki-Müller theory (Laskowski and Pugh 1992; Gregory and Nelson 1984). If the collision efficiency is represented by E_c, the rate of disappearance of the number particles by coagulation in a suspension is given by

$$\frac{dn}{dt} = -k_c n^2, \text{ where } k_c = \frac{4E_c kT}{3\mu} \tag{4.24}$$

where k is the Boltzman constant, μ is the liquid viscosity, and T is the absolute temperature. The collision efficiency, with values between 0 and 1, depends on the suspension stability, where $E_c = 1$ corresponds to a completely destabilized system. The sizes of fine particles in mineral processing are relatively big compared with colloidal particles; therefore, the suspension may be considered completely destabilized. Integrating Equation 4.21 results

$$n = n_0 \frac{1}{1 + n_0 k_c t} \quad (4.25)$$

where n_0 is the initial particle concentration.

4.3.2 Flocculation

The aggregation of particles from dispersions by adsorption of large polymer chains on several particles is called flocculation. Depending on the form in which these polymers act on the particle surface, several forms of flocculation may be distinguished. The most important flocculation type is bridging flocculation. Here, small quantities of large chain flocculant are adsorbed over many particles simultaneously producing strong flocs. At higher flocculant concentrations than those needed, they adsorb completely in one particle, leaving less opportunity for adsorbing simultaneously at other particles which restabilizes the suspension. For this reason, optimal doses of polymeric flocculant exist. The action of cationic polymers on negatively charged particles is a process similar to coagulation, where the surface charge neutralization is predominant over the formation of polymer bridges. This aggregation is called electrostatic flocculation and should be distinguished from the sequence of coagulation followed by bridging flocculation. Flocculants may be applied either after destabilizing the suspension via coagulation or without prior destabilization. The flocculants are known to be not very effective in treating stable suspensions and so the first option which involves prior destabilization by coagulation is always better.

The mining industry uses anionic over cationic flocculant more frequently on negatively charged particles. In this case, the adsorption mechanism is by covalent bonds or by chemical reaction on the particle surface. This aggregation is called *flocculation by salt bond*. From all these aggregation mechanisms, bridging flocculation is the most important.

4.3.2.1 Orthokinetic Theory of Flocculation

The optimum flocculation conditions for a given concentrate or tailings refers to the highest settling velocity of the aggregate, that is, to the size and density of the flocs and to the agitation time to obtain them. The aggregation is formed by collision of the primary particles in a hydrodynamic field. Since collision between particles is produced by mechanical energy in a shear field, it is essential that the means of producing the agitation permits the calculation of the shear rate of the flow. The magnitude of shear rate and the application time is important since it, on one side, improves the contact between particles to produce agglomeration and, on the other, destroy the flocs due to the strong agitation. The balance of these two effects is fundamental to obtain flocs of big size and density.

Traditionally it was thought that during flocculation the shear rates $\dot{\gamma}$ should be less than $\dot{\gamma} \approx 100\left[s^{-1}\right]$ for several minutes to avoid the rupture of the formed flocs (Hogg et al. 1966), but Rulyov et al. (2009) showed that values much greater than these, $\dot{\gamma}$ up to $3000\left[s^{-1}\right]$, could be used successfully for a very short time, less than 10 s. This method received the name of *ultra-flocculation*.

Flocculation is highly affected by the shape of the size distribution of the particles. Recent analysis of the flocs in the cement industry indicated that the multi-fractal spectrum of flocculation growth can be used to describe the growth of flocs (Han et al. 2020). The influence of the particle size distribution can be measured directly performing a water mass balance around thickener. Details are provided in Chapter 11 and Chapter 12. When the Gaudin Schumann model shape parameter is higher than 0.37 the flocculation and the settling rate improve considerably (Bascur and Retamal 2021, 2023; Bascur et al. 2022). An increase of 40% water is obtained, enabling to maximize the copper production by better control of the size distribution shape in the grinding circuit. The P80 indicator is not a good performance measure for either the flotation or flocculation performance.

The process of flocculation and sedimentation can be modeled using population balance modeling (PBM), as described in Chapters 7 and 9, and computational fluid dynamics (CFD). The changes in particle or floc size and their settling velocities over time can be described. Flocculation is a dynamic and nonlinear process containing aggregation, breakage, reaggregation, and rebreak age between particles, microflocs, and macroflocs.

Recently due to necessity to improve the efficiency of water recovery from water treatment and tailings, scientific studies have looked for a detailed analysis of this process. Egarr et al. (2016) have developed a population balance model for the flocculation process. Their research focus on the mixing performance of a flocculation tank. The population balance model has proven to be an advanced model that is capable of approximating flocculation and sedimentation of particles than previous research. The PBM can simultaneously simulate both the aggregation and breakage processes between microflocs and macroflocs with irregular shape (Shi et al. 2019).

4.3.2.2 Degree of Flocculation

The rate at which flocculation occurs is principally governed by the frequency of collision of particles, but also by the fraction of surface covered by the flocculant. When collision is the result of mechanical or hydrodynamics agitation, it is called *orthokinetic*. According to Healy and La Mer (1966) and Gregory and Nelson (1984), orthokinetic flocculation is produced at this rate:

$$\frac{dn}{dt}=-k_1\left(\theta\right)\times k_2\left(E_c,\dot{\gamma},d_p\right)n^2 \tag{4.26}$$

Where n is the number of primary particles remaining in the suspension at time t and $k_1\left(\theta\right)$ and $k_2\left(E_c,\dot{\gamma},d_p\right)$ are the kinetic constant with values dependent on the flocculant coverage θ, the collision efficiency E_c, shear rate $\dot{\gamma}$, and particle size d_p respectively:

$$k_1(\theta)=\theta(1-\theta)\ \text{and}\ k_2\left(E_c,\dot{\gamma},d_p\right)=\frac{8}{3}E_c\dot{\gamma}d_p^3 \tag{4.27}$$

The maximum value of $k_1(\theta) = 0.25$ is obtained when $\theta = 0.5$, therefore for this flocculant coverage, Equation 4.23 becomes

$$\frac{dn}{dt} = -\frac{2}{3}\dot{\gamma} E_c d_p^3 n(t)^2 \tag{4.28}$$

This equation is the starting point of most flocculation kinetic studies and was first developed by Smoluchowski in 1917 (Bhattacharjee et al. 1998). To a primary particle of size d_p other primary particles will be added to form and grow an aggregate. Therefore, while the numbers of primary particles vanish, the concentration of primary particles in the suspension diminishes, and the concentration of flocs increases in the system. Equation describes the rate of change of primary particles with time, where $n(t)$ is the number of primary particles left in the suspension at time *t* and n_0 is the number of original primary particles present in the suspension.

For constant values of $\dot{\gamma}$, d_p, and E_c, the number of primary particles remaining in suspension at time and the number of flocs at time t is as follows:

$$\frac{\boldsymbol{n_p(t)}}{\boldsymbol{n_0}} = \frac{1}{1+\left(\frac{2}{3}\right)\boldsymbol{n_0\dot{\gamma} E_c d_p^3 t}} \tag{4.29}$$

$$n_F = \frac{\left(\frac{2}{3}\right) n_0^2 \dot{\gamma} E_c d_p^3 t}{n_{pF}\left(1+\left(\frac{2}{3}\right) n_0 \dot{\gamma} E_c d_p^3 t\right)} \tag{4.30}$$

The difference between the initial and the actual concentration of primary particles represents the concentration of flocs $\varphi_F = \varphi_0 - \varphi$, then:

$$\varphi_F(t) = \frac{4\varphi_0^2 \dot{\gamma} E_c t}{\pi + 4\varphi_0 \dot{\gamma} E_c t} \tag{4.31}$$

4.3.2.3 Ultra-Flocculation

In general flocculation is performed at shear rates lower than $100\left[s^{-1}\right]$ to avoid the rupture of flocs. Lower values of the sear rate imply longer reaction time (Rulyov et al. 2009) demonstrate that excellent flocculation is possible at very high shear rates avoiding the rupture of the flocs if very short times of reaction are considered. In this way the contact between the flocculant and the solid is optimized. For example, a suspension of copper tailings was flocculated at less than $10[s]$.

The flocculation efficiency of a pulp depends on four factors: (1) the pulp concentration; (2) the flocculant dosage; (3) the strength of the agitation of the pulp and flocculant, measured in terms of the shear rate; and (4) the time. For example, for a copper flotation tailing, Figure 4.23 gives the efficiency of flocculation in terms of these four variables.

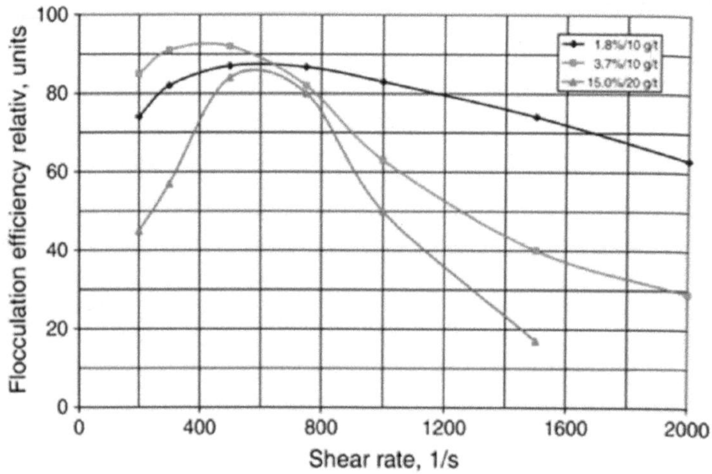

FIGURE 4.23 Effect of the suspension concentration, the flocculant dosage and the shear rate for a $10[s]$ treatment.

4.3.3 Pelletizing

Pelletizing is the particle-enlargement process, in which tumbling of moist fines in drums, disks, or conical devices leads to deformation of agglomerates called *pellets*. The pelletizing process involves three consecutive processes: nucleation, coalescence, layering, and abrasion transfer (Meyer 1980).

- Nucleation occurs whenever a powder is wetted with solvent system and is drawn together to form three-phase air-water-liquid nuclei system which are held together by liquid bridges. The rate and the extent of nuclear formation depends upon the size of the particles, the moisture content, the viscosity of the binding particles, the wettability of the substrate, and the tumbling and drying rates.
- Nucleation is followed by coalescence and layering. Coalescence is defined as the formation of large-sized particles by random collision of well-formed nuclei. This mechanism requires slightly excess moisture on the surface of the nuclei. Layering is the successive addition of fragments and fines where the number of particles remains constant while the total mass of the system increases due to increasing particle size as a function of time.
- Abrasion transfer involves the transfer of materials from one granule to another without any preference in either direction, without any change in the total number or mass of the particles. The particles undergo a continuous change in their size as long as the tumbling continue.

The population balance for a well-mixed batch system undergoing coalescence alone is described in the literature (Kapur and Fuerstenau 1969; Kapur and Runkana 2003; Dhawan et al. 2001) as Equation 4.1.

4.4 PARTICLE SIZE ANALYSIS

The size and size distribution of rocks play an important if not decisive role in the metallurgical efficiency of mineral processing of metallic and nonmetallic ores. Crushing and grinding equipment and their comminution circuits depend basically on the size of the ore fed to the circuit. In concentration processes such as flotation and gravity separation, the size of the mineral particles establishes the boundary of applicability of each technique, and in solid-liquid separation processes the particle size and shape influences their performance. Note that in mineral processing we call "*particle*" to a macroscopic piece of material that in mechanics we would call "*body*."

The problem of characterizing the size of a population of particles is difficult because the particles are generally irregular in shape and span a wide range in size, as shown in Figure 4.24 (Wills and Napier-Munn 2011). Typical mineral processing material include particles with sizes from 1 meter to several microns (Allen 1994; Herbst 1979; Bahar and Khalilib 2015).

In this section we discuss the methodology required to quantify a particulate system encountered in mineral processing. The analysis includes the measurement of size, size distribution, and shape. A summary of the most common empirical distribution equations is presented, and the most important techniques to measure them are surveyed.

4.4.1 Configuration of a Particle

In mechanics the properties of a body P is described in relation to its *configuration*. We define configuration as a one to one mapping of the body onto the space. This means that the configuration represents the extension of space occupied by the body. Geometry enters mechanics through the volume $V(P)$ and its surface $S(P)$. Therefore,

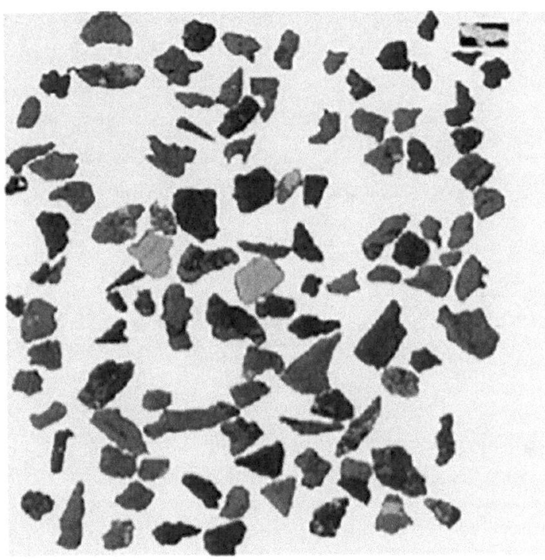

FIGURE 4.24 Mineral particulate system.

for a body of arbitrary shape, a set of two variables are necessary to characterize the configuration, its size, volume, and surface area. Only for bodies of regular geometrical shape is one variable, the size, sufficient to define the geometry of the body.

4.4.1.1 The Size of a Particle

It is fortunate that minerals present cleavage when subjected to fracture, which make the daughter fragments have the same or similar shapes. This property permits the definition of shape factors for each mineral from which the lineal dimension, the volume and surface of the particles can be described in terms of a *nominal size*. Define the linear, surface, and volume shape factors by $\alpha_l\alpha_l$, $\alpha_s\alpha_s$, and $\alpha_v\alpha_v$ such that for a particle P of nominal size xx the linear dimension, surface, and volume of the particle are given by

$$L(P) = \alpha_k x, S(P) = \alpha_s x^2 \text{ and } V(P) = \alpha_v x^3 \tag{4.32}$$

The set of one variable and two constant parameters $(x, \alpha_l, \alpha_s, \alpha_v)$ replaces with advantage the original set of two variables $(L, V(P), S(P))$ in the description of the configuration of the particle (Concha et al. 1973a, 1973b). The nominal size xx is arbitrary and can be chosen conveniently according to the measurement technique used. It is important however that once a nominal size has been chosen, the same should be used throughout the case on hand.

Two of the most common nominal sizes are the *Feret* and the *Martin* sizes. Feret size is the distance between two parallel tangents to the particle always drawn in the same direction. Figure 4.25 shows both Feret and Martin sizes. For example, in microcopy the direction should be perpendicular to the perpendicular to the scanning direction.

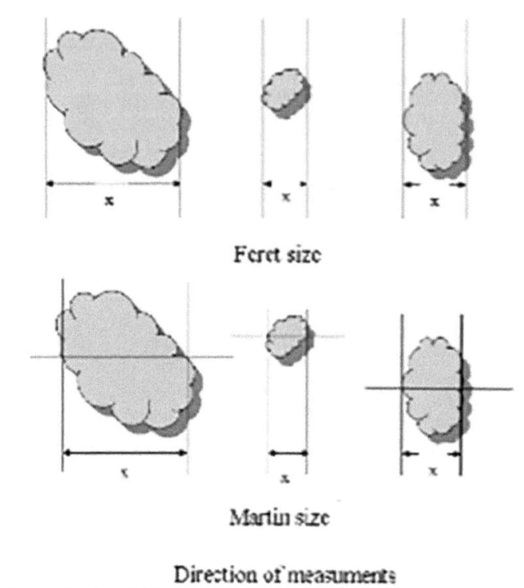

FIGURE 4.25 Feret and Martin sizes.

4.4.1.2 Equivalent Diameters

Very often is convenient to describe the size of particles in term of other variables induced by the method of the particle size determination such as the volume, surface, or its settling velocity. Figure 4.26 shows the relationship between the diameter of a sphere to its volume V, surface S, specific $\hat{S}$ surface, cross-sectional area A, and its sedimentation velocity u_∞.

4.4.1.2.1 Equivalent Surface and Volume Diameters

Using the expression for sphere diameters given previously, we can define the properties of particles in relationship with spheres. Define *equivalent* of the circular diameter d_C, the surface of the diameter d_S, volume diameter d_V, and equivalent specific surface diameter d_{SV} are the diameters of a sphere having the same surface, the same volume, and the same surface by unit volume d_{SV} respectively. By definition, and using Figure 4.26 to determine the nominal size x we have

$$d_C = \left(\frac{P_p(P)}{\pi}\right)^{\frac{1}{2}} \text{ and } d_C = \left(\frac{1}{\pi}\alpha_l\right)^{\frac{1}{2}} x \tag{4.33}$$

$$d_s = \left(\frac{1}{\pi}S(P)\right)^{\frac{1}{2}} \text{ and } d_s = \left(\frac{1}{\pi}\alpha_s\right)^{\frac{1}{2}} x \tag{4.34}$$

$$d_V = \left(\frac{6}{\pi}V(P)\right)^{\frac{1}{3}} \text{ and } d_V = \left(\frac{6}{\pi}\alpha_V\right)^{\frac{1}{3}} x \tag{4.35}$$

$$d_{SV} = \left(6\frac{V(P}{S(P)}\right)\Bigg) \text{ and } d_{SV} = 6\frac{\alpha_V}{\alpha_S} x \tag{4.36}$$

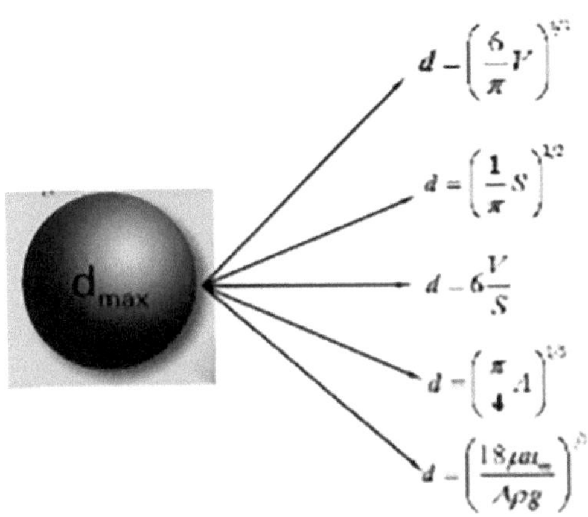

FIGURE 4.26 Property of a sphere.

It is important to remark that the volume diameter is the same as the hydrodynamic diameter since the drag force on a sphere is a function of the volume of the particle.

4.4.1.3 Stokes Diameter

In some cases, it is convenient to use other properties of the particles for their size definition, notably the Stokes diameter $d_S(P)d_S(P)$ defined as the diameter of a sphere that has the same settling velocity than the particle in Stokes regime.

$$d_S(P)=\left(\frac{18\mu u_\infty(P)}{\Delta\rho g}\right)^{\frac{1}{2}} \tag{4.37}$$

Where $\Delta\rho, \mu u_\infty$ are the solid-fluid density difference, the fluid viscosity, and the particle terminal settling velocity.

4.4.1.4 Particle Shape

In general particles have an irregular shape, and the particle size analysis should not be restricted to the determination of the size and size distribution but should consider their shape. In the same way as for the sizes, several parameters measure the shape of particles, but most of them can be determined from the fundamental set $(x,\alpha_l,\alpha_s,\alpha_v)$. In what follows we will mention only the sphericity.

When a more accurate description of the shape of a particle is desired, the Fourier's analysis is used. This method (Schwarz and Shane 1969; Ehrlich and Weinberg 1970) consists of describing the surface of a particle by means of a vector originating in the particle's center of mass, which depends on the two angles of the spherical coordinates θ and ϕ. For a spherical particle, the radius is a constant, and for irregular particles it can be expressed as the sum of terms called spherical harmonics. Since the determination of the coefficients in this series, called the Fourier Series, is very difficult, one can resort to the two-dimensional representation of the projected image of the particle on a plane. In this case, the periphery of the image is represented by a vector formed by a Fourier Series in terms of the angle θ of the cylindrical coordinates only.

4.4.1.5 Sphericity

The sphericity, designated by the symbol ψ, is according to Wadell (1835) the ratio between the surface of a sphere S_e and the surface of the particle S_p both having the same volume V_p. Since both have the same volume, it can also be defined as the ratio of the specific surface of the sphere and the specific surface of the particle $\frac{\overline{S}_e}{\overline{S}_p}$. This shape is useful since the specific surface of particles can be measured experimentally.

$$\psi=\frac{S_e}{S_p}\equiv\frac{\frac{S_e}{V_p}}{\frac{S_p}{V_p}}\equiv\frac{\overline{S}_e}{\overline{S}_p}=\frac{\frac{6}{d_V}}{\frac{6}{dV_{SV}}}=\frac{d_{SV}}{d_V} \tag{4.38}$$

From (4.35) and (4.36)

$$\psi = (36\pi)^{\frac{1}{3}} \frac{\alpha_V^{\frac{2}{3}}}{\alpha_S} \tag{4.39}$$

4.4.2 Frequency and Distribution Function

4.4.2.1 Frequency Function

Once the nominal particle size has been defined, a function should describe the "quantity" of particles of a given size range present in the population. The frequency, or density function $f(x)$, is defined in the form:

$$f(x)dx = \text{or}\begin{cases} \text{fraction in the population with sizes in the range } x \text{ and } x + dx, \text{ or the} \\ \text{probability to find in the population a particle of size } x \text{ between } x + dx \end{cases}$$

This fraction can be represented by the shaded area of Figure 4.27. Adding all the fractions from the minimum size $x_{\min}$ to the maximum size $x_{\max}$ the normalization condition is defined:

$$\int_{x_{min}}^{x_{max}} f(x)dx = 1 \tag{4.40}$$

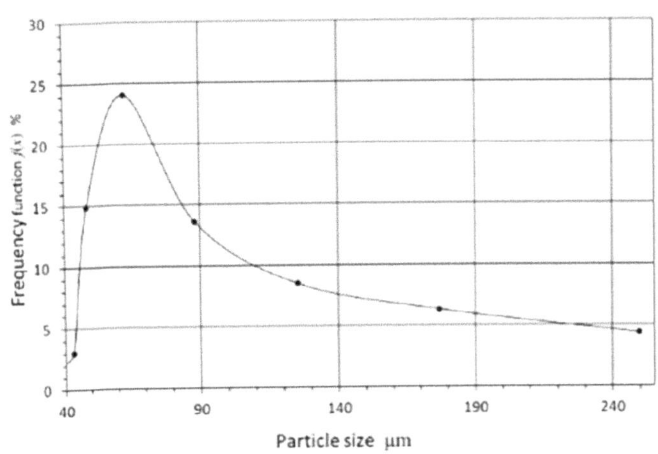

FIGURE 4.27 Frequency function.

4.4.2.2 Distribution Function

Adding all the fractions from the minimum size $x_{\min} x_{\min}$ to the size xx, we obtain the *passing size distribution* function $F(x)F(x)$ and adding from the size xx to the maximum size the retained distribution.

$$F(x) = \text{or} \begin{cases} \text{fraction in the population with sizes less than } x, \text{or the} \\ \text{probability to find in the population a particle with less than } x \end{cases}$$

$$F(x) = \int_{x_{min}}^{x} f(\xi)d\xi \quad R(x) = \int_{x}^{x_{max}} f(\xi)d\xi \tag{4.41}$$

Obviously, the sum of $F(x) + R(x) = 100$.
Figure 4.28 shows such distribution functions.

4.4.3 Metrics of the Particle Size Distribution

The principal parameters to quantify a particle size distribution are the *range*, the *mode*, the *quantiles*, the *mean value*, and the *variance*.

4.4.3.1 Range

The range x_{range} of the population is the domain of the variable $x_{\min} \leq x_{\text{range}} \leq x_{max}$.

4.4.3.2 Mean and Variance

The distribution of sizes in a particulate system can be characterized by two parameters, the mean and the variance. Mean value is the expected value of the particle distribution, variance is the second moment about the mean and describes the dispersion of the population. They are defined by

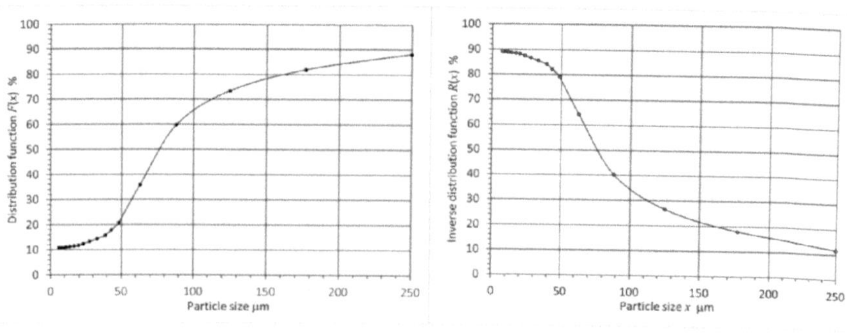

(a) % passing. (b) % retained

FIGURE 4.28 Distributions functions for a particle population.

$$\overline{x}=\int_{x_{\min}}^{x_{\max} f} xf(x)dx \text{ and } \sigma^2=\int_{x_{\min}}^{x_{\max}} \overset{2}{\int f(x)dx}\,(x-\overline{x}) \tag{4.42}$$

The square root of the variance receives the name of *standard deviation* $\sigma\sigma$ and the ratio of the standard deviation and the mean is the *coefficient of variation* C_v.

4.4.3.3 Mode

The mode $x_{mod}x_{mod}$ is the most frequent value. The mode is represented by the maximum of the curve:

$$\left.\frac{df(x)}{dx}\right|_{x=x_{med}} = 0 \text{ and } \left.\frac{d^2 f(x)}{dx^2}\right|_{x-x_{mad}} < 0 \tag{4.43}$$

4.4.3.4 Quantiles

Quantile x_{ij} is the size for which the distribution function is ij. Figure 4.29 shows the quantiles x_{25}, x_{50}, and x_{80} corresponding to $F(25)$, $F(50)$, and $F(80)$.

4.4.3.5 Change of Independent Variable

In some cases, it is convenient to work with the distribution of a different property of the distribution instead of the size. For example, in some cases the volume $V(x)$ could be more convenient than the particle size as an independent variable. In these cases, a simple transformation of variable is possible.

Consider x as the original variable and y, a function of x, the new independent variable, then $y=\tilde{y}(x)$ so that $x=\tilde{y}^{-1}(x)$. As before $f(x)$ is the initial frequency function and $g(y)$ the new one. The normalization condition holds for the two functions so that $f(x)dx = g(y)dy$ and:

$$f(x)=\left(\frac{dy}{dx}\right)g(y) \tag{4.44}$$

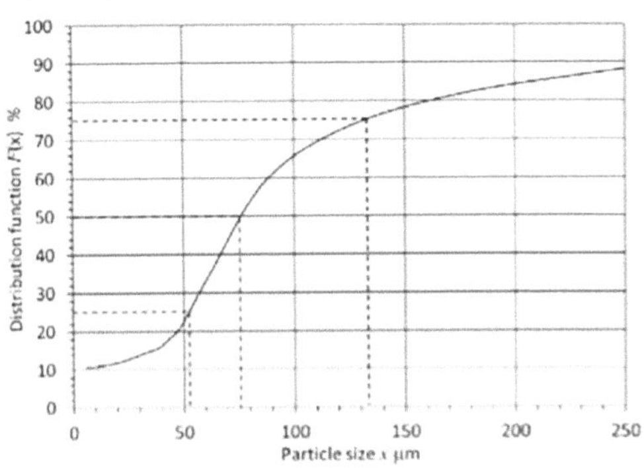

FIGURE 4.29 Distribution function and quartiles $x_{25}=10,\ x_{50}=16,\ x_{75}=19\mu\text{m}$.

Consider the following example. If the new independent variable is the volume $yy = V$ with $V = \alpha_V x^3$, then $\frac{dV}{dx =} 3\alpha_V x^2$, since $x = \left(\frac{V}{\alpha_V}\right)^{\frac{1}{3}}$, $\frac{dV}{dx =} 3\alpha_V^{\frac{1}{3}} V^{\frac{2}{3}}$

$$f(x) = \left(3\alpha_V^{\frac{1}{3}} V^{\frac{2}{3}}\right) g(V) \tag{4.45}$$

For second example let the new independent variable by $y = \ln(x)$, $\frac{dy}{dx} = \exp(-y)$

$$f(x) = \exp(-y) g(y) \tag{4.46}$$

4.4.4 Experimental Representation of a Particle Size Distribution

It is convenient to describe the size distribution of a population with mathematical functions which would provide a convenient representation of the model of mineral processing operations. These functions should also permit a statistical analysis calculation of the distribution parameters mentioned previously.

4.4.4.1 Frequency Functions Induced by the Method of Measurement

Experimental method to measure the frequency function of a particulate system, such as counting particles or measuring length or area by microscopy and screening present the problem that each of these methods yield the frequency in different units. For example, microscopic measurements deliver number, length, or area of particles and screening give result by weight. For this reason, to be able to convert the results from one form to the other, define the elative frequencies in the form, which are valid for homogeneous and uniform in shape.

$$f_0(x)dx = \text{number of particles with sizes between } x \text{ and } x + dx$$

$$f_1(x)dx = \text{length of particles with sizes between } x \text{ and } x + dx$$

$$f_2(x)dx = \text{area of particles with sizes between } x \text{ and } x + dx$$

$$f_3(x)dx = \text{volume of particles with sizes between } x \text{ and } x + dx$$

Each of these functions must obey the normalization condition for $j = 0,1,2,3$:

$$max\int F_j(x) = \int_{x_{min}}^{x\int} f_j(\xi)d\xi, R_j(x) = \int_{x}^{xmax\int} f_j(\xi)d\xi, f_j(x) = \frac{dF_j}{dx}$$

$$\int_{x_{min}} f_j(s)dx = 1 \tag{4.47}$$

The relationship between the various frequency functions results from the fact that the length $f_1(x)dx$ is proportional to β_1 the number function $f_0(x)dx$ multiplied by the length $L(x)$ of the size x particles. The same reasoning for surface or volume functions results in

$$\text{Length}: \quad f_1(x)dx = \beta_1 L(x) f_0(x)dx$$

$$\text{Surface}: \quad f_2(x)dx = \beta_2 S(x) f_0(x)dx$$

$$\text{Volume}: \quad f_3(x)dx = \beta_2 V(x) f_0(x)dx \tag{4.48}$$

The β_j parameters are obtained from (5.45)

$$\begin{aligned} \int_{x_{\min}}^{x_{\max}} \beta_1 L(x) f_0(x)dx \equiv 1 \rightarrow \beta_1 &= \frac{1}{\int_{x_{\min}}^{x_{\max}} L(x) f_0(x)dx} \\ \int_{x_{\min}}^{x_{\max}} \beta_2 S(x) f_0(x)dx \equiv 1 \rightarrow \beta_2 &= \frac{1}{\int_{x_{\min}}^{x_{\max}} S(x) f_0(x)dx} \\ \int_{x_{\min}}^{x_{\max}} \beta_3 S(x) f_0(x)dx \equiv 1 \rightarrow \beta_3 &= \frac{1}{\int_{x_{\min}}^{x_{\max}} V(x) f_0(x)dx} \end{aligned} \tag{4.49}$$

Then substituting Equations 4.48 in 4.49 and considering that $L(x)$, $SS(x)$, and $V(x)$ can be expressed in terms of α_l, α_s, and α_v gives

$$f_1(x)dx = \frac{xf_0(x)dx}{\int_{x_{min}}^{x_{max}} xf_0(x)dx, f_2(x)dx = \frac{x^2 f_0(x)dx}{\int_{x_{min}}^{x_{max}} x^2 f_0(x)dx, f_3(x)dx = \frac{x^3 f_0(x)dx}{\int_{x_{min}}^{x_{max}} x^3 f_0(x)dx}}} \tag{4.50}$$

These expressions may by generalized by

$$f_j(x) = \frac{x^i f_0(x)}{\int_{x_{min}}^{x_{max}} x^i f_0(x)dx} \quad \text{also expressed as} \quad f_j(x) = \frac{x^{j-k} f_k(x)}{\int_{x_{min}}^{x_{max}} x^{j-k} f_k(x)dx} \tag{4.51}$$

4.4.4.2 Average Sizes and Moments of the Distribution

From Equation 4.51, we obtain the average sizes $\overline{L}$, $\overline{S}$, and $\overline{V}$, that we will designate by $\overline{x}_{10}$, $\overline{x}_{20}$, and $\overline{x}_{30}$ respectively.

$$\frac{1}{\beta_1} = \int_{x_{\min}}^{x_{\max}} L(x) f_0(x)dx = \overline{L} \equiv \alpha_\ell \overline{x}_{10}$$
$$\frac{1}{\beta_2} = \int_{x_{\min}}^{x_{\max}} S(x) f_0(x)dx = \overline{S} \equiv \alpha_s \overline{x}_{20}^2 \qquad (4.52)$$
$$\frac{1}{\beta_3} = \int_{x_{\min}}^{x_{\max}} V(x) f_0(x)dx = \overline{V} \equiv \alpha_v \overline{x}_{30}^3$$

$$\overline{x}_{10} = \int_{x_{\min}}^{x_{\max}} x f_0(x)dx, \overline{x}_{20}^2 = \int_{x_{\min}}^{x_{\max}} x^2 f_0(x)dx, \overline{x}_{30}^3 = \int_{x_{\min}}^{x_{\max}} x^3 f_0(x)dx$$
$$\overline{x}_{12} = \int_{x_{\min}}^{x_{\max}} x f_2(x)dx, \overline{x}_{13}^2 = \int_{x_{\min}}^{x_{\max}} x f_3(x)dx, \overline{x}_{30}^3 \qquad (4.53)$$

The subscript 0, 1, 2, and 3 correspond to the type of frequencies number length, surface, and volume. Equation 4.53 can be generalized leading to the terms called moment of the distribution.

$$\overline{x}_{lj}^i = \left(\int_{x_{min}}^{x_{max}} x^i f_j(x) dx \right) \qquad (4.54)$$

Using (4.54) the relationship between the several averages can be written in the form

$$\overline{x}_{ij}^i = \frac{\int_{x_{\min}}^{x_{\max}} x^{i+j-k} f_k(x)dx}{\int_{x_{\min}}^{x_{\max}} x^{j-k} f_k(x)dx} \equiv \frac{\overline{x}_{o+j-k,k}^{i+j-k}}{\overline{x}_{j-k,k}^{j-k}} \qquad (4.55)$$

As example let's observe the averages $\overline{x}_{12}\overline{x}_{12}$ and $\overline{x}_{13}\overline{x}_{13}$:

$$\overline{x}_{12} = \int_{x_{min}}^{x_{max}} x f_2(x) dx \equiv \frac{\overline{x}_{30}^3}{\overline{x}_{20}^2} \text{ and } \overline{x}_{13} = \int_{x_{min}}^{x_{max}} x f_3(x) dx \equiv \frac{\overline{x}_{40}^4}{\overline{x}_{30}^3} \qquad (4.56)$$

4.4.4.3 Equivalent Diameter of a Distribution of Particles

In the same way as we defined the equivalent diameter of a particle, we can define the equivalent diameter (average) of the distribution:

$$\overline{d}_v = \left(\frac{6}{\pi} \alpha_v \right)^{\frac{1}{3}} \overline{x}_{13}, \overline{d}_s = \left(\frac{\alpha_v}{\pi} \right)^{\frac{1}{2}} \overline{x}_{20} \text{ and } \overline{d}_{sv} = \left(\frac{6}{\pi} \frac{\alpha_v}{\alpha_s} \right) \overline{x}_{12} \qquad (4.57)$$

4.4.4.4 Variance of a Distribution of Particles

According to the definition of variance in (4.42) and making use of (4.48), two versions of the frequency function $f_j(x) f_j(x)$ can be written:

$$\sigma_j^2 = \int_{x_{\min}}^{x_{\max}} \left(x - \bar{x}_{1j}\right)^2 f_j(x)dx$$

$$\sigma_j^2 = \frac{\int_{x_{\min}}^{x_{\max}} \left(x - \bar{x}_{1j}\right)^2 x^{j-k} f_k(x)dx}{\int_{x_{\min}}^{x_{\max}} x^{j-k} f_k(x)dx} \tag{4.58}$$

4.4.4.5 Specific Surface of a Particle Size Distribution

Since $\bar{S} = \frac{\bar{S}(x)}{\bar{V}(x)}$ we can write

$$\bar{S} = \frac{\alpha_s \bar{x}_{20}^2}{\alpha_v \bar{x}_{30}^3} \equiv \frac{\alpha_s}{\alpha_v} \frac{1}{\bar{x}_{12}} \tag{4.59}$$

4.4.4.6 Representation of Frequency and Distribution Functions

Among the most popular empirical functions are the Schumann, Rosin-Rammler, and Weibull. Three parameters' equations, normal and log-normal and the gamma functions. There is no theoretical justification for these equations, and usually they are chosen only by the quality of their fitting to experimental data.

4.4.4.6.1 Schumann Distribution Function

The Schumann distribution function, also known as the Gates-Gaudin-Schumann function, GGS (Tarján 1974a), the most used in America, is a potential function of the particle size. The GGS is the two-parameter distribution function given in Figure 4.8 where the volume distribution is used.

$$F_3(x) = \left(\frac{x}{k}\right)^m \quad \text{and} \quad f_3(x) = \frac{dF_3}{dx} = \frac{m}{k}\left(\frac{x}{k}\right)^{m-1} \tag{4.60}$$

Where the parameters k and m are known as the size modulus and distribution modulus. The normalization condition is $\int_{x_{min}}^{x_{max}} f_3(x)dx = 100$.

The logarithm of Equation 4.60 yields

$$\log F_3(x) = m\log(x) - m\log(k) \tag{4.61}$$

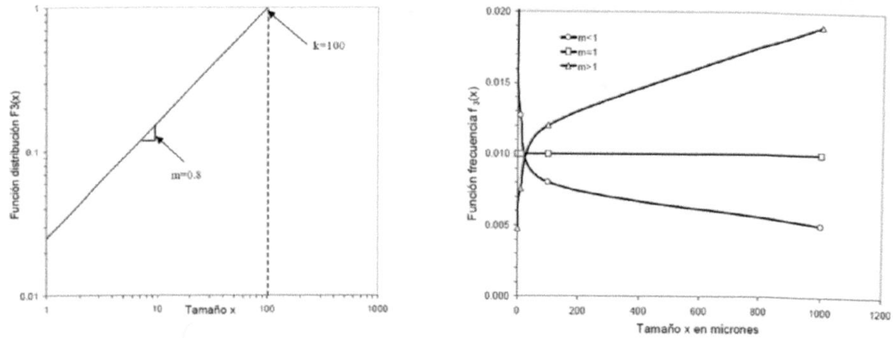

(a) Distribution. (b) Frequency.

FIGURE 4.30 Gaudin-Schumann's functions for k = 100 and m = 0.8.

This plot of Equation 4.61 yields a straight line with m as the slope and k as the interception of the line with the value of $F_3(x) = 1$. Figure 4.30 shows a Schumann's log-log plot using $\frac{x}{k}$ as independent variable. From Figure 4.30, we see that the Schumann distribution has no mean value, which is a drawback of this function. This particle size distribution model provided by GGS has shown to be extremely for real time estimation of the P80 and the particle size distribution shape m from online particle size distribution measurements in industrial grinding circuits. The availability of the size distribution shape allows to control the creation of fines in the grinding circuits, as discussed in Chapter 7, 11, and 12 (Bascur 1990, 1991, 2023; Bascur et al. 2022).

4.4.4.6.2 Rosin-Rammler Distribution Function

The Rosin-Rammler (Tarján 1974b) size distribution function is the one most frequently used in Europe.

$$F_3(x) = 1 - \exp\left\{-\left(\frac{x}{x_0}\right)^n\right\} \quad \text{and} \quad R_3(x) = 1 - F_3(x) = \exp\left\{-\left(\frac{x}{x_0}\right)^n\right\} \tag{4.62}$$

$$\text{Writing} \quad \ln\left(1 - F_3(x)\right) = -\left(\frac{x}{x_0}\right)^n \rightarrow \frac{1}{\ln\left(1 - F_3(x)\right)} = \left(\frac{x}{x_0}\right)^n$$

$$\log\left\{\frac{1}{\ln\left(1 - F(x)\right)}\right\} = n\log(x) - n\log(x_0) \tag{4.63}$$

where x_0 is the size modulus and nn the distribution modulus. Figure 4.31 shows the plot of Equation 4.63. The distribution modulus is obtained from the slope of the straight line, and the size modulus is the quantile $x_0 = x_{63}$.

The frequency function leads to

$$f_3(x) = \frac{n}{x_0}\left(\frac{x}{x_0}\right)^{n-1} \exp\left\{-\left(\frac{x}{x_0}\right)^n\right\} \tag{4.64}$$

The Rosin-Rammler distribution system may be expanded in a potential series

$$F_3(x) = 1 - \left[1 - \left(\frac{x}{x_0}\right)^n + \frac{1}{2}\left(\frac{x}{x_0}\right)^{2n} - \frac{1}{3!}\left(\frac{x}{x_0}\right)^{3n} + \cdots\right] \tag{4.65}$$

For small values of x, it is possible to neglect the terms with greater order which, making $x_0 = k$ and $n = m$, leave Schuman's function $F_3(x) = 1 - \left(\frac{x}{x_0}\right)^n$.

Plotting Equations 4.63 and 4.64 yields Figure 4.31.

4.4.4.6.3 *Weibull Functions*

The Weibull distribution function

$$F(x,\lambda,k) = 1 - \exp\left[-\left(\frac{x}{\lambda}\right)^k\right] f(x;\lambda,k) = \frac{k}{\lambda}\left(\frac{x}{\lambda}\right)^{k-1} \exp\left\{-\left(\frac{x}{\lambda}\right)^k\right\} \tag{4.66}$$

$k > 0$ is a shape parameter and $\lambda > 0$ is a scale parameter (Figure 4.32) (NIST n.d.).

4.4.4.7 Normal and Log-Normal Distribution Functions

Experimental data from particle size analysis from microscopy, which measures the number of particles, gives a normal distribution. That is not the case with surface or volume measurements.

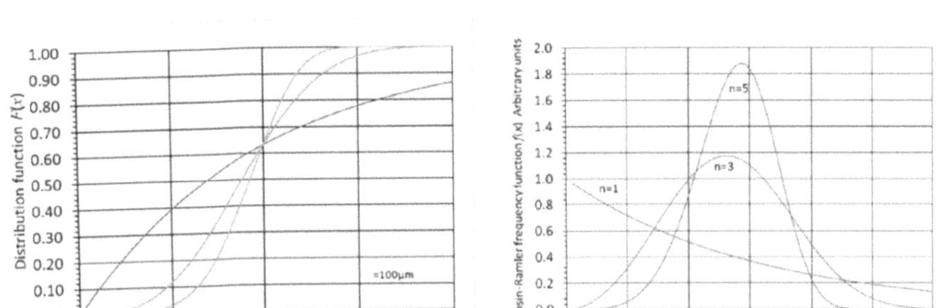

FIGURE 4.31 Rosin-Rammler distribution and frequency function for several values of n and with $x_0 = 100\mu m$.

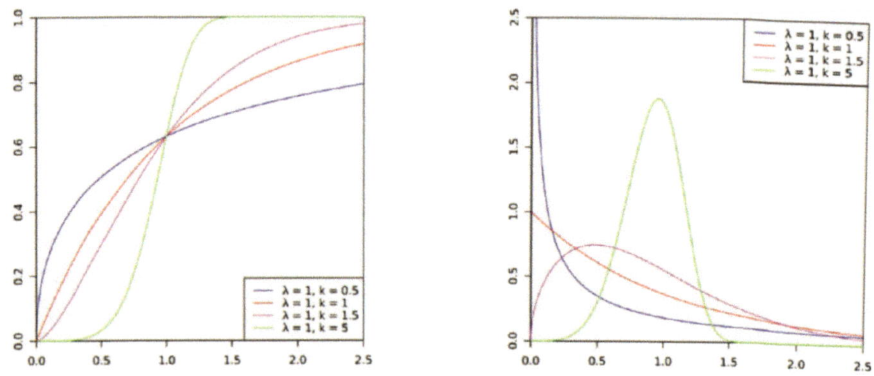

(a) Distribution function. **(b) Frequency function.**

FIGURE 4.32 Weibull functions (NIST).

4.4.4.7.1 Normal Distribution

The normal frequency and distribution functions are given by

$$f_0(x) = \frac{1}{\sqrt{2\pi\sigma}} \exp\left\{-\left(\frac{1}{\sqrt{2}}\frac{\xi-\mu}{\sigma}\right)^2\right\} \quad \text{and}$$

$$F_0(x) = \int_{-\infty}^{x} \frac{1}{\sqrt{2\sigma}} \exp\left\{-\left(\frac{\xi-\mu}{\sqrt{2\sigma}}\right)^2\right\} d\xi \tag{4.67}$$

where the parameters $\mu\mu$ and $\sigma\sigma$ are the mean value and variance of the population. Figure 4.33 shows a normal or probabilistic frequency curve. The value of the integral in Equation 4.67 is shown in Table 4.2 for several values $\frac{(x-\mu)}{\sigma}\frac{(x-\mu)}{\sigma}$

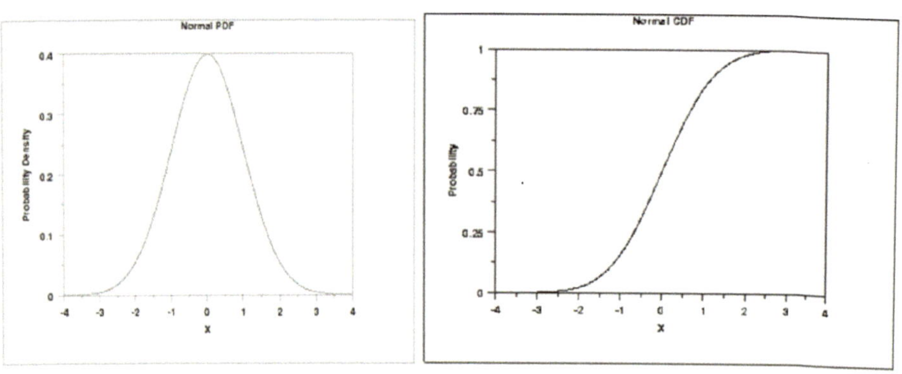

FIGURE 4.33 Normal distribution.

TABLE 4.2
Normal Distribution Function

$\frac{\xi-\mu}{\sigma}$	$F_o(x)=\int_{-\infty}^{x}\frac{1}{\sqrt{2\sigma}}\exp\left\{-\left(\frac{\xi-\mu}{\sqrt{2\sigma}}\right)^2\right\}d\xi$
0	0.5000
0.5	0.69150
1.0	0.84130
1.5	0.93320
2.0	0.97720
3.0	0.99870
4.0	0.99997

From Table 4.2, for $\frac{(x-\mu)}{\sigma}=0,\ F_0(x)=0.5$, which means that

$$\mu = x_{50}\mu = x_{50} \tag{4.68}$$

That represents the median of the distribution. Moreover for $\frac{(x-\mu)}{\sigma}=1, F_0(x)=0.8413$, and we can write $\frac{(x_{84.31}-x_{50})}{\sigma}=1$ so that

$$\sigma = x_{84.31}-x_{50} \tag{4.69}$$

The average sizes are as follows:

$$\left(\frac{\bar{x}_{10}}{\mu}\right)=1,\left(\frac{\bar{x}_{20}}{\mu}\right)^2=1+\left(\frac{\sigma}{\mu}\right)^2,\ \left(\frac{\bar{x}_{30}}{\mu}\right)^3=1+3\left(\frac{\sigma}{\mu}\right)^2$$

$$\left(\frac{\bar{x}_{40}}{\mu}\right)^4=1+6\left(\frac{\sigma}{\mu}\right)^2+3\left(\frac{\sigma}{\mu}\right)^4$$

$$\left(\frac{\bar{x}_{12}}{\mu}\right)=\frac{1+3\left(\frac{\sigma}{\mu}\right)^2}{1+\left(\frac{\sigma}{\mu}\right)^2},\left(\frac{\bar{x}_{13}}{\mu}\right)=\frac{1+6\left(\frac{\sigma}{\mu}\right)^2+3\left(\frac{\sigma}{\mu}\right)^4}{1+\left(\frac{\sigma}{\mu}\right)^2} \tag{4.70}$$

The median and variance are

$$x_{50} = \bar{x}_{10} \text{ and } \sigma^2 = \bar{x}_{20}^2 - \bar{x}_{10}^2 \tag{4.71}$$

The defect of the normal distribution function is that it has a minimum value of $-\infty$ instead of zero.

4.4.4.7.2 Log-Normal

Several authors (Hatch and Choate 1929; Epstein 1947, 1948) have demonstrated that the products of the size reduction of ores produce logarithm-normal distributions, that is, the logarithm of the size is normally distributed.

It has been demonstrated that if a particle system has a log-normal distribution in number, it will also have it in any of the other ways of measuring probability, length, surface, or volume. The frequency and distribution functions are given by

$$f_j(x) = \frac{1}{\sqrt{2\pi}\ln\sigma}\exp\left\{-\left(\frac{\ln\frac{x}{\mu}}{\ln\sqrt{2}}\right)^2\right\} \quad \text{and}$$

$$F_j(x) = \int_0^x \frac{1}{\sqrt{2\pi}\ln\sigma}\exp\left\{-\left(\frac{\ln\left(\frac{\xi}{\mu}\right)}{\ln\sqrt{2\sigma}}\right)^2\right\}d\xi \tag{4.72}$$

A plot of the log normal function is shown in Figure 4.34.

The average sizes are

$$\ln\left(\frac{\bar{x}_{10}}{\mu}\right) = \frac{1}{2}\times(\ln\sigma_0)^2, ln\left(\frac{\bar{x}_{20}}{\mu}\right) = 1\times(\ln\sigma_0)^2, \ln\left(\frac{\bar{x}_{30}}{\mu}\right) = \frac{3}{2}\times(\ln\sigma_0)^2,$$

$$\ln\left(\frac{\bar{x}_{40}}{\mu}\right)^4 = 2\times(\ln\sigma_0)^2$$

$$\ln\left(\frac{\bar{x}_{13}}{\mu}\right) = -\frac{5}{2}\times(\ln\sigma_0)^2, \ln\left(\frac{\bar{x}_{13}}{\mu}\right) = \frac{5}{2}\times(\ln\sigma_0)^2 \tag{4.73}$$

The median and variance are $x_{50} = \bar{x}_{10}$ and $\sigma_0^3 = \bar{x}_{20}^2 - \bar{x}_{10}^2$

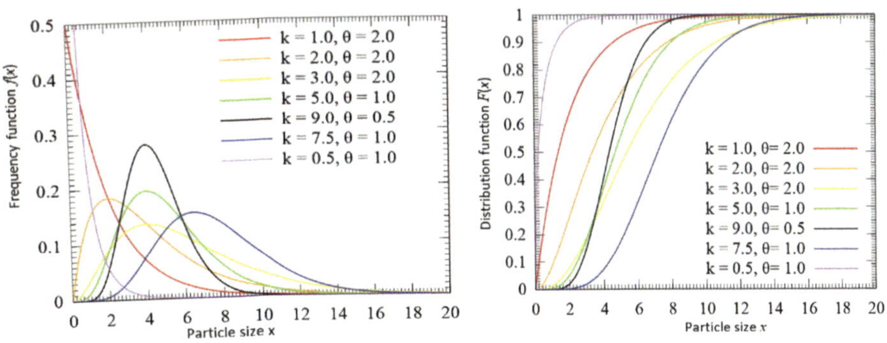

FIGURE 4.34 Log-normal distribution.

4.4.4.7.3 *Gamma Distribution Functions*

Randolph and Larson (1971) proposed the gamma distribution as an adequate representation of products of the size reduction of ores. Defining the new variable $z = \left(\frac{a}{b}\right)x$, where a is a parameter unknown until the experimental data are fitted to the incomplete gamma function, Equation 4.72, which is widely tabulated, and b is the size mode, the frequency and distribution functions of the normalized are

$$f_3(z) = \frac{z^a}{\Gamma(a+1)}\exp(-z), \bar{z}_{1.0} = \frac{b}{a}(a+1), c.v = \frac{1}{(a+1)^{(\frac{1}{2})}} \tag{4.74}$$

Where $c.v$ is the coefficient of variation. The distribution function is

$$F(z) = \int_0^z \frac{z^a}{\Gamma(a+1)}\exp(-z)\,dz$$

$$= \frac{z^a}{\Gamma(a+1)} \times \left[\frac{z^{a+1}}{a+1} - \frac{z^{a+2}}{a+2} + \frac{z^{a+3}}{(a+3)2!} + - \cdots\right] \tag{4.75}$$

Figure 4.35 shows the Gamma frequency and distribution functions (k = a)

The average sizes are

$$\bar{x}_{12} = b, \bar{x}_{13} = \frac{b}{a}(a+1) \tag{4.76}$$

From which we have

$$a = \frac{\bar{x}_{12}}{\bar{x}_{13} - \bar{x}_{12}} \tag{4.77}$$

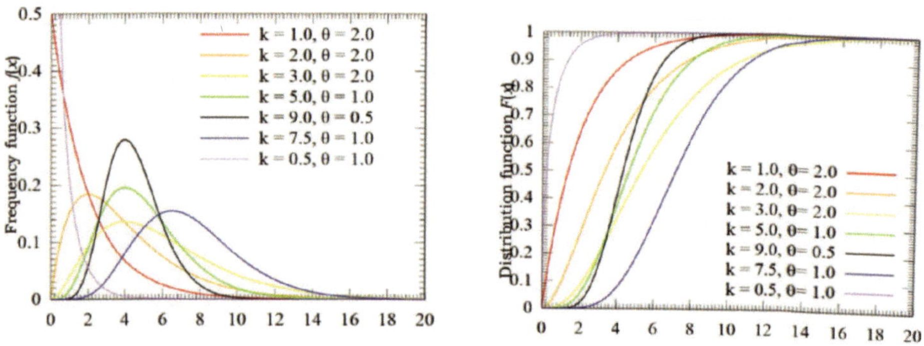

FIGURE 4.35 Gamma frequency and distribution functions ($k = a$).

This means that parameters a and b can be obtained directly from experimental data by calculating averages.

4.4.5 Discrete Size Distribution

In the previous sections we have implicitly assumed that size is a continuous variable, and we have used this property to integrate or derive the frequency or size distribution functions, respectively. However, the methods available for experimentally measuring the size distribution of a particulate system deliver data discreetly, that is, they measure the fraction of particles contained in successive finite intervals.

Consider fractions of particles with sizes between x_a and x_b. If these intervals are close enough, a discrete relative frequency variable f_{ab} may defined by

$$f_{ab}(x) = \int_{x_a}^{x_b} f(x)\,dx \equiv F(x_b) - F(x_a) \tag{4.78}$$

If we evaluate the integral using the mean value theorem, where $\bar{x}_{ab}$ is the average value of the interval $\Delta x = x_b - x_a$. The function $f(\bar{x}_{ab})$ and the relative frequency f_{ab} are given by

$$f(\bar{x}_{ab}) = \frac{F(x_b) - F(x_a)}{x_b - x_a} \text{ and } f_{ab} = f(\bar{x}_{ab})\Delta x \tag{4.79}$$

Assume that a particle size analysis method separates the particulate system into nn fractions of sizes, using the index $k = i$ for the lower value of the range of size "i" and f_i for discrete frequency functions, and $k = i + 1$ for the upper value of the range of size "i" with $k = 0, 1, 2 \cdots n$. See Figures 4.36 and 4.37.

As shown in Figure 4.37, all sizes smaller than the size x_i receive the designation F_i and all sizes greater than x_i as designated by R_i. The relative f_i discrete frequency and relative F_i discrete distribution are then defined by

TABLE 4.3
Incomplete Gamma Function Values

a	Γ(a)	a	Γ(a)	a	Γ(a)	a	Γ(a)
1.00	1.00000	1.25	0.90640	1.50	0.88623	1.75	0.91906
1.01	0.99433	1.26	0.90440	1.51	0.88659	1.76	0.92137
1.02	0.98884	1.27	0.90250	1.52	0.88704	1.77	0.92376
1.03	0.98355	1.28	0.90072	1.53	0.88757	1.78	0.92623
1.04	0.97844	1.29	0.89904	1.54	0.88818	1.79	0.92877
1.05	0.97350	1.30	0.89747	1.55	0.88887	1.80	0.93138
1.06	0.96874	1.31	0.89600	1.56	0.88964	1.81	0.93408
1.07	0.96415	1.32	0.89464	1.57	0.89049	1.82	0.93685
1.08	0.95973	1.33	0.89338	1.58	0.89142	1.83	0.93969
1.09	0.95546	1.34	0.89222	1.59	0.89243	1.84	0.94261
1.10	0.95135	1.35	0.89115	1.60	0.89352	1.85	0.94561
1.11	0.94740	1.36	0.89018	1.61	0.89468	1.86	0.94869
1.12	0.94359	1.37	0.88931	1.62	0.89592	1.87	0.95184
1.13	0.93993	1.38	0.88854	1.63	0.89724	1.88	0.95507
1.14	0.93642	1.39	0.88785	1.64	0.89864	1.89	0.95838
1.15	0.93304	1.40	0.88726	1.65	0.90012	1.90	0.96177
1.16	0.92980	1.41	0.88676	1.66	0.90167	1.91	0.96523
1.17	0.92670	1.42	0.88636	1.67	0.90330	1.92	0.96877
1.18	0.92373	1.43	0.88604	1.68	0.90500	1.93	0.97240
1.19	0.92890	1.44	0.88581	1.69	0.90678	1.94	0.97610
1.20	0.91817	1.45	0.88566	1.70	0.90864	1.95	0.97988
1.21	0.91558	1.46	0.85560	1.71	0.91057	1.96	0.98374
1.22	0.91311	1.47	0.88563	1.72	0.91258	1.97	0.98768
1.23	0.91075	1.48	0.85575	1.73	0.91467	1.98	0.99171
1.24	0.90850	1.49	0.85595	1.74	0.91683	1.99	0.99581

$$f_{\text{ab}} = f\left(\overline{x}_{\text{ab}}\right)\Delta x \quad \text{with} \quad \Delta x = x_b - x_a \tag{4.80}$$

The F_i function is called *inverse cumulative distribution* because it sums all fractions less than x_i and the *direct cumulative distribution* function R_i to the sum of all fractions greater than x_i, such that

$$F_i = \sum_{k=i}^{n} f_k = f_i + f_{i+1} + f_{i+2} + \cdots + f_{n-1} + f_n$$

$$R_i = \sum_{k=0}^{i-1} f_k = f_0 + f_1 + f_2 + \cdots + f_{i-2} + f_{i-1} \tag{4.81}$$

Of course, all measurements, whether in number, linear, superficial, or in volume (weight) can use the discrete distribution system and can be described by the empirical functions developed in the previous section. Figure 4.38 shows the discrete frequency and distribution functions.

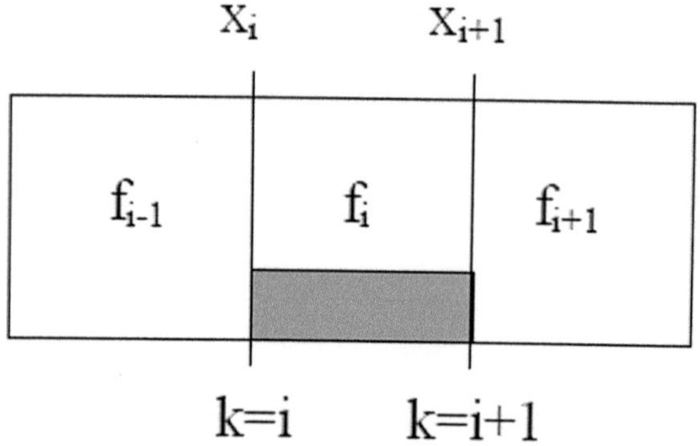

FIGURE 4.36 Definition of "f_i."

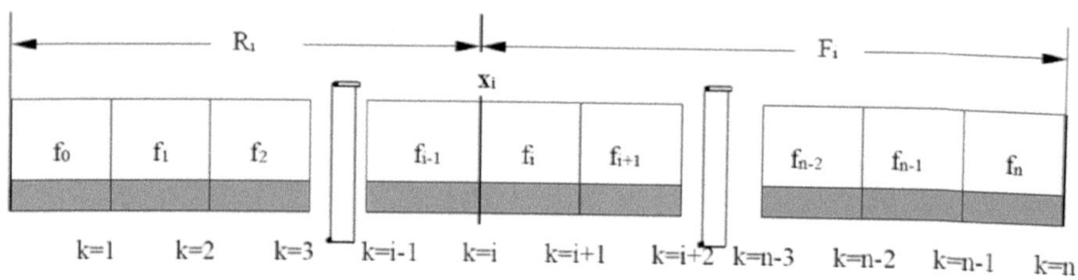

FIGURE 4.37 Particulate system divided in k sizes from 1 to n.

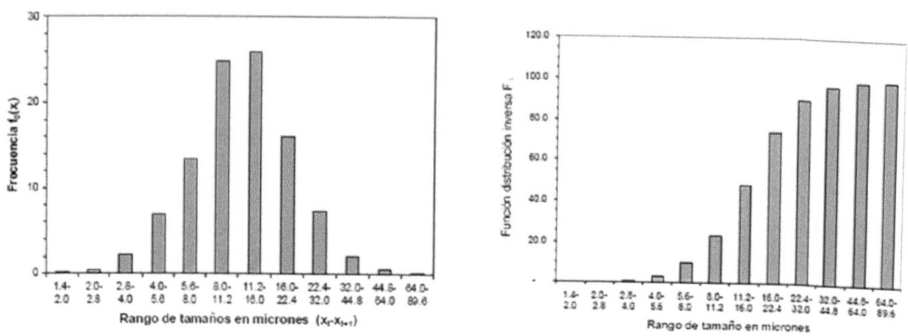

FIGURE 4.38 Discrete frequency and distribution functions: (a) discrete frequency function, (b) discrete distribution function.

4.4.6 Mesh Size

The most common way to perform a particle size analysis is to subject the particulate system to the action of the force of gravity in a series of sieves placed on top of the other. The largest sieve is at the top, and each sieve that follows has a mesh with

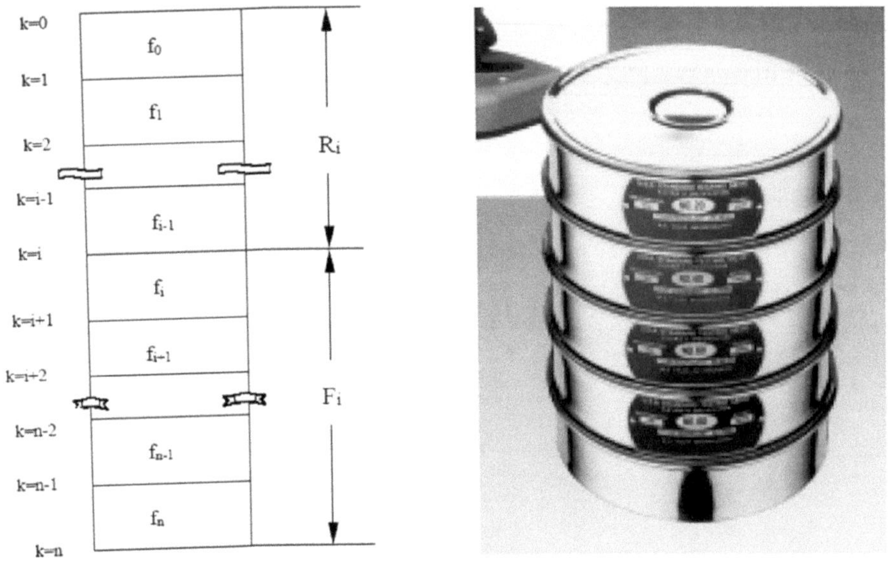

FIGURE 4.39 Discrete distribution functions for sieving.

smaller opening than the previous one. Particles pass through or are retained on the sieves according to their size. If "n" sieves are used, the particle system is separated into "n" sizes. The Equations 4.80 and 4.81, associated with Figure 4.39, are valid for this measurement method.

The sieves used in particle size analysis are standardized as to the relationship between the openings of successive sieves. The Normal Series is called that series of sieves in which the area of the opening of a sieve is twice the area of the sieve opening that follows. As the openings of these sieves are square, the distance between the wires that form a screen is $\sqrt{2}$ times that of the next sieve. Because of the geometric relationship between the linear distances of the openings of successive sieves, the average size of the particles trapped between two sieves is defined as the geometric mean between these distances:

$$\overline{x}_i = \sqrt{x_i \times x_{i+1}} \tag{4.82}$$

Since in a *Normal Sieves Series* $x_i = \sqrt{2}x_{i+1}$ and for the *Double Sieves Series* is $x_i = \sqrt{\sqrt{2}}\,x_{i+1}$, in this case sieve opening *area* of $\sqrt{2}$ the area of the sieve that follows it, the average size is

$$\overline{x}_i = \frac{1}{4\sqrt{2}}x_i = 0.841x_i \quad \text{and} \quad \overline{x}_i = \frac{1}{6\sqrt{2}}x_i = 0.917x_i \tag{4.83}$$

TABLE 4.4
Standard International Normal Series

TYLER Standard Screen 1910 mesh	ASTM E 11–87 1987 mesh	INTERNACIONAL ISO 565 (TBL 2) 1983 µm
3.0	-	6.70
4.0	4.0	4.75
6.0	6.0	3.35
8.0	8.0	2.36
10	12	1.70
14	16	1.18
20	20	850
28	30	600
35	40	425
48	50	300
65	70	212
100	100	150
150	140	106
200	200	75
270	270	53
400	400	38

Source: Gilson Company. 2023. ASTM test sieves. www.globalgilson.com/astm-test-sieves. Accessed August 2023.

Size x_i can be expressed in millimeters, microns, or any other unit. It is common to use the "*mesh size*" as a measure of the opening of a sieve, defined as the number of openings contained in an inch (linear) of sieve. Because this number depends on the thickness of the wire used to manufacture the sieve, the designation is standardized by ISO, ASTM, Tyler, and other international standards. Tables 4.4 and 4.5 show the mesh size and size in microns of the Normal and Double Sieve International Series.

Broadbent and Calcott (1956a, 1956b) demonstrated that a particle size analysis can be represented in a vectorial form, in which the base is the sequence of sizes $\boldsymbol{i}$, and the magnitudes consist of the frequency and distribution functions:

$$\boldsymbol{f}^T = \left[\boldsymbol{f}_0, \boldsymbol{f}_1, \boldsymbol{f}_2, \cdots \boldsymbol{f}_{n-1}, \boldsymbol{f}_n\right]$$

$$\boldsymbol{F}^T = \left[\boldsymbol{F}_0, \boldsymbol{F}_1, \boldsymbol{F}_2, \cdots \boldsymbol{F}_{n-1}, \boldsymbol{F}_n\right]$$

$$\boldsymbol{R}^T = \left[\boldsymbol{R}_0, \boldsymbol{R}_1, \boldsymbol{R}_2, \cdots \boldsymbol{R}_{n-1}, \boldsymbol{R}_n\right] \tag{4.84}$$

TABLE 4.5
Equivalence of Standard Double Sieve Series

TYLER Standard Screen 1910 mesh	ASTM E 11–87 1987 mesh	INTERNACIONAL ISO 565 (TBL 2) 1983 mm
2.5	-	8.00
3.0	-	6.70
-	3.0	6.30
3.5	3.5	5.60
4.0	4.0	4.75
5.0	5.0	4.00
6.0	6.0	3.35
7.0	7.0	2.80
8.0	8.0	2.36
9.0	10	2.00
10	12	1.70
12	14	1.40
14	16	1.18
16	18	1.00
20	20	850
24	25	710
28	30	600
32	25	500
35	40	425
42	45	355
48	50	300
60	60	250
65	70	212
80	80	180
100	100	150
115	120	125
150	140	106
170	170	90
200	200	75
250	230	63
270	270	53
325	325	45
400	400	38

4.4.7 Representation of Experimental Data

Experimental determination of the size distribution of a particulate system generates a list of data to be sorted into tables and charts. From this data the parameters of the distribution can be calculated, the required average values, and other properties such as the specific surface. Since all experimental methods give discrete data, their representation using discrete functions is the most appropriate. Tables must contain the following information:

Column 1: Size range $x_i - x_{i+1}$ (in length units or in meshes).
Column 2: Size:

- Designation of size x_i (top of the size of the range).
- Average size of the range $\overline{x}_i$ (arithmetic or geometric).

Column 3: Number of particles in the $x_i - x_{i+1}$ range (in fraction or %).
Column 4: Frequency function f_i (in fraction or %).
Column 5: Inverse distribution function F_i (in fraction or %).
Column 6: Direct distribution function R_i (in fraction or %).

The following are two examples of data tables. Table 4.6 corresponds to an analysis under the microscope, in which particles are counted, and Table 4.7 corresponds to analysis by screening, in which fractions are weighed.

The disadvantage of the plot in Figure 4.40 is that the points are grouped in the region of fine sizes. This is because the sizes are in geometric progression. For this reason, it is convenient in these cases to choose logarithmic scales for the coordinate axes, which is equivalent to plotting the logarithm of the distribution versus the logarithm of the size. The result is shown in Figure 4.42, where the points are evenly spaced.

TABLE 4.6
Microscopy Analysis Data

Range microns	Size microns		Number of particles	Frequency	Distribution inverse	Distribution direct
x_i-x_{i+1}	x_i	$\overline{x}_i = \sqrt{x_i x_{i+1}}$	N	f_i %	F_i %	R_i %
1.4–2.0	1.4	1.7	1	0.1	**100.0**	**0.0**
2.0–2.8	2.0	2.4	4	0.4	99.9	0.1
2.8–4.0	2.8	3.3	22	2.2	99.5	0.5
4.0–5.6	4.0	4.7	69	6.9	97.3	2.7
5.6–8.0	5.6	6.7	134	13.4	90.4	9.6
8.0–11.2	8.0	9.5	249	24.9	77.0	23.0
11.2–16.0	11.2	13.4	259	25.9	52.1	47.9
16.0–22.4	16.0	18.9	160	16.0	26.2	73.8
22.4–32.0	22.4	26.8	73	7.3	10.2	89.8
32.0–44.8	32.0	37.9	21	2.1	2.9	97.1
44.8–64.0	44.8	53.5	6	0.6	0.8	99.2
64.0–89.6	64.0	75.7	2	0.2	0.2	9938
Total			**1000**	**100.0**	**0.0**	**100.0**

TABLE 4.7
Screening Analysis Data

Range mesh	Size *i* microns		Mass of particles	Frequency	Distribution inverse	Distribution direct
$x_i - x_{i+1}$	x_i	$\sqrt{x_i x_{i+1}}$	m	f_i%	F_i%	
8–10	2360	2003.0	2.0	1.84	100.00	0.00
10–14	1700	1416.3	2.2	1.95	98.16	1.84
14–10	1180	1001.5	4.6	4.15	96.21	3.79
20–35	850	714.1	8.2	7.44	92.06	7.94
28–35	600	505.0	13.2	11.95	84.62	15.38
35–48	425	357.1	22.6	20.48	72.67	27.33
48–65	300	252.2	17.1	15.51	52.19	47.81
65–100	212	178.3	14.8	13.40	36.68	63.32
100–150	150	126.1	8.5	7.71	23.28	76.72
150–200	106	89.2	7.4	6.66	15.57	84.43
200–270	75	63.0	1.7	1.57	8.91	91.09
270–400	53	44.9	1.0	0.90	7.34	92.66
400–fondo	38	19.5	7.1	6.44	6.44	93.56
Total	(10)		110.5	**100.00**	**0.00**	**100.00**

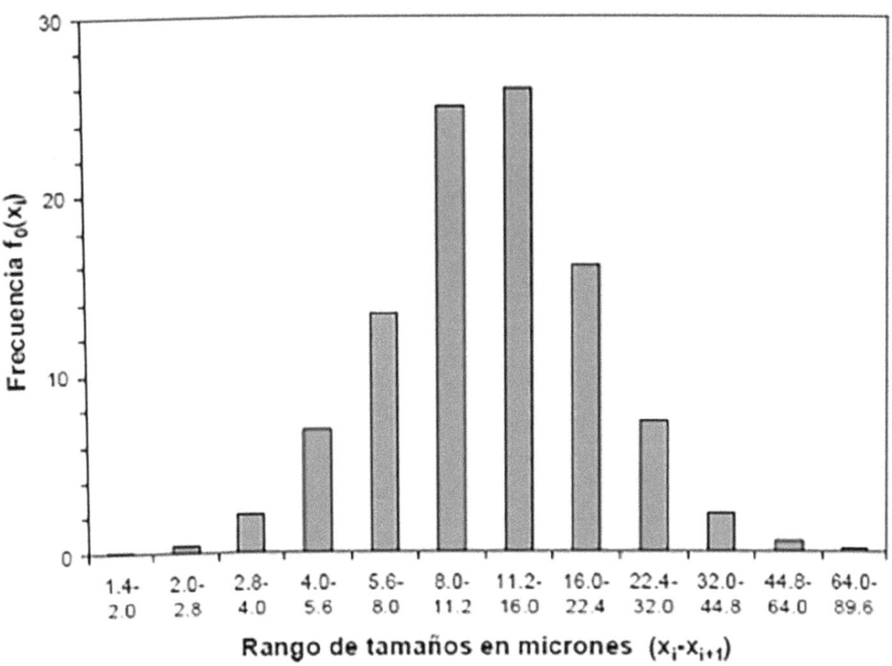

FIGURE 4.40 Number frequency function for the data of Table 4.6.

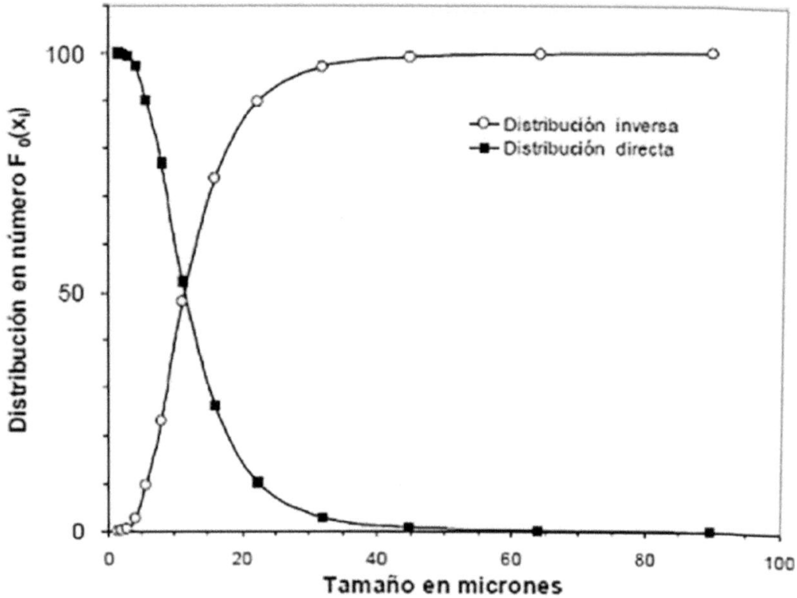

FIGURE 4.41 Number distribution function, data from Table 4.6.

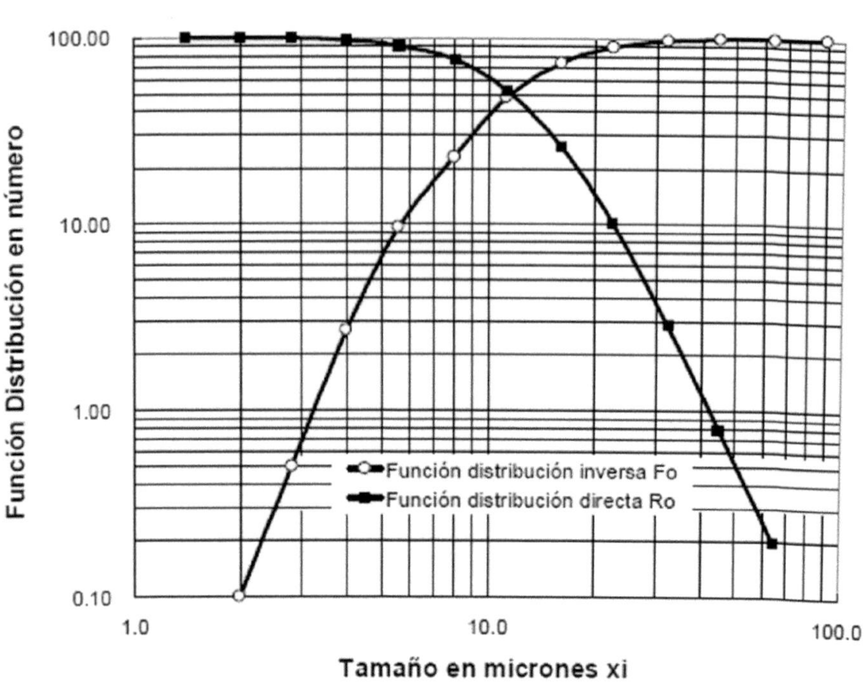

FIGURE 4.42 Direct and reverse distribution functions from data in Table 4.6 on a log-log scale.

Figures 4.43, 4.44, and 4.45 show the results of the size distribution using the standard mesh range distributions.

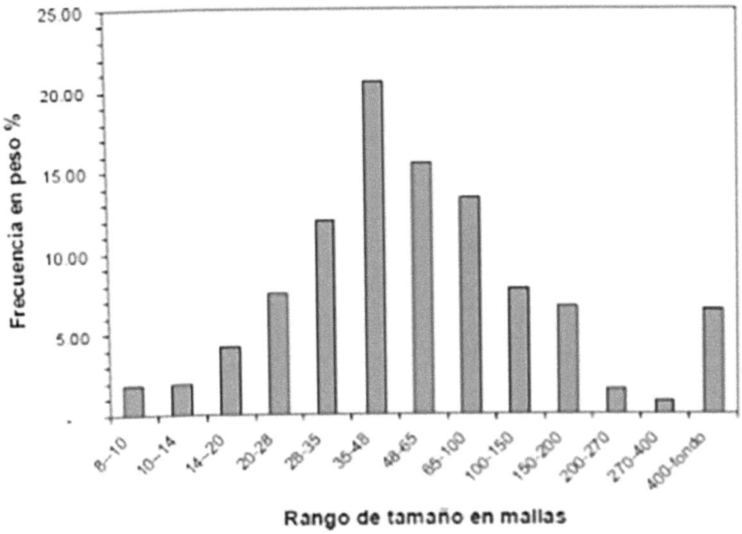

FIGURE 4.43 Mass frequency function.

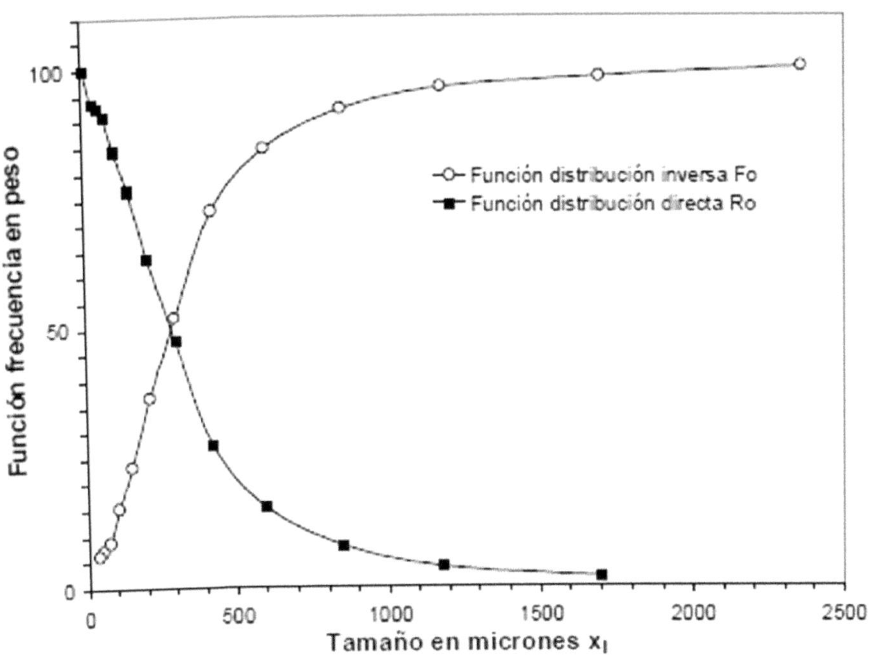

FIGURE 4.44 Direct and inverse volume distribution function.

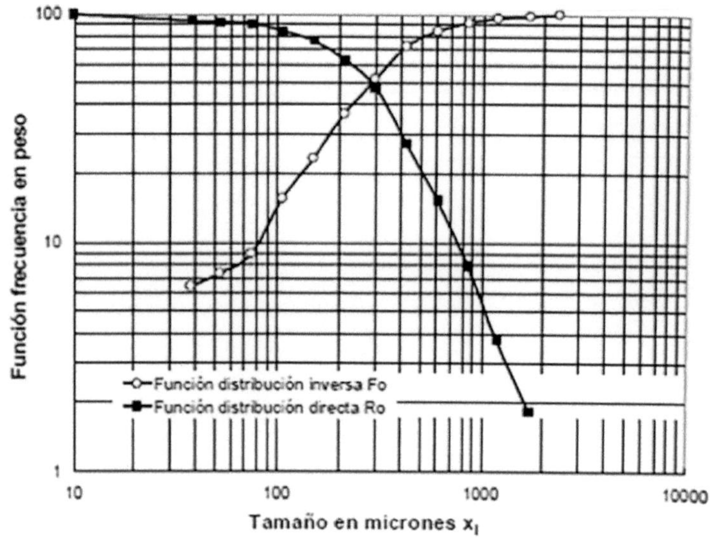

FIGURE 4.45 Direct volume log-log distribution function.

4.4.8 Experimental Determination of the Size and Shape of a Particulate System

The wide range of sizes of a particulate system and the specific properties of it explains the existence of the large number of methods for the experimental determination of its size distribution. To decide about which method to use, the analyst must first consider the objective of the analysis, that is, to know which variable is really of interest to know about the particulate system.

For example, if the particle size of a powder is needed to predict the air pollution it will produce the rate of sedimentation of the particles is the variable to use, and the proper method of measurement is the one that determines the "Stokes size." Alternatively, if you want to know the size of a catalyst to calculate its reactivity, the interest is the specific surface of the particle, so the method to be used is the measurement of the "surface volumetric size."

If instruments that precisely measure the variable of interest is not available, the choice is often made based on particle size. For the range of particles greater than 75 microns, the most used method is screening, while for particles less than 1 micron, it is to use a microscope. In the intermediate range, there are many instruments based on different methods, such as classification, sedimentation, optical and electrical methods, and the flow in porous media. In this section we will briefly describe the main existing methods and give examples of some available instruments using these methods.

4.4.8.1 Methods Based on Particle Classification

One of the most common ways to determine particle size is to separate them into categories according to some property or properties. These methods include screening, which classifies particles according to their size and shape, subjecting the particulate system to the action of the force of gravity in a series of sieves placed on top of each other. The largest sieve is at the top, and each sieve that follows has a mesh with a smaller opening than the previous one. Particles pass through or are retained in the openings of the sieves according to their size and shape. If "*n*" sieves are used, the particle system is separated into *n* sizes. Another method very commonly used in the mining-metallurgical field for particles between 10 and 50 microns is based on sedimentation of the particles in a fluid.

4.4.8.1.1 Sieving

In Section 4.4.6, we described the concept of mesh size and the standardized series of sieves used to perform sieving. We should add that there are micro sieves with meshes between 1 and 50 microns, which are mainly used in particle calibration, or when very narrow ranges of fine particles are needed. Figure 4.46 shows a set of standardized sieves and micro sieves.

Mechanical screening is done by placing a series of five or six sieves, one on top of the other with a closed bottom and a lid, inside a sieve, which is printed a horizontal oscillating motion and a vertical sharp movement. The first allows the mesh surface to all particles, and the second prevents the blockage of it. About 50 to 100 g of the material to be analyzed is placed in the upper sieve, which corresponds to the mesh of greater opening. It is usually sifted for 20 minutes. Upon completion of this time interval, the bottom is changed and sifted for an additional 10 minutes, until no more than 0.5% of the original load passes. This procedure follows ASTM D452. If material losses exceed 0.5% of the original material, the test should be discarded. The pre-treatment material with the bottom of the last screening should be combined.

When fines tend to agglomerate or, when the original material is a suspension in water, the sifting can be done wet by a pre-treatment to remove the fines. The suspension or material, containing 50 to 100 g of solids, is placed in a suitable sieve, for example 200 meshes, (ASTM D-313 and C117), or a sieve of 325 meshes (ASTM D185–72). It is sifted manually, helping the passage of particles by means of a brush with camel hair.

The suspension passed by the last sieve is collected in a bucket. 50 to 100 g of the material to be analyzed is placed in the upper sieve, which corresponds to the mesh

FIGURE 4.46 Set of standardized sieves and micro sieves.

of greater opening. It is usually sifted for 20 minutes. Upon completion of this time interval, the bottom is changed and sifted for additional 10 minutes, until no more than 0.5% of the original load passes. This procedure follows ASTM D452. If material losses exceed 0.5% of the original material the test should be discarded. The pre-treatment material with the bottom of the last screening should be combined.

The sieving can be done dry or wet manually or mechanically. Factors that influence the result of a screening are the following:

- material load
- the amount of fine material
- the shape of the particles
- screening time
- the cohesiveness of the dust
- the fragility of particles

Based on these factors, the following recommendations can be made:

- The lower the load of a sieve, the faster the screening, as long as the amount is sufficient to perform a precise weighing.
- It is recommended to remove the fine in the dry screening to prevent it from becoming very long.
- The movement of the sieve should be such that it minimizes the locking of the mesh.
- Materials that clump should be sifted in wet.
- Degradable or very soft materials should be treated gently.

There are many types of sieves on the market. We will only mention one because of the widespread use and the good quality of the screening it achieves. It is Tyler Ro-Tap (Tyler n.d.). This equipment, which we can see in Figure 4.47, combines a horizontal rotating motion of 270 to 300 rotations per minute with 140 to 160 strokes on the top cover per minute (ASTM B-214).

FIGURE 4.47 Sieve shaker and ultrasonic micro sieve.

For very small particles, between 1 and 50 microns, wet screening in microsieves is recommended. Approximately 1 g of material dispersed in a low surface tension liquid, such as acetone, is used. Microsieves are placed inside an ultrasonic bath, which performs screening. The liquid can be recovered by distillation.

4.4.8.1.2 Sedimentation-Based Methods

Sedimentation-based size analysis techniques determine the Stokes diameter defined as the diameter of a sphere that has the same velocity of sedimentation as the particle. As seen in Section 4.4.1.3, Stokes diameter d_S and the volume equivalent diameter of a particle of arbitrary shape d_V are given by (4.34 and 4.35)

$$d_S(P) = \left(\frac{18\mu u_\infty}{\Delta\rho g}\right)^{\frac{1}{2}} \quad \text{and} \quad d_S(P) = \left(\frac{18\mu u_V}{\Delta\rho g}\right)^{\frac{1}{2}} \tag{4.85}$$

Pettyjohn y Christiansen (1948) established that the ratio between the velocity u_V of a particle P and that of its equivalent sphere $d_S(P)$ is

$$\frac{d_S}{d_V} = \left(0.843\log\left(\frac{\psi}{0.065}\right)\right)^{\frac{1}{2}} \tag{4.86}$$

Experimental determination of the size distribution using the sedimentation method is based on the dependence of the settling rate on the concentration of the suspension. The concentration of the suspension at a certain level of the settling column, initially homogeneous, diminish its concentration as the sedimentation progresses. Measuring the change in concentration, or weight, of the particles induces various methods of sedimentation analysis such as the classic Andreasen Pipette the Micromeritics Sedigraph, the Quantachrome Microscan, the Retsch Lumosed, the Analysette of Fritsch and the Gallenkamp, Sartorius and Shimadzu, sedimentation balances, among others. These instruments have as their main element a sedimentation column, which is filled with a homogeneous suspension of concentration which is measured over time. Consider a small constant volume at a certain height z of the column. In the first few moments of sedimentation, each particle of a certain size that leaves the small volume in z by sedimentation is replaced by the same number of particles of the same size that come from above. However, particles that have a sedimentation rate greater than $u(d_{Si}) = \frac{(L-z)}{t_i}$ will have passed the z position in the time interval t_i so that there will be no particles of that size that are coming from above to replace those that came out of the small volume. Therefore, the concentration of the suspension in the volume in z in time t_i will have only particles less than size d_{Si}, that is, its size distribution will be $F(d_{Si})$. If the initial concentration small volume is φ_0 at z and it is sampled for increase time increments, preferably in geometric progression, and the solid material obtained is weighed or its concentration is measured by light extinction or X-ray, the weight distribution can be calculated by the expression

$$F_3(d_i) = \frac{\varphi(d_{Si})}{\varphi_0} \tag{4.87}$$

4.4.8.1.2.1 Andreasen Pipette The Andreasen's Pipette (Allen 1981), shown in Figure 4.48, consists of a 500 cm^3 cylindrical vase graduated in centimeters capacity from z = 0 at a certain distance from the bottom up to $z = 20$ cm at the top. The pipette frosted lid has a thin tube welded through its center which leads to the $z = 0$ graduation inside the vase. As seen previously, the tube ends in a two-way valve, which leads to a widening of the tube with a capacity of 10 cm^3. Suspension can be sucked into the receptacle through this tube and, by manipulating the valve, can be transferred to a 25 cm^3 beaker.

The measurement method is based on the sedimentation in a column as discussed previously. A suspension is placed in the pipette up to the top mark of $z = 20$cm, $z = z = 20\,cm$ is homogenized and left to settle. If at each time $t_1 = t_1, t_2, t_3$ a sample is drawn at the *z*-level, the successive samples will lack the sizes $d_{Si} = d_{S1}, d_{S2}, d_{S3}$ and the sizes corresponding to the other sampling times. From these samples the particle size distribution can be obtained according to Equation 4.85.

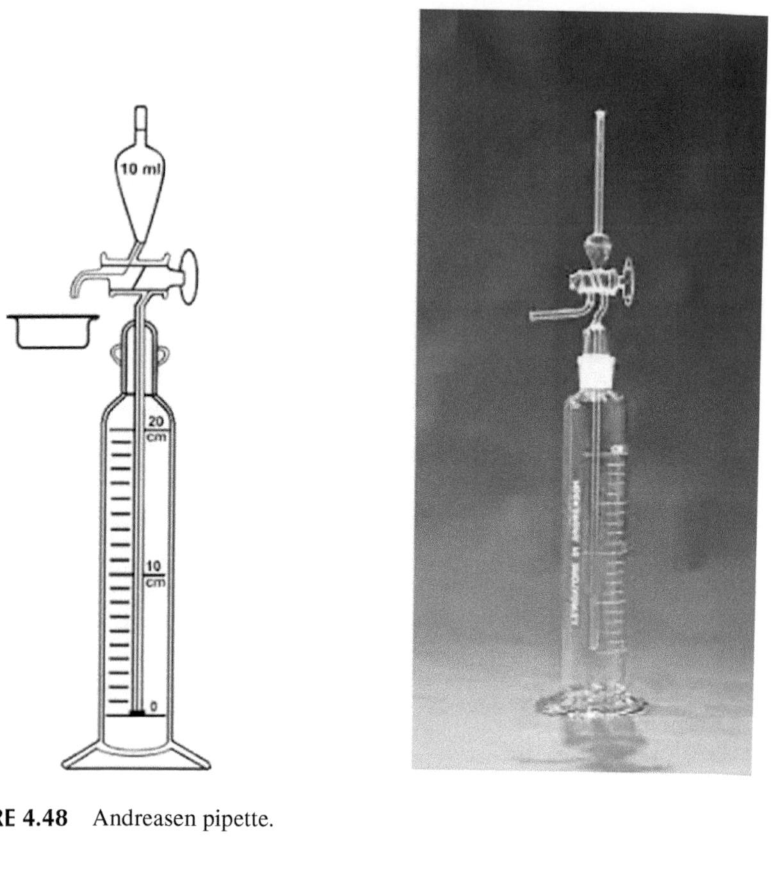

FIGURE 4.48 Andreasen pipette.

4.4.8.1.2.2 Cyclosizer The Cyclosizer is an elutriator (Michaud 2015), that is, a device that separates the particles by a continuous current of fluid, making use of the different terminal rate of sedimentation of particles of different size. The device makes use of the separation properties of a hydrocyclone. This consists of a cylindrical tube, with its axis in the vertical direction, followed by an inverted cone, which ends in a small diameter tube called apex. The cylindrical part has a lid on top with a second tube in the center, called vortex, and has a tangential entrance in its mantle as shown in Figure 4.49.

The Cyclosizer consists of five inverted hydrocyclones of equal size but different combination of apex and vortex arranged in series. See Figure 4.50.

The choice of apex/vortex is such that successive hydrocyclones classifies smaller particles. For example, for a quartz suspension under normal conditions, each hydrocyclone classifies (accumulate in the apex) fractions of particle with sizes 44 μm, 33μm, 23μm, 15μm, and 13μm which correspond to the standard size series.

Because the classification on each hydrocyclone is not perfect, the equipment was arranged in reverse form (Figure 4.50), with a closed cylindrical chamber with larger diameter added to the end of the apex so that the coarse particles that come out of the apex are retained in each hydrocyclone along with impurities finer than the separation size. Because of gravity this material can reenter the equipment and reclassify. After a short time, the particles in the apex chamber become free of contaminant and can be recovered, dried, and weighed.

The finest material that comes out from the overflow of the last hydrocyclone is collected in a pond with more than 180 liters of water; it is practically impossible to recover the fine solid material. For this reason, it is necessary to remove the initial

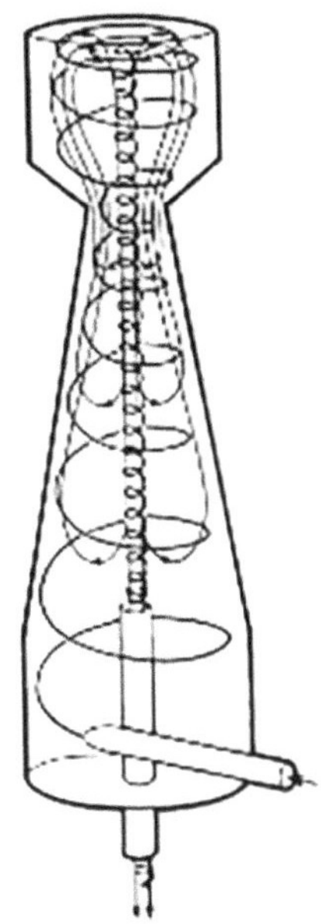

FIGURE 4.49 Inverted hydrocyclone with closed apex.

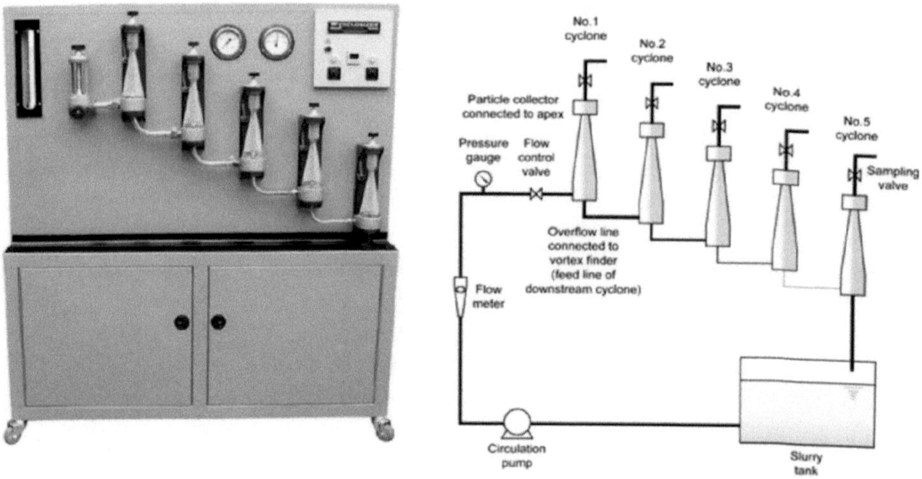

FIGURE 4.50 Cyclosizer particle size analyzer.

Source: Michaud, D. 2015. Cyclosizer. *Metallurgist 911*, June 18. www.911metallurgist.com/blog/cyclosizer. Accessed August 2023.

sample in a pre-treatment to separate the material less than 1 $m\mu$. This pre-treatment consists of the following:

- Disperse between 25 and 200 g of the sample of solid powder in water inside a 2 liter filled beaker up to a height of 16 cm and shake until the solid is completely dispersed.
- Leave to settle for 60 minutes and then remove the supernatant water up to a height of 1.5 cm.
- The contents of the specimen are again completed with water up to the same level, shaken, and the procedure is repeated four times.
- The content of the four supernatant suspension are combined, flocculated, left to settle, dried, and weighed.
- The residue of the sample is wetly sieved on a 400 meshes screen, and the undersize material is introduced to the Cyclosizer.

The procedure for operating the Cyclosizer consists of the following:

- Start the equipment and regulate the water flow by adjusting the working pressure to 40 psi.
- Measure water temperature.
- Insert the deslimed sample into the Cyclosizer and leave it running for 30 minutes.
- After this time interval, the suspension retained in each chamber is removed in a 50 cm3 beaker, dried on a stove, and the solids content calculated.

- The sedimentation diameter is obtained by adjusting the solid and fluid properties according to the temperature (density and viscosity) to the result given by the instrument, which is calibrated for quartz.

4.4.8.1.3 *Sedimentometers*

There are several instruments that measure the concentration in a sedimentation column by light attenuation or X-ray. Among the firsts, Topas, Retsch, Wagner, EEL, Bound Brooke, Seishin, and others. While for Photo-Sedimentometers, Quantachrom's Micromeritics, and Microscan Sedigraph, use X-rays to measure concentration. The basics of X-rays attenuation and light attenuation will be given in the next section.

4.4.8.1.3.1 X-Ray Sedimentometer The Micromeritics Sedigraph is an example of an instrument using X-Ray to determine the solid particle size distribution in a suspension (Micromeritics n.d.). It is an instrument for particle size analysis using the sedimentation method. Particle mass is measured directly via X-ray absorption. By measuring the rate at which particles settle as described by Stokes' law under gravity through a liquid having known properties, the SediGraph determines the equivalent spherical diameter of particles ranging from 300 to 0.1 micrometers.

4.4.8.1.4 *Microscopy*

Microscopy is one of the most direct methods of measuring particle size because it is the only method in which particles are individually observed and measured. For this reason, it is the absolute method of particle size analysis. It has the additional advantage that it allows us to observe and measure the shape of the particles and determine their composition. Since particles are considered individually, the quality of the sample becomes important. It is necessary to measure at least 10,000 particles in order to

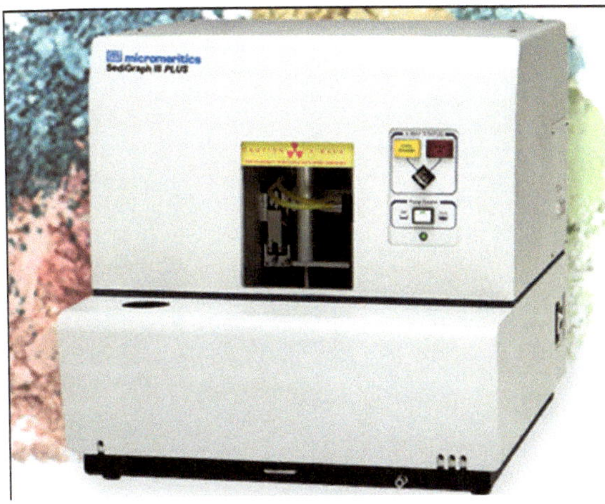

FIGURE 4.51 X-ray SediGraph from micromeritics.

have a sampling error of 1%, while if only 100 particles are measured, the error rises to 10% (as explained in the next section). It is clear then, that the only way to perform a particle measurement under the microscope is by means of an automatic system since otherwise it would be excessively time-consuming. The result of a microscopic study of a particulate system results in frequency and distribution in number.

4.4.8.1.4.1 Optical Microscopy Optical microscopy is used to measure particles with sizes between 0.8< x <150 μm. Its biggest limitation is its small focal depth, which is around 10μm for an expansion of 100× and 0.5μm for a 1000×. The resolution limit for an optical microscope is given by

$$x = \frac{f\lambda}{2NA} \tag{4.88}$$

where x is the resolution limit, λ the wavelength of the light illuminating, NA is the numerical aperture of the target, and f is a factor of about 1.3. Particles less than 5 μm should be studied with transmission microscopes, while larger particles may use reflection microscopy. Images produced by an optical microscope can be observed directly or on their projection on a surface (photograph). It is preferable to make the observation directly to avoid the problem of those particles that were not left in focus. Observation can be made from samples dispersed homogeneously on a glass slide or embedded in a briquette. In the latter case, one of the problems is how to consider those particles that were cut and what relationship exists between the flat microscope image and the three-dimensional shape of the particle. This phenomenon is treated in the discipline called stereology.

Several companies offer automatic optical microscopic particle analysis systems. Figure 4.52a shows an outline of an optical microscope and an automatic analysis system using this principle (Gianfrancesco 2017; Evident 2016). Figure 4.52b shows an example of optical microscopy images in reflected light of a Pb-Zn-Cu-Ag-Bi-W polymetallic ore.

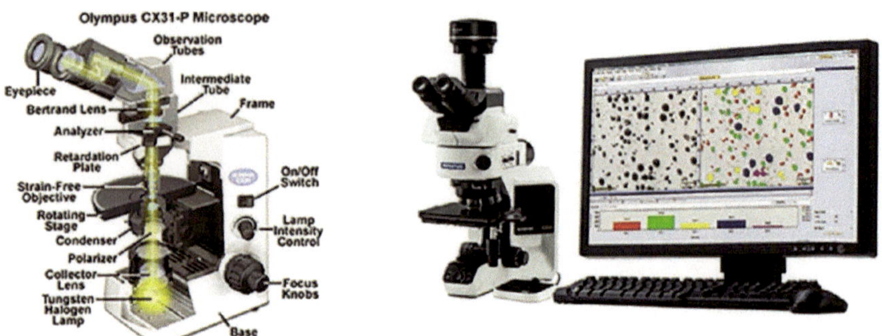

FIGURE 4.52 Olympus polarized optical microscopic system.

4.4.8.1.4.2 Electronic Microscopy There are two types of electron microscopy used for particle observation, transmission electron microscopy (TEM) (Figure 4.53), and scanning electron microscopy (SEM).

TEM is used to directly examine particles in the range from 0.001 to 5μm in the image produced on a fluorescent screen or generally on the photographic image. Relatively inexpensive instruments can reach increases of 4000× while more sophisticated equipment can reach up to 10000× with a resolution of 0.3 to 0.5 nm. Great effort should be used in producing good sample preparation.

SEM is a technique in which a fine ray of medium energy electrons (5–50 keV) is swept over the sample in parallel paths. These electrons interact with the sample producing a secondary emission of electrons (SEE), diffracted electrons (BSE), light and X-rays. Each of these signals can be detected and displayed on the screen of a cathode tube as a television image. Observation is usually done in a photographic reproduction of the screen. SEM is much faster and gives more three-dimensional detail than TEM due to its large depth of focus, which is about 300 times larger than that of an optical microscope. Sample up to 25 × 25 mm with partial amplifications of 20 to 100,000 and with a resolution of 15 to 20 nm can be observed. Figure 4.54 shows a scanning electron microscope (SEM) (Fandrich et al. 2007).

4.4.8.2 Methods Based on the Interaction of Light

When a ray of light passes through a particle suspension, part of the beam is absorbed by the particles, transforming it into another type of energy, and the other part is dispersed as diffracted light. This part is emitted in all directions of space, although

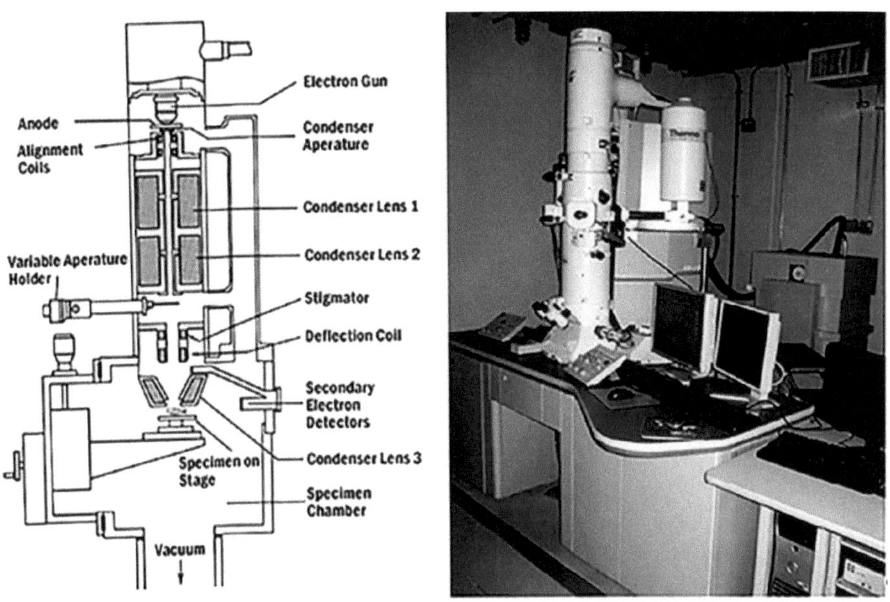

FIGURE 4.53 TEM electron microscope: (a) Outline of a TEM System, (b) TEM System Instrument.

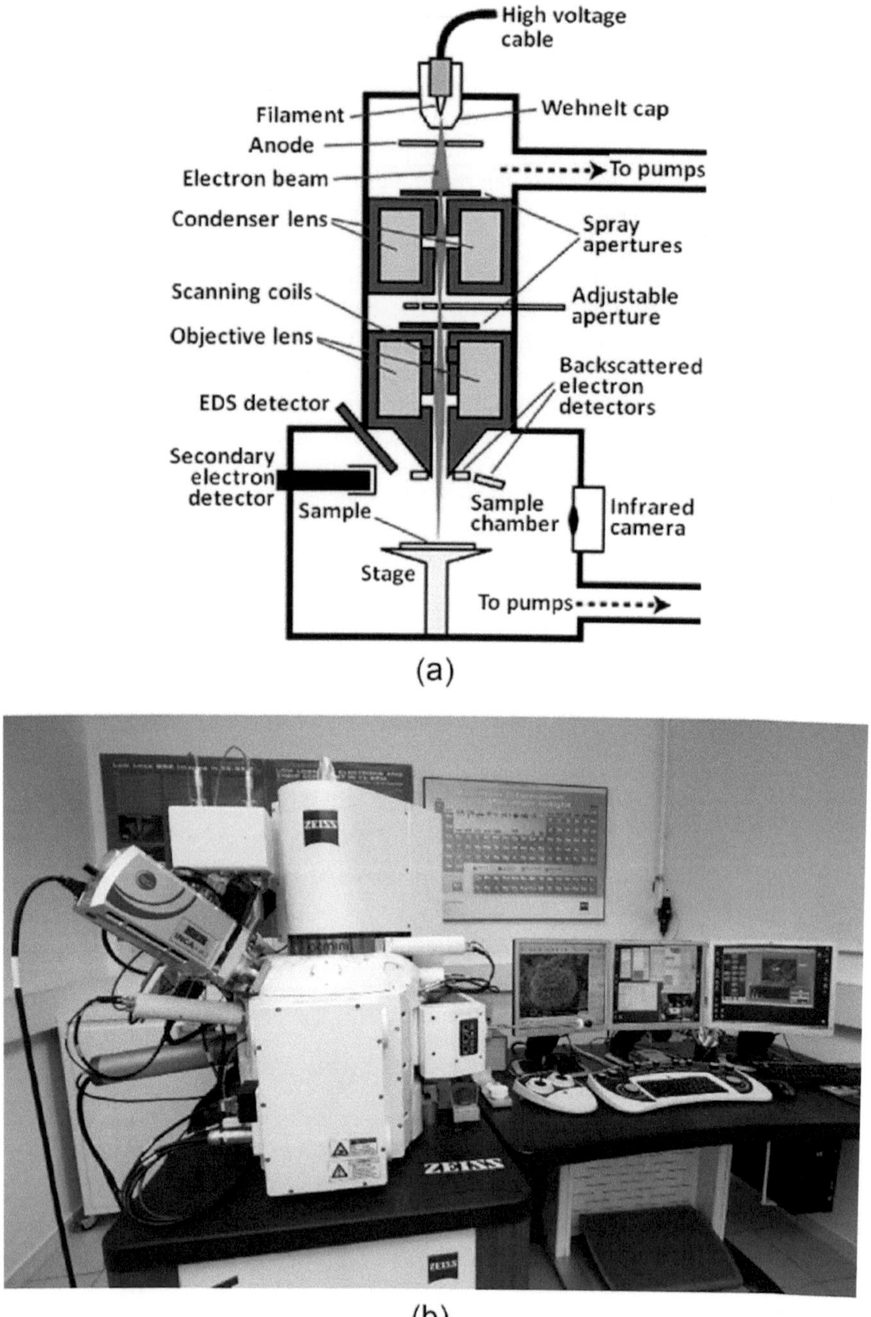

FIGURE 4.54 Scanning electron microscope (SEM). (a) SEM Figure System Scheme (b) SEM System Instrument.

the intensity is higher in some directions than others. Diffracted light produces a characteristic diffraction pattern for particles which can be used to measure particle size and concentration.

4.4.8.2.1 Light Scattering

Light scattering is the deviation of luminous rays from their original direction by small particles. Light is scattered in all directions of space, but its intensity depends on particle size and dispersion angle. Figure 4.55 shows the intensity of light dispersed by two particles of different sizes. It can be observed that the intensity of scattered light increases significantly with the increase in particle size and that it is maximum for a dispersion angle of $\theta = 0°$.

Figure 4.56 shows the form of dependence on the intensity of the scattered light with the particle size and the angle of dispersion. Only the data of the angle $\theta = 0$ are soft enough to be used as a correlation.

Mie's theory describes dependence on particle sizes less than or equal to the wavelength of the ray $(x \leq \lambda)$. From the figure you see that in this range, called Rayleigh's range, it can extend up to $x < 1\mu m$ with $I \propto x^6$.

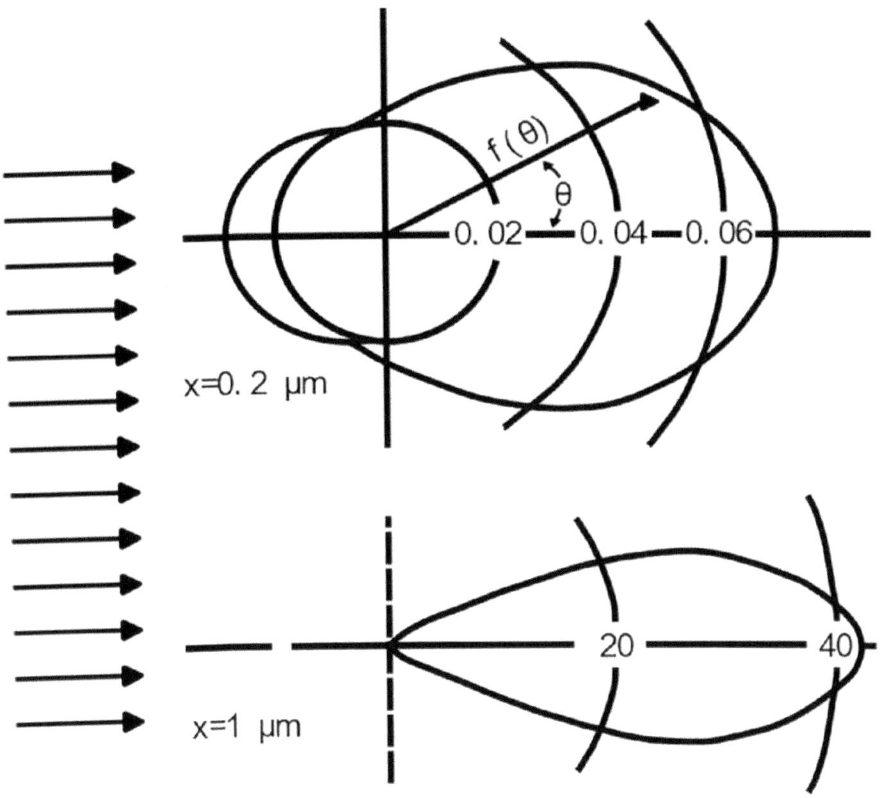

FIGURE 4.55 Intensity of light dispersed by two particles of different sizes.

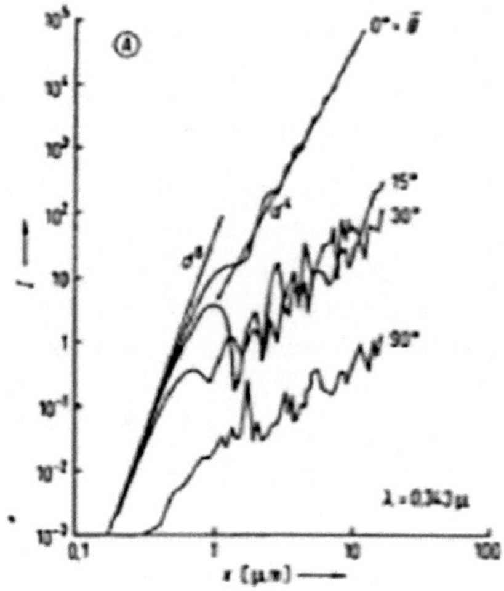

FIGURE 4.56 Intensity of light dispersed by a particle of 0.2 and another of 1 μm for a ray with wavelength 0.63 μm and a refraction index of $n = 1.2$.

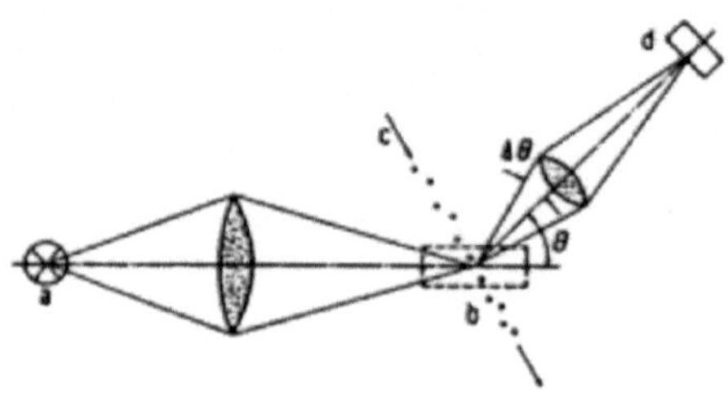

FIGURE 4.57 Optical system to measure light dispersion.

For sizes larger than $x < 1\mu m$ and up to $10\mu m$, called the Mie effect, the dependency is $I \propto x^4$.

Instruments that use scattering to determine particle size have an optical arrangement similar to that in Figure 4.57.

A light source sends a beam through an optical system into a small volume through which a flow of particles passes at very small concentration. Scattered light rays are captured by a second optical system located at an angle θ of the first. Figure 4.58 shows an instrument that uses this principle.

4.4.8.2.2 Light Extinction

To measure the extinction due to particles, which is the intensity of the transmitted light, a detector is placed at the $\theta = 0$ angle. The scattered light intensity that is detected at an angle δ, around $\theta = 0$ (Figure 4.59), depending on the particle size, the wavelength of the incident light, the refractive index of the optical system and the δ angle.

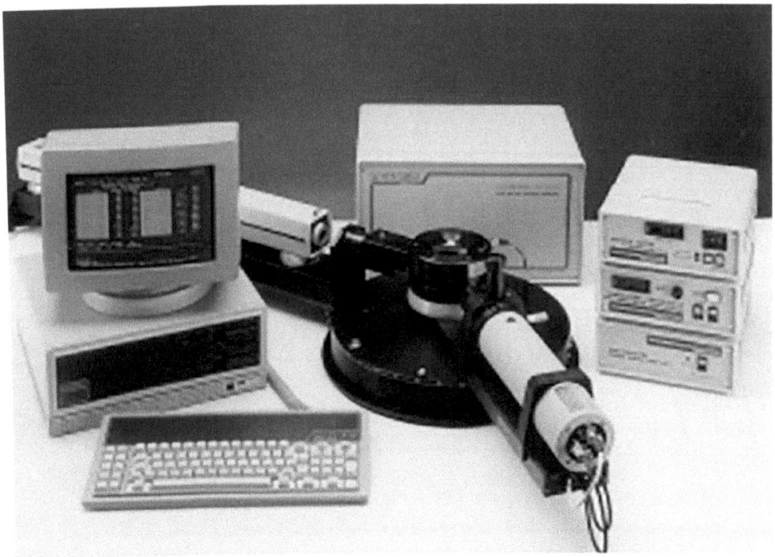

FIGURE 4.58 Instrument for measuring light scattering.

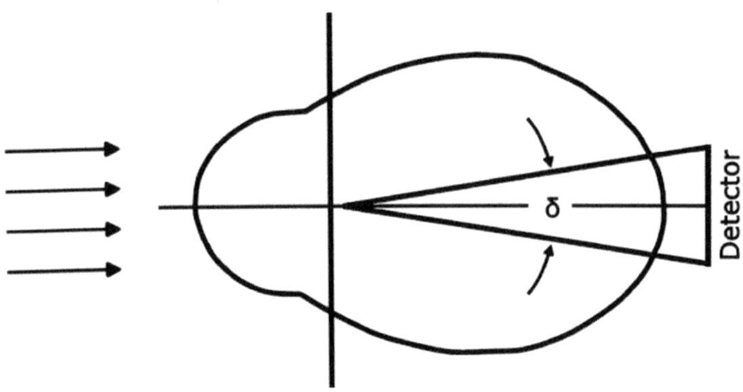

FIGURE 4.59 Part of the scattered light falling into the detector.

For diluted particle suspensions, the decrease in light intensity from incident I_0 the transmitted light I is given by

$$I = I_0\left(1 - NK(x,n,\delta)\right)\frac{\pi x^2}{4} \tag{4.89}$$

where N is the concentration of particles in number and $K(x,n,\delta)$ is the extinction coefficient, which depends on the particle size, its refractive index, and the angle of detection. Figure 4.60 shows the variation of the extinction coefficient with the particle size.

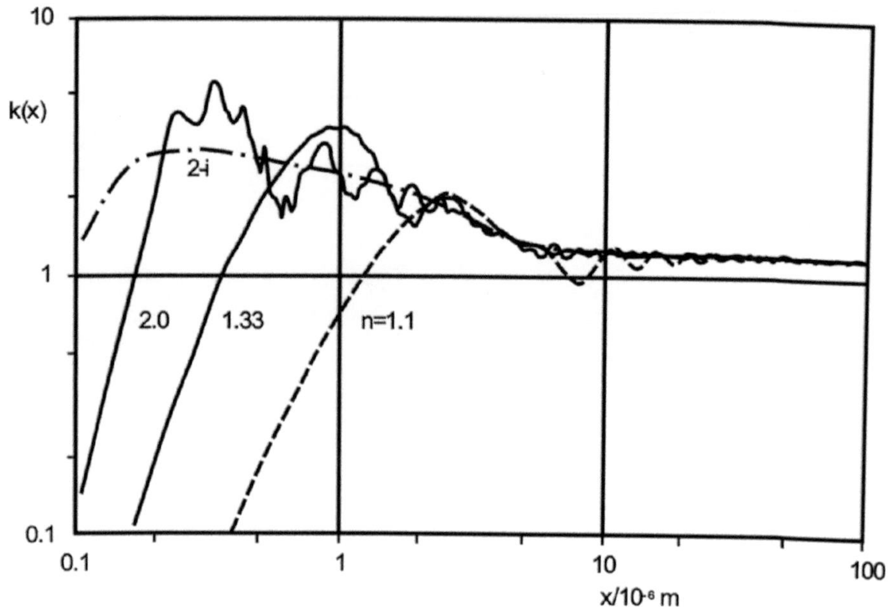

FIGURE 4.60 Extinction coefficient K based on particle size and wavelength, for various refractive indices.

4.4.8.2.3 Light Photosedimentometer

Light photosedimentometers are equipment that uses the extinction of light scattered throughout particles. For example, the US-Sedimentometer USS 791 applies the sedimentation principle in suspensions to determine the particle size distribution of samples. The operating principle is based on measuring the time in which certain quantities of sample accumulate in a calibrated tube. The particle size distribution is then calculated by applying Stokes' law. A light barrier system automatically follows the altering height of material in the collection tube by applying Stokes' law. All parts of the sedimentometer are made in the glass workshop at Topas. Figure 4.61 shows the US-Photosedimentometer.

4.4.8.2.4 Light Diffraction

When a parallel, monochromatic ray of light strikes a small opening, light transmitted through a lens forms a diffraction pattern on a flat wall. This pattern, called Fraunhofer diffraction pattern, consists of a strongly illuminated circle surrounded by dark and light rings respectively, as shown in Figure 4.62.

If the ray of light illuminates a particle suspension, it also forms a Fraunhofer diffraction pattern, which is not disturbed by particle movement. Figure 5.36 shows a simple optical system for measuring diffraction patterns. A He/Ne laser beam targets a lens that widens it and transforms it into parallel beam. This beam passes through the suspension and is focused on another lens on a wall, where its intensity is measured by a photoelectric detector.

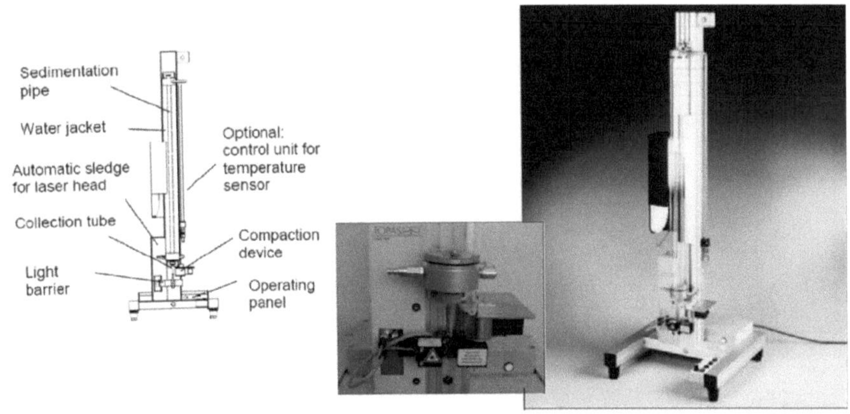

FIGURE 4.61 US-Photosedimentometer 791.

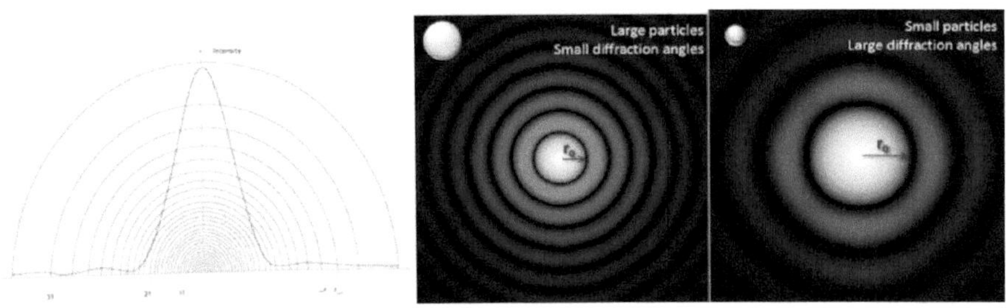

FIGURE 4.62 Fraunhofer diffraction pattern.

Source: Sympatec. n.d. Helos/BR, Rodos/L, Vibri/L. www.sympatec.com/en/particle-measurement/sensors/laser-diffraction/helos/helos-br-rodos-l-vibri-l/. Accessed August 2023.

When the particle size is much larger than the wavelength of the laser beam, $x > 1\mu m$ the light intensity measured in the detector is independent of the optical properties of the material then, according to Fraunhofer's theory, the light intensity of the radius circle r, produced by a spherical particle of diameter x, depends on the radius r of the image and the particle size:

$$i(r,x) = i_0 \left(k \frac{x^4}{4} \right) \left(\frac{J_1(k,r,x)}{k,r,x} \right)^2 \text{ with k=} \frac{\pi}{\lambda f} \tag{4.90}$$

where i_0 is the intensity and is the wavelength of the incident, x is the size of the particles, r the radius of the pattern circles, f is the focal length, and J_1 is the first-class Bessel function. The radius of the center circle is given by

$$r_0 = 1.22 \frac{\lambda f}{x} \tag{4.91}$$

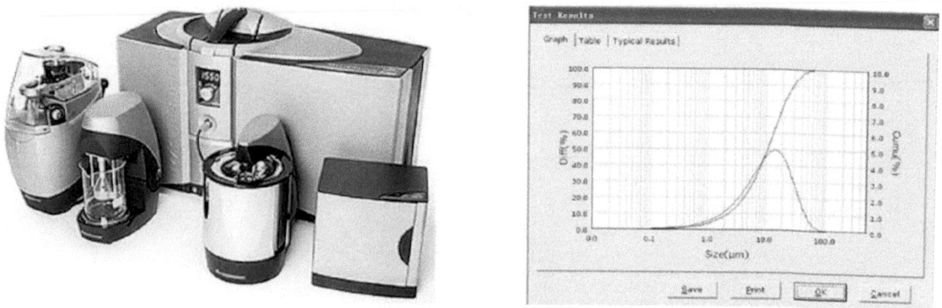

FIGURE 4.63 Mastersizer 3000 particle size analyzer.

If the particle is part of a suspension, the intensity is measured at the angle. The integration results from

$$I(r) = \frac{\int_0^\infty i(x)\frac{f_3(x)}{x^3}dx}{\int_0^\infty \frac{f_3(x)}{x^3}dx} \tag{4.92}$$

There are numerous instruments for particle size analysis in the market that use this principle. The best known are Helos-Rodos from Sympatec Germany, Malvern, England, Cilas Laser Granulometer, France and Leads and Northrup Microtrac, USA. Figure 4.63 shows a Mastersizer 3000 particle size analyzer.

4.4.8.3 Electric Resistance Methods

Particle electrical resistance counters count particles that, as they pass through a small hole, change the electrical resistance of the fluid.

Consider a cylinder-shaped control volume immersed in a homogeneous electric field. Through this volume flows an electrolyte in which solid particles are suspended. Particles pass through the volume one by one. When a particle is in the control volume, the control volume's electrical resistance increases and therefore for equal current increases the voltage $V = I \times R$. It can be shown that the change in electrical resistance is proportional to the volume of the particle:

$$V_p = V_0 \frac{\Delta R}{R_0} \equiv V_0 \frac{\Delta V}{V} \tag{4.93}$$

Where V_p is the volume of the particle, V_o the control volume, R_o is the resistance in the electrolyte only, ΔR is the resistance increment, ΔV is the voltage increase, and V is the original voltage.

There are several instruments that use this principle, the best known of which is the *Coulter particle counter* (Figure 4.66).

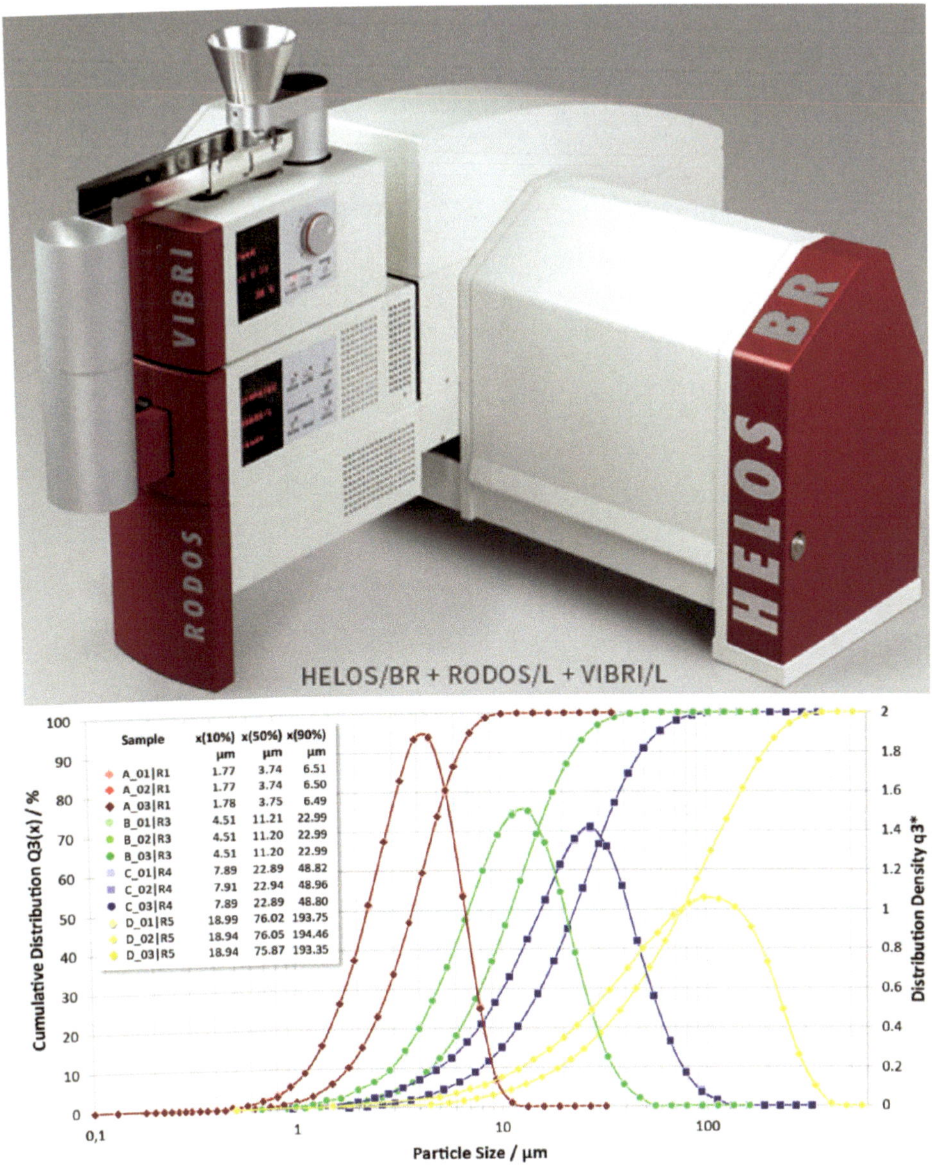

FIGURE 4.64 Particle size analyzer Helos-Rodos by Sympatec.

4.4.8.4 Permeability-Based Method

4.4.8.4.1 Fisher Test

The Fisher Sub Sieve Sizer consists primarily of an air pump, a pressure regulator, a sample tube, a standardized dual-range flowmeter, and a chart to give the results. Figure 4.67 shows an outline of the instrument (Isizer 2023).

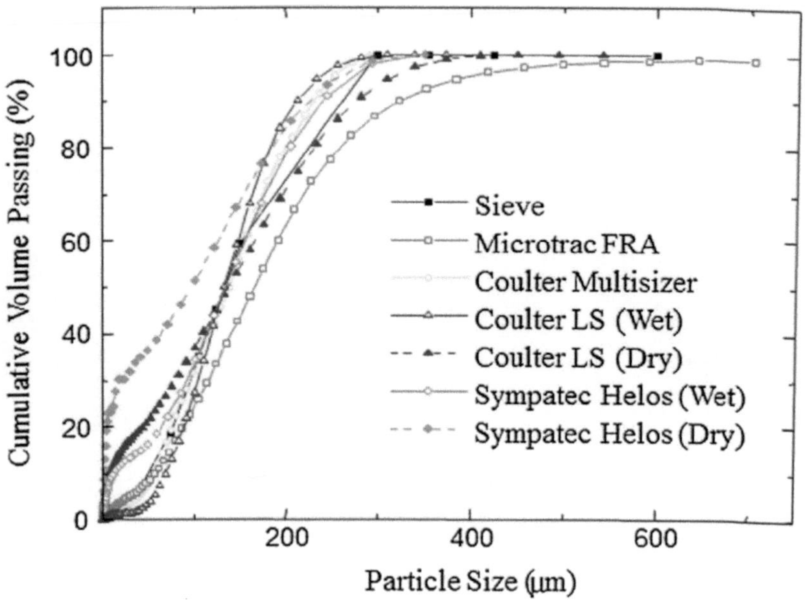

FIGURE 4.65 Comparison of several methods of particle size analysis (Basim and Khalilib 2015).

The Fisher Sub Sieve Sizer is an instrument that uses air permeability, at constant pressure, as a method to determine the average size of a particulate system using the expression

$$d_{sv} = 6\times10^4\left(k\frac{(1-\varepsilon)^2}{g\varepsilon^3}\frac{\mu_a Lq}{\Delta p}\right)^{\frac{1}{2}} \tag{4.94}$$

where μ_a the viscosity of the air, which depends on the temperature, q is the airflow, ε is the porosity of the particle bed and is its L thickness.

Among the explicit factors that affect the outcome, the main one is porosity that depends on how the porous bed is packed. It is recommended that the entire body be inserted at once into the cell and that the cell be hit briefly, or vibrated, in a standardized manner for each test. Factors that are not explicitly covered in the procedure are those on which Darcy, Carman, and Kozeny's theories are based. First, the airflow must be slow so that the linear dependence on the pressure drop flow is met. This is true for a Reynolds number, defined in the form (Concha 2001),

$$R_e^* = \frac{0.142\rho_a c(\varepsilon)\sqrt{k(\varepsilon)q}}{\mu_a\varepsilon^{\frac{2}{3}}} \tag{4.95}$$

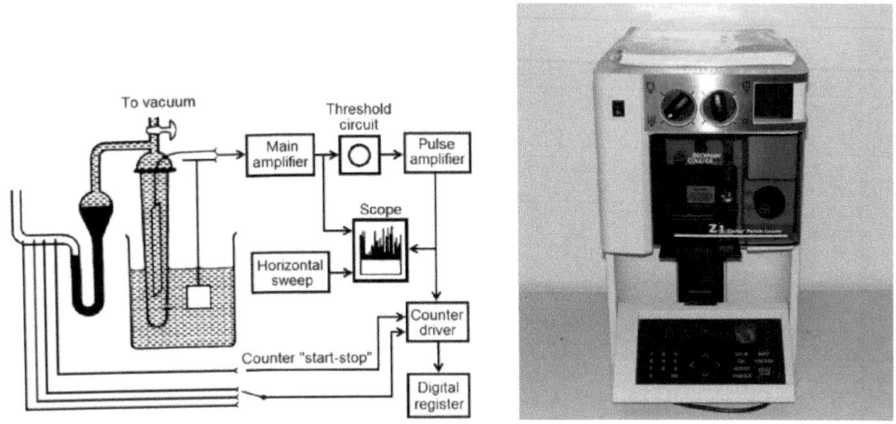

FIGURE 4.66 Coulter sizer principle and Beckman Coulter Z1 D Particle Counter 15230.

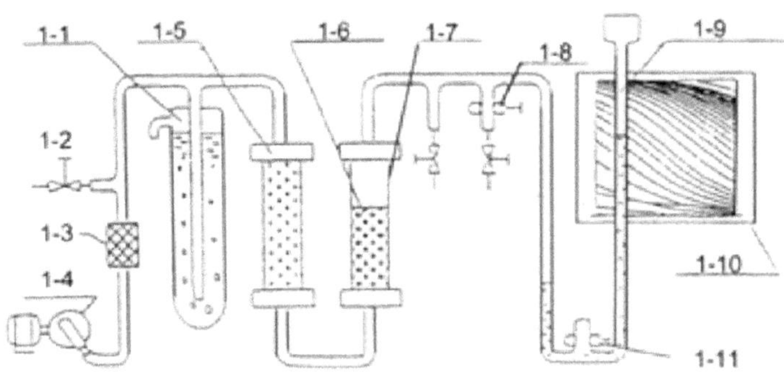

FIGURE 4.67 Fisher-sub-sieve-sizer-principle, where 1–1 Constant Pressure Regulator, 1–2 Pressure Control, 1–3 Air Filter, 1–4 Air Pump, 1–5 Dryer, 1–6 Packed Powder Sample, 1–7 Sample Tube, 1–8 Range Control, 1–9 Manometer, 1–10 Calculator Chart, and 1–11 Manometer Level Control.

Source: Isizercimadmin. 2023. Fisher sub sieve sizer principle 201–6. Particle Size Analyzers, July 26. www.isizer.com/fisher-sub-sieve-sizer-principle/. Accessed August 2023.

Second, the capillary model of a Carman-Kozeny porous bed assumes that the bed is made up of a very large set of straight capillaries and that the tortuosity parameter is $\beta = 5$. Finally, we will comment that the permeability method allows to also determine the shape factors of the particles.

Figures 4.68 and 4.69 present the permeability particle size analysis and the comparison between Micromeritics SAS and Fisher FSSS instruments.

4.4.8.5 Image-Based Methods

QEMSCAN (Quantitative Evaluation of Minerals by Scanning Electron Microscopy) is a fully automated micro-analysis system that enables quantitative chemical analysis

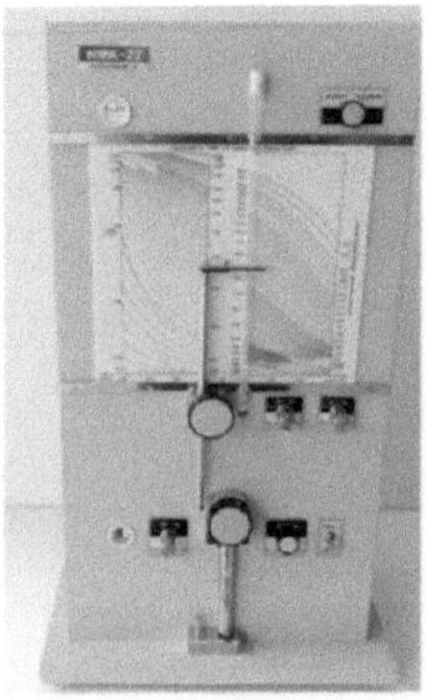

Fisher sub-sieve Sizer.

Micromeritics Sub-Sieve AutoSizer II.

FIGURE 4.68 Permeability sub-sieve instrument for particle size analysis.

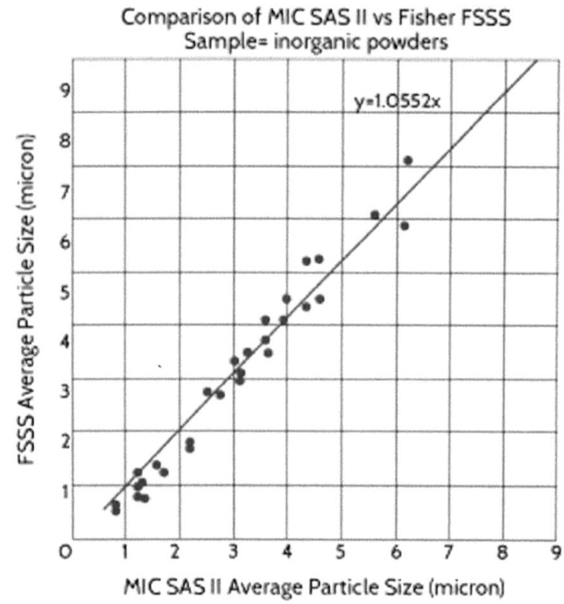

FIGURE 4.69 Comparison of results from of the Micromeritics SAS and Fisher FSSS.

of materials and generation of high-resolution mineral maps and images as well as porosity structure (Ayling et al. 2012; Grammatikopoulos 2018). It is an important tool in supporting mineral processing test work by quantitative and statistical analysis of the size distribution of the ore and its liberation and association.

Figure 4.70 and 4.71 show the QEMSCAN particle and species analyzer.

A good sample preparation is vital for a QEMSCAN Analysis (Figure 4.72).

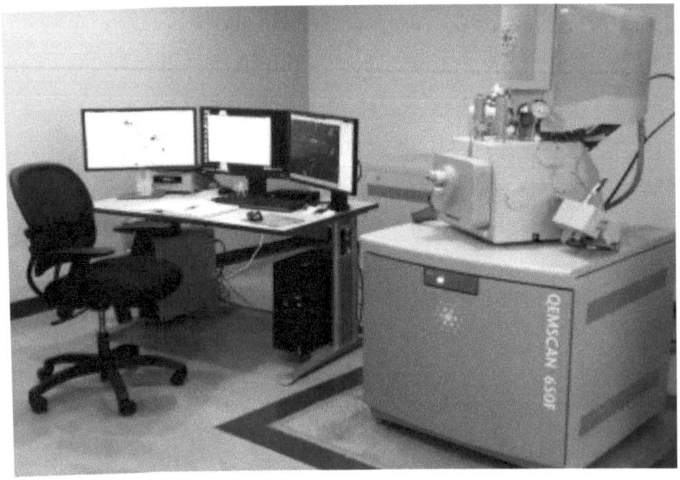

FIGURE 4.70 QUEMSCAN Zeiss EVO 50 SEM with four silicon detectors.

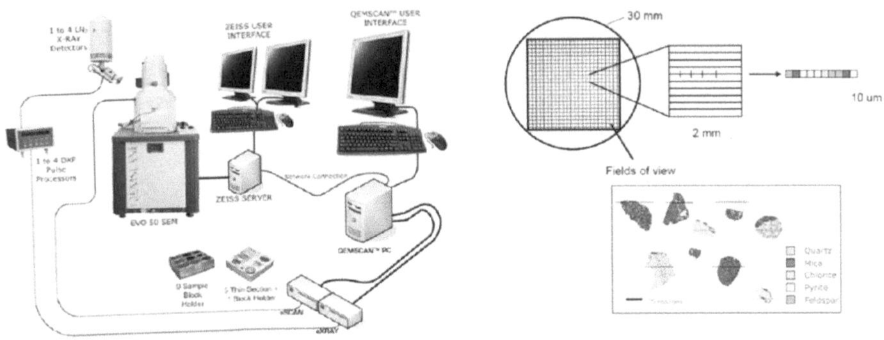

FIGURE 4.71 QEMSCAN system and modal analysis.

FIGURE 4.72 Preparation of a sample for QEMSCAN and final briquette.

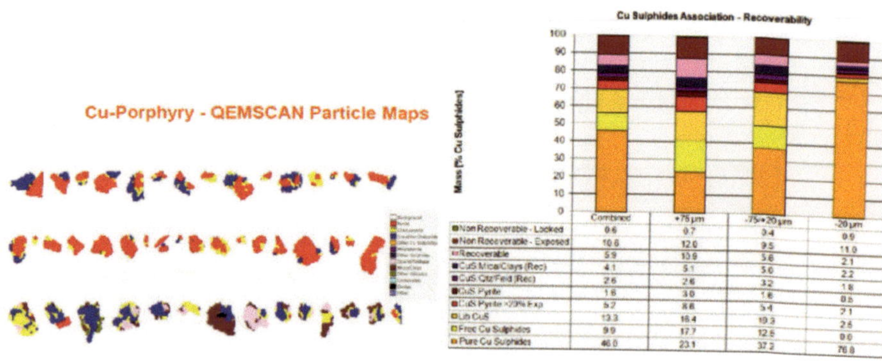

FIGURE 4.73 Porphyry copper QEMSCAN Maps and copper sulfides recoverability.

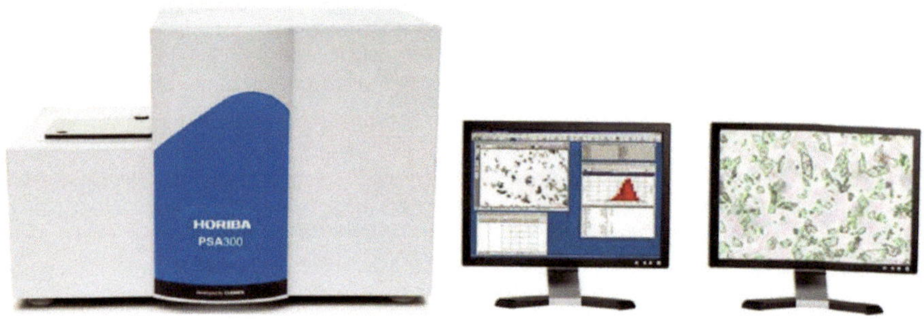

FIGURE 4.74 Compact laser diffraction particle size analyzer PSA-300 from Horiba.

The HORIBA's CAMSIZER uses a two-camera system with digital image processing to measure particle size distribution and particle shape, density of powders, number and particles simultaneously in the size range between 30 μm and 30 mm.

4.4.9 Particle Sampling

Any particle size analysis is performed on a sample of the particulate system. For this sample to be useful, a fundamental requirement for the measurement is that the sample is representative of the original system.

4.4.9.1 Sampling Methods

A laboratory sample for a particle size analysis can be obtained in various ways. For all of them it is necessary that the material is previously well mixed. The most used manual way to mix a material is by means of a cloth, such as the one shown in Figure 4.75a.

The size of the rolling cloth varies depending on the size of the material to be mixed. If this material is several dozen kilograms, two people must make the role. The square-shaped cloth rests on the floor, and each person takes two ends by rolling the material from one corner to the other. The operation is repeated for five minutes.

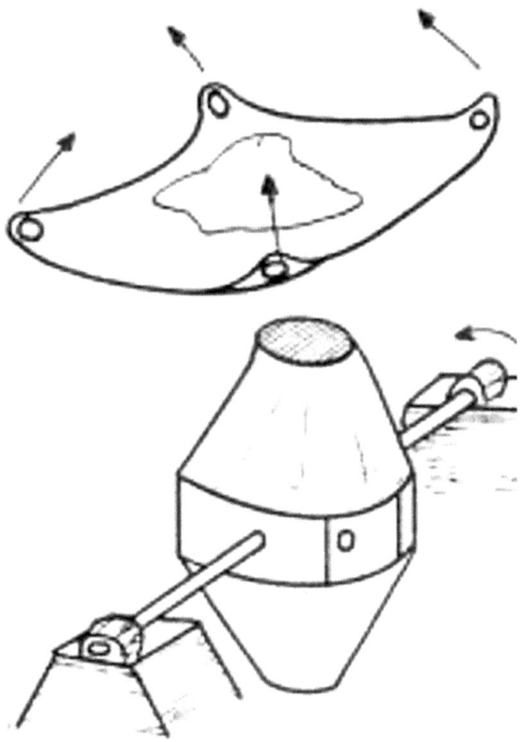

FIGURE 4.75 Mechanisms for mixing a sample: (a) roll cloth and (b) mechanical mixer.

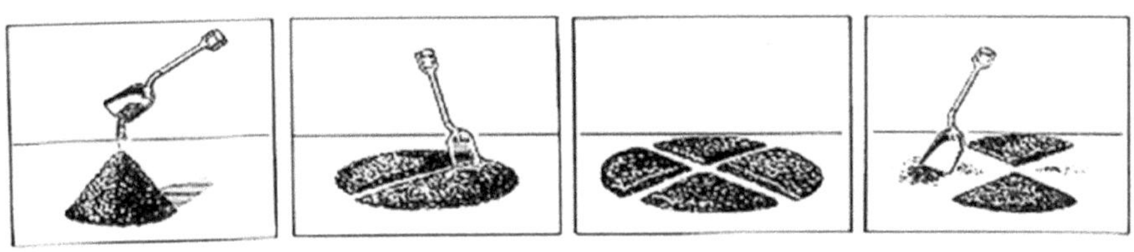

FIGURE 4.76 Coning and quartering sampling technique.

When the sample is small, less than three kilograms, the operation can be done by a person on a table.

When the amount of material to be mixed is very large, a mechanical mixer is used as the schematic in Figure 4.75b.

4.4.9.1.1 *Coning and Quartering*

The method of coning and quartering to reduce a sample is widely used due to its simplicity and because it does not require special equipment. The method, applicable to materials with particles smaller than 1 inch in size, is to form a cone with the material, on the floor or on a table, and crush it forming a circular cake. See Figure 4.76.

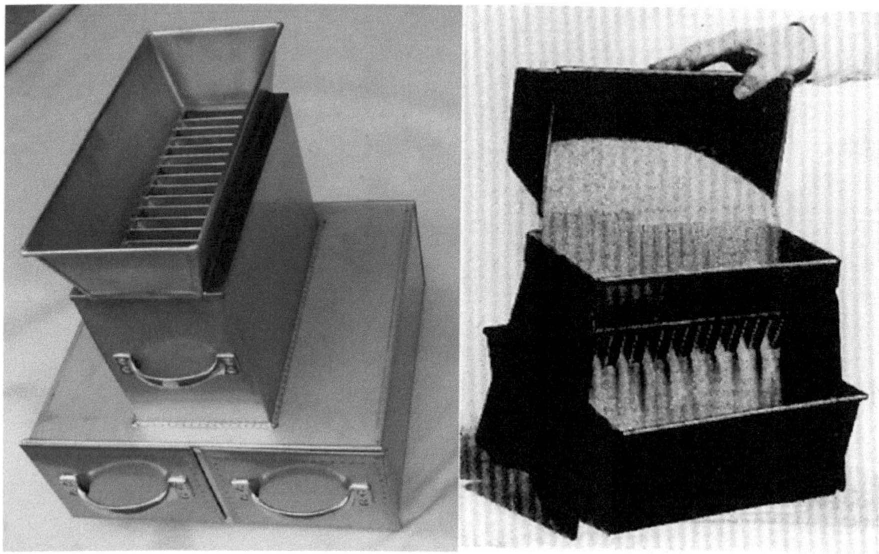

FIGURE 4.77 Chute splitter.

The symmetry of the cone is very important. The resulting cone trunk is divided into four equal parts. Two opposite quarters are carefully removed, gathering the remaining two fractions and building a new cone. The procedure is repeated until a sample is obtained of the desired size.

4.4.9.1.2 Riffle or Chute Splitter

A good method of reducing a sample is to pour the particles evenly onto a cutter of shots or riffles, such as the one in Figure 4.77. This device divides the original ample into two roughly equal parts. The repeated use of the cutter allows to obtain fractions of original.

4.4.9.1.3 Binomial Sample Reducer

Another piece of equipment used in the laboratory to reduce a sample is the binomial reducer. This device allows you to obtain one of the original sample in a single pass, see Figure 4.78. The particle size limit is 12 inches.

4.4.9.1.4 Spinning Riffle

An excellent sampling method is the rotary sampler. The material is placed in the mailbox of a vibrating feeder, which discharges into a rotating system of tanks, for example, from bottles. The system moves at a constant speed, allowing each bottle to receive the same sample amount shown in Figure 4.79.

4.4.9.2 Determination of Sample Size

The analysis of a particulate system requires two consecutive operations both subject to error, sampling, and sizing. As sampling and analysis operations are independent,

FIGURE 4.78 Binomial reducer.

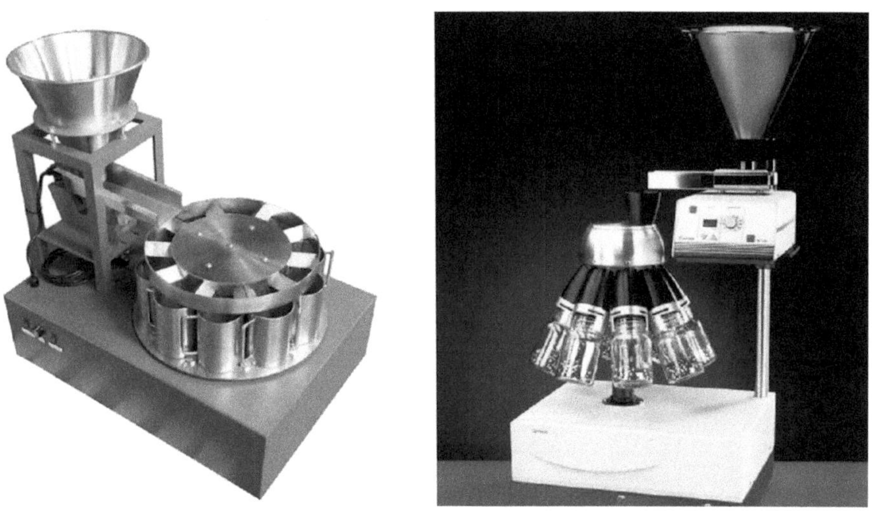

FIGURE 4.79 Rotary samplers.

errors are additive, and the total standard error (SE) associated with the determination will be determined by the following:

$$SE^2 = SE_m^2 + SE_a^2 \quad (4.96)$$

where SE_m^2 is the standard error associated with sampling and SE_a^2 the standard error associated with the analysis.

4.4.9.3 Sample Size

Consider the true value of the inverse distribution of a particulate system is $F(x)F(x)$, then we will designate by $F^*(x)$ the value of the distribution in the sample. As sampling is subject to an error SE_m, the relationship between these two values is as follows:

$$F(x) = F^*(x) \pm SE_m \tag{4.97}$$

On the other hand, let us designate by $F^{**}(x)$ the distribution obtained by the experimental analysis, then

$$F^*(x) = F^{**}(x) \pm SE_a \tag{4.98}$$

and therefore, for the particulate system we have the following:

$$F(x) = F^{**}(x) \pm SE \tag{4.99}$$

In practice, we can accept 1% sampling errors and 2% analysis errors; this means $SE_m = 0.010$, $SE_a = 0.020$, so that $SE = 0.022$.

$$\text{Then} \quad F(x) = F^{**}(x) \pm 0.022 \tag{4.100}$$

The question is then, what is the sample size so that the sampling error is within 1%?

The way samples are cut, the details of which will be described later, suggest that sampling errors are distributed in binomial form, that is,

$$\sigma_m^2 = \frac{1}{n} F^*(x)\left(1 - F^*(x)\right) \tag{4.101}$$

Where n is the number of particles in the sample. As

$$SE_m = t_{0.95}\sigma_m \tag{4.102}$$

where $t_{0.95} = 2.326$ is the statistical parameter of t-student test for one sample, from which we infer that

$$\sigma_m = \frac{SE_m}{2.326} = \frac{0.022}{2.326} = 0.0095 \tag{4.103}$$

Then the number of particles needed to obtain a standard deviation of $\sigma_m = 0.0095$ will be as follows:

$$n = \frac{F^*(x)\left(1 - F^*(x)\right)}{(0.0095)^2} \tag{4.104}$$

From this expression we see that the number of particles depends on the value of $F(x)$ and that it will be maximum for the following:

$$\frac{dn}{dF} = \left(1 - 2F^*(x)\right) = 0, \rightarrow F^* 0.5 \tag{4.105}$$

From where it follows that the maximum error will be made for $F^*(x) = 0.5$. Substituting this value in Equation 4.101, the required number of particles n for the sampling error to be less than 1%, and from Equation 4.100, is

$$n = \frac{0.25}{\sigma_m^2} = \frac{0.25}{(0.0095)^2} = 2770 \tag{4.106}$$

It is possible to calculate that if only 100 particles are taken, the sampling error increases to 10%.

The number of 2770 particles represent different problems for the various methods of analysis, depending on the size of the particles to be analyzed. The sample size depends on the average particle system size. The mass sample size will be

$$W = 2770 \times \rho_s \alpha_v \bar{x}_{30} \tag{4.107}$$

Assuming a density of the material of $\rho_s = 2.65$ a shape factor of $\alpha_v = 0.5$ (corresponding to quartz), and several different sizes, the total mass of the sample are given in Table 4.8.

TABLE 4.8
Sample Mass

Particle size		Density	Sphericity	Number par.	Total mass	
Mesh	cm	g/cm3	alfa	n	W (g)	kg
	10	2.6	0.5	2770	36010.	36.010.
3	0.67	2.6	0.5	2770	2413.	2.413.
8	0.236	2.6	0.5	2770	850.	.850.
100	0.015	2.6	0.5	2770	54.	.054.
150	0.0106	2.6	0.5	2770	38.	.038.

4.5 POPULATION BALANCE

Population balance was introduced in the early 1970s by Randolph and Larson (1971) and Hulburt and Katz (1964) in response to the impossibility of the theory of continuous media to describe the evolution of particles characterized, not only by space-time coordinates, but also by a set of properties closely related to the discrete characteristic of particulate systems. “It is a predictive multidimensional particle distribution theory, which in the case of certain well-defined regular process of particle formation, is useful in the, a priory prediction of the form and often the magnitude of the particle distribution” (Randolph and Larson 1971).

A *particle phase space* is defined which describes the evolution of the properties $\psi(\xi_1 \ldots \xi_N)$ of the system at a point in the three physical space of coordinate $(x, y\, z)$ at time t.

$$\underbrace{\frac{\partial}{\partial t}\int_{\xi(t)}\left(\int_{V(t)}\psi(\xi)dV\right)d\xi}_{\substack{\text{Rate of change of the number of}\\ \text{particles with property } \psi(\xi) \text{ in}\\ \text{the phase space } V(t)}} = \underbrace{-\int_{\xi(t)}\left(\int_{\partial V(t)}(\boldsymbol{Jn})dS\right)d\xi}_{\substack{\text{Net flux of number of particles}\\ \text{with property } \psi(\xi) \text{ through the}\\ \text{boundaries } \partial V \text{ of } V(t)}}$$
$$\underbrace{+\int_{\xi(t)}\left(\int_{V(t)}\left(B(\psi,t)-\left(D(\psi,t)\right)\right)d\xi\right)}_{\substack{\text{Net rate of generation of particles with}\\ \text{property } \psi(\xi) \text{ in the phase space } V(t)}} \tag{4.108}$$

Where $\boldsymbol{J}$ is the flux density of the total number of particles with property $\psi(\xi)$ crossing the boundary $\partial V(t)$ of the active volume of $V(t)$ at time t and A and D are the rate of appearance and disappearance of total number of particles of particles with property $\psi(\xi)$ per unit time.

The flux density of number of particles $\boldsymbol{J}$ can be separated in two terms, the convection across the boundary $\oint_{\partial V}\psi\boldsymbol{v}_c\boldsymbol{n}dS$, and the internal convection of particles $\oint_{\partial V}\psi\boldsymbol{v}_i\boldsymbol{n}dS$, a function of the internal state of the particles in $V(t)$.

$$\frac{d}{dt}\int_{\xi(t)}\left(\int_{V(t)}\psi\, dV\right)d\xi = -\int_{\xi(t)}\left(\int_{\partial V(t)}\left(\psi\left(\boldsymbol{v}_e\boldsymbol{n}\right)+\psi\left(\boldsymbol{v}_i\boldsymbol{n}\right)\right)dS\right)d\xi$$
$$+\int_{\xi(t)}\left(\int_{V(t)}\left(B\psi(\xi)-D(\psi(\xi))\right)\right)d\xi \tag{4.109}$$

Using Leibnitz differentiation rule to differentiate defined integrals whose limits are variable over time, the derivative of the first term in Equation 4.109 may be expressed as

$$\int_{\xi(t)} \left(\begin{array}{l} \dfrac{\partial}{\partial t}\left(\displaystyle\int_{V(t)} \psi(\xi)\,dV + \int_{\partial V(t)} \big(\psi(\xi)(\boldsymbol{v}_e\boldsymbol{n}) + \psi(\xi)(\boldsymbol{v}_i\boldsymbol{n})\big)\,dS \right) \\ \qquad - \displaystyle\int_{V(t)} \big(B\psi(\xi) - D(\psi(\xi)\big)\,dV \end{array} \right) d\xi = 0 \quad (4.110)$$

Since the control region of the internal coordinate subspace were chosen arbitrarily, the sufficient condition for Equation 4.110 to be satisfied is that the integrant be zero, then

$$\begin{array}{l} \dfrac{\partial}{\partial t}\left(\displaystyle\int_{V(t)} \psi(\xi)\,dV + \int_{\partial V(t)} \big(\psi(\xi)(\boldsymbol{v}_e\boldsymbol{n}) + \psi(\xi)(\boldsymbol{v}_i\boldsymbol{n})\big)\,dS \right) \\ - \displaystyle\int_{V(t)} \big(B\psi(\xi) - D\psi(\xi)\big)\,dV \end{array} \quad (4.111)$$

$$\underbrace{\frac{\partial \psi(t)}{\partial t}}_{\substack{\text{Rate of change}\\ \text{of the number of}\\ \text{particles with}\\ \text{property}\,\psi}} = -\underbrace{\nabla(\psi \boldsymbol{v}_e)}_{\substack{\text{Convective flux of}\\ \text{number of particles}\\ \text{with property}\,\psi}} - \underbrace{\nabla(\psi \boldsymbol{v}_i)}_{\substack{\text{Convective flux of}\\ \text{internan property}\,\psi}} + \underbrace{(B\psi - D\psi)}_{\substack{\text{Net rate of generation of}\\ \text{particles with property}\,\psi}} \quad (4.112)$$

This is the *general population balance* for a particulate system. This model or other similar have been successfully applied to processes such as crystallization (Randolph and Larson 1971), leaching (Sepúlveda and Herbst 1978), comminution (Herbst et al. 1985; Sastry and Concha 1985; Bürger et al. 2018), ball wear (Menacho and Concha 1985), flotation (Bascur and Herbst 1982; Bascur 2013), and floculation (Egarr et al. 2016; Shi et al. 2019; Quezada et al. 2020).

4.5.1 Comminution

Consider the case of grinding in a tumbling mill with a batch or continuous feeding of a mass of coarse particles of several sizes and a discharge of a distribution of finer particles. In this case and in many other processes, it is unnecessarily a detailed spatial position of the particles, thus a *macroscopic population balances* using the physical average of the population properties in an arbitrary control region of space (the mill) is appropriate. Consider only one property in the mill $\bar{\psi} = w_i$, that is the mas of particle size i as the volume average of ψ, and define the total volume and holdup of mass of solid particles in the mill by $V(t) = V_H(t)$ and $W(t) = \rho_s V_H(t)$, the total mass of particles of size i in the holdup by $W(t)w_i(t) = \rho_s V_B(t)\bar{\psi}$.

$$\frac{d}{dt}\left(W(t)w_i(t)\right) = \left(F_i(t)w_i(t) - O_O(t)w_O(t)\right) + W(t)w_i(t) - A\bar{\psi} + C\bar{\psi} \tag{4.113}$$

Where A is the rate if breakage of the particles of size i, $-S_i w_i$, and C is the rate of generation of size of the other sizes $B_{ij}S_i w_i$.

$$\frac{d}{dt}\left(W(t)w_i(t)\right) = \left(F_i(t)w_i(t) - O_O(t)w_O(t)\right) + W(t)w_i(t) - WS_i w_i + WB_{ij}S_j w_j \tag{4.114}$$

Finally, the macroscopic population balances for a system of particles in a ball mill can be written as

$$\underbrace{\frac{d}{dt}\left(W(t)w_i(t)\right)}_{\substack{\text{Rate of change of mass} \\ \text{of particles of size } x_i \text{ in } W(t).}} = \underbrace{\left(F_i(t)w_i(t) - O_O(t)w_i(t)\right)}_{\substack{\text{Net rate of flow of particles with property} \\ \text{of size } x_i \text{ in and nout from } W(t).}} + \underbrace{W(t)w_i}_{\substack{\text{Mass of particles} \\ \text{of size } x_i \text{ in the} \\ \text{hodlup } W(t)}}$$

$$\underbrace{-WS_i(t)w_i + W\sum_{j=1}^{I-1} b_{ij}S_j(t)}_{\substack{\text{Net rate w of generation of particles} \\ \text{of size } x_i \text{ in } W(t).}} \tag{4.115}$$

4.5.2 Size Enlargements

The reduction of size is necessary to achieve the appropriate fine particle sizes for their recovery by concentration from metalliferous ores due to the fine dissemination in which the minerals are found. As these fine sizes are not suitable for subsequent processes, such as thickening when there is a need to recover water from the tailings or when oxidized minerals are lixiviated, it is necessary to increase the size of these fines again by agglomeration. When agglomerating, particle of a suspension increases in size and acquire a greater sedimentation velocity essential to obtain a good separation by thickening. The agglomerates also form more permeable cakes accelerating filtration processes. Different methods are available to increase the size of solid particles. The most common are coagulation, by reduction of the interparticle electrostatic repulsion, flocculation by bridging particles with polymeric agents (flocculants) which are water-soluble, and oil agglomeration, in which an oily hydrocarbon is used to agglomerate hydrophobic particles suspended in water, hence producing hydrophobic agglomerate (Concha 2014).

4.5.3 Coagulation and Flocculation

In Chapter 7 we saw that the DLVO theory consider that suspension of particles in water possess electrical charges, mainly negative, that maintain them in suspension, avoiding sedimentation. To make them settle, we must set the physicochemical condition in the ionic solution; electrostatic attraction, for particles to collide with another and form coarse entities that, subjected to gravity, will settle. This is the critical coagulation concentration (CCC), condition of neutralization of the surface charge by providing salts with ionic charge opposite to that of the particles. The aggregation efficiency follows Schulze-Hardy rule that propose that the critical coagulation concentration decreases as the inverse sixth power of the counterion valence (Trefalt et al. 2020). However, for diffuse potentials multivalent counterions, the CCC increases with the surface charge density, inducing the strong decrease of the CCC with valence leading to a scaling of CCC with z is stronger than z^2 depending on the nature of the counterion (Lyklema 2013; see Chapter 6).

Particle suspensions present in the several stages of mineral processing are much larger than those of a colloidal system. The particle range for the fines in these processes is from 1 to 10 microns, therefore the dominant forces are inertial and convective forces, which are produced by an agitator, a pump, or the action of gravity. These flow conditions may impose enormous energy to the particles, permitting them to overcome the energy barriers of electrostatic repulsive forces without the necessity of neutralizing the charge. An excessively violent agitation may re-disperse the particles, and coagulation would not occur. The results of these two types of coagulation, by charge neutralization or by hydrodynamic surpassing the energy barrier, are different, and in the first case a neutral coagulum is obtained, while in the second the aggregate has an electric charge.

4.5.3.1 Coagulation Kinetics

Orthokinetic coagulation of colloidal systems is described by the Smoluchowski-Müller theory (Laskowski and Pugh 1992; Gregory and Nelson 1984). If the collision efficiency is represented by E_c, the rate of disappearance of the number particles by coagulation in a suspension is given by

$$\frac{dn}{dt} = -k_c n^2, \text{where } k_c = \frac{4E_c kT}{3\mu} \tag{4.116}$$

where k is the Boltzman constant, μ is the liquid viscosity, and T is the absolute temperature. The collision efficiency, with values between 0 and 1, depends on the suspension stability, where $E_c = 1$ corresponds to a completely destabilized system. The sizes of fine particles in mineral processing are relatively big compared with colloidal particles; therefore, the suspension may be considered completely destabilized. Integrating Equation 4.113 results in the following:

$$n = n_0 \frac{1}{1 + n_0 k_c t} \tag{4.117}$$

where n_0 is the initial particle concentration.

4.5.3.2 Flocculation

The aggregation of particles from dispersions by adsorption of large polymer chains on several particles is called flocculation. Depending on the form in which these polymers act on the particle surface, several forms of flocculation may be distinguished. The most important flocculation type is bridging flocculation. Here, small quantities of large chain flocculant are adsorbed over many particles simultaneously producing strong flocs. At higher flocculant concentrations than those needed, they adsorb completely in one particle, leaving less opportunity for adsorbing simultaneously at other particles, which restabilizes the suspension. For this reason, optimal doses of polymeric flocculant exist. The action of cationic polymers on negatively charged particles is a process similar to coagulation, where the surface charge neutralization is predominant over the formation of polymer bridges. This aggregation is called electrostatic flocculation and should be distinguished from the sequence of coagulation followed by bridging flocculation. Flocculants may be applied either after destabilizing the suspension via coagulation or without prior destabilization. The flocculants are known to be not very effective in treating stable suspensions, and so the first option which involves prior destabilization by coagulation is always better.

The mining industry uses anionic over cationic flocculant more frequently on negatively charged particles. In this case, the adsorption mechanism is by covalent bonds or by chemical reaction on the particle surface. This aggregation is called flocculation by salt bond. From all these aggregation mechanisms, bridging flocculation is the most important.

4.5.3.3 Orthokinetic Theory of Flocculation

The optimum flocculation conditions for a given concentrate or tailings refer to the highest settling velocity of the aggregate, that is, to the size and density of the flocs and to the agitation time to obtain them. The aggregation is formed by collision of the primary particles in a hydrodynamic field. Since collision between particles is produced by mechanical energy in a shear field, it is essential that the means of producing the agitation permits the calculation of the shear rate of the flow. The magnitude of shear rate and the application time is important since it, on one side, improves the contact between particles to produce agglomeration and, on the other, destroy the flocs due to the strong agitation. The balance of these two effects is fundamental to obtain flocs of big size and density.

4.5.3.4 Degree of Flocculation

The rate at which flocculation occurs is principally governed by the frequency of collision of particles, but also by the fraction of surface covered by the flocculant. When collision is the result of mechanical or hydrodynamics agitation, it is called *orthokinetic*. According to Healy and La Mer (1966) and Gregory and Nelson (1984), orthokinetic flocculation is produced at the rate:

$$\frac{dn}{dt} = -k_1\left(\theta\right) \times k_2\left(E_c, \dot{\gamma}, d_p\right) n^2 \tag{4.118}$$

Where n is the number of primary particles remaining in the suspension at time t and $k_1(\theta)$ and $k_2(E_c, \dot{\gamma}, d_p)$ are the kinetic constant with values dependent on the flocculant coverage θ, the collision efficiency E_c, shear rate $\dot{\gamma}$, and particle size d_p respectively:

$$k_1(\theta) = \theta(1-\theta) \text{ and } k_2(E_c, \dot{\gamma}, d_p) = \frac{8}{3} E_c \dot{\gamma} d_p^3 \tag{4.119}$$

The maximum value of $k_1(\theta) = 0.25$ is obtained when $\theta = 0.5$; therefore, for this flocculant coverage, Equation 4.115 becomes the following:

$$\frac{dn}{dt} = -\frac{2}{3}\dot{\gamma} E_c d_p^3 n(t)^2 \tag{4.120}$$

This equation is the starting point of most flocculation kinetic studies and was first developed by Smoluchowski in 1917 (Bhattacharjee et al. 1998). To a primary particle of size d_p, other particles primary particles will be added to form and grow an aggregate. Therefore, while the numbers of primary particles vanish, the concentration of primary particles in the suspension diminishes, and the concentration of flocs increases in the system. Equation 4.117 describes the rate of change of primary particles with time, where $n(t)$ is the number of primary particles left in the suspension at time t, and n_0 is the number of original primary particles present in the suspension.

For constant values of $\dot{\gamma}$, d_p, and E_c, the number of primary particles remaining in suspension at time and the number of flocs at time t is as follows:

$$\frac{n_p(t)}{n_0} = \frac{1}{1 + \left(\frac{2}{3}\right) n_0 \dot{\gamma} E_c d_p^3 t} \tag{4.121}$$

$$n_F = \frac{\left(\frac{2}{3}\right) n_0^2 \dot{\gamma} E_c d_p^3 t}{n_{pF}\left(1 + \left(\frac{2}{3}\right) n_0 \dot{\gamma} E_c d_p^3 t\right)} \tag{4.122}$$

The difference between the initial and the actual concentration of primary particles represents the concentration of flocs $\varphi_F = \varphi_0 - \varphi$, then:

$$\varphi_F(t) = \frac{4\varphi_0^2 \dot{\gamma} E_c t}{\pi + 4\varphi_0 \dot{\gamma} E_c t} \tag{4.123}$$

4.5.3.5 Ultra-Flocculation

In general flocculation is performed at shear rates lower than $100\left[s^{-1}\right]$ to avoid the rupture of flocs. Lower values of the sear rate imply longer reaction time. Rulyov (2004, 2008) demonstrates that excellent flocculation is possible at very high shear rates, avoiding the rupture of the flocs if very short times of reaction are considered. In this way the contact between the flocculant and the solid is optimized. For example, a suspension of copper tailings was flocculated at less than $10[s]$.

4.5.3.5.1 *Pelletizing*

The continuous and size continuous population balance model of a well-mixed batch system undergoing coalescence is as follows (Dhawan et al. 2001):

$$\frac{\partial N(v,t)}{\partial t} = -\frac{1}{N_T^{\alpha}}\int_0^{\infty}\lambda(u,v,t)N(ut)N(v,t)\,du + \frac{1}{2N_T^{\alpha}}\int_0^{\infty}\lambda(u,v-u,t)N(ut)N(v-u,t)\,du \tag{4.123}$$

where u and v refers to volume of pellet, $N(v,t)$ is the number density function of particles, $\lambda(u,v,t)$ is coalescence kernel rate, N_T is the total number of particles at time t, and α is space system parameter. The total volume should be conserved in the system since pellets of volume v_i and v_j combine together to form an agglomerate of volume $v_i + v_j$.

Coalescence. The basic postulate for random coalescence is that the particle mass is well mixed, all agglomerates are spherical, and equal probability of coalescence for all size classes, that independent of size. The specific rate constant or coalescence kernel λ, that is, the degree of agglomeration, is independent of the size of the interacting agglomerates, then Equation 5.120 reduces to

$$\frac{dN_i}{dt} = -\lambda N_i(t) + \frac{\lambda}{2N(t)}\sum_{j=1}^{i-1}N_{i-j}(t)N_j(t) \tag{4.124}$$

Summing this equation over all sizes, the rate of depletion of the total number of agglomerates $N(t)$ shows a first order decay kinetics:

$$\frac{dN_i}{dt} = -\frac{\lambda}{2}N_i(t) \text{ with solution } N(t) = N(0)\exp\left(\frac{\lambda}{2}t\right) \tag{4.125}$$

4.6 THE THEORY OF MIXTURES

The system of a small rigid or a deformable particle suspended in a fluid, such as the case of sedimentation and transport of suspensions, may be considered a continuous media. The theory of mixtures is the appropriate tool to study this system (Truesdell and Toupin 1960; Truesdell 1965, 1984; Atkin and Crain 1976; Bowen 1976, 1984; Bedford and Drumheller 1983; Bustos et al. 1999; Concha 2014). This theory postulates that the solid-fluid system may be represented as a superposition of two continuous media, one for the solid and another for the fluid, each following its own movement with restrictions imposed by the interaction between the two components. This idea can be generalized for mixture of many different materials. Each component of the mixture should obey the laws of conservation of mass and momentum, incorporating terms to account for the interchange of mass and momentum between components and that the mixture behaves as a single component body. For the case of a set of solid particles thoroughly mixed with a fluid forming a suspension under the following conditions:

1. All particles are small, with respect to the container, and of the same size form and density.
2. The individual particles and the fluid are incompressible.
3. There is no mass transfer between the solid and the fluid.
4. The only body force is gravity.

The motion of each of the components of the mixture may be described through the local mass and linear momentum balances in all regions where the field variables are continuous:

$$\text{Mass balance of the component:} \frac{\partial}{\partial t}(\rho_\alpha \varphi_\alpha) + \nabla(\rho_\alpha \varphi_\alpha v_\alpha) = \bar{g}_\alpha \tag{4.126}$$

$$\text{Mass balance of the mixture:} \nabla \cdot q = \sum_{2}^{\alpha=1} \frac{\bar{g}_\alpha}{\rho_\alpha} \tag{4.127}$$

Momentum balance of the component:

$$\rho_\alpha \dot{\boldsymbol{v}}_\alpha = \nabla \bullet \boldsymbol{T}_\alpha + \boldsymbol{b}_\alpha + \boldsymbol{m}_\alpha \tag{4.128}$$

$$\text{Restriction:} \sum_{2}^{\alpha=1} (\boldsymbol{m}_\alpha + \bar{g}_\alpha \boldsymbol{v}_\alpha) = 0 \tag{4.129}$$

In these expressions the field variables $\varphi(\boldsymbol{r},t)$, $\boldsymbol{v}_\alpha(\boldsymbol{r},t)$, and $\mathbf{T}_\alpha(\boldsymbol{r},t)$ represent the volume fraction, the velocity and the stresses of the α component, $\boldsymbol{b}_\alpha(\boldsymbol{r},t)$ is the body force and $\boldsymbol{m}_\alpha(\boldsymbol{r},t)$ represent the interaction force between components, and $\bar{g}_\alpha(\boldsymbol{r},t)$ measures the rate at which the other components deliver mass per unit volume to the component α. The operator $\nabla \bullet ()$ represents the divergence of the function ().

At discontinuities, Equations 5.125 and 5.126 must be replaced by the corresponding jump conditions:

$$\sigma[\varphi_\alpha] = [\varphi_\alpha \mathbf{v}_\alpha \mathbf{e}_I] \tag{4.130}$$

$$\sigma[\varphi_\alpha \mathbf{v}_\alpha] = [\varphi_\alpha \mathbf{v}_\alpha (\mathbf{v}_\alpha \mathbf{e}_I)] - [\mathbf{T}_\alpha \mathbf{e}_I] \tag{4.131}$$

Where $\sigma = v_s \bullet e_I$ is the displacement velocity of the discontinuity and $\boldsymbol{e}_I$ is the unit vector in the i direction.

If the solid component is $\alpha = 1 = s$ and the fluid component $\alpha = 2 = f$, and if there is no mass transfer between components and the only body force is gravity, then the momentum balance becomes the following:

$$\text{Volume balance for the solid: } \frac{\partial \varphi}{\partial t} + \nabla\left(\varphi v_s\right) = 0 \quad (1.132)$$

$$\text{Volume balance for the mixture: } \nabla \boldsymbol{q} = 0, \text{with } \mathbf{q} = \varphi \mathbf{v}_s - \left(1-\varphi\right)\boldsymbol{v}_f \quad (1.133)$$

$$\text{Momentum balance for the solid: } \rho_s \varphi \dot{\boldsymbol{v}}_s = \nabla \boldsymbol{T}_s + \rho_s \varphi \boldsymbol{g} + \boldsymbol{m} \quad (1.134)$$

Momentum balance for the fluid:

$$\rho_f\left(1-\varphi\right)\dot{\boldsymbol{v}}_f = \nabla \boldsymbol{T}_f + \rho_f\left(1-\varphi\right)\boldsymbol{g} - \boldsymbol{m} \quad (4.135)$$

In these equations $\varphi(\boldsymbol{r},t)$ is the volume fraction of solids at the position r and time t and $\boldsymbol{v}_s(\boldsymbol{r},t), \boldsymbol{v}_f(\boldsymbol{r},t), \boldsymbol{T}_s(\boldsymbol{r},t), \boldsymbol{T}_f(\boldsymbol{r},t), \boldsymbol{m}(\boldsymbol{r},t)$ are the solid and fluid velocities, the solid and fluid partial stress tensors, the solid-fluid interaction force, and $\boldsymbol{g}$ is the gravitational constant.

In most cases of practical interest, the acceleration terms are negligible (see Bustos et al. 1999 and Concha 2014) and can be disregarded in Equations 4.131 and 4.132. The interaction force $\boldsymbol{m}$ may be separated into a static force $\boldsymbol{m}_E$ and a dynamic force $\boldsymbol{m}_D$ that depends on the motion, so that the Equations 4.131 and 4.132 are reduced to the following:

$$\text{Momentum balance for the solid: } 0 = \nabla \boldsymbol{T}_s + \rho_s \varphi \boldsymbol{g} + \boldsymbol{m}_E + \boldsymbol{m}_D \quad (4.136)$$

$$\text{Momentum balance for the fluid: } 0 = \nabla \boldsymbol{T}_f + \rho_f\left(1-\varphi\right)\boldsymbol{g} - \boldsymbol{m}_E - \boldsymbol{m}_D \quad (4.137)$$

4.6.1 Fluid Component

For all types of fluids, the stresses may be divided into an equilibrium isotropic stress $p_f \boldsymbol{I}$ (the pressure) and an extra stress $\boldsymbol{T}_f^E$ (viscous stress):

$$\mathbf{T}_f = -p_f \mathbf{I} + \mathbf{T}_f^E \quad (4.138)$$

where p_f is the fluid pressure and $\boldsymbol{T}_f^E$ is the viscous stress in the fluid. Replacing Equation 4.135 into the momentum balance of the fluid, we obtain the following:

$$0 = \nabla \boldsymbol{T}_f + \rho_f\left(1-\varphi\right)\boldsymbol{g} - \boldsymbol{m}_E - \boldsymbol{m}_D \quad (4.139)$$

In flow of a solid-fluid mixture, there are two variables related to friction. One is the viscous stress tensor $\boldsymbol{T}_f^E$, representing the friction within the fluid, and the other is the interaction force m corresponding to the friction between the solid and the fluid. Experience has shown (Whitaker 1986) that the friction within the fluid is much smaller than the friction between solids and fluids, and therefore may be neglected in Equation 4.136 leading to:

$$\nabla p_f = \rho_f (1-\varphi)\mathbf{g} - \mathbf{m}_E - \mathbf{m}_D \tag{4.140}$$

Neglecting the viscous term in Equation 4.135 is equivalent to consider the fluid as an elastic fluid.

The fluid pressure p_f is a variable defined over the entire surface of the mixture (remember the solids and fluids are superimposed continuous media), and therefore it is not experimentally measurable. The experimental pressure associated with the flow in a porous media is the pore pressure.

For a solid-fluid mixture, the fluid pressure is continuous across semi-permeable surfaces, that is, surfaces that let the fluid pass but retain the solids. For this reason, the pore pressure $p(z,t)$ may be measured with a manometer. Figure 4.81 shows the way to do it.

$$p(z,t) = \rho_f g\left(h - z(t)\right) \tag{4.141}$$

Dynamic processes involving particulate systems generally depend on the pressure in excess to the hydrostatic pressure, since at the hydrostatic pressure they come to end. Therefore, the concept of excess pore pressure $p_e(z,t)$ is introduced:

$$p_e(z,t) = p(z,t) - \rho_f g_z\left(L - z(t)\right) \tag{4.142}$$

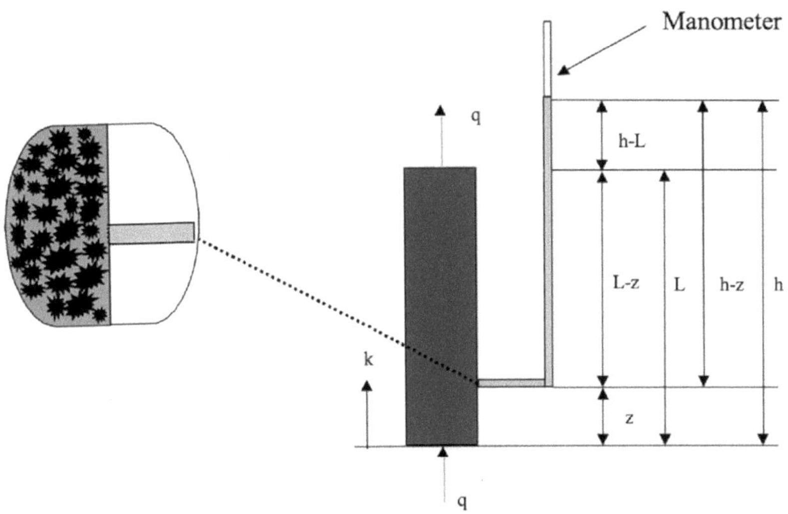

FIGURE 4.80 Measurement of pore pressure.

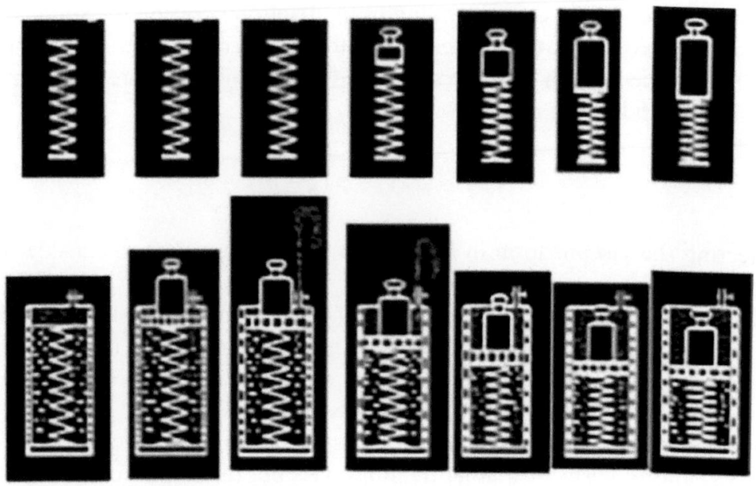

Peso total	0	20	20	20	20	20	20
Presión poros	0	20	20	15	10	5	0
Presión sólido	0	0	0	5	10	15	20
Figura	1	2	3	4	5	6	7

FIGURE 4.81 Mechanical analog of consolidation.

4.6.2 Solid Component

The properties of the solid component in particulate systems depend strongly on its concentration. In diluted systems particles have greater liberty of movement as opposed to more concentrated suspensions, where the motion of one particle is hindered by the presence of other particles. The characteristic concentration at which this change occurs is known as critical concentration φ_c, defined as that concentration at which the particles form a network capable of transmitting compression contact forces. At concentration less than the critical, the mixture is called suspension, and all contact forces are transmitted from particle to particle through the fluid. At concentration greater than the critical, the mixture is called a porous medium, porous bed or sediment, and the stresses are transmitted through both the fluid and from particle to particle by direct contact.

If we consider that the solid component also behaves as an elastic fluid, the tress tensor $\boldsymbol{T}_s$ becomes a pressure that can be expressed by the following:

$$\boldsymbol{T}_s = \begin{cases} -\beta I & \text{for } \varphi \le \varphi_c \\ -\left(\beta + p_s\left(\varphi,\varphi_c\right)\right)\boldsymbol{I} & \text{for } \varphi > \varphi_c \end{cases} \tag{4.143}$$

Where φ_c is the critical concentration, β is a constant, and $\boldsymbol{I}$ is the second order unit tensor. For concentrations less than the critical, the solid stress is constant, and above it, it is a function of the concentration and of the critical concentration. This result (but not the basic principle) coincides with more elaborate constitutive equation of the solid stress tensor, for example, for isotropic elastic solids (Bustos et al. 1999) or a plastic solid (Buscall and White 1987; Green et al. 1996).

From Equation 4.133, the local momentum balance becomes as follows:

$$\nabla p_s = \rho_s \varphi \boldsymbol{g} + \boldsymbol{m}_E + \boldsymbol{m}_D \tag{4.144}$$

4.6.2.1 Solid Effective Stress

For the same reason that in the case of the fluid component, the solid pressure p_s is not an experimentally measurable variable. When a compressive force is applied to a porous medium, the fluid filling the pores between the particles immediately supports the total force exerted, increasing its pore pressure. The excess pore pressure initiates the fluid flow out of the pores, which is accompanied by a reduction of the pore pressure and a progressive transfer of the stress to the solid structure. The rate at which the porous medium deforms depends on the constitutive equation of the solid stress and on the solid-fluid interaction force opposing the fluid flow.

The transient process of stress transfer from the fluid filling the pores to the solid skeleton is called consolidation. We will take the analog to consolidation used in civil engineering. In Figure 4.81 the length of a given spring is shown when it is subjected to different weights, 5, 10, 15, and 20 kg. In the bottom figure, the same spring is submerged in water contained in a frictionless piston having a valve.

(1) Initially the system is in equilibrium with the valve open and without any weight; (2) the valve is closed, and a 20 kg weight is put on the piston. Since water is incompressible, it supports the whole weight, increasing its "pore pressure." The spring conserves its total length. (3) Represents the instant when the valve is open and before the water starts to flow out; (4) with the valve open, the water under pressure starts to flow out through the valve. The rate of outflow is controlled by the amount of excess pore pressure and by the friction through the valve (permeability). As the water leaves the piston, it diminishes the level inside the piston, the spring compresses until it supports 5 kg. At this moment, the water supports the difference of 15 kg; (5) more water flows out, and the piston continue dropping. When the spring has compressed to the size shown in the figure, the spring and the water support 10 kg each; (6) the water continues flowing out, and the spring supports 15 kg while the water supports 5 kg; (7) finally, the spring supports the 20 kg, and the water has hydrostatic pressure (zero excess pore pressure). The complete process consists in transferring the 20 kg load from the water (pore pressure) to the spring (solid effective stress).

4.6.2.2 Total Pressure

In a particulate system, the measurable total pressure p_t is the sum of the pore pressure p and the solid effective stress σ_e. Then, the relationship between the theoretical and the experimental variables are as follows:

$$p_t = p_f + p_s \equiv p + \sigma_e \tag{4.145}$$

The force exerted by the fluid on a surface S of the porous medium is as follows:

$$\int_S p_f dS = \int_S p dS = \int_S p\left(\varepsilon_S dS\right) \tag{4.146}$$

Where S is a section of the porous medium, S_f is the section of the porous medium including the fluid only, and ε_S is the surface porosity (fraction of the total surface formed by the fluid). If the surface porosity ε_S is assumed to be equal to the volume porosity ε (fraction of the total volume of the porous medium occupied by the fluid), $\varepsilon_S \equiv \varepsilon = 1-\varphi$, Equation 4.143 may be written in this form:

$$\int_S p_f dS = \int_S p\left(1-\varphi\right) dS \tag{4.147}$$

From which

$$p_f = \left(1-\varphi\right) p \tag{4.148}$$

Using Equation 4.142 the following is obtained for the solid pressure:

$$p_s = \varphi p + \sigma_e \tag{4.149}$$

Substituting the theoretical pressures p_f and p_f by its experimental equivalents p and σ_e from Equations 4.145 and 1.149 into the local balances of momentum (Equations 4.141 and 4.137) results in the following:

$$\nabla\left(\varphi p\right) + \nabla\sigma_e = \rho_s \varphi \boldsymbol{g} + \boldsymbol{m}_E + \boldsymbol{m}_D \tag{4.150}$$

$$\nabla\left(1-\varphi\right) p = \rho_f \left(1-\varphi\right) \boldsymbol{g} - \boldsymbol{m}_E - \boldsymbol{m}_D \tag{4.151}$$

4.6.2.3 Interaction Force at Equilibrium

Consider the local momentum balance of the fluid at equilibrium, Equation 4.148, where the dynamic interaction force $\boldsymbol{m}_D = 0$ and the hydrostatic pressure is given by $p(z) = \rho_f g\left(L-z\right)$, where L is the height of the water level. Substituting into Equation 4.148 yields the following:

$$\left(1-\varphi\right)\rho_f \boldsymbol{g} - p\nabla\varphi\Big|_{\text{equilibrium}} = \rho_f\left(1-\varphi\right)\boldsymbol{g} - \boldsymbol{m}_E$$

$$\boldsymbol{m}_E = p\nabla\varphi\Big|_{\text{equilibrium}} \tag{4.152}$$

Assuming that the functional form of Equation 4.149 is always valid, we can write for the static interaction force:

$$\boldsymbol{m}_E\left(\boldsymbol{r},t\right) = p\nabla\varphi \tag{4.153}$$

Substituting into the local momentum balances (Equations [4.148] and [4.147]) results in the following:

$$\nabla p = \rho_f \boldsymbol{g} - \frac{\boldsymbol{m}_D}{1-\varphi} \tag{4.154}$$

$$\nabla \sigma_e = \Delta\rho\varphi \boldsymbol{g} + \frac{\boldsymbol{m}_D}{1-\varphi} \tag{4.155}$$

In terms of the excess pore pressure p_e the local momentum balance for the fluid is as follows:

$$\nabla p_e = -\frac{\boldsymbol{m}_D}{1-\varphi} \tag{4.156}$$

Combining Equations 4.152 and 4.153, we can substitute Equation 4.152 by the following:

$$\nabla p_e + \nabla \sigma_e = \Delta\rho \boldsymbol{g} \tag{4.157}$$

4.6.3 Discontinuities

It is well-known that suspensions develop discontinuities. Therefore, the local field equations must be substituted at discontinuities by jump conditions:

$$\sigma[\varphi] = [\boldsymbol{v}_s \boldsymbol{e}_I] \tag{4.158}$$

$$\sigma[\varphi \boldsymbol{v}_s] = [\varphi \boldsymbol{v}_s (\boldsymbol{v}_s \bullet \boldsymbol{e}_I)] - [(\varphi\{p_e + \rho_f g(L-z)\} + \sigma_e)\boldsymbol{e}_I] \tag{4.149}$$

4.6.4 Dynamic Process

Collecting the previous results, we can say that the flow of a particulate systems may be represented by the following field variables: the volume fraction of solids $\varphi(\boldsymbol{r},t)$, the solid and fluid velocities $v_s(\boldsymbol{r},t)$ and $v_f(\boldsymbol{r},t)$, the excess pore pressure $p_e(\boldsymbol{r},t)$, the solid effective stress $\sigma_e(\boldsymbol{r},t)$, and the solid-fluid dynamic interaction force $\boldsymbol{m}_D(\boldsymbol{r},t)$. These variables constitute a *dynamical process* if, for regions where the variables are continuous, they obey the following:

1) The field equations:

$$\frac{\partial\varphi}{\partial t} + \nabla(\varphi \boldsymbol{v}_s) = 0 \tag{4.160}$$

$$\nabla \boldsymbol{q} = 0 \text{ with } \boldsymbol{q} = \varphi \boldsymbol{v}_s + (1-\varphi)\boldsymbol{v}_f \tag{4.161}$$

$$\nabla \sigma_e = \Delta\rho\varphi \boldsymbol{g} + \frac{\boldsymbol{m}_D}{1-\varphi} \tag{4.162}$$

$$\nabla p_e = -\frac{\boldsymbol{m}_D}{1-\varphi} \tag{4.163}$$

2) The jump conditions at discontinuities:

$$\sigma[\varphi] = [\varphi(\boldsymbol{v}_s \bullet \boldsymbol{e}_I)] \tag{4.164}$$

$$\sigma[\varphi \boldsymbol{v}_s] = [\varphi \boldsymbol{v}_s(\boldsymbol{v}_s \bullet \boldsymbol{e}_I)] - \left[\left(\varphi\{p_e + \rho_f g(L-z)\} + \sigma_e\right)\boldsymbol{e}_I\right] \tag{4.165}$$

3) Constitutive equations for the solid-fluid interaction force and for solid effective stress:

$$\boldsymbol{m}_D = \boldsymbol{m}_D(\varphi, v_s, \boldsymbol{q}) \tag{4.166}$$

$$\sigma_e = \sigma_e(\varphi, v_s, \boldsymbol{q}) \tag{4.167}$$

REFERENCES

Allen, T. 1981. *Particle Size Measurement*, 3rd ed. London: Chapman & Hall.

Allen, T. 1994. *Particle Size Measurement*, 4th ed. London: Chapman & Hall. pp. 192–216.

Atkin, R.J., and Crain, R.E. 1976. Continuum theories of mixtures: Basic theory and historical development. *Quarterly Journal of Applied Mathematics* 29:209–244.

Ayling, B., Rose, P., Petty, S., Zemach, E., and Drakos, P. 2012. QEMSCAN® (Quantitative Evaluation of Minerals by Scanning Electron Microscopy): Capability and application to fracture characterization in geothermal systems. In *37th Workshop on Geothermal Reservoir Engineering*. Stanford Geothermal Program Workshop Report SGP-TR-194, vol. 1. Stanford, CA: Stanford Geothermal Program. pp. 596–606.

Bahar, G., and Khalilib, B.M. 2015. Particle size analysis on wide size distribution powders; effect of sampling and characterization technique. *Advanced Powder Technology* 26(1):200–207.

Bakalarz, A., Duchnowskal, M., and Luszczkiewicz, A. 2017. Influence of liberation of sulphide minerals on flotation of sedimentary copper ores. *E3S Web of Conferences* 18:01025. https://doi.org/10.105/e3conf/20171801025

Barrios, G.K.P., Jiménez-Herrera, N., and Tavares, L.M. 2020. Simulation of particle bed breakage by slow compression and impact using a DEM particle replacement model. *Advanced Powder Technology* 31(7):2749–2758.

Bascur, O.A. 1990. Profit-based grinding controls. *Minerals and Metallurgical Processing*:9–15.

Bascur, O.A. 1991. Integrated grinding/flotation controls and management. Presented at the Copper 91/Cobre 91 International Symposium, Ottawa, Ontario.

Bascur, O.A. 2013. A dynamic flotation framework for performance management. In *Separation Technologies for Minerals, Coal and Earth Resources*. Edited by C. Young and G. Luttrel. Englewood, CO: SME. pp. 40–45.

Bascur, O.A. 2023. Digital twins in mining and mineral processing. In *26th World Mining Conference Proceedings, Resourcing Tomorrow, Creating Value for Society, Artificial Intelligence invited lecture*, Brisbane. www.wmc2023.org

Bascur, O.A., and Herbst, J.A. 1982. Dynamic modelling of a flotation cell with a view toward automatic control. Presented at XIV IMPC, Toronto, Canada. www.researchgate.net/

publication/284604831_Dynamic_Modelling_of_a_Flotation_Cell_with_a_view_towards_Automatic_Control
Bascur, O.A., and Retamal, F. 2021. Integrated rock processing strategies in the digital age. *Presented at Minería Digital*, Gecamin, Santiago, October 17–19.
Bascur, O.A., and Retamal, F. 2023. Increasing water recovery to maximize copper production in low grade ores. In *Proceedings of the 62 Annual Conference of Metallurgists, COM 2023*, Springer Nature, Ontario, Canada.
Bascur, O.A., Retamal, F., and Espinosa, S. 2022. Treating low grades ores using a digital in mining and mineral processing plants. In *Proceedings of the Copper 2022 International Conference*, IIMChile, Santiago. www.copper2022.cl
Basim, G.B., and Khalilib, M. 2015. Particle size on wide distribution powders; effect of sapling and characterization technique. *Advanced Power Technology* 26:200–207.
Baumbgardt, S., Buss, B., May, P., and Schubert, H. 1975. On the comparison of results in single grain crushing under different load. In *Proceedings of the XI International Mineral Processing Congress*. Westmount, QC: Canadian Institute of Mining, Metallurgy and Petroleum. p. 3.
Bedford, A., and Drumheller, D.S. 1983. Theories of immiscible and structured mixtures. *International Journal of Engineering Science* 21(8):863–960.
Bergstrom, B.H., and Sollenberger, C.L. 1961. Kinetic energy effect in single particle crushing. *AIME Transactions* 220:373–379.
Bhattacharjee, S., Elimelech, M., and Borkovec, M. 1998. DLVO interaction between colloidal particles: Beyond Derjaguin (Invited Paper for a Special Issue in Honor of Professor Egon Matijevic). *Croatica Chemica Acta* 71:883–903.
Bourgeois, F. 1993. Microscale modelling of comminution processes. PhD thesis, University of Utah, Salt Lake City, UT.
Bourgeois, F., King, R.P., and Herbst, J.A. 1992. Low impact energy single-particle fracture. In *Comminution: Theory and Practice*. Edited by S. Kawatra. Littleton, CO: SME.
Bowen, R.M. 1976. Theory of mixtures. In *Continuum Physics*, vol. 3. Edited by A.C. Eringen. New York: Academic Press.
Broadbent, S.R., and Calcott, T.G. 1956a. A matrix analysis of processes involving particle assemblies. *Philosophical Transactions of the Royal Society A* 249:960.
Broadbent, S.R., and Calcott, T.G. 1956b. A matrix representation of breakage. *Journal of the Institute of Fuel* 29:528–539.
Brough, C., Strongman, J., Fletcher, J., Zając, M., Garside, R., Garner, C., and Rose, L. 2019. Operational mineralogy: An overview of key practices in sample analysis, sample preparation and statistics. In *Life with Ore Deposits on Earth: 15th SGA Biennial Meeting*, vol. 4. Geneva: Society for Geology Applied to Mineral Deposits.
Bürger, R., Bustamante, O., Fulla, M.R., and Rivera, I.E. 2018. A population balance model of ball wear in grinding mills: An experimental study. *Minerals Engineering* 128: 288–293.
Buscall, R., and White, L.R. 1987. The consolidation of concentrated suspensions. *Journal of the Chemical Society, Faraday Transactions* 1(83):873–891.
Bustos, M.C., Concha, F., Bürger, R., and Tory, E.M. 1999. *Sedimentation and Thickening: Phenomenological Foundation and Mathematical Theory*. Dordrecht: Kluwer. pp. 46–47.
Concha, F. 2001. *Manual de Espesamiento y Filtración*. Concepción, Chile: Cettem.
Concha, F. 2014. *Solid Fluid Separation in the Mining Industry*. Cham: Springer.
Concha, F., Aravena, J., and Tovar, P. 1973a. Descripción del tamaño y tamaño promedio de un sistema de partículas. In *Proceedings of the I Congreso Latinoamericano de Ingeniería de Minas y Metalurgia Extractiva*, Santiago. pp. 1–11.
Concha, F., Reyes, J., and López, J. 1973b. Determinación experimental de factores de forma de partículas. In *Proceedings of the I Congreso Latinoamericano de Ingeniería de Minas y Metalurgia Extractiva*, Santiago. pp. 13–18.

Cutting, G.W. 1979. Development of a multicomponent-multisize liberation model. *The Chemical Engineer* 351:845.

Cutting, G.W., and Barber, S.P. 1980. Characterization and prediction of mineral release during grinding processes: Application to a lead/zinc ore. In *European Symposium on Particle & Technology Preprint*. Edited by K. Schoenert. Kalruhue, Germany.

Dhawan, N., Rashidi, S., and Rajamani, R.K. 2001. Population balance model for crushed ore agglomeration for heap leach operations. *KONA Powder and Particle Journal* 31:200–213.

Egarr, D.A., Horton, L., Rice, H., et al. 2016. Development of flocculation models for improving water treatment. 10th European Waste Water Management Conference, Manchester, October 11–12. https://eprints.whiterose.ac.uk

Ehrlich, R., and Weinberg, B. 1970. An exact method for the characterization of grain shape. *Journal of Sedimentary Petrology* 40(1):205–212.

Epstein, B. 1947. The mathematical description of certain breakage mechanism leading to the logarithm-normal distribution. *Journal of the Franklin Institute* 244:471–477.

Epstein, B. 1948. Logarithm-normal distribution in breakage of solids. *Journal of Industrial and Engineering Chemistry* 40:2289.

Evident. 2016. Olympus Stream 2.2 [software]. www.olympus-ims.com.

Fandrich, R., Ying Gu, Debra Burrows, Kurt Moeller. (2007). Modern SEM-based mineral liberation analysis. *International Journal of Mineral Processing* 84(1–4):310–320. http://dx.doi.org/10.1016/j.minpro.2006.07.018. Accessed 19 October 2007.

Gaudin, A.M. 1939. *Principles of Mineral Processing*. New York: Mc Graw-Hill, Book.

Gianfrancesco, A. 2017. *Materials for Ultra-Supercritical and Advanced Ultra-Supercritical Power Plants*. London: Woodhead.

Gilson Company. 2023. ASTM test sieves. www.globalgilson.com/astm-test-sieves. Accessed August 2023.

Gilvarry, J.J. 1961. Fracture of brittle solids. I. Distribution function for fragment size in single fracture (theoretical). *Journal of Applied Physics* 32(3):391–399.

Gilvarry, J.J., and Bergstrom, B.H. 1961a. Fracture and comminution of brittle solids, theory and experiment. *AIME Transactions* 220:380–389.

Gilvarry, J.J., and Bergstrom, B.H. 1961b. Fracture of brittle solids. II. Distribution function for fragment size in single fracture (experimental). *Journal of Applied Physics* 32:400.

Grammatikopoulos, T. 2018. High-definition mineralogy by QEMSCAN: From exploration to focus on automated/process mineralogy. https://www.e4m.fsg.ulaval.ca/fileadmin/documents/Evenements/T_Grammatikopoulos2018.pdf

Green, M.D., Eberl, M., and Landman, K.A. 1996. Compressive yield stress of flocculated suspensions: Determination via experiment. *AIChE Journal* 42(8):2308–2318.

Gregory, J., and Nelson, D.W. 1984. A new optical method for flocculation monitoring. In *Solid-Liquid Separation*. Chichester: Society of Chemical Industry (E. Horwood). pp. 172–184.

Griffith, A.A. 1921. The phenomena of rupture and flow in solids. *Philosophical Transactions of the Royal Society A* 6:163–197.

Gutsche, O., Kapur, P.C., and Fuerstenau, D.W. 1993. Comminution of single particles in a rigidly-mounted roll mill. Part 3: Particle interaction and energy dissipation. *Powder Technology* 76:271.

Han, K., Xiao, J., Zhang, Z., Wang, J., and Hu, X. 2020. Effect of particle size distribution on flocculation and its growth in cement-ground limestone suspensions. *Construction and Building Materials* :1–15. https://doi.org/10.1016/j.conbuildmat.2020.120047

Hatch, T., and Choate, S.P. 1929. Statistical description of the size properties of non-uniform particulate substances. *Journal of the Franklin Institute* 207:369–387.

Healy, T.W., and La Mer, V.K. 1966. *J. Phys Chem.* New York: Elsevier. 68:3535.

Herbst, J.A. 1979. Particle size analysis. In *Mineral Processing Handbook*. New York: AIME.

Herbst, J.A., Lo, Y.C., and Rajamani, K. 1985. Population balance model prediction of the performance of large-diameter mills. *Mineral and Metallurgical Engineering* :1–31.

Herbst, J.A., Rajamani, R.K., Lin, C.L., and Miller, J.D. 1988. Development of a multicomponent multisize mineral liberation model. *Minerals Engineering* 1(2):97–111.

Hiramatsu, Y., and Oka, Y. 1966. Determination of the tensile strength of rock by a compression test of an irregular test piece. *International Journal of Rock Mechanics and Mining Sciences* 3:89.

Hoffmann, N. 1976. The size distribution during the fracture of binary and tertiary mixture. *Chemie Ingenieur Technik* 48(4):329.

Hogg, R., Healy, T.W., and Fuerstenau, D.W. 1966. Mutual coagulation of colloidal dispersions. Transactions of the Faraday Society 62.

Hulburt, H.M., and Katz, S. 1964. Some problems in particle technology: A statistical mechanical formulation. *Chemical Engineering Science* 19:555–574.

Isizer. 2023. Fisher sub sieve sizer principle 201–6. *Particle Size Analyzers*, July 26. www.isizer.com/fisher-sub-sieve-sizer-principle/. Accessed August 2023.

Kapur, P.C., and Fuerstenau, D.W. 1969. A coalescence model for granulation. *Industrial & Engineering Chemistry Process Design and Development* 8:56–62.

Kapur, P.C., and Runkana, V. 2003. Balling and granulation kinetics revisited. *International Journal of Mineral Processing* 72(1–4):417–427.

Kapur, P.C., Gutsche, O., and Fuerstenau, D.W. 1993. Comminution of single particles in a rigidly-mounted roll mill. Part 3: Particle interaction and energy dissipation. *Powder Technology* 76:271–276.

Karpinski, M.M., and Tervo, R.M. 1964. Fragmentation analyses—fundamental processes. *AIME Transactions* 229:126.

King, R.P. 2001. *Modeling and Simulation of Mineral Processing Systems*. Oxford: Butterworth-Heinemann. p. 140.

Klotz, K., and Shubert, H. 1982. Crushing of single irregularly shaped particles by compression: Size distribution of progeny particles. *Powder Technology* 32:129.

Koch, P., and Rosenkranz, J. 2017. *Texture-Based Liberation Models for Comminution*. Lulea: Diva, Luleå University of Technology Division of Minerals and Metallurgical Engineering.

Lamberg, P., and Vianna, S. 2009. *A Technique for Establishing Multiphase Particle Balance in Flotation Circuits*. Indooroopilly, Brisbane: Julius Kruttschnitt Mineral Research Centre (JKMRC).

Laskowski, J., and Pugh, R. 1992. Dispersions stability and dispersing agents. *Developments in Mineral Processing* 12:115–171.

Laslett, G.M., Sutherland, D.N., Gottlieb, P., and Allen, N.R. 1990. Graphical assessment of a random breakage model for mineral liberation. *Powder Technology* 60(2):83–97. https://doi.org/10.1016/0032-5910(90)80135-L

Liu, J., and Schönert, K. 1996. Modelling of interparticle breakage. *International Journal of Mineral Processing* 44(5):101.

Lyklema, J. 2013. Coagulation by multivalent counterions and the Schulze-Hardy rule. *Journal of Colloid and Interface Science* 392:102–104.

Menacho, J.M., and Concha, F. 1985. Ball wear model and ball size optimization in tumbling mills. *Advances in Mineral Processing* 1:478.

Meyer, K. 1980. *Pelletizing of Iron Ores*. Düsseldorf: Springer Verlag.

Michaud, D. 2015. Cyclosizer. *Metallurgist 911*, June 18. www.911metallurgist.com/blog/cyclosizer. Accessed August 2023.

Micromeritics. n.d. SediGraph III Plus 5125. www.micromeritics.com/sedigraph-iii-plus/. Accessed August 2023.

Morrell, S., and Morrison, R.D. 1989. Ore charge, ball load and material flow effects on a energy based SAG mill model. In *Proceeding Advance in Autogenous and SAG*

Technology. Edited by A.L. Mular and G.E. Agar. Vancouver: University of British Columbia. pp. 697–712.

Napier-Munn, T.J., Morrell, S., Morrison, R.D., and Kojovic, T. 1996. *Mineral Comminution Circuits. Their Operation and Optimization*. Brisbane: Julius Kruttschnitt Mineral Research Center.

Nied, R. 2007. Rotor impact mills. In *Handbook of Powder Technology. Vol. 12: Particle Breakage*. Edited by A.D. Salman, M. Ghadiri, and M.J. Hounslow. New York: Elsevier.

NIST (National Institute of Standards and Technology). n.d. Weibull distribution. In *Engineering Statistics Handbook*. www.itl.nist.gov/div898/handbook/eda/section3/eda3668.htm. Accessed August 2023.

Oliveira, G.J.R. 2021. A Aplicabilidade da Microtomografia de Rayos-X para a Determinação de Liberação Mineral. PhD thesis, Federal University of Minas Gerais, Brazil.

Pettyjohn, E.S., and Christiansen, E.B. 1948. Drag coefficient of particles. Chemical Engineering Progress 44:157.

Powell, M., Hildenh, M., Ballantyne, G., Liu, J., and Tavares, L.M. 2014. The appropriate, and inappropriate, application of the JKMRC t_{10} relationship. Presented at IMPC XXVIII, Santiago.

Quezada, G.R., Ayala, L., Leiva, W.H., Toro, N., Toledo, P.G., Robles, P., and Jeldres, R.I. 2020. Describing mining tailing flocculation in seawater by population balance models: Effect of mixing intensity. *Metals* 10(2):240.

Randolph, A.D., and Larson, M.A. 1971. *Theory of Particulate Processes*. New York: Academic Press. pp. 12–40.

Rulyov, N. 2004. Ultra-flocculation: Theory, experiment, applications conference: "Particle size enlargement in mineral processing" at Hamilton, Ontario, Canada. In *Proceedings of the 5-th UBC—McGill Biennial International Symposium on Fundamentals of Mineral, Hamilton, Ontario, Canada, August 22–25*. Edited by J.S. Laskowski. Montreal, Canada: Canadian Institute of Mining, Metallurgy and Petroleum.

Rulyov, N.N. 2008. Turbulent microflocculation of ultrafine minerals. *Mineral Processing and Extractive Metallurgy IMM Transaction Section C* 117(1):32–37. https://doi.org/10.1179/174328507X198744

Rulyov, N.N., Korolyov, V.J., and Kovalchuk, N.M. 2009. Ultra flocculation of quarts suspensions: Effect of shear rate, particle size distribution and solids content. *Mineral Processing and Extractive Metallurgy IMM Transactions Section C* 118(3):175–181. https://doi.org/10.1179/174328509X408243

Rumpf, H. 1973. Physical aspects of comminution and new formulation of a law of comminution. *Powder Technology* 7:145.

Saeidi, F. 2010. New approach for characterizing a breakage event as a multi-stage process. PhD thesis, University of Queensland, Australia.

Saho, R. 2006. An investigation of single particle breakage tests for coal handling system of the Gladstone port. *Powder Technology* 161:158.

Sastry, K.V., and Concha, F. 1985. Population balance with applications in mineral processing. In *Proceedings of the International Symposium in Mineral Processing*. Concepción: University of Concepción, Chile.

Schönert, C. 1980. *Zerkleinern, Mechanishe Verfahrenstechnik*, 4th ed. Karlsruhe: Universität Karlsruhe.

Schönert, K. 1972. Role of fracture physics in understanding comminution phenomena. *AIME Transactions* 252:21.

Schönert, K. 1991. Advances in comminution fundamentals and impacts on technology. *Aufbereitungs Technologie* 32(9):487–494.

Schönert, K. 1996. The influence of particle bed configurations and confinements on particle breakage. *International Journal of Mineral Processing* 44–45:1–16.

Schwarz, M.P., and Shane, K.C. 1969. Measurements of the particle shape with Fourier analysis. *Sedimentology* 13:213–231.

Sepúlveda, J., and Herbst, J.A. 1978. A population balance approach to the modeling of multistage continuous leaching systems. In *AIChE Symposium Series: Fundamental Aspects of Hydrometallurgical Processes*. New York: American Institute of Chemical Engineers. pp. 41–65.

SGS Minerals Services. 2013. JK drop weight test. T3 SGS 151. https://es.scribd.com/document/455245524/SGS-MIN-WA232-JK-Drop-Weight-Test-EN-11-pdf.

Shi, Z., Zhang, G., Zhang, Y., He, T., and Pei, G. 2019. Modeling of flocculation and sedimentation using population balance equation. *Journal of Chemistry* 2019:9187204. https://doi.org/10.1155/2019/9187204

Sikong, L. 1990. Breakage behavior of fine particles of brittle minerals and coals. *Powder Technology* 61(1):51–57.

Sympatec. n.d. Helos/BR, Rodos/L, Vibri/L. www.sympatec.com/en/particle-measurement/sensors/laser-diffraction/helos/helos-br-rodos-l-vibri-l/. Accessed August 2023.

Tarján, G. 1974a. Gates-Gaudin-Schuman distribution function: A contribution to particle size distribution functions. *Powder Technology* 10(1–2):73–77.

Tarján, G. 1974b. Rosin-Rammler distribution function: A contribution to particle size distribution functions. *Powder Technology* 10(1–2):73–77.

Tavares, L.M. 2000. Role of particle microstructure in comminution. *Developments in Mineral Processing* 13:C4-106.

Tavares, L.M. 2007. Breakage of single particles: Quasi-static. In *Handbook of Powder Technology. Vol. 12: Particle Breakage*. Edited by A.D. Salman, M. Ghadiri, and M.J. Hounslow. New York: Elsevier. pp. 3–66.

Tavares, L.M., and King, R.T. 1998. Single-particle fracture under impact loading. *International Journal of Mineral Processing* 54(1):1–28.

Trefalt, G., István Szilágyi, I., and Borkovec, M. 2020. Schulze-Hardy rule revisited. *Colloid and Polymer Science* 298:961–967.

Truesdell, C. 1965. Sulle basi de la termomecanica. *Red. Acad. Lincei* 22:33–88 (1957 English Translation in Rational mechanics of materials. *Int. Sci. Rev.* :292–305).

Truesdell, C. 1984. *Rational Thermodynamics*, 2nd ed. New York: Springer Verlag.

Truesdell, C., and Toupin, R.A. 1960. The classical field theories of mechanics. In *Handbook of Physics*, vol. III-1. Edited by S. Flugge. New York: Springer.

Tyler, W.S. n.d. Ro-Tap sieve shaker. https://wstyler.com/particle-analysis/ro-tap-sieve-shaker/. Accessed August 2023.

Wadell, A. 1835. Volume, shape and roundness of quartz particles. *Journal of Geology* 43:250–280.

Weichert, R., and Herbst, J.A. 1986. An ultrafast load cell device for measuring particle breakage. In *Proceedings, 1st World Congress on Particle Technology*, Elsevier, Nuremberg, Germany. pp. 3–14.

Whitaker, S. 1986. Flow in porous media: 1 a theoretical derivation of Darcy's law. *Transport in Porous Media* 1:3–23. http://dx.doi.org/10.1007/BF01036523

Wiegel, R.L. 1976. Integrated size reduction-mineral liberation model. *Trans. Soc. Mining Eng.* 260:147–152.

Wills, B., and Napier-Munn, T. 2011. *Mineral Processing Technology: An Introduction to the Practical Aspects of Ore Treatment and Mineral Recovery*. Maryland Heights, MO: Elsevier.

Yashima, S., Kanda, Y., and Sano, S. 1987. Relationships between particle size and fracture energy or impact velocity required to fracture as estimated from single particle crushing. *Powder Technology* 5:277.

FURTHER READING

Kubik, R. 2018. Creation and development of Qemscan analysis procedure based on Polish copper ores investigation. In *Proceedings of the 18th International Multidisciplinary Scientific GeoConference (SGEM)*. https://www.researchgate.net/publication/327970529

LabX. n.d. Beckman coulter Z1 particle counter. www.labx.com/product/beckman-coulter-z1-particle-counter. Accessed August 2023.

Truesdell, C., and Noll, W. 1965. The non-linear theory of mechanics. In *Encyclopedia of Physics*, vol. 111/3. Edited by S. Fiogge. New York: Springer Verlag.

5 Physical Chemistry of Interfaces

5.1 INTRODUCTION

In the previous chapters, we have seen that particulate systems are a constant in every operation of mineral processing. In each of these operations, the particles are mixed with water at different concentration in which they exposed their new surface originated by breakage and grinding. As particle size diminishes, the exposed surface area increases, and every factor affecting the surface become important. During the whole process pH is important, and reagents are added to regulate this parameter to the correct value in each operation. During flotation, collectors and frothers are added and in solid-liquid separation, coagulant and flocculant are extensibility used. All the phenomena brought about by the effect of pH and the addition of chemical reagent belong to the area of physical chemistry of surfaces, and the knowledge of this science is important to the scientist and engineers studying an operating mineral processing plant.

The mixture of small particles, solid, liquid, or gaseous, in a solvent are called a *colloidal dispersion* if their sizes are in the range of 1 nanometer to 1 micron. Smaller particles are called *solutions* and bigger are *suspensions*. Colloids systems can be reversible with stable products or irreversible if they cannot be returned to its original components. Colloidal systems are important for subject both from academics and industry. These systems include solid-liquid, liquid-liquid (emulsions), and gas-liquid dispersions (foams). Interactions of suspensions of solid particles with other solids, liquids, or gases, such as minerals, liquid drops, and gas bubbles in the presence of chemical reagents, are very important in the mineral and other important industries. Figure 5.1 shows some engineering applications of different types of interactions between two phases.

When fluids come in close contact with a solid forming a suspension, the interaction forces between the molecules of these materials govern the possibility of certain phenomena to occur such as adsorption, wetting, spreading, and coating, which acquire importance in several mineral recovery processes. These forces are *attractions* and *repulsion* forces, the importance of which depend on the closeness between the space separating them. The stability of those systems depends on the balance of the forces, the principal of which are the *van der Waals attraction*, the *boundary layer repulsion, the disjoining pressure*, and the *steric and hydrophobic interaction forces* (Ruylov 2020).

DLVO (named after Boris Derjaguin, Lev Landau, Evert Verwey, and Theodoor Overbeek) interactions (Derjaguin and Landau 1941; Verwey and Overbeek 1948) include the *van der Waals attraction VDW and the electrostatic interactions EDL*. Non-DLVO interactions, those that are not included in the original DLVO theory, include *steric and bridging interaction* due to polymers adsorption, solvation forces, hydrophobic and hydration and oscillatory structural interactions.

DOI: 10.1201/9781032614212-5

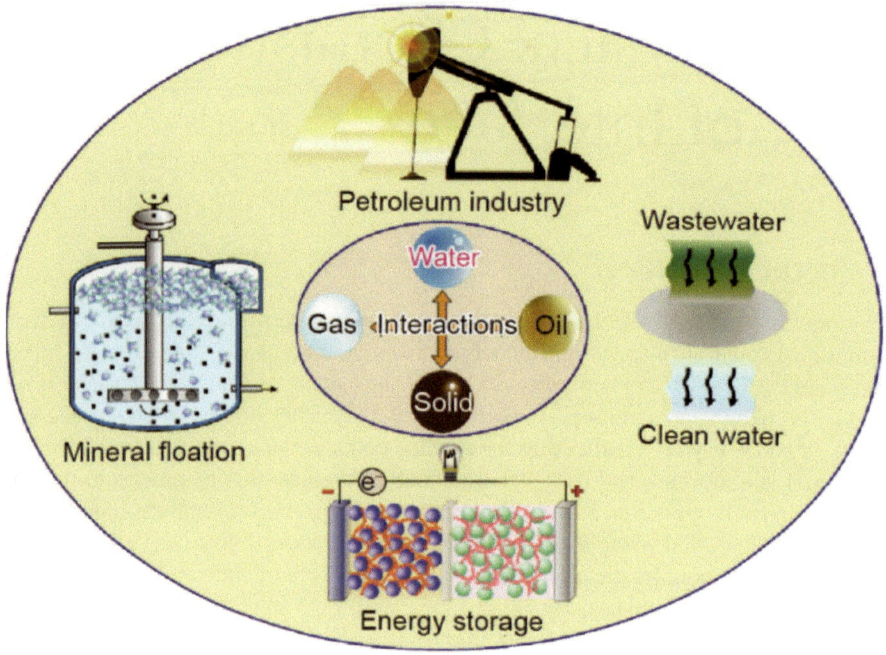

FIGURE 5.1 Schematic of surface interactions involved in various engineering processes.

Source: Zhang, J., and Zeng, H. 2021. Intermolecular and surface interactions in engineering processes. *Engineering* 7(1):63–83.

5.2 INTERACTION FORCES

The distances between solid and the fluid particles in a suspension, which is thickness of the layer that separates them, and their electrical charges determine the type of forces that attracts or repulse them.

5.2.1 DLVO Theory

The classical DLVO theory (Derjaguin and Landau 1941; Verwey and Overbeek 1948) has been widely applied to describe the stability of colloidal particles in aqueous media. The DLVO theory is a theory of colloidal dispersion stability in which zeta potential is used to explain that as two particles approach one another, their ionic atmospheres begin to overlap, and a repulsion force is developed. The balance of these forces results in the electrostatic stability of a colloid. It suggests that the interface between two particles at a close distance h in a suspension is separated by a thin liquid film where the pressures are greater than at the bulk (Lopez et al. 2015). This excess pressure in the film is called *disjoining pressure* that represents the change in the Gibbs free energy in the liquid film (Derjaguin 1932). The force components are the van der Waals attraction F_{VDW} and the electrostatic repulsive interaction Electric Double Layer EDL.

5.2.1.1 Van der Waals Interactions

The van der Waals (VDW) interactions originate from the fluctuation of the electric dipole moments of two molecules while they are approaching each other, including induction, orientation, and dispersion forces, thus VDW forces are always present at molecules and surfaces, where attractions forces $F_{DWV}(D)$ and energies W related by $F_{DWV}(D) = -dW / dD$ are given by

$$F_{VDW}(D) = -\frac{6}{D^7}\underbrace{\left[\frac{3kT}{4}\left(\frac{\varepsilon_1 - 1}{\varepsilon_1 + 1}\right)^2 + \frac{3h_p v_e}{16\sqrt{2}}\frac{\left(n_1^2 - 1\right)^2}{\left(n_1^2 + 1\right)^{3/2}}\right]}_{A_{131}} \tag{5.1}$$

Where A_{131} is the DLVO Hamaker constant for two equal spheres of the same size; D is the dimensions of the particles; ε_1 and ε_3 are the dielectric permittivity; n_1 and n_3 are refractive indices respective (for vacuum/air, $\varepsilon_3 = 1$ and $n_3 = 1$); h_p, k, and T are the Plank constant, the Boltzmann constant, and the absolute temperature; v_e is the maximum electronic ultraviolet adsorption frequency; and e is the electrostatic charge. The VDW interactions dominate the wetting behavior of nonpolar films on solid surfaces.

5.2.1.2 Electrostatic Interactions F_{EDL}

When the media between the particles are aqueous solutions, the long-range electrostatic repulsion forces (EDL) are present together with the VDW attraction forces at the surface. Different models were developed to describe the EDL, such as the Helmholtz, Gouy-Chapman, and Stern models (Gouy 1917; Stern 1924) introduced the *double-layer theory* formulating the *dilute layer* concept, which was later modified by Stern with the introduction of the concept of *specifically adsorbed counter ions*, which received the name of the *Stern plane* (Kopeliovich 2019). Figure 5.2 shows the electric double layer schematics, as described by Wang et al. (2011).

The double layer theory indicates that most particles suspended in water possess surface charge due to the dissociation of surface groups such as OH^-, $COOH^+$, $NaSO_4^+$ and adsorption of certain ionic species such as surfactants. Specific ions adsorbed at the surface form a charged surface ψ_0, while an unequal distribution of counter and co-ions compensate the surface charge. The potential at the surface decreases linearly to the value ψ_d at the center of the specifically adsorbed counter ions and then decrease exponentially with distance from the surface, reaching zero at the bulk solution.

The *critical coagulation concentration* (CCC) is that condition when salts with ionic charge opposite to that of the particles neutralize the particle surface charge. The aggregation efficiency follows Schulze-Hardy rule that propose that the critical coagulation concentration decreases as the inverse sixth power of the counterion valence (Trefalt et al. 2017).

$$CCC = \frac{K}{A^2 z^6}\left(\tanh\left(z\frac{e\psi^d}{kT}\right)\right)^4 \tag{5.2}$$

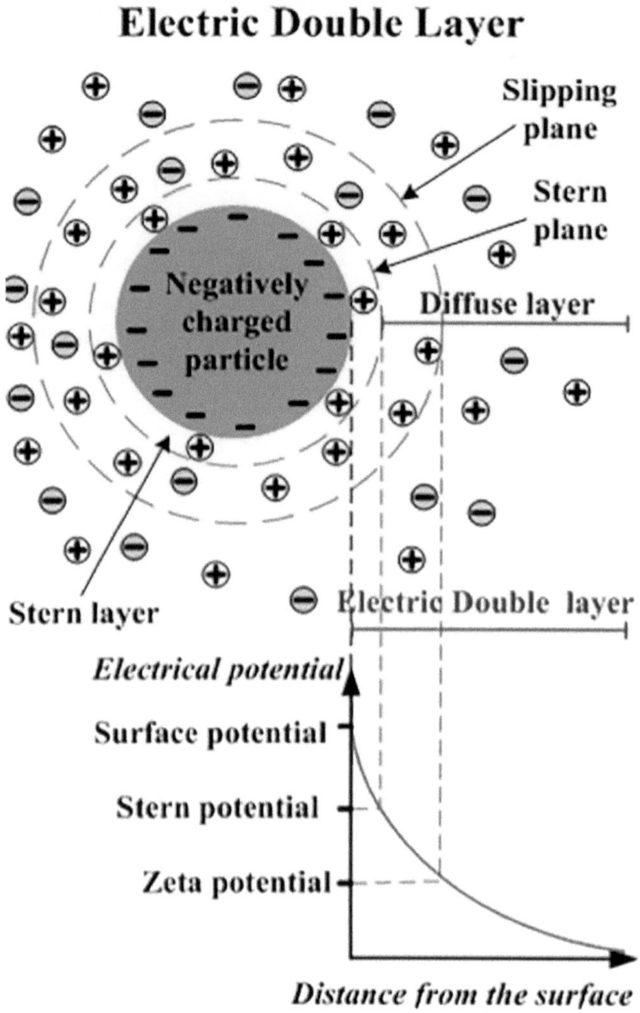

FIGURE 5.2 Electrical double layer (EDL) potential distribution near a particle surface submerged in an electrolyte.

Source: Wang, R-M., Zheng, S-R., and Zheng, Y-P. 2011. Chapter 4: Interface of polymer matrix composites. In *Polymer Matrix Composites and Technology.* Edited by R-M. Wang, S-R, Zheng, and Y-P. Zheng. Woodhead Publishing Series in Composites Science and Engineering. Cambridge, UK: Woodhead. pp. 169–209, 547–548.

Where A is the Hamaker constant and ψ the surface potential.

When a solution flows along the particle, it becomes stagnant at a certain distance from the particle surface. This distance is called the shear plane. The difference of electrical potential between the shear plane and the bulk of the solution is called *zeta potential* ζ, which can be measured experimentally by several commercial instruments.

For diffuse potentials multivalent counterions, the CCC increases with the surface charge density, inducing the strong decrease of the CCC with valence z.

$$CCC = \frac{K}{A^2 z^2}\left(z \exp\frac{\zeta}{kT}\right)^4 \tag{5.3}$$

Quantitatively, the extent of this decrease in z is generally system dependent. Hence, the general conclusion is that the scaling of CCC with z is stronger than z^2 depending on the nature of the counterion (Lyklema 2005).

For two equal spheres of the same size, the EDL interaction forces F_{EDL} is given by

$$F_{EDL} = \frac{z_1 z_2 e^2}{4\pi\varepsilon\varepsilon_0 D^2}\frac{1+\kappa D}{1+\kappa\sigma}\exp^{-\kappa(D-\sigma)} \tag{5.4}$$

Where κ is the reciprocal of the distances between the surfaces given by

$$\frac{1}{\kappa} = 64\pi\varepsilon_0\varepsilon\left(\frac{kT}{z_i e}\right)^2 \tanh^2\left(\frac{ze\psi_0}{4kT}\right) \tag{5.5}$$

Where ε_0 and ε are the dielectric permittivity of vacuum and medium if the medium respective, z_i is the electrolyte ions valence, ψ_0 the surface potential and T the absolute temperature, e is the electrostatic charge, ψ_0 is the surface potential, and σ are the sphere diameters.

For a monovalent electrolyte, such as NaCl:

$$\frac{1}{\kappa} = 9.22\times 10^{-11}\tanh^2\left(\frac{\psi_0}{130}\right), \text{ in}[\text{J}/\text{m}] \tag{5.6}$$

The total DLVO energy G_T of interaction between two particles is given by the sum of attraction forces $G_A = F_{VDW}$ at every distance h and the repulsive electrical forces $G_{el} = F_{EDL}$.

$$G_T = G_A + G_{el} \tag{5.7}$$

This expression represents the kinetic stability of colloidal dispersions and is pictured schematically in Figure 5.3.

This curve shows that increasing the ion concentration, which would diminish the distance h between the particle and increasing the energy barrier, would turn coagulation or flocculation more difficult. Fast flocculation would occur in the absence of an energy barrier and low flocculation in the presence of a barrier (Chemistry Stack Exchange 2020).

The results of the DLVO theory, with the VDW and EDL interaction forces and energies between two bodies of different geometries with separation D in terms of the Hamaker constant are shown in Table 5.1.

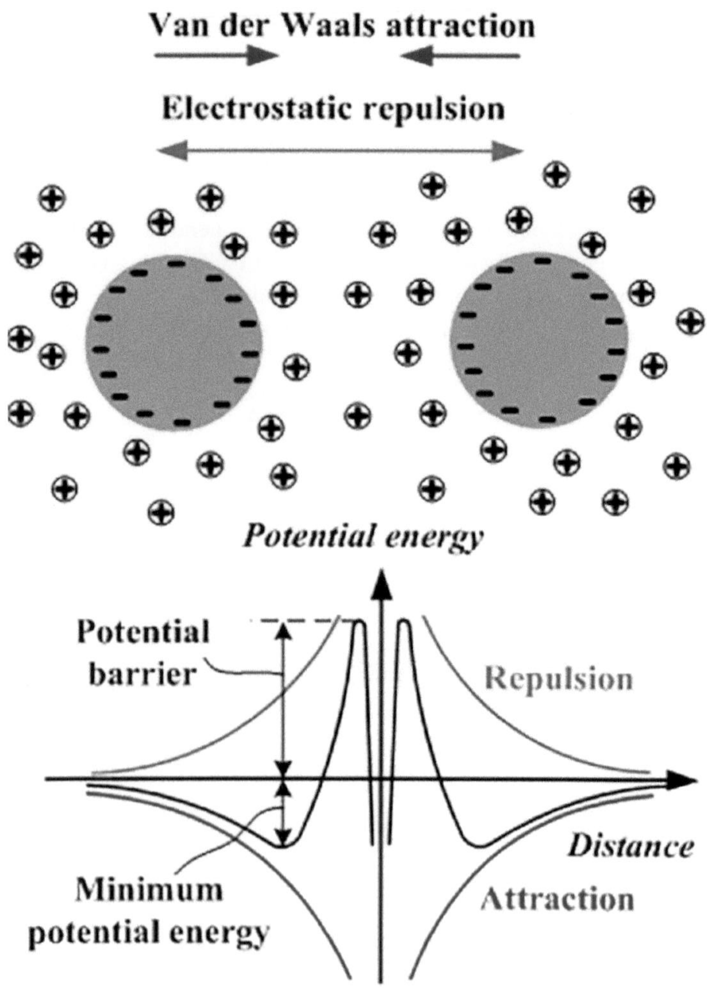

FIGURE 5.3 Energy-distance curve for electrostatically stabilized systems.

Source: Kopeliovich, D. 2019. Stabilization of colloids. www.substech.com/dokuwiki/doku.php?id=stabilization_of_colloids. Accessed August 2023.

5.2.2 Non-DLVO Interactions

The DLVO interactions, described in the previous paragraphs, cannot describe several other interactions that significantly influence the interactions of molecules and surfaces in engineering processes such as hydrophobic effects, polymer bridging, and hydration force. These interactions are known as non-DLVO (Trefalt et al. 2017; Zhang and Zeng 2021). Israelachvili (2011) measured for the first time long-range attractive force much stronger than the VDW interaction between two hydrophobic surfaces in aqueous solutions that was found to decay exponentially and the characteristic of 1 nm, but the origin is still not fully understood.

TABLE 5.1
Results of the DLVO Theory for Different Geometries

Item	Two atoms or small molecules	Two flat surfaces	Sphere (radius R) and a flat surface	Two orthogonal rods/cylinders (radii R_1 and R_2)
Geometry of bodies with separation D ($D \ll R$)	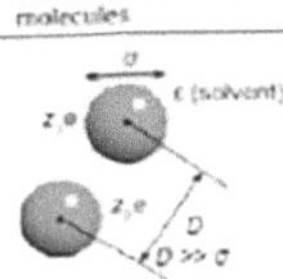	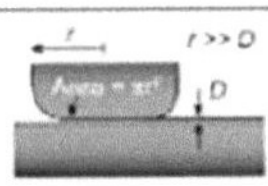	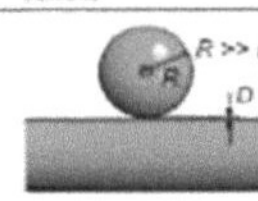	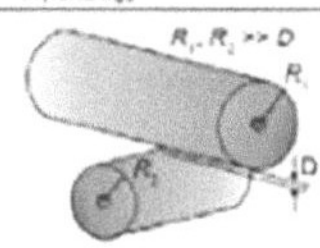
VDW interaction $W_{VDW}(D)$	$-C_{VDW}/D^6$	$W_{VDW\,plane} = -A/(12\pi D^2)$	$-AR/(6D)$	$-A\sqrt{R_1R_2}/(6D)$
$F_{VDW}(D)$	$-6C_{VDW}/D^7$	$-A/(6\pi D^3)$	$-AR/(6D^2) = -2\pi R W_{VDW\,plane}$	$-A\sqrt{R_1R_2}/(6D^2) = -2\pi\sqrt{R_1R_2}W_{VDW\,plane}$
EDL interaction $W_{EDL}(D)$	[illegible]	$W_{EDL\,plane} = (\kappa/(2\pi))Ze^{-\kappa D}$	$RZe^{-\kappa D}$	$\sqrt{R_1R_2}Ze^{-\kappa D}$
$F_{EDL}(D)$	[illegible]	$(\kappa^2/(2\pi))Ze^{-\kappa D}$	$\kappa RZe^{-\kappa D} = \pi RW_{EDL\,plane}$	$\kappa\sqrt{R_1R_2}Ze^{-\kappa D} = 2\pi\sqrt{R_1R_2}W_{EDL\,plane}$

Source: Zhang, J., and Zeng, H. 2021. Intermolecular and surface interactions in engineering processes. *Engineering* 7(1):63–83.

5.2.2.1 Steric Forces

When polymers are adsorbed at solid-liquid interfaces, a repulsive entropic force occur due to the compression of polymer chains during the approach of the two surfaces. This is known as the steric force. When two surfaces approach at solid-liquid interfaces, polymers adsorb at the surfaces, and a repulsive force occurs due to the thermal fluctuation force that the interfaces undergo. When the amount of adsorbed polymer is small, the free end of the polymer bound to a surface may attach to the other surface, bridging both surfaces, as shown in Figure 5.4. This force is called *steric force*.

5.2.2.2 Hydration Force

Hydrophilic surfaces in aqueous media show short distance range hydration force between the interacting surfaces due to the hindrance of the hydrated ions attached to the charged surfaces. They depend on the ion species and concentrations in the liquid medium, the surface roughness, and the location in the diffusive boundary layer and could be repulsive but showing oscillatory behavior at shorter distances.

Trefalt et al. (2017) indicates that the classical DLVO theory represents an accurate description of the interface force even in the presence of multivalent ions, typically down to distances of few nanometers. However, the corresponding Hamaker constants and diffuse layer potentials must be obtained from the force profiles. At low salt concentrations, double layer forces remain repulsive and can be long ranged.

At short distances, additional short-range non-DLVO interactions may become important. Such an interaction is particularly relevant in the presence of multivalent counterions. In concentrated salt solutions, the double layer forces are fully screened, and the particle interactions are dominated by dispersion forces.

Some surfaces may maintain a constant charge (CC) upon approach, but the magnitude of the surface potential will then increase, and in other situations the magnitude of the charge density decreases, while the surface maintains a constant a constant potential (CP), but in real systems the situation is often intermediate (CR). Measurements of such forces are shown in Figure 5.5. The salt concentrations were chosen to be sufficiently high to fully eliminate the contribution from double layer forces.

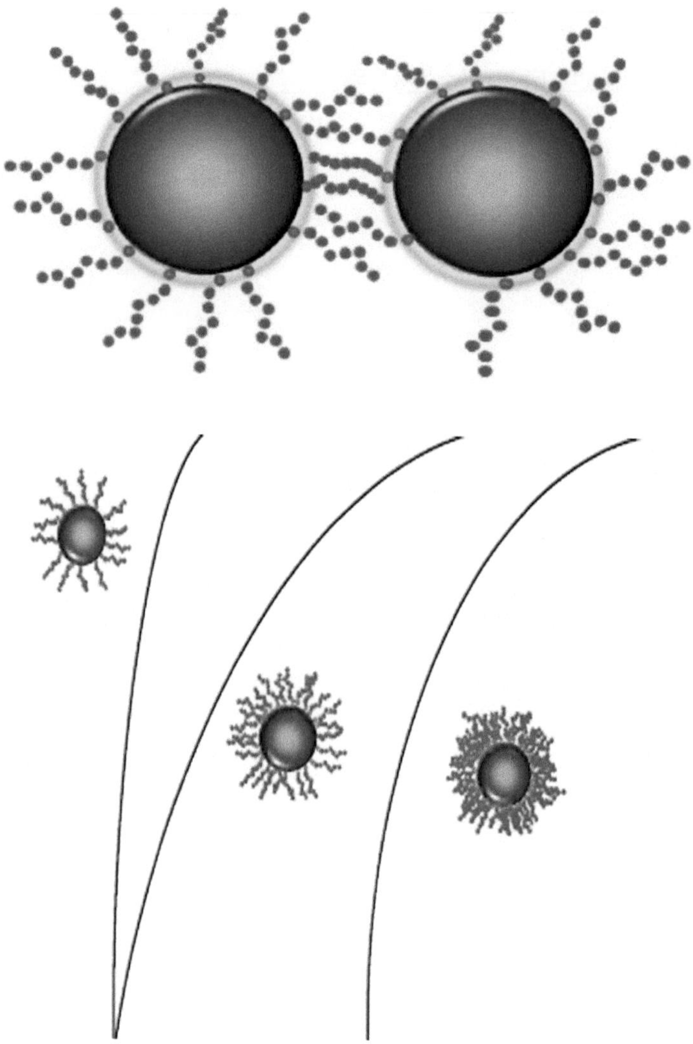

FIGURE 5.4 Steric forces due to adsorption of polymers: (a) colloid stabilization, (b) effect of increase in polymerization.

Source: Lopez, R., Balderas, M.A., Pérez, E., and Gama, A. 2015. Importance of molecular interactions in colloidal dispersion. ***Advances in Condensed Matter Physics*** **683716:1–8.**

5.2.3 Augmented DLVO Theory

Figure 5.6 shows a new augmented-DLVO for the force-distance between surfaces and particles including the non-DLVO interactions.

5.3 EXAMPLES IN FLOTATION

Derjaguin and Dukhin (1993) and Dukhin et al. (1995) suggested that the collection efficiency of solid particles is the product of three probability functions, quantifying

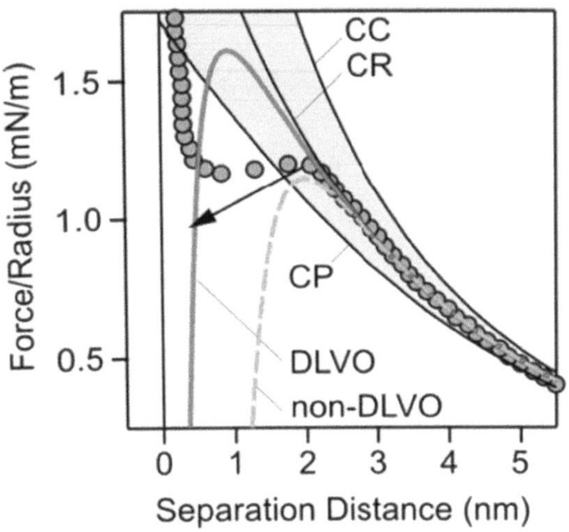

FIGURE 5.5 Comparison of the experimental data with DLVO theory for forces between sulfate latex particles of 3.0 [μm] in 10 [mM] KCl solution adjusted to pH 4.0.

Source: Trefalt, G., Palberg, T., and Borkovec, M. 2017. Forces between colloidal particles in aqueous solutions containing monovalent and multivalent ions. *Current Opinion in Colloid & Interface Science* 27:9–17.

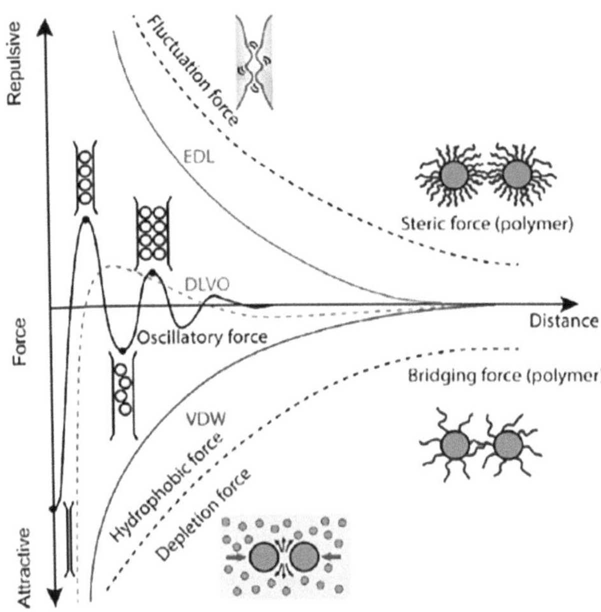

FIGURE 5.6 Force-distance between surfaces and particles with various possible interactions forces, attractive or repulsive.

Source: Zhang, J., and Zeng, H. 2021. Intermolecular and surface interactions in engineering processes. *Engineering* 7(1):63–83.

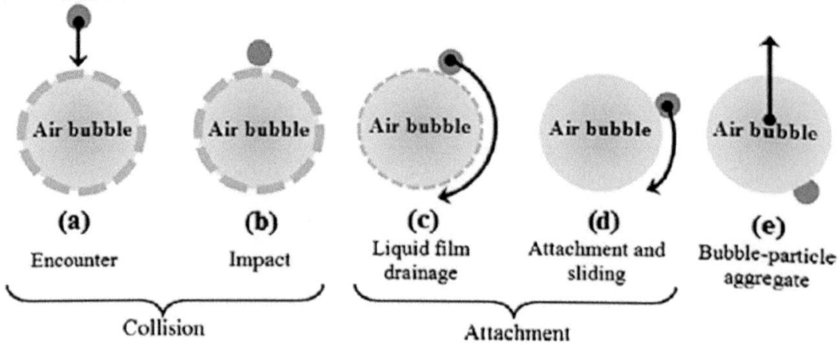

FIGURE 5.7 Schematic representation of the flotation subprocesses.

Source: Karimi, N., Akdogan, G., and Bradshaw, S.M. 2014. A CFD-kinetic model for the flotation rate constant. Part II: Model validation. *Minerals Engineering* 69:205–213.

the collision, attachment, and detachment efficiencies. Therefore, the general form of the flotation rate model can be represented in Figure 5.7.

5.3.1 Bubble-Particle Collision

According to Ruylov (2020) the particle-bubble collision energy and its dissipation play an essential role in the formation of particle-bubble aggregates, as shown in Figure 5.8. If the particle-bubble bond energy is less than the collision energy, then the formation of a particle-bubble aggregate requires at least partial dissipation of the collision energy. The dissipation of the collision energy may occur by overcoming the viscous liquid interlayer that separates a particle and a bubble or, alternatively, through the excitation of capillary waves on a bubble surface and through the formation and extension of a three-phase wetting perimeter.

The capture efficiency of particles by bubbles during the flotation process depends on several factors, the particles and bubble sizes, and velocities and the physicochemical condition of the process ambient, that is, on its physicochemical hydrodynamics shown in Figure 5.9. With this in mind, Ruylov (2020) classified the capture efficiency in four modes, (1) Stokes surface with retarded bubble surface, transient mode, and transient modes with (2) retarded and (3) Ffee bubble surface and (4) potential mode with free bubble surface (Table 5.2).

Duan et al. (2003) presented the following equation for the particle-bubble collision:

$$Z_{pb} = 2.39\left(\frac{G_{fr}}{d_b \times V_{\text{cell}}}\right)\left(\frac{0.33\epsilon^{4/9} d_b^{7/9}}{\nu^{1/3}}\left(\frac{\Delta\rho_b}{\rho_f}\right)^{2/3} \times \frac{1}{v_B}\right) E_c \times E_a \times E_s \qquad (5.8)$$

Where k is the rate constant of flotation $[1/T]$; G_{rf} is the gas flow rate $\left[L^3/T\right]$; d_b is the bubble diameter [cm]; V_{cell} is the volume of the flotation cell; $\in$ is the turbulent

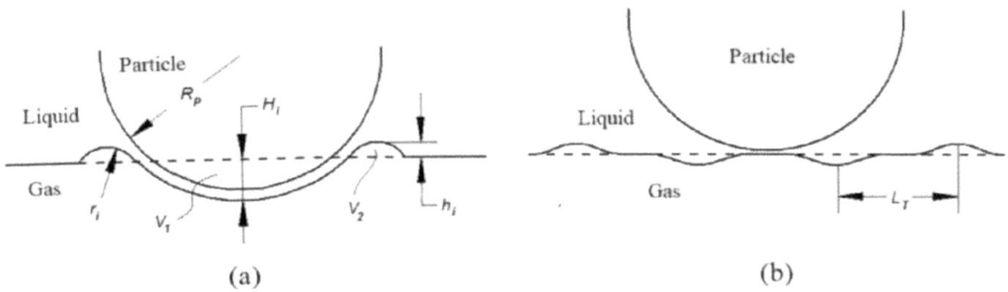

FIGURE 5.8 Impact of a spherical particles on a bubble surface (a) whit first circular capillary wave, (b) particle rebooting and causing a secondary capillary wave.

Source: Ruylov, N. 2020. Chapter 1: Theory of mineral-particle adhesion to air–water interface. In *Flotation: Physical Theory and Applications*. Kyiv: National Academy of Sciences of Ukraine.

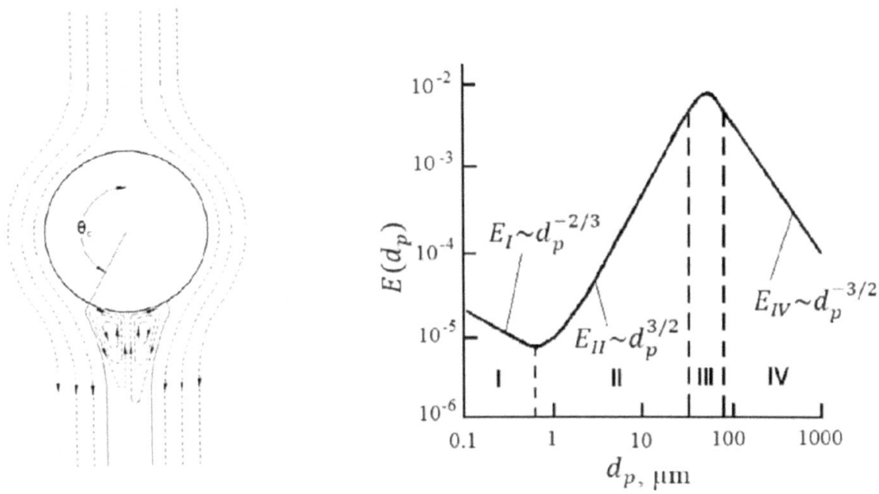

FIGURE 5.9 Capture efficiency $E(d_p)$ versus particle diameter d_P for a 1[mm] bubble and a $7.5\left[g/cm^3\right]$.

Source: Ruylov, N. 2020. Chapter 1: Theory of mineral-particle adhesion to air–water interface. In *Flotation: Physical Theory and Applications*. Kyiv: National Academy of Sciences of Ukraine.

dissipation rate $\left[L^2/T^3\right]$; v is the fluid kinematic viscosity $\left[L^2/T\right]$; $\Delta\rho$ is the bubble-fluid density difference; ρ_f is the fluid density, the numbers of solid particles and bubbles respectively; v_B is the bubble velocity; E_c, E_a, and E_s are the efficiencies of collision, attachment, and stability respectively; and Z_{pb} is the number of bubble-particle collisions.

TABLE 5.2
The Four Modes of Particles Capture from a Bubble

Hydrodynamic streamline flow mode around bubble	Flow function $\Psi_H(r,\theta)$ of fluid around bubble	Capture efficiency with no account of particle sedimentation $E_0\left(R_p / R_b\right)$	Capture efficiency with the account of particle sedimentation $E_{0g}\left(R_p / R_b\right)$
Stokes mode $(Re < 1)$, retarded bubble surface	$\frac{1}{2}r^2 \sin^2\theta\left[1-\frac{3}{2}\left(\frac{R_b}{r}\right)+\frac{1}{2}\left(\frac{R_b}{r}\right)^3\right]$	$\frac{3}{2}\left(\frac{R_p}{R_b}\right)^2$	$\left[\frac{3}{2}\left(\frac{R_p}{R_b}\right)^2+G_s\right]/\left(1+G_s\right)$
Transient mode $(1 < Re < 100)$, retarded bubble surface	$\frac{1}{2}r^2 \sin^2\theta\left[1+2\left(\frac{R_b}{r}\right)^2 \sum_{n=1}^{4}\left(A_n - B_n\cos\theta\right)\left(\frac{R_b}{r}\right)^n\right]$	$\left(\frac{R_p}{R_b}\right)^2\left[1+\sum_{n=1}^{4} n(n+1)\left(A_n - B_n\cos\theta_c\right)\right]$	$\frac{1}{\left(1+G_s\right)}\left\{G_s+\left(\frac{R_p}{R_b}\right)^2\left(1+\sum_{n=1}^{4} n(n+1)\left(A_n - B_n\cos\theta_c\right)\right)\right\}$
Transient mode $(1 < Re < 100)$, free bubble surface	$\frac{1}{2}r^2 \sin^2\theta\left[1+2\left(\frac{R_b}{r}\right)^2 \sum_{n=1}^{4}\left(A_n - B_n\cos\theta\right)\left(\frac{R_b}{r}\right)^n\right]$	$\left(\frac{R_p}{R_b}\right)^2\left[1+\sum_{n=1}^{4} n(n+1)\left(A_n - B_n\cos\theta_c\right)\right]$	$\frac{1}{\left(1+G_s\right)}\left\{G_s+\left(1-\sum_{n=1}^{4} n(n+1)\left(A_n - B_n\cos\theta_c\right)\right)\right\}$
Potential mode $(100 < Re < 500)$, free bubble surface	$\frac{1}{2}r^2 \sin^2\theta\left[1-\left(\frac{R_b}{r}\right)^3\right]$	$3\frac{R_p}{R_b}$	$\left[3\frac{R_p}{R_b}+G_s\right]/\left(1+G_s\right)$

Source: Ruylov, N. 2020. Chapter 1: Theory of mineral-particle adhesion to air–water interface. In *Flotation: Physical Theory and Applications*. Kyiv: National Academy of Sciences of Ukraine.

$$E_c = E_{c-su} \times \sin^2\theta_t \times \exp\left\{\left(3\frac{v_f\left(\rho_p - \rho_f\right)d_p^2}{9v}\left(\frac{d_B}{d_p} - 1.8\right)\right) - 4\left(\frac{\frac{2}{3} + \frac{\cos^3\theta_t}{3} - \cos\theta_t}{\sin^4\theta_t}\right)\cos\theta_t\right\} \quad (5.9)$$

Where θ_t is the maximum possible collision angle of the particle on the surface of the bubble beyond which collision is prevented.

If we call θ_a the adhesion angle, F_{att} and F_{det} are the attachment and detachment forces, respectively:

$$E_a = \frac{\sin^2\theta_a}{\sin^2\theta_t} \text{ and } E_s = 1 - \exp\left(1 - \frac{F_{\text{att}}}{F_{\text{det}}}\right) \quad (5.10)$$

5.3.2 Bubble Attachment-Detachment

Schubert (2005) pointed out that the long-range attractive interaction forces between hydrophobic surfaces in aqueous systems are caused by the capillary forces of gas bridges which form at the coalescence of nanobubbles adhering on the surfaces. The coalescence of nanobubbles on selectively hydrophobized particles with coarser bubbles initiates the jump into the three-phase contact at the attachment events in flotation. Therefore, one should no longer speak of hydrophobic forces in these events but of hydrophobic effects because they are caused by capillary forces. However, in the selective hydrophilization of particles by long-chain collectors or surfactants, the association of nonpolar groups in the adsorption layers plays an important role in the energy balance of adsorption. Appling Young's equation $\gamma_{SG} = \gamma_{SL} + \gamma_{LG}\cos\theta$, the change in the Gibbs free energy ΔG is

$$\Delta G = \gamma_{LG}(\cos\theta - 1) \quad (5.11)$$

Where, γ_{LG} is the interfacial liquid gas tension and θ is the contact angle. Figure 5.10 shows the interfacial energy and that for adsorption to occur $\Delta G < 0$; therefore, the contact angle must be $\theta > 0$.

5.4 FLOTATION PHENOMENOLOGICAL MODEL

A phenomenological flotation model is based on performing mass particle, mass balances, and defining the constitutive formulation for the interactions between particles species of many types and particle size distribution, the water and the bubbles created by a bubble generator using energy to create the bubbles to assist in mixing and attaching the hydrophobic particles to the bubbles.

5.4.1 Bascur and Herbst Dynamic Flotation Framework

In 1982, Bascur and Herbst developed a flotation model involving the interaction of three phases, solid, liquid, and gas governed by the physicochemical and

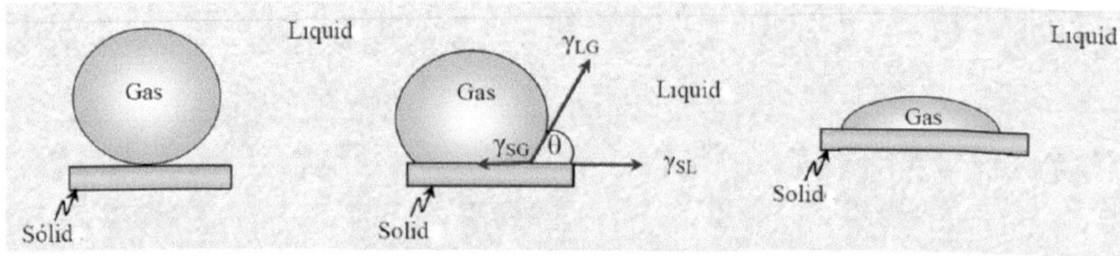

FIGURE 5.10 Interfacial energy γ of the three-phase in equilibrium.

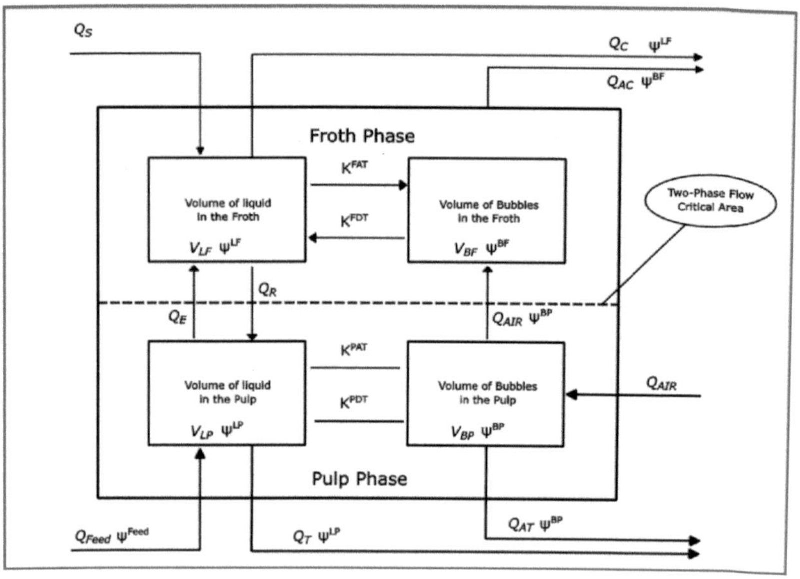

FIGURE 5.11 Schematic diagram of the four states for particles in a flotation cell according to the population balance model.

Source: Bascur 1982, 2000, 2013.

hydrodynamic environment where each particle, existing in several mineralogical species each with a number of particle sizes, which can exist in four states (phases): (1) free in the pulp, (2) attached to bubbles in the pulp, (3) free in the froth, and (4) attached to bubbles in the froth (Bascur and Herbst 1982).

The most important aspect of the phenomenological model is the modeling of the suspension with bubbles and then the froth created after the constraint created by the nature of the two-phase continuity equation (Bascur 1976, 1991).

The dynamic model was developed with the intention of providing a basis for testing process control strategies and training of process engineers. It has been favorably distinguished in a recent advanced in flotation technology review by Fuerstenau (1999).

Figure 5.11 shows the four phases with information to calculate the overall flotation recovery.

This dynamic flotation framework is based on the two-phase continuity equation model of Bascur (1976), Concha and Bascur (1977), the Abrahamson (1975) collision theory together with the Bischofberger and Schubert (1979) and Schubert et al. (1979), considering the Kolmogoroff turbulence phenomena as a key factor in the attachment and detachment processes of bubbles and particles and the Mika and Fuerstenau (1968) conceptual population balance flotation integrated microprocesses. We discuss the dynamic model development in the next section and the dynamic flotation framework for dynamic simulation and online predictive flotation behavior in Chapter 9. Implicit in the process model described in Chapter 7 all these fundamentals apply, and particles in the slurries require collectors, depressants, modifiers for the process hydrodynamic, and the bubbles and particles interactions to occur.

The Bascur and Herbst model has been used to develop online models to predict the behavior of the flotation process online using the measurements provided by the plant control system. It has been used by several institution to dynamically simulate the flotation process and integrate the grinding and flotation to discover new ways to improve both the grinding and flotation process to optimize industrial flotation plants (Conradie et al. 2003; Bascur and Soudek 2019; Bascur and Gorrie 2021; Bascur et al. 2022; Bascur and Retamal 2021).

REFERENCES

Abrahamson, J. 1975. Collision rates of small particles in a vigorously turbulent fluid. *Chemical Engineering Science* 30:1371–1379.

Bascur, O.A. 1976. Phenomenological model of suspensions in sedimentation. Engineering thesis, University of Concepción (in Spanish).

Bascur, O.A. 1982. Modeling and computer simulation of a flotation cell. PhD dissertation, University of Utah, Salt Lake City.

Bascur, O.A. 1991. A unified solid liquid separation framework. *Fluid Particle Separation Journal* 4:117–122.

Bascur, O.A. 2000. An interactive dynamic flotation model framework. In *Proceedings of the 14th Proceedings of the International Mineral Processing Congress: Development in Mineral Processing*. pp. 21–31. https://doi.org/10.1016/S0167-4528(00)80064-2

Bascur, O.A. 2013. A dynamic flotation framework for performance management. In *Separation Technologies for Minerals, Coal and Earth Resources*. Edited by C. Young and G. Luttrel. Englewood, CO: SME. pp. 40–45.

Bascur, O.A., and Gorrie, T. 2021. Maximizing copper recovery using digital twins plant. In *Proceedings of Procemin Geomet 2021: 17th International Conference on Mineral Processing and Geometallurgy*. Santiago: Gecamin.

Bascur, O.A., and Herbst, J.A. 1982. Dynamic modelling of a flotation cell with a view toward automatic control. Presented at the 14th International Mineral Processing Congress, Toronto, Canada.

Bascur, O.A., and Retamal, F. 2021. Integrated rock processing strategies in the digital age. Presented at Minería Digital, Gecamin, Santiago, Chile, October 17–19.

Bascur, O.A., Retamal, F., and Espinosa, S. 2022. Treating low grades ores using a digital in mining and mineral processing plants. In *Proceedings of the Copper 2022 International Conference: IIMChile. Vol. 7: Process Optimization through Innovation, Technology, Automation and Design*. Santiago: Gecamin.

Bascur, O.A., and Soudek, A. 2019. Grinding and flotation optimization using operational intelligence. *Mining, Metallurgy and Exploration* 36:139–149. https://doi.org/10.1007/s42461-018-0036-4.

Bischofberger, C., and Schubert, H.C. 1979. On the optimization of hydrodynamics in flotation processes. In *Proceedings of the 13th International Mineral Processing Congress*, vol. 2. Edited by J. Laskowski. New York: Elsevier Scientific. pp. 1261–1287.

Chemistry Stack Exchange. 2020. Electric double layer formation. https://chemistry.stackexchange.com/questions/126773/electric-double-layer-formation.

Concha, F., and Bascur, O.A. 1977. A phenomenological model of sedimentation. In *Proceedings of 12th International Mineral Processing Congress: Special Publication* 1:29–46.

Conradie, A.V.E., Bascur, O.A., Aldrich, C., and Nieuwoudt, I. 2003. Integrated comminution and flotation neurocontrol using evolutionary reinforcement learning. In *Proceedings of the Application of Computers and Operations Research in the Mineral Industries (APCOM)*. Johannesburg: Southern African Institute of Mining and Metallurgy. pp. 209–216.

Derjaguin, B., and Landau, L.D. 1941. Theory of stability of strongly charged lyophobic sols and of the adhesion of strongly charged particles in solutions of electrolytes. *Acta Physicochimica U.S.S.R.* 17:633–662.

Derjaguin, B.V. 1932. Elastic properties of thin layers of water. *Russian Journal of Physical Chemistry A* 3(13):29–41.

Derjaguin, B.V., and Dukhin, S.S. 1993. Theory of flotation of small and medium-size particles. *Progress in Surface Science* 43(1):241–266. https://doi.org/10.1016/0079-6816(93)90034-S

Duan, J., Fornasiero, D., and Ralston, J. 2003. Calculation of the flotation rate constant of chalcopyrite particles in an ore. *International Journal of Mineral Processing* 72:227–237.

Dukhin, S.S., Kretzschmar, G., and Miller, R. 1995. *Dynamics of Adsorption at Liquid Interfaces: Theory, Experiment, Application*. The Netherlands: Elsevier Science B.V.

Fuerstenau, D.W. 1999. The froth flotation century. In *Advances in Flotation Technology*. Edited by B.K. Parek and J.D. Miller. Littleton, CO: SME. pp. 3–21.

Gouy, G. 1917. Sur la function electrocapillary. *Annals of Physics* 9(7):129.

Israelachvili, J.N. 2011. *Intermolecular and Surface Forces*, 3rd ed. New York: Academic Press.

Kopeliovich, D. 2019. Stabilization of colloids. www.substech.com/dokuwiki/doku.php?id=stabilization_of_colloids. Accessed August 2023.

Lopez, R., Balderas, M.A., Pérez, E., and Gama, A. 2015. Importance of molecular interactions in colloidal dispersion. *Advances in Condensed Matter Physics* 683716:1–8.

Lyklema, J. 2005. *Fundamentals of Interface and Colloid Science: Solid-Liquid Interfaces*. Amsterdam: Academic Press, Elsevier.

Mika, T.S., and Fuerstenau, M. 1968. A microscopic model of the flotation process. In *Proceedings of the Eight International Mineral Processing Congress*, vol. 2. Leningrad: Mechanobr. pp. 246–269.

Ruylov, N. 2020. Chapter 1: Theory of mineral-particle adhesion to air–water interface. In *Flotation: Physical Theory and Applications*. Kyiv: National Academy of Sciences of Ukraine.

Schubert, H. 2005. Nanobubbles, hydrophobic effect, hetero-coagulation and hydrodynamics in flotation. *International Journal of Mineral Processing* 78(1):11–21.

Schubert, H., Heidenreich, E., Liepe, F., and Neese, T. 1979. *Mechanische Verfarenstechnick I*. Leipzig: Deutscher Verlag. p. 75.

Stern, O.Z. 1924. On the theory of the electrolytic double layer. *Electrochemistry* 30(21–22):508–516.

Trefalt, G., Palberg, T., and Borkovec, M. 2017. Forces between colloidal particles in aqueous solutions containing monovalent and multivalent ions. *Current Opinion in Colloid & Interface Science* 27:9–17.

Verwey, E.W., and Overbeek, J.Th.G. 1948. *Theory of Stability of Lyophobic Colloids*. Amsterdam: Elsevier.

Wang, R.-M., Zheng, S.-R., and Zheng, Y.-P. 2011. Chapter 4: Interface of polymer matrix composites. In *Polymer Matrix Composites and Technology*. Edited by R.-M. Wang, S.-R. Zheng, and Y.-P. Zheng. Woodhead Publishing Series in Composites Science and Engineering. Cambridge: Woodhead. pp. 169–209, 547–548.

Zhang, J., and Zeng, H. 2021. Intermolecular and surface interactions in engineering processes. *Engineering* 7(1):63–83.

6 Suspension Rheology

6.1 INTRODUCTION

Continuum mechanics has been very successful when using linear constituent equations, such as Hooke's law of elasticity, in the quantitative description of the deformation of metals and other materials for the construction of industrial equipment parts and bridges. Its treatment, using Newton's viscosity law, of the fluid flow (see Chapter 3) such as water in pipes, canals, and rivers, and objects such as aircraft in the air and gases in pipes and chimneys are excellent. However, the advent of new materials, such as polymers and composite substances such as suspensions, emulsions, and foams, make it necessary to clarify the concepts of elasticity and viscosity since many materials can simultaneously or sequentially present these properties. Sometimes a material will behave as solid when deformed quickly (small time scale) and as a fluid when slowly deformed (major time scale). These evidences have led some researchers to say that all material flows if expected long enough. Bingham (1922) studied the deformation and flow of this type of materials, which he called rheology. By definition rheology covers the whole field of continuous dynamics; however, by convention, it has been decided to limit it to that field not covered by linear theories of elasticity and viscosity, that is, the study of Hookean solids and Newtonian fluids does not belong to rheology. The modern thinking of rheology is the description of a set of material behavior instead of the description of a set of materials. Typically, rheological behaviors include the following:

- Viscosity dependent on shear rate. Many fluids exhibit a decrease in shear viscosity with shear rates. This decrease in viscosity may reach values of 10^{-3} or 10^{-4} times the viscosity at rest, zero shear rate.
- Normal stresses in stationary shear flows. Some fluids possess different values of their normal stresses σ_{11}, σ_{22}, and σ_{33} producing strange behaviors, such as climbing the stem of an agitator, changing the direction of flow in an agitated pond and convex curvature of the surface in a flow in a channel.
- Transient responses in shear flow. There are materials that experience a variety of transient phenomena in shear flows, such as stress relaxation and creep. Many of these phenomena are used in the determination of parameters of rheological materials.
- Other phenomena. In addition, they exhibit nonconventional rheological behaviors such as the siphon without a tube.

To quantify the effects described, it is necessary to establish constitutive equations other than the Hookean solid and Newtonian fluid laws and conduct experimental research to measure the associated parameters. A rheological study usually includes the formulation of constitutive equations to describe the behavior of a certain material in more complex geometries and flow measurements in more complex geometries.

DOI: 10.1201/9781032614212-6

The axioms of mechanics, mass, and momentum equations are valid for all macroscopic bodies considered as a continuous medium. The field equations are as follows:

$$\begin{aligned} &\text{Mass balance} && \rho + \rho(\nabla \cdot \boldsymbol{v}) = 0 \\ &\text{Momentum balance} && \rho \boldsymbol{v} = -p\boldsymbol{I} + \nabla \bullet \boldsymbol{T}^E + \rho \boldsymbol{b} \end{aligned} \tag{6.1}$$

Where ρ, p, $\boldsymbol{v}$, $\boldsymbol{T}^E$, and $\boldsymbol{b}$ are the fluid density, pressure, velocity field, extra stress, and body force. Since our focus is centered in the transport of concentrates and tailings in the mining industry, we will restrict our discussion to the flow of incompressible fluids in cylindrical tubes. For incompressible fluids, from Chapter 3 of this book we can write the following:

$$\begin{aligned} \frac{\partial v_z}{\partial z} &= 0 \\ T^E(r) &= -\frac{1}{2}\frac{\Delta p}{L} r \end{aligned} \tag{6.2}$$

Since we still have not defined a constitutive equation for the shear stress, Equation 6.2 is valid for any type of fluids.

6.2 NEWTONIAN FLUIDS

As we have already mentioned, a Newtonian fluid is that fluid whose shear stress $T^E(r)$ is a linear function of the shear rate $\dot{\gamma}$. The constant of relating them is the shear viscosity μ.

$$T^E(r) = -\mu\dot{\gamma} \tag{6.3}$$

The best example of a Newtonian fluid is water with a viscosity of 1 [mPa-s] (Figure 6.1).

A nice review of viscosity called the flow of wet water can be found in the Richard Feynman notes (Gottier and Pfeiffer 2013).

6.3 NON-NEWTONIAN FLUIDS

A non-Newtonian fluid is a fluid for which the viscous stress, at a given temperature, is a nonlinear function of the shear rate. An example of non-Newtonian behavior in a fluid is observed in concentrated suspensions of solids in fluids (concentrates and tailings), in high molecular weight materials such as polymers and in solutions of small portions of polymers in a liquid (collectors, flocculants). These responses include among others viscous and elastic behavior, or both, dilatancy or shear thinning at shear rates, which is an increase or decrease in viscosity when sheared. Among the

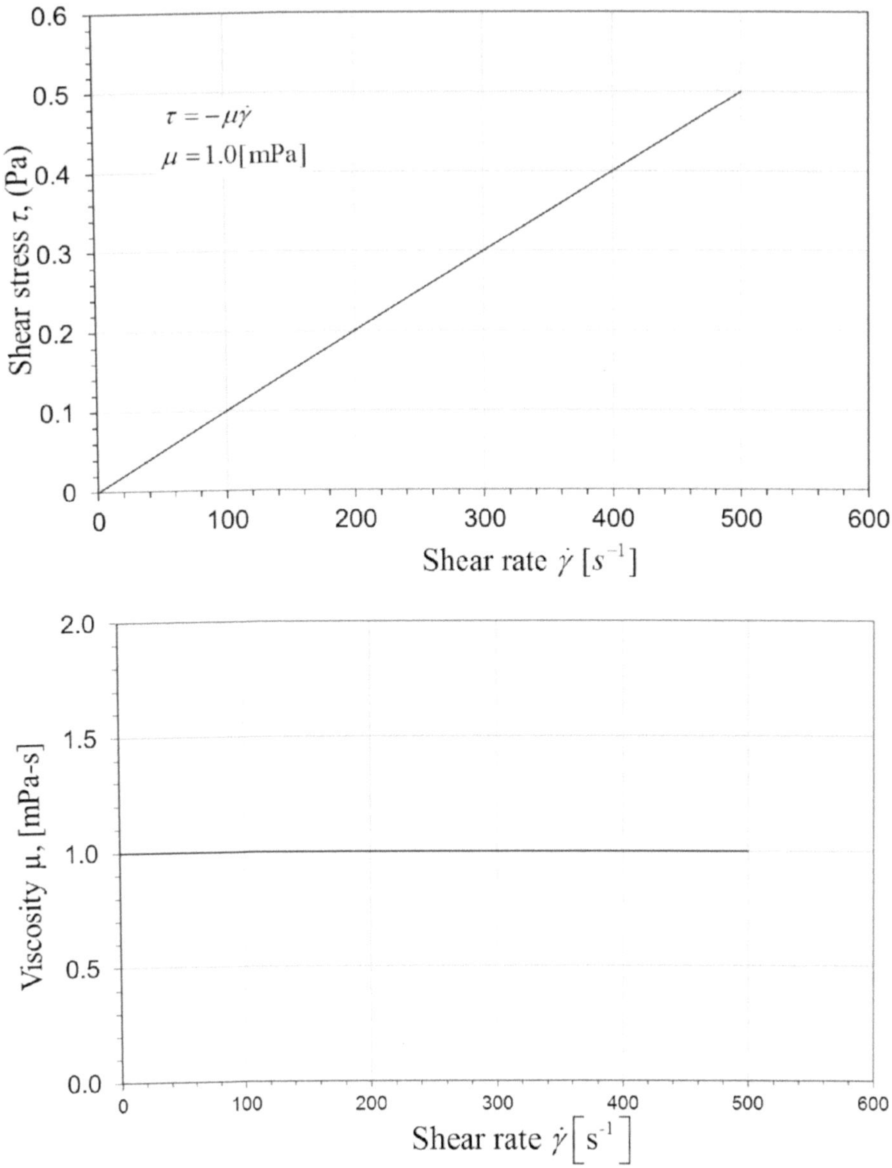

FIGURE 6.1 Rheogram for a Newtonian fluid: (a) shear stress; (b) shear viscosity.

non-Newtonian fluids, the most interesting examples in mineral processing are concentrate and tailing suspensions and polymer solutions as mentioned (Chhabra and Richardson 1999; Concha et al. 1991; Concha 2014).

The shear stress-shear rate relationship may be written in a similar way than that for Newtonian fluid, with the extra shear stress. $T^E(r)$ written in the form

$$T_{rv}^{E} = -\eta\dot{\gamma} \tag{6.4}$$

Where T_{rv}^{E} is the extra shear stress, $\dot{\gamma}$ is the shear rate, and η the shear viscosity.

6.3.1 Rheological Models

There are several models to describe the behavior of non-Newtonian fluid, among the best known those of Bigham plastic, power law, Herschel-Buckley, Cross, Casson, and Sisko (Macosko 1994). It is important to emphasize that it is not possible to select one model by statistical methods as the best for all application. For our purpose, the Bingham plastic and the power law models are the simplest and best models for mineral processing, each for different applications, the Bingham model suitable for suspension transport in pipes and the power law for the development of process models such as grinding, classification, flotation, and thickening (Barnes et al. 1989). Many experimental determinations of materials such as flotation concentrates tailings show the typical rheograms shown in Figure 6.2.

6.3.1.1 Bingham Model

There are materials that behave like solids for low deformations (elastic deformation) and as a fluid for higher shear rates. These materials receive the name of *viscoelastic*. The elastic limit of deformation is called *yield stress* and its constitutive equation is called Bingham model and have the form

$$\tau = \tau_y + K\dot{\gamma} \text{ for } \dot{\gamma} > 0 \tag{6.5}$$

Where $\tau = T_{rv}^{E}$, τ_y is the yield stress, K is a constant, the *plastic viscosity* and $\dot{\gamma}$ is the shear rate. Figure 6.3 shows that Bingham model is useful only for flow with shear $\dot{\gamma} \geq 300\left[s^{-1}\right]$ where the slope K represents the shear rate.

The concentration of a suspension has a great effect on its viscosity, while the temperature has a lower effect, since the viscosity decrease exponentially with temperature but increases slightly with pressure. For mining companies this combination is therefore small since they usually are located at high altitudes above sea level. Several equations have been proposed to describe the effect of the concentration on the yield stress and plastic viscosity such as the well-known expression of Krieger and Doherty (1959). It is interesting to note that the increase in the suspension density does not require necessarily an increase in the amount of solid; the increase in density may well come from surface active reagents producing coagulation or flocculation.

In the Bingham model the yield stress is constant, and the plastic viscosity is also constant and similar to the Newtonian fluid as shown in Figure 6.3 but depends on the suspension concentration. They can be represented by

$$K(\varphi) = \left(1 - \frac{\varphi}{\varphi_k}\right)^{-2.5\varphi_k} \tag{6.6}$$

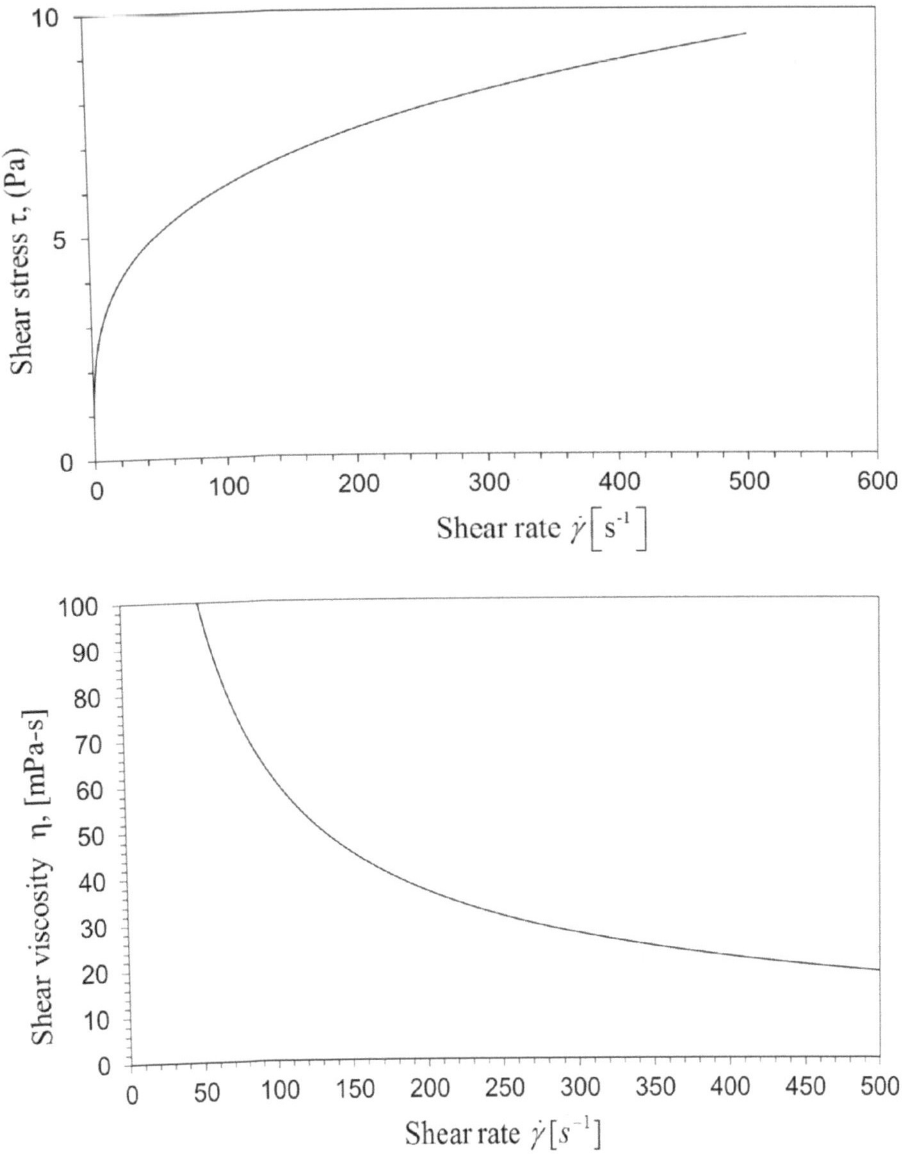

FIGURE 6.2 Rheogram: shear-stress versus shear-rate relationship for a copper flotation tailing, a non-Newtonian fluid.

$$\tau_y(\varphi) = \tau_k \left(\frac{\varphi^3}{\varphi_k} \right)$$

$$\tau_y(\varphi) = \tau_k \left(\varphi^3 / \varphi_k \right)$$

where φ_k is the maximum value φ can attain.

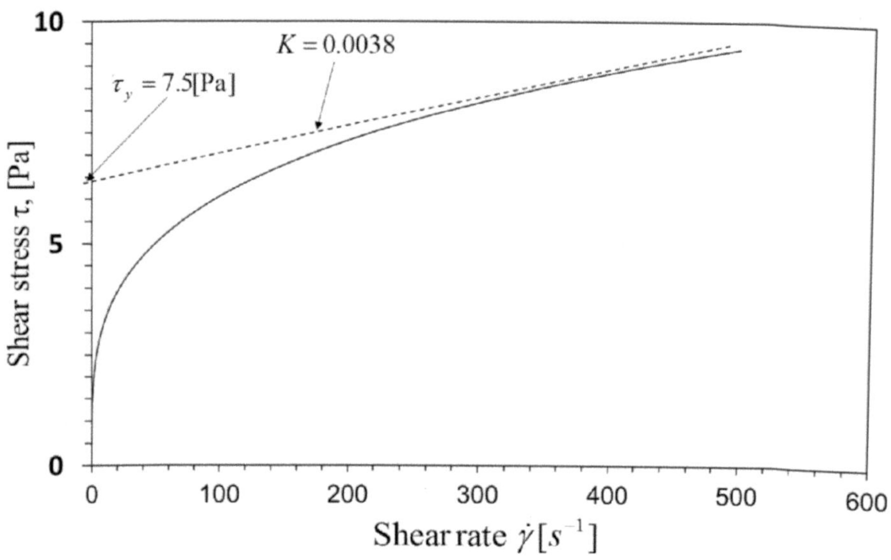

FIGURE 6.3 Plastic viscosities in the Bingham model.

The plastic viscosity $K(\varphi)$ and the dimensionless yield stress $\tau_y / \tau_K(\varphi)$ versus concentration φ as shown in Figure 6.4.

6.3.1.2 Power Law Model

Figure 6.2 shows that the shear viscosity is a function of the shear rate and can be written in the form

$$\tau = \eta(\dot{\gamma}) \tag{6.7}$$

If the viscosity in this figure is plot in a log-log scale, a straight line is obtained (Figure 6.5), so that the shear viscosity as a function of the shear rate can be written in the form

$$\eta(\gamma) = m\dot{\gamma}^{n-1} \tag{6.8}$$

Substituting in Equation 6.8 in Equation 6.7 yields

$$\tau = m\dot{\gamma}^{n} \tag{6.9}$$

In this expression m is called *consistency index* and n is the *power law index*. In Figure 6.5, $m = 1723.4$, $n - 1 = -0.727$, so that $n = 0.273$.

In a similar way to the case of the Bingham model, the shear viscosity in the power law model is a function of the viscosity and the concentration $\eta(\dot{\gamma}, \varphi)$. Experimental data of the effect of shear rate and concentration on the viscosity of a copper flotation tailing was plotted in Figures 6.6 and 6.7.

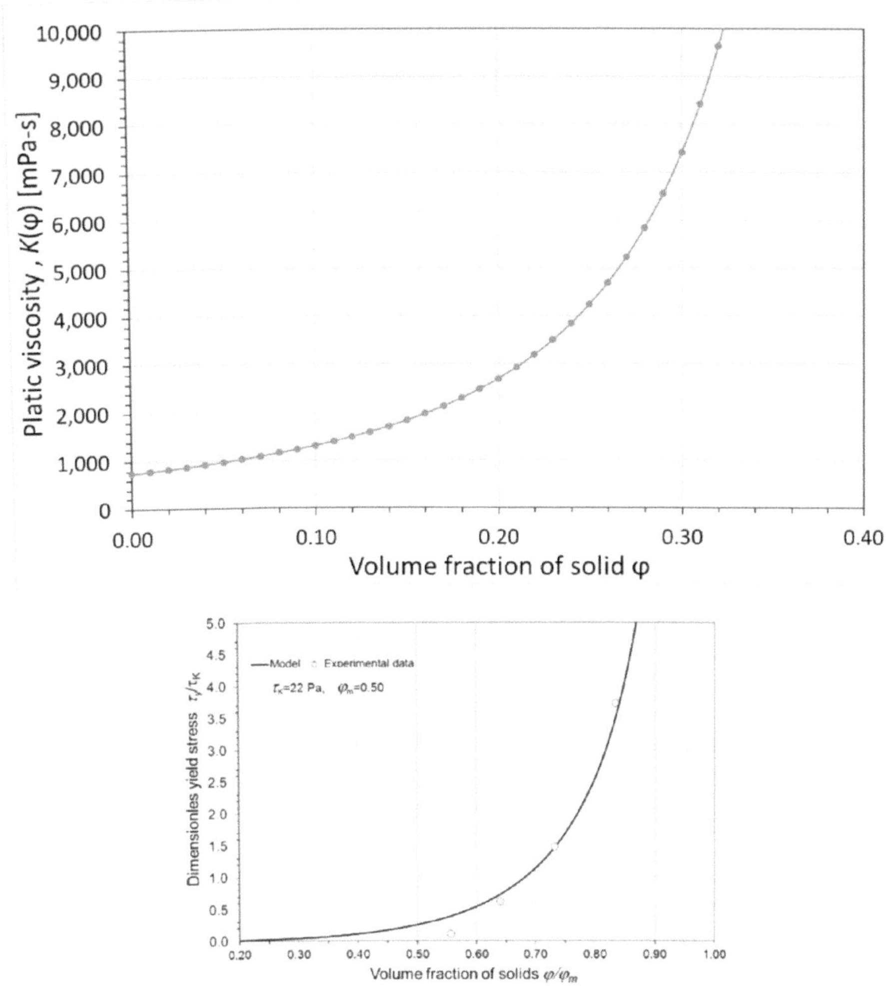

FIGURE 6.4 Plastic viscosity $K(\varphi)$ and dimensionless yield stress $\tau_y / \tau_K(\varphi)$ versus concentration φ for a copper flotation tailing.

Assuming that the relationship of the shear viscosity of a suspension with the shear rate and concentration $\eta(\dot{\gamma},\varphi)$ may be written in the form:

$$\eta(\dot{\gamma},\varphi) = a\dot{\gamma}^c \exp\left(b\dot{\gamma}^d\varphi\right) \tag{6.10}$$

The following experimental parameters were obtained:

$$a = 5.333, b = 0.152, c = -0.467 \text{ and } d = -0.048 \tag{6.11}$$

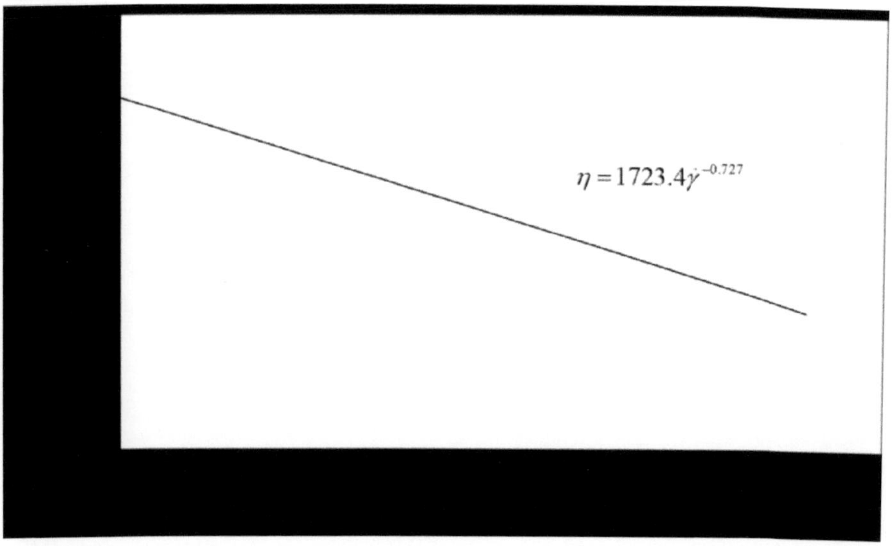

FIGURE 6.5 Viscosities in the Power law model.

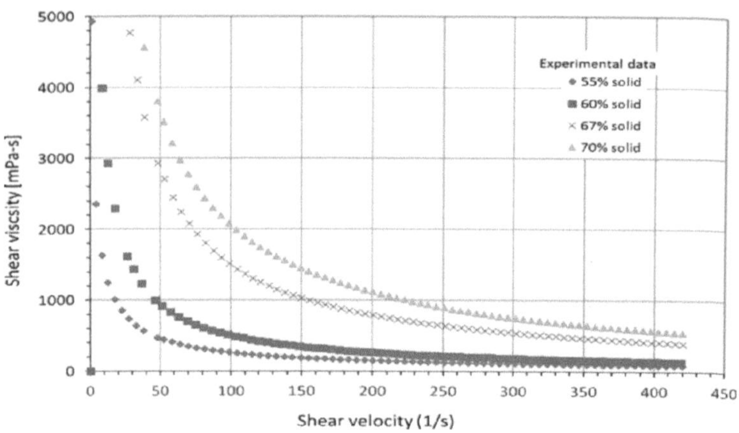

FIGURE 6.6 Shear viscosity versus the shear rate of a suspension with the concentration as a parameter.

Source: Concha, F., Barrientos, A., Torres, V., and Alejo, B. 1991. Suspension rheology. Universidad de Concepción [in Spanish].

6.4 TRANSPORT OF CONCENTRATED SUSPENSIONS IN CYLINDRICAL TUBES

Concentrated suspension, as opposite to dilute suspensions, are non-settling, that is, they behave as homogeneous fluids, and the shear stress is a linear function of the radius of the tube as shown in Figure 6.8:

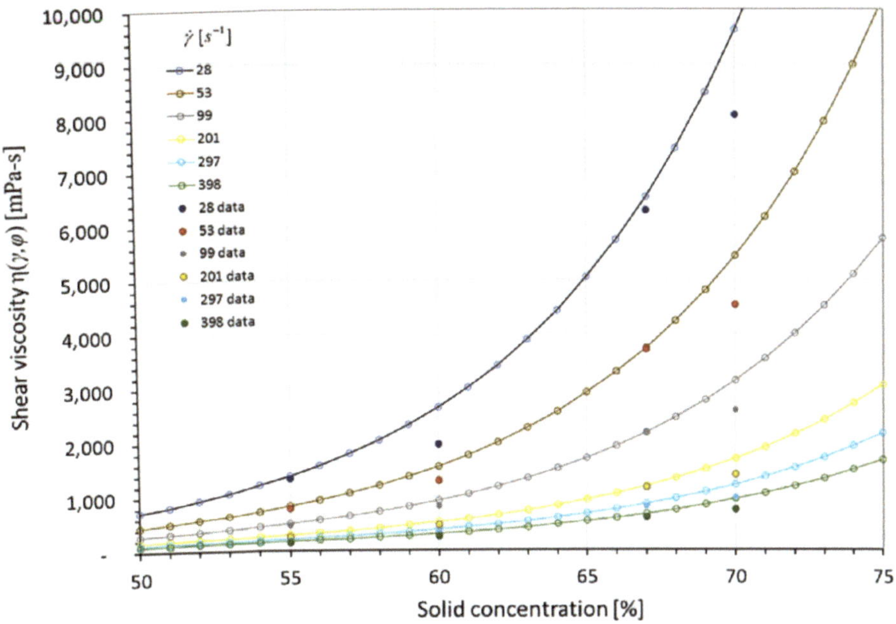

FIGURE 6.7 Plot of the experimental data and simulation of the shear viscosity as a function of solid concentration with shear rate as a parameter.

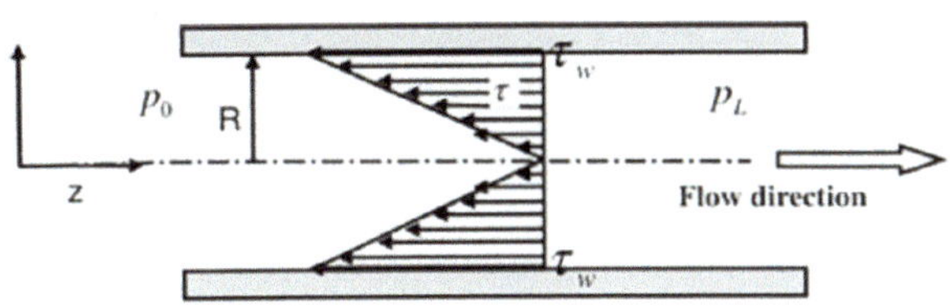

FIGURE 6.8 Shear stresses in the flow in tubes.

Source: Concha, F.A. 2014. Solid Liquid Separation in the Mining Industry. Cham, Switzerland: Springer.

$$T_{rz}^{E}(r) \equiv \tau(r) = -\frac{1}{2}\frac{\Delta p}{L}r, \quad \tau_y = -\frac{1}{2}\frac{\Delta p}{L}R_y, \quad \tau_w = -\frac{1}{2}\frac{\Delta p}{L}R \tag{6.12}$$

Where the linear parameter is the gradient of pressure $\Delta p / L$, with p the pressure and L the length of the tube and the minus sign indicates the stress τ increases in magnitude opposite to the direction of the flow, with $\tau_w = \tau(R)$.

6.4.1 Bingham Fluids Transport

Equation 6.12 shows that the shear stress for all fluid is a linear function of the pressure gradient. Combining this with the constitutive equation for a Bingham fluid given by Equation 6.5 yields the following:

$$\frac{\partial v_z}{\partial r} = -\begin{cases} \frac{1}{2}\frac{\Delta p}{LK} r - \frac{\tau_y}{K}, & r < R_y \\ 0 & r \geq R_y \end{cases} \tag{6.13}$$

$$\text{For} : \gamma > 0 \qquad \frac{\partial v_z}{\partial r} = -\left(\frac{1}{2}\frac{\Delta p}{KL} r - \frac{\tau_y}{K}\right) \tag{6.14}$$

Integrating $\quad v_z(r) = \frac{1}{2}\frac{\Delta p}{KL}\left(R_y r - \frac{1}{2}r^2\right) + C$

For $v_z(R) = 0 \quad C = -\frac{1}{2}\frac{\Delta p}{KL}\left(R_y R - \frac{1}{2}R^2\right)$

$$v_z(r) = \frac{1}{2}\frac{\Delta p}{KL}\left(R_y r - \frac{1}{2}r^2\right) - \frac{1}{2}\frac{\Delta p}{KL}\left(R_y R - \frac{1}{2}R^2\right) \tag{6.15}$$

Using Equation 6.12, Equation 6.15 may be written as

$$\text{Velocity distribution} \quad v_z\left(R_y\right) = \frac{1}{4}\frac{\Delta p R^2}{KL}\left(\left(\frac{\tau_y}{\tau_w}\right)^2 - \left(\frac{\tau_y}{\tau_w} - 1\right)\right) \tag{6.16}$$

$$\text{Volume flow rate} \quad Q = \frac{\pi}{8}\frac{\Delta p R^4}{KL}\left(\left(\frac{\tau_y}{\tau_w}\right)^2 - \left(\frac{\tau_y}{\tau_w} - 1\right)\right) \tag{6.17}$$

$$\text{Average velocity} \quad \bar{v}_z = \frac{1}{8}\frac{\Delta p R^2}{KL}\left(\left(\frac{\tau_y}{\tau_w}\right)^2 - \left(\frac{\tau_y}{\tau_w} - 1\right)\right) \tag{6.18}$$

$$\text{Maximum velocity} \quad v_m = v_z(r) = \frac{1}{4}\frac{\Delta p R^2}{KL}\left(\left(\frac{\tau_y}{\tau_w}\right)^2 - \left(\frac{\tau_y}{\tau_w} - 1\right)\right) \tag{6.19}$$

Figure 6.9 shows the velocity profiles for three Bingham fluids.

6.4.2 Power Law Fluids Transport

The combination of Equations 6.12 and 6.9:

$$\frac{\partial v_z}{\partial r} = -\left(\frac{1}{2m}\frac{\Delta p}{L} r\right)^{1/n} \tag{6.20}$$

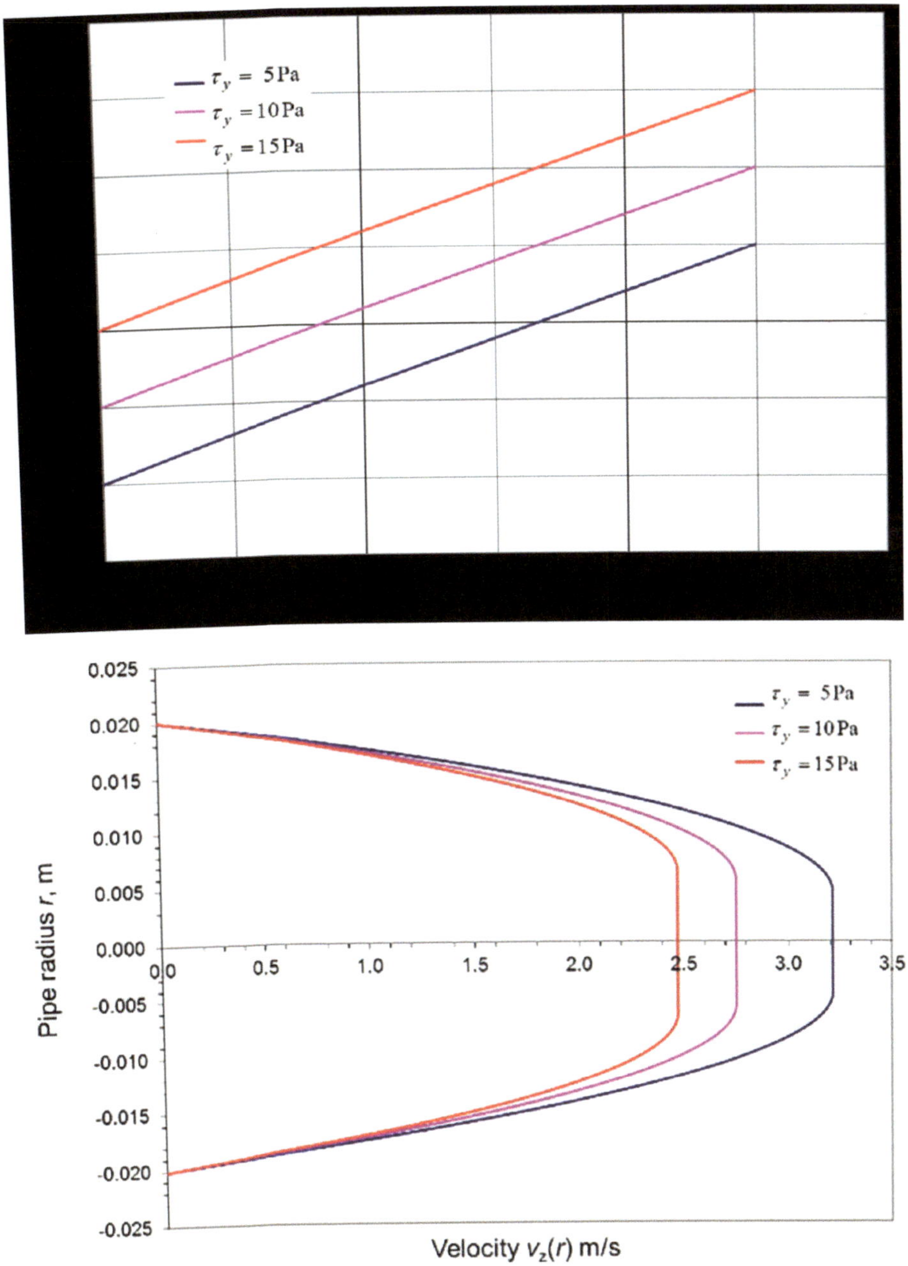

FIGURE 6.9 Velocity profiles for three Bingham fluids: (a) rheogram, (b) velocity profiles.

Integrating and using the boundary condition yields the following:

$$\frac{\partial v_z}{\partial r} = \left(\frac{-\Delta p}{2mL} r \right)^{1/n} \tag{6.21}$$

$$v_z(r) = \left(\frac{\Delta p}{2mL} \right)^{1/n} \frac{n}{n+1} \left(-r^{n+1/n} \right) + C$$

Using the boundary condition $v_z(R) = 0$ results in the following:

$$\text{Velocity} \quad v_z(r) = \frac{nR}{n+1} \left(\frac{\Delta pR}{2mL} \right)^{1/n} \left(1 - \left(\frac{r}{R} \right)^{(n+1)/n} \right) \tag{6.22}$$

$$\text{Volume flow rate} \quad Q = \pi R^2 \frac{n}{(3n+1)} \left(\frac{\Delta pR^{n+1}}{2mL} \right)^{1/n} \tag{6.23}$$

$$\text{Average velocity} \quad \bar{v}_z = \frac{n}{(3n+1)} \left(\frac{\Delta pR^{n+1}}{2mL} \right)^{1/n} \tag{6.24}$$

$$\text{Maximum velocity} \quad v_m = \frac{n}{n+1} \left(\frac{\Delta pR^{n+1/n}}{2mL} \right)^{1/n} \tag{6.25}$$

$$\text{Wall shear rate} \quad \dot{\gamma}_w = \frac{(3n+1)}{4n} \frac{8\bar{v}_z}{D} \tag{6.26}$$

$$\text{Wall shear stress} \quad \tau_w = m \left(\frac{(3n+1)}{4n} \frac{8\bar{v}_z}{D} \right)^n \tag{6.27}$$

$$\text{Reynolds number} \quad \text{Re}_{PL} = \frac{\rho \bar{v}_z^{2-n} D^n}{8^{n-1} m \left(\frac{3n+1}{4n} \right)^n} \tag{6.28}$$

Figure 6.10 shows the velocity distribution for the flow of a power low fluid in a tube for different values of the power index.

6.5 EFFECT OF RHEOLOGY IN MINERAL PROCESSING

Particle size and concentration of a suspension affect the efficiency in all the process in a concentrator due to its effect on suspension rheology. The viscosity of the

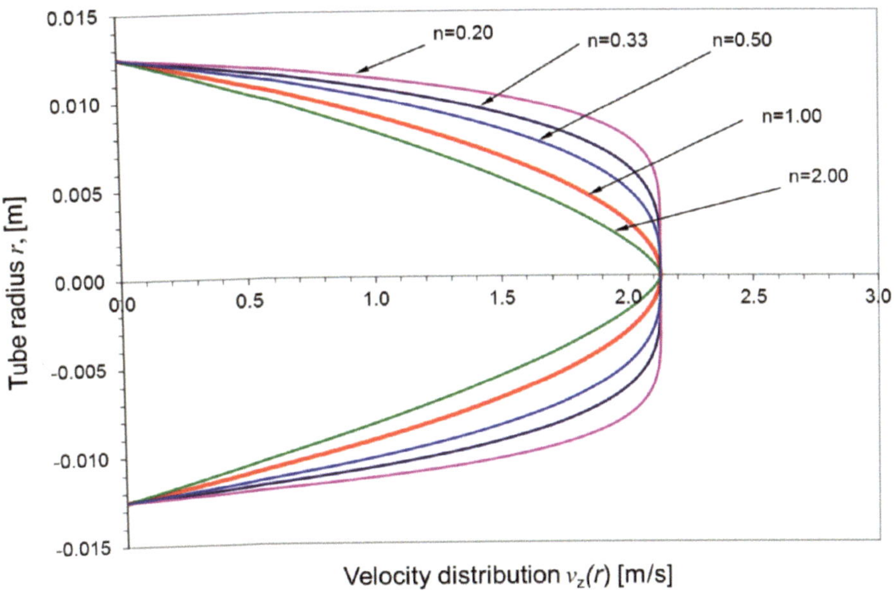

FIGURE 6.10 Velocity distribution of power low fluids with consistency index $m = 3\text{Pa} - \text{s}^{\text{n}}$ and power low indices 0.20; 0.33; 0.50; 1 and 3.

Source: Concha, F.A. 2014. Solid Liquid Separation in the Mining Industry. Cham: Springer.

suspension of a non-Newtonian fluid depends on these two variables in the form given by Equation 6.10:

$$\eta(\gamma,\ \varphi)=m\bar{\gamma}^{\text{n}-1}\times\exp\left(a\gamma^{-b\varphi}\right) \tag{6.29}$$

Where a and b are parameters, and n is the parameter of the power law model.

6.5.1 Comminution

In a grinding mill full of balls, the feed passes through the ball charge to reach the outlet. The ball charge is in rotational movement forming a porous bed through which the pulp moves (Ergun 1952). Ergun considered the static problem with no movement of this porous bed.

A single-phase flow, water-solid particles, through a porous medium can be represented by Darcy's law.

$$q(t)=\left(\frac{k(\varepsilon)}{\eta(\gamma,\varphi)}\right)\frac{\Delta p}{L} \tag{6.30}$$

Where $q(t)$ is the volume average or percolation velocity, $k(\varepsilon)$ is the permeability of the porous bed, ε is the porous bed formed by the balls of diameter d, μ is the fluid viscosity given by, and $\Delta p / L$ is the pressure gradient.

$$q = \left(\frac{k(\varepsilon)}{\eta(\gamma,\varphi)} \right) \frac{\Delta p}{L} \tag{6.31}$$

Tupper et al. introduced the slip velocity between the fluid and the balls to take into account the dynamics of the system, but since the balls are considered moving mainly in circular form, its axial velocity in the mill can be considered zero, and therefore the simple static model would be applicable.

The pulp fluid velocity is related to the volume average velocity by $v_f = q\varepsilon \equiv q(1-\varphi)$, where φ is the solid volume fraction in the pulp, and the pressure gradient through the mill is proportional to the level height difference of the pulp between the feed and discharge side of the mill, therefore the solid feed rate in the mill can be expressed as follows:

$$F(t) = \frac{6}{180} \left\{ \frac{\rho_s}{\varphi^2(t)} \right\} \left(\frac{\rho g D \Delta h}{L} \right) \left\{ \frac{1}{m\bar{\gamma}^{n-1} \times \exp\left(a\bar{\gamma}^{b\varphi}\right)} \right\} \tag{6.32}$$

Appendix 1 shows the derivation of equation 6.32.

6.5.2 Flotation

The small particles, especially clay minerals, in a suspension have several deleterious effects on flotation. They have small inertia and follow the streamlines in the flow evading interception with bubbles lowering the flotation efficiency (Miettinen et al. 2010); they transfer large quantities of slimes into the concentrate, lowering the concentrate grade; they adsorb a large amount of reagents due to their large surface area increase reagent consumption, increasing pulp viscosity producing a decrease in froth stability (Farrokhpay and Bradshaw 2012).

6.5.3 Solid-Liquid Separation

High viscosity of a pulp produced by slimes hinder the water recovery from thickeners and tailing dams and increase the dosage of flocculant. Filtration becomes more difficult with flocculated feed pulps.

6.6 PRACTICAL CONCLUSIONS

In all these suspensions systems, there are nonlinearities caused by the concentration of solids, particle size distributions, and density of the particles. As a practical advice for rock of density 2.7, the cyclone feed maximum is 65% solids by weight, the

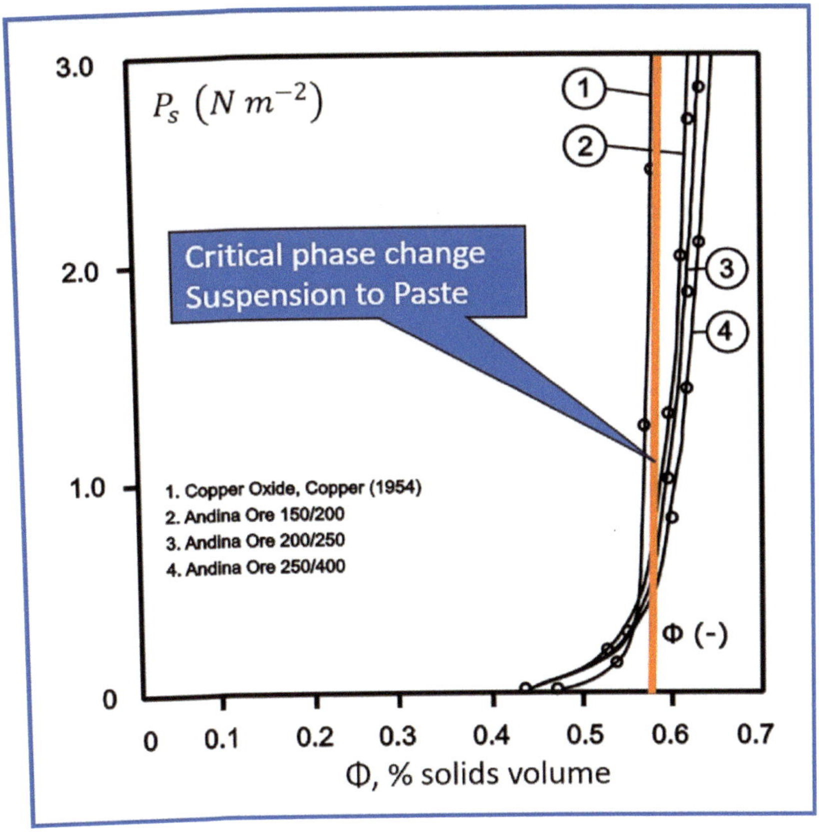

FIGURE 6.11 Compression data for copper sulfide and copper oxide ores.

cyclone overflow is in the range of 34%–45% solids by weight, cyclone underflow is 70%–80% solids, in a ball mill the maximum percent solids are in the order of 75%–77% solids by weight, and a thickener underflow is in the order of 60%–65% solids by weight. In aerated pulps such as the flotation cells, everything changes to a typical % solids by weight of 35% solids with an air holdup of about 7%–30% range depending on the scale of the turbulence available, as we will discuss in the particulate flotation chapter.

Bascur and Barrientos (1977) and Bascur (1991) have generalized the effect of the percent solids, density of the ore, and different particle sizes in the suspension. As shown in Figure 6.11 in the suspension compression study, the drastic change from suspension happens very sharply at 0.57 to 0.60 volume fraction of solids in a suspension. This critical solid volume fraction tells us the suspension is passing a fluid suspension into a paste state with very low flowing capabilities ("like water going into ice"). This study was made for copper sulfide ore with a specific density of 2.7 for a range of particle size between 150 to 400 mesh. These results are used to define process constraints for online constraint predictions. The key is to avoid costly constraint violations that might interrupt industrial operations or that damage the process equipment. These slurry changes are also detected by filtering the power amps measurements in the slurry pumps feeding the hydrocyclone banks and the thickener underflow streams.

The regression equation of this set of data is helpful in predicting the change from suspension to paste in grinding, classification, and thickening to avoid costly pumping, mill overloads, or cyclone roping problems in the mineral processing industry. From the data we can obtain an Apparent Viscosity = Viscosity (%SV, PSDS (M), RHO Rock) with a form Apparent Viscosity = K * (% solids)$^{-19}$. This empirical correlation is obtained from the data available in an industrial site.

REFERENCES

Barnes, H.A., Hutton, J.F., and Walters, K. 1989. *An Introduction to Rheology*. Amsterdam: Elsevier.

Bascur, O.A. 1991. A unified solid liquid separation framework. *Fluid Particle Separation Journal* 4:117–122.

Bascur, O.A., and Barrientos, A. 1977. Phenomenological model of sedimentation: Experimental parameters. Presented at the IV Chilean Chemical Engineering Congress, University de Concepcion.

Bingham, E.C. 1922. *Fluidity and Plasticity*. New York: McGraw-Hill. p. 215.

Chhabra, R.P., and Richardson, J.F. 1999. *Non-Newtonian Flow in the Process Industry*. London: Butterworth-Heinemann. p. 436.

Concha, F., Barrientos, A., Torres, V., and Alejo, B. 1991. *Suspension Rheology*. Concepcion, Chile: Universidad de Concepción (in Spanish).

Concha, F.A. 2014. *Solid Liquid Separation in the Mining Industry*. Cham: Springer.

Ergun, S. 1952. Fluid flow through packed columns. *Chemical Engineering Progress* 48:89–94.

Farrokhpay, Saeed and Bradshaw, Dee J. 2012. *Effect of clay minerals on froth stability in mineral flotation: a review*. XXVI International Mineral Processing Congress – IMPC 2012, New Delhi, India, 24–28 September 2012. New Delhi, India: Technowrites. https://espace.library.uq.edu.au/view/UQ:282993

Gottier, M.A., and Pfeiffer, R. 2013. The Flow of Wet Water, California Institute of Technology, The Feynman Lecture on Physics, California. www.feynmanlectures.caltech.edu/II_41.html#Ch41-S1

Krieger, I.M., and Doherty, T.J. 1959. A mechanism for non-Newtonian flow in suspensions of rigid spheres. *Transactions of the Society of Rheology* 3:137–152.

Macosko, C.W. 1994. *Rheology: Principles, Measurements and Applications*. New York: CCH Publishers.

Miettinen, T., Ralston, J., and Fornasiero, D. 2010. The limits of fine particle flotation. *Minerals Engineering* 23:420–437.

APPENDIX 6.1

FORMULATION FOR EQUATION 6.32

Shear velocity	γ
Shear rate	$\tau = \eta(\gamma, \varphi)\gamma$
Shear viscosity	$\eta(\gamma, \varphi) = m\bar{\gamma}^{n-1} \times \exp\left(a\bar{\gamma}^{b\varphi}\right)$
Porosity	$\varepsilon = 1 - \varphi$
Solids volume fraction	φ
Pulp density	$\rho = 100 \times f = \dfrac{\rho_s \times \rho_f}{\rho_s \times (100 - w) + \rho_{f^w}}$
Percent solids by weight	ω
Percolation volumetric flow	$q = \left(\dfrac{k(\varepsilon)}{\eta(\gamma, \varphi)}\right)\dfrac{\Delta p}{L}$
Pulp velocity	$v_f = \varepsilon q$
Mill flux area	$S = \dfrac{\pi}{4} D^2$
Volumetric pulp flow	$Q_F = v_f S$
Mass flow rate	$F = \rho_s Q_F \times \dfrac{\varphi}{1 - \varphi}$
Pressure gradient	$\dfrac{\Delta p}{L} = \rho g \varepsilon \times \dfrac{\Delta h}{L}$
Porous media (balls) permeability	$k(\varepsilon) = \dfrac{1}{180}\left(\dfrac{6}{S}\right)\dfrac{\varepsilon^3}{(1-\varepsilon)^3}$

$$\frac{v_f S}{\varepsilon} = \left(\frac{1}{\eta(\gamma, \varphi)}\right)\left(\frac{1}{180}\right)\left(\frac{6}{S}\right)\left(\frac{\varepsilon 2}{(1 - \varepsilon 3)}\right)\rho g \varepsilon\left(\frac{\Delta h}{L}\right)$$

$$v_f S = \frac{6}{180}\left(\frac{1-\varphi}{\varphi^3}\right)\left(\frac{\rho g D \Delta h}{L}\right)\left(\frac{1}{m\bar{\gamma}^{n-1} \times \exp\left(\pm\bar{\gamma}^{b\varphi}\right)}\right)$$

$$Q_F = \frac{6}{180}\left(\frac{1-\varphi}{\varphi^3}\right)\left(\frac{\rho g D \Delta h}{L}\right)\left(\frac{1}{m\bar{\gamma}^{n-1} \times \exp\left(\pm\bar{\gamma}^{b\varphi}\right)}\right)$$

$$F(t) = \frac{6}{180}\left\{\frac{\rho_s}{\varphi^2(t)}\right\}\left(\frac{\rho g D \Delta h}{L}\right)\left(\frac{1}{m\bar{\gamma}^{n-1} \times \exp\left(\pm\bar{\gamma}^{b\varphi}\right)}\right)$$

7 Particulate Comminution

7.1 INTRODUCTION

Comminution is the operation of *size reduction* of a material with the objective to obtain a material with particle size distribution suitable for its direct use or for its subsequent processing by the metallurgical industry.

As we have indicated in Chapter 1, size reduction is accomplished in two stages, *crushing* for reducing larger block sizes to a size of approximately 50 to 10 mm (3/8 in), a job that takes up to four steps, and *grinding* for particle size less than 50 mm, which is usually performed in a dry or wet environment. This chapter will provide an overview of the fundamentals of the grinding processes (Austin et al. 1984; Austin and Concha 1994; Herbst et al. 1973, 1985) and present recent implementations of software for optimizing grinding circuits using online process data.

7.1.1 Tumbling Mills

The crushed product with 50 mm is reduced to values of up to 10 μm. Depending on the fineness of the product required, the process is divided into *primary*, *secondary*, and *tertiary* grinding. The most used equipment in grinding is the *tumbling mill* also known as *rotary mill*. Primary mills use steel bars as grinding charge media *bar mills* and secondary and tertiary grinding are done in *ball mills* using steel balls as grinding charge or grinding media. Figure 7.1 shows typical rod mills and Figure 7.2 ball mills.

FIGURE 7.1 Rod mills.

Source: SolidsWiki Contributors. 2013. Rod mills. *SolidsWiki*. www.solidswiki.com/index.php?title=Rod_Mills&oldid=15378. Accessed July 2023.

DOI: 10.1201/9781032614212-7

FIGURE 7.2 Ball mills.

Source: Jankovic, A. 2015. Chapter 8: Developments in iron ore comminution and classification technologies. In *Mineralogy, Processing and Environmental Sustainability.* Edited by B. Lu. New York: Woodhead.

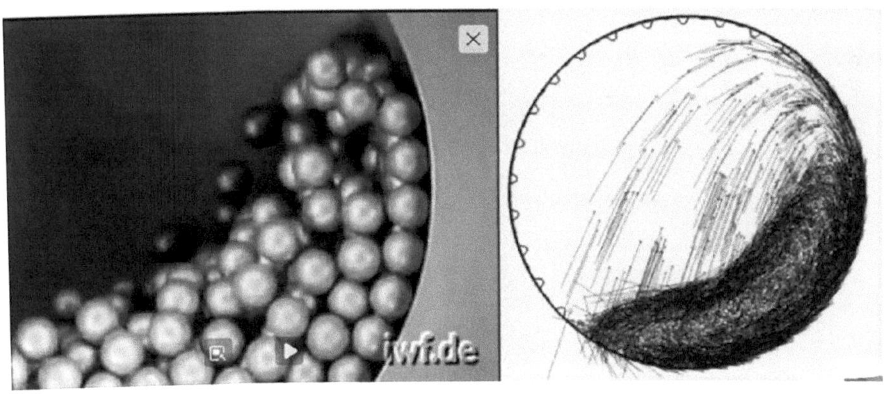

FIGURE 7.3 Motion of the ball charge in a ball mill.

The reduction ratio, which is the ratio between the initial particle size broken to the final particle size, is higher in mills than in crushers. The primary mills are of the order of 5/1, while for secondary and tertiary mills, it increases to values of up to 30/1, and sometimes even more than 100/1.

7.1.2 Ball Charge Motion

The grinding mechanism, the same for primary, secondary, or tertiary mills, is based on the rotation of its outer cylinder wall that lifts the bars or balls which fall on the load of ore fracturing it by impact and friction. The movement of the grinding media can be seen in Figure 7.3.

Critical speed is that in which a ball becomes attached to the top mill's wall against gravity (see Figure 7.4a) so that a momentum balance reads as follows:

$$\underbrace{m\omega_c^2 D/2}_{\text{Centripetal force}} - \underbrace{mg}_{\text{Ball nweight}} = 0 \tag{7.1}$$

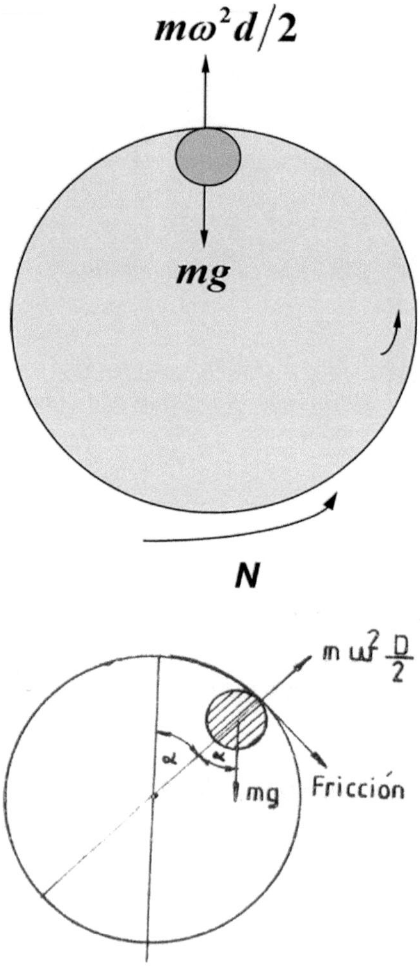

FIGURE 7.4 Ball position at critical velocity: (a) at critical velocity, (b) at detachment velocity.

Considering that the relationship between the angular velocity ω and revolution per minute N (rpm) is $\omega \equiv 2\pi N_c \cdot [\mathrm{m/s}]$,

$$\omega = \sqrt{\frac{2g}{D}}\,[\mathrm{rad}/s] \qquad N_c = \frac{60}{\pi}\sqrt{\frac{2g}{D}}\,[\mathrm{rpm}] \tag{7.2}$$

since $g = 9.81\left[\mathrm{m/s^2}\right]$.

$$\text{for: D [m]} \quad N_c = \frac{42.3}{\sqrt{D}}[\mathrm{rpm}] \text{ and for: D [in] } N_c = \frac{76.6}{\sqrt{D}}[\mathrm{rpm}] \tag{7.3}$$

Since a mill should operate with the ball charge sliding and falling over the particles, the actual velocity should be between 50% and 90% of the critical, which should be determined requiring maximum energy on impact.

7.1.2.1 Optimal Balls Trajectory

We said that ball is dragged along the wall as the mill rotates. At critical speed, the ball reaches the top of the mill without falling. At lower speeds it will detach when it reaches an angle α, as shown in Figure 7.4b. It is important for us to know at what angle the trajectory of the ball will produce the greatest kinetic energy of impact on the wall of the mill, what is its trajectory, how long it will take to get there, and at what percentage of the critical speed will it happen? See Figure 7.5.

The mass balance in this case is

$$m\omega^2 D/2 - mg\cos\alpha = 0 \tag{7.4}$$

Then

$$\omega = \left(\frac{2g\cos\alpha}{D}\right)^{0.5} \tag{7.5}$$

The fraction of critical velocity is

$$N^* = N/N_c = \omega/\omega_c \quad \omega = N^*\omega_c \tag{7.6}$$

From (7.4),

$$\cos\alpha = \omega^2 D/2g \tag{7.7}$$

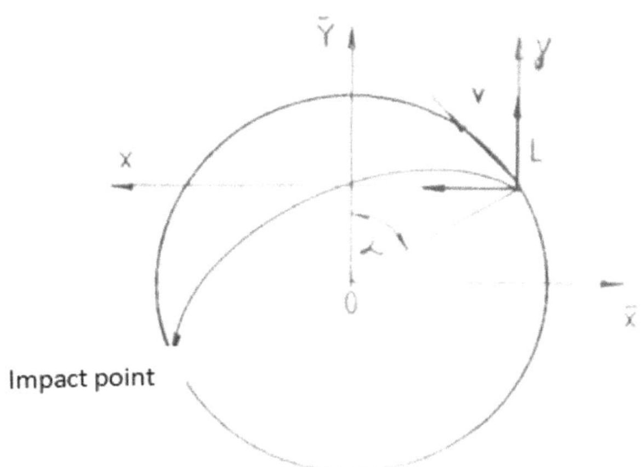

FIGURE 7.5 Ball trajectory.

so that substituting the last expressions of Equation 7.6 in Equation 7.7 gives $\cos\alpha = N^{*2}\omega_c^2 D/2g$, and replacing ω_c from Equation 7.6 gives

$$N^{*2} = \cos\alpha \tag{7.8}$$

Calling $v_0 = \omega R$ the ball initial speed

$$v_0 = \left(\frac{D}{2} g\cos\alpha\right)^{0.5} \tag{7.9}$$

The components of the initial velocity are $v_x = v_0\cos\alpha$ and $v_y = v_o sen\theta - gt$, eliminating v_0 by combining both equations. The component of the trajectory of the ball is $x = v_0\cos\alpha \times t$ and $y = v_0(sen\alpha)\times t - (g/2)\times t^2$. Removing t from these two equations results in the trajectory

$$y = x\tan\alpha - \frac{x^2}{D\cos^3\alpha} \tag{7.10}$$

The ball will hit the mill's wall when its trajectory cuts the circle $\overline{x}^2 + \overline{y}^2 = R^2$ where $\overline{x} = x - Rsen\alpha$ and $\overline{y} = y + R\cos\alpha$. Introducing Equation 7.10 and the values of $\overline{x}$ and $\overline{y}$, we find the equation $x^3 \times \left(\frac{x}{2D\cos^2\alpha} - \text{sen}\alpha\right) = 0$ with roots 0 at point L and $x_T = 2D\text{sen}\alpha\cos^2\alpha$. The time to reach point T is $t_T = (x_T / v_{xT}) = \frac{2D}{v_0}\text{sen}\alpha\cos\alpha$, and the velocity at the point of contact T is $v_T = \left(v_{xT}^2 + v_{yT}^2\right)^{0.5}$ with $v_{xT} = v_0\cos\alpha$ and $v_{yT} = v_o sen\theta - gt_T$. Replacing the terms in v_T with v_0 given in Equation 7.9 yields for the impact velocity

$$v_T = \left(\frac{D}{2} g\cos\alpha \times \left(108\text{sen}^2\alpha\right)\right)^{0.5} \tag{7.11}$$

The kinetic energy at impact is $K \equiv (1/2)v_T^2$

$$K = \frac{mDg}{4}\left(\cos\alpha \times 8\text{sen}^2\alpha\cos\alpha\right) \tag{7.12}$$

The maximum energy is obtained for $dK/d\alpha = 0$, then

$$\frac{dK}{d\alpha} = \frac{mDg}{4} sen\alpha\left(15 - 24sen^2\alpha\right) = 0 \tag{7.12a}$$

Then

$$\sin\alpha = 0.79;\ \alpha = 52.24^{\circ} \text{ and } \cos\alpha = 0.61$$

Substituting in Equation 7.8 gives the value of N^* and the percent of the critical value

$$100N^* = 100(\cos\alpha)^{0.5} = 0.78\% \tag{7.13}$$

7.1.3 Ball Wear in a Tumbling Mill

Consider the case of the wear of balls in a tumbling mill, with a continuous feeding of new balls and discharge of the worn balls, as shown in Figure 7.6.

Ball wear is a very simple case of a population balance. With respect to the balls, the mill is considered a fully mixed continuous process reactor where steel is lost from the balls by wear, without no fractures of balls and where they are transported from the input to the output, therefore the model based on the space-averaged population balance is shown in Equation 7.14. Using Figure 7.7 we apply the Population Balance to the Process (Sastry and Concha 1985).

We will assume that no breakage of balls occurs and that the change of size of the balls is a linear continuous function of the ball size. Consider the volume average number of ball size $\bar{\psi} = N(d)$ as the only coordinate in a tumbling mill, and define the of total number of balls in the holdup of balls by $V(t) = V_B(d,t)$, the convection of balls equal to the net in-outflow of balls in the mill $\phi_I(d_I) - \phi_O(d_O)$, and the internal convection flux, wear of the balls in $V_B(t)$ with the term $g(d) = \mathrm{d}(d)/\mathrm{d}t < 0$, where $g(d)$ is the rate of decrease in diameter of a ball of size d. Then, the evolution of the ball charge sizes in the mill is given by

$$\frac{\partial}{\partial t}N(d,t) = -\sum_{j=1}^{n}\frac{\partial}{\partial d}(N(d,t)g(d)) + \phi_I(d_I,t) - \phi_O(d_O,t) \tag{7.14}$$

Since the ball wear in a tumbling mill is a very slow process, the usual industrial practice is charging balls once a day or once a shift, which may be considered continuous in time regarding the long time-intervals involved in the process. On the other hand, a limited number of ball size classes are charged to the mill, and approximately one size class is discharged at the outflow. Figure 7.8 shows the distribution of balls at the inlet and outlet of an industrial grate discharge mill.

The inflow and outflows of balls to a mill may be regarded as a continuous function of sum of several impulses in the sizes d_k, with $k = 1, 2, \ldots p$:

$$\phi_I(d,t) = \Phi_I \sum_{k=1}^{p} m_o^I(d)\delta(d - d_k) \tag{7.15}$$

$$\phi_o(d_o,t) = \Phi_o m_o^o(d)\delta(d - d_o) \tag{7.16}$$

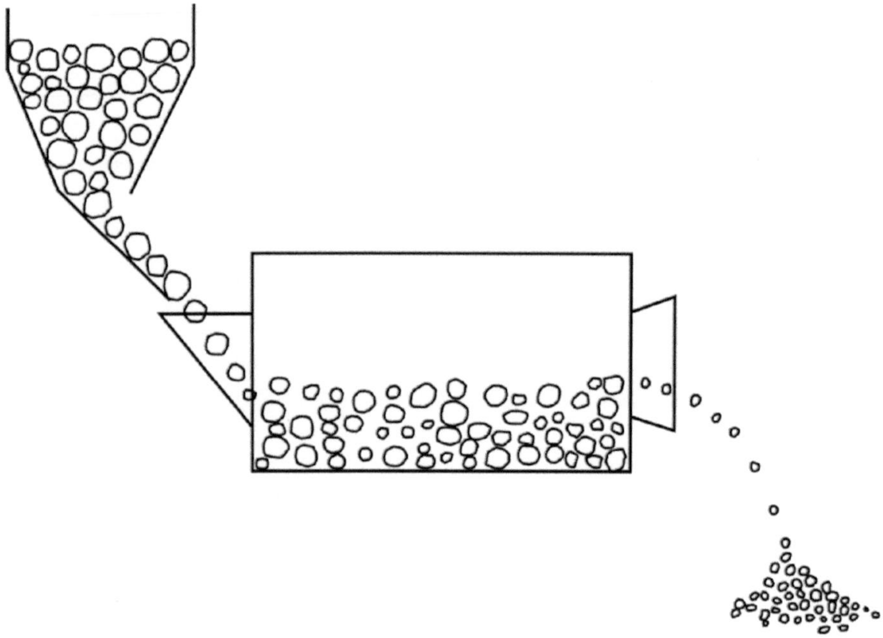

FIGURE 7.6 Balls inflow and outflow in the mill.

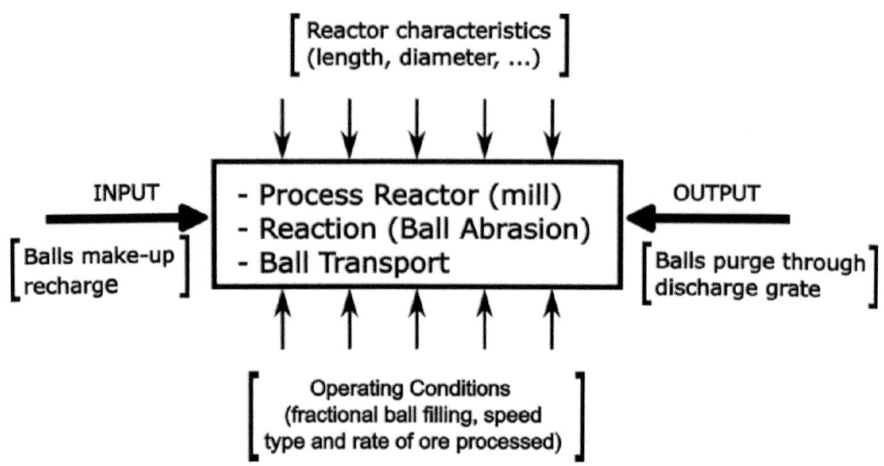

FIGURE 7.7 Ball wear represented as a rate process.

Where Equations 7.15 and 7.16 represent respectively the ball flow rate in sizes d to $\mathrm{d}+d(d)$ at time t at the inlet and outlet of the mill and $\sum_{k=1}^{p} m_o^I(d)$ is the relative frequency number of the balls at the inflow and δ is the Dirac Delta function defined by

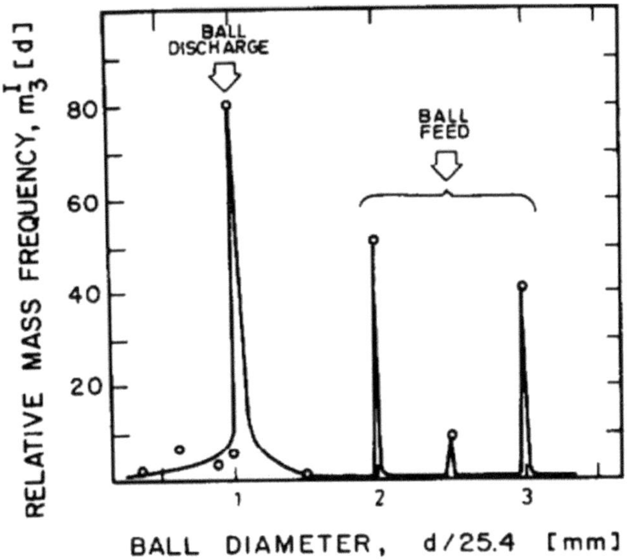

FIGURE 7.8 Numbers of sized d_I and d_O in the feed and discharge ball size distribution in a 3.7 m by 4.9 m ball mill.

Source: Menacho, J., and Concha, F. 1986. Mathematical model of ball wear in grinding mills. *Powder Technology* 47:87–96.

$$\begin{cases} \delta\left(d-d_k\right)=0, \text{for } d \neq d_k \\ \int_{-\infty}^{+\infty} \delta\left(d-d_k\right) \mathrm{d}(d)=1 \end{cases} \tag{7.17}$$

Substituting Equations 7.15 and 7.16 in Equation 7.14 and under the initial condition $N(d,0)=N_0(d)$ and $N\left(d_0,t\right)=0$ yields the general spaced average population balance for the ball wear in a tumbling mill (Menacho and Concha 1986):

$$\underbrace{\frac{\partial N(d,t)}{\partial t}}_{\substack{\text{Rate of change of number of balls} \\ \text{in the mill with sizes between d} \\ \text{and } d+\mathrm{d}(d) \text{ in the ball charge.}}} + \underbrace{\frac{\partial}{\partial d}[N(d,t)g(d)]}_{\substack{\text{Rate of change of the ball with} \\ \text{sizes between } d \text{ and } d+\mathrm{d}(d) \text{ in} \\ \text{the ball charge due to the ball} \\ \text{wear kinetics g(d).}}} = \underbrace{\Phi_I(d,t)\sum_{k=1}^{p} m_o^I(d)\delta\left(d-d_k\right) - \Phi_o\left(d_o,t\right)\delta\left(d-d_o\right)}_{\substack{\text{Total number of balls with size} \\ \text{between d and d+d}(d)\text{ flowing in or} \\ \text{out of the ball charge.}}} \tag{7.18}$$

In this equation, g is the rate of decrease in diameter of the ball of sized d.

Many industrial mills show a first order rate of decrease in diameter such as shown in Figure 7.9.

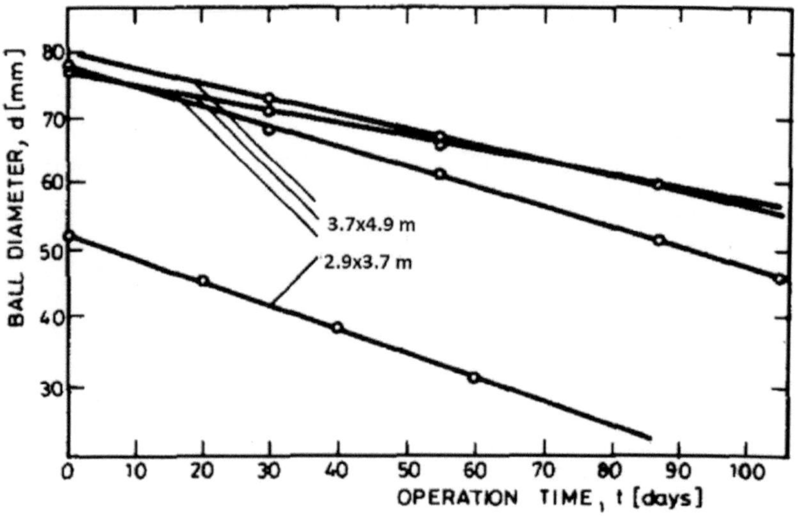

FIGURE 7.9 Data obtained from three industrial ball mills with three different ball qualities.
Source: Menacho, J., and Concha, F. 1986. Mathematical model of ball wear in grinding mills. *Powder Technology* 47:87–96.

When $g = \alpha$, this is zero order, and the equation becomes

$$\underbrace{\frac{\partial N(d,t)}{\partial t}}_{\substack{\text{Rate of change of number of balls} \\ \text{in the mill with sizes between d} \\ \text{and } d+\mathrm{d}(d) \text{ in the ball charge.}}} + \underbrace{\alpha \frac{\partial}{\partial d}[N(d,t)]}_{\substack{\text{Rate of change of the ball with} \\ \text{sizes between } d \text{ and } d+\mathrm{d}(d) \text{ in} \\ \text{the ball charge due to the ball} \\ \text{wear kinetics if zero order.}}} = \underbrace{\Phi_I(d,t)\sum_{k=1}^{p} m_o^I(d)\delta\left(d-d_k\right) - \Phi_o(d_o,t)\delta\left(d-d_o\right)}_{\substack{\text{Total number of balls with size} \\ \text{between d and } \mathrm{d}+\mathrm{d}(d) \text{ flowing in or} \\ \text{out of the ball charge.}}} \quad (7.19)$$

Equation 7.19 was solved with the Lagrange method of characteristic with initial condition $N(d,0) = N_0(d)$ giving (Menacho and Concha 1986)

$$N(d,t) = N_0(d-\alpha t) + \frac{\phi_1}{\alpha}\sum_{k=1}^{p} m_0^I\left(d_k\right)\times\left(U\left(d-d_k\right) - U\left(d-\alpha t-d_k\right)\right) \quad (7.20)$$

where

$$U\left(d-d_k\right) = \begin{cases} 1 \text{ for } d > d_k \\ 0 \text{ for } \mathrm{d} < \mathrm{d}_k \end{cases} \text{ and } U\left(d-\alpha t-d_k\right) = \begin{cases} 1 \text{ for } d-\alpha t > d_k \\ 0 \text{ for } \mathrm{d}-\alpha t < \mathrm{d}_k \end{cases} \quad (7.21)$$

7.1.4 Ball Mass Size Distribution in the Mill at Steady State

Defining the mass size distribution $M_3^{ss}(d)$ in the mill at steady state by

$$M_3^{ss}(d) = \frac{\int_{d_0}^{d} N(\xi)\xi^3 d\xi}{\int_{d_0}^{d_1} N(\xi)\xi^3 d\xi} \quad (7.22)$$

Substituting Equation 7.21 in Equation 7.22 yields the final equation for the steady state mill ball mass distribution

$$M_3^{ss}(d) = \frac{\left(d^4 - d_0^4\right) - \sum_{k=1}^{p} d_k^{-3} m_3^I(d_k)\left(d^4 - d_k^4\right)U(d - d_k)\left[\sum_{k=1}^{p} d_k^{-3} m_3^I(d_k)\right]^{-1}}{\left(d^4 - d_0^4\right) - \sum_{k=1}^{p} d_k^{-3} m_3^I(d_k)\left(d^4 - d_k^4\right)\left[\sum_{k=1}^{p} d_k^{-3} m_3^I(d_k)\right]^{-1}} \quad (7.23)$$

where $m_3^I(d_k)$ is the relative mass frequency function.

It is interesting to point out that the shape modulus of $m = 3.8$ for the Gaudin Schumann plot presented by Bond (1960) is close to that obtained with single-size ball recharges in an overflow mill, that is, for $p = 1$ and $d_0 = 0$

$$M_3^{ss}(d) = \left(\frac{d}{d_1}\right)^4 \quad (7.24)$$

7.1.4.1 Balls Holdup

A good industrial practice is to maintain a constant ball charge level in a mill. If the inflow of balls is known, and the size of the balls discharged and the wear kinetics are also known, the ball holdup W_B may be calculated. Since the mass frequency function of a ball of size d at the feed is $m_3^I = \rho(\pi/6)d^3$,

$$W_B = \int_{d_0}^{d_1} m_3^I N(d,t)d(d) \quad (7.25)$$

Integrating between the smallest ball size d_0 to the size of the balls fed to the mill d_1, the holdup at steady state results in

$$W_B^{ss} = \frac{\pi\rho\phi_1}{24\alpha} \frac{\sum_{k=1}^{p} d_k^{-3} m_3^I(d_k)\left(d_0^4 - d_k^4\right)}{\sum_{k=1}^{p} d_k^{-3} m_3^I(d_k)} \quad (7.26)$$

Where ϕ_1 is the inflow of balls, ρ is the ball density, α is the rate of decrease of a ball diameter (Menacho and Concha 1986).

7.1.4.2 Steel Consumption

Based on Equation 7.26, the total steel consumption for a grate discharge mill at steady state C_T^{ss} is

$$C_T^{ss} = \frac{4\alpha W_B^{ss}}{\sum_{k=1}^{p} d_k^{-3}\left(d_0^4 - d_k^4\right) m_3^I\left(d_k\right)} [\text{ton}/\text{h}] \tag{7.27}$$

For the special case of an overflow mill at steady state, Equation 7.27 simplifies to

$$C_T^{SS} = 10^6 \frac{C_T^{SS}}{Q} [\text{g}/\text{ton}] \tag{7.28}$$

7.1.5 Optimization of the Make-Up Ball Charge

The efficient grinding of the different particle sizes in a tumbling mill requires the existence in the mill of a variety of ball sizes, where each ball size maximizes the breakage of each particle size range. Due to wear in the abrasive environment in the mill, it is necessary to charge periodically new balls to maintain a steady balanced ball charge in the mill. The industrial practice of charging balls to a tumbling mill is a matter of experience. From the models of a ball mill and ball charge wear that we have seen in this chapter and the model of classification found in Chapter 9, the optimum choice of make-up balls with regards to both the steel and energy consumption may be estimated by an adequate optimizing procedure (Concha et al. 1992).

If the objective of the grinding process is to maximize the output rate of the grinding circuit to produce a material with 80% less than 65 mesh and steel consumption from ball wear restricted to a maximum of 500 g/ton, that is

$$Max\left(I = Q\left(m_k\right)\right) \tag{7.29}$$

With restrictions:

$$0 \le m \le \text{with}\,\text{k} = 1, \cdots p\,\text{and} \sum_{k=}^{p} m_k = 1, \quad F\left(x_j\right) = F^*\left(x_j\right), C_T^{ss} \le C_T^{ss^*} \tag{7.30}$$

For the case of $F^* = 0.8$ and $C_T^{ss^*} = 500[\text{g}/\text{ton}]$, and several values for the grinding and ball wear parameters (Concha et al. 1992), the results predicted that an increase in 12% in the circuit capacity could be obtained by changing a mono-sized ball charge of 2 [in] to an optimum make-up ball charge of 1.5 [in]. The result of the simulation is shown in Figure 7.10. A stricter mathematical version that includes the transient process of ball wear was presented by Bürger et al. (2018).

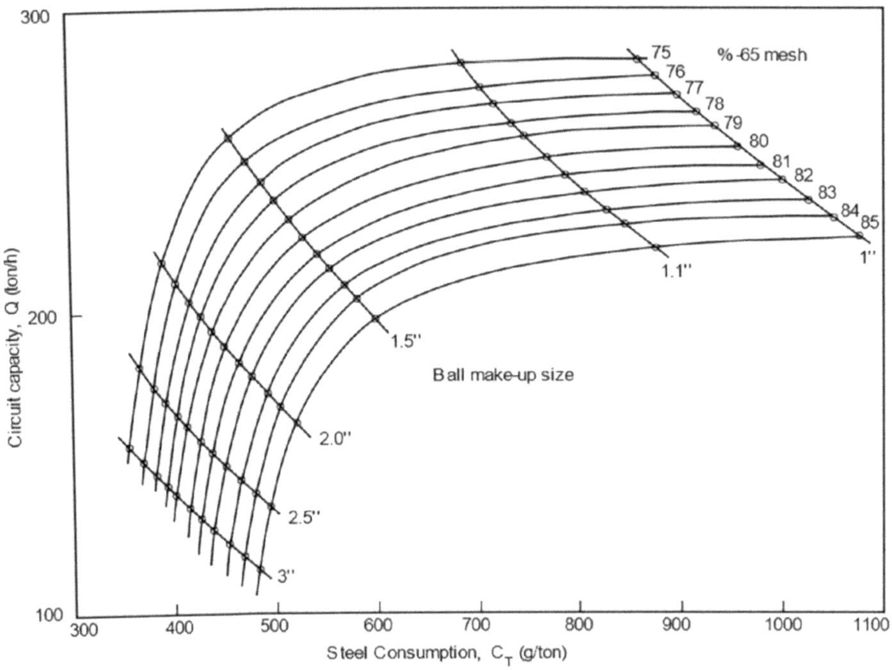

FIGURE 7.10 Circuit capacity versus steel consumption for mono-sized ball make-up meeting the product specification.

7.1.6 Batch Grinding in a Ball Mill

Consider a discontinuous laboratory ball mill as a well-mixed reactor containing a mass W of coarse ore that will experience fracture for a time t. If the feed to the mill is restricted to one particle if size i, experience has shown that the rate of disappearance by fracture of the mass of the fraction of this size is governed by a law of the first order (see Figure 7.11).

Applying the population balance to the fracture in a batch mill, we have

$$W(t)w_i(t) = -D\overline{\psi} + C\overline{\psi} \tag{7.31}$$

where $A = WS_i$ and $C = WB_{ij}$. The rate of breakage S_i may be fitted with the expression (Austin et al. 1984)

$$S_i = a_T\left(\frac{x_i}{x_j}\right)^a \tag{7.32}$$

The progeny obtained from the fracture of mass size w_j to size w_i will be designated by b_{ij}. According to Austin et al. (1984) a useful representation of the breakage distribution function in terms of three parameters, Φ_j, β and γ

$$B_{ij} = \Phi_j \left(\frac{x_{i-1}}{x_j} \right)^{\gamma} + \left(1 - \Phi_j\right) \left(\frac{x_{i-1}}{x_j} \right)^{\beta} \quad \text{for } 0 \leq \Phi_j \leq 1,\ i > j \tag{7.33}$$

where

$$B_{ij} = \sum_{j=1}^{i-1} b_{ij} \tag{7.34}$$

With the two parameters S_i and b_{ji}, and dividing by the total mass W, the rate of grinding of particles of size i will be

$$\underbrace{\frac{\partial w_i}{\partial t}}_{\substack{\text{Rate of change of} \\ \text{particle of size}\, i}} = \underbrace{-S_i w_i(t)}_{\substack{\text{Rate of dissapearance} \\ \text{of particle of size}\, i}} + \underbrace{\sum_{j=1}^{i-1} b_{ij} S_j w_j(t)}_{\substack{\text{Rate of generation of size}\, i \\ \text{from breakage of size}\, j}} \quad , \text{ for } n \geq i \geq j \geq 1 \tag{7.35}$$

Figure 7.12 shows a plot of the primary breakage distribution function B_{ij} versus the particle size x_i in log-log scale and the resulting particle size distribution.

7.1.7 Effect of Ball Size on the Rate of Breakage

Austin et al. (1984) presented the effect of the ball size on the primary rate of breakage in the following expression:

$$S_i = a(d) x_i^{\alpha} \frac{1}{1 + \left(x_i / \mu(d)\right)} \tag{7.36}$$

where a and μ are parameter functions of the ball diameter d, a correction factor that consider a decline of the rate of breakage when the size of the particles are too big in size for a given ball size to efficiently break. See Figure 7.13.

The figure shows that small balls break small particles faster while the big balls break faster the bigger particles. This conclusion implies that, for a wide range of particles, a mixtures of balls sizes is required.

7.2 CONTINUOUS GRINDING

The characteristic of a ball mill is the good mixture of the load of the mill that prevents defining several stages related to a physical phenomenon of the operation. On the other hand, the behavior as a perfect mixer implies that the mill product has at every moment the same particle size composition of the load, and there is no sorting effect inside a ball mill.

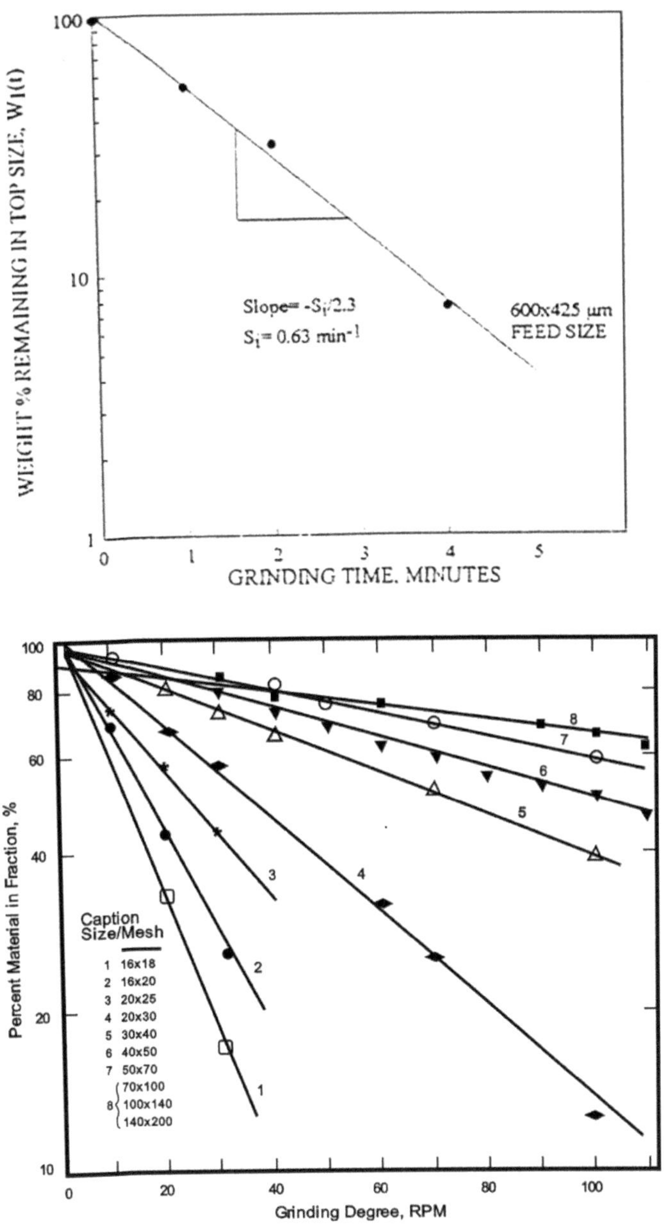

FIGURE 7.11 Rate of fracture of particles percent w_i.

(a) Source: Yekeler M., Ozkan, A., and Austin, L.G. 2001. Kinetics of fine wet grinding in a laboratory ball mill. *Powder Technology* 114:224–228. (b) Source: Austin, L.G., Menacho, J.M, and Pearcy, F. 1987. A general model for semi-autogenous and autogenous milling. In *APCOM 87: Proceedings of the Twentieth International Symposium on the Application of Computers and Mathematics in the Mineral Industries, vol. 2: Metallurgy.* Johannesburg: Southern African Institute of Mining and Metallurgy. pp. 107–126.

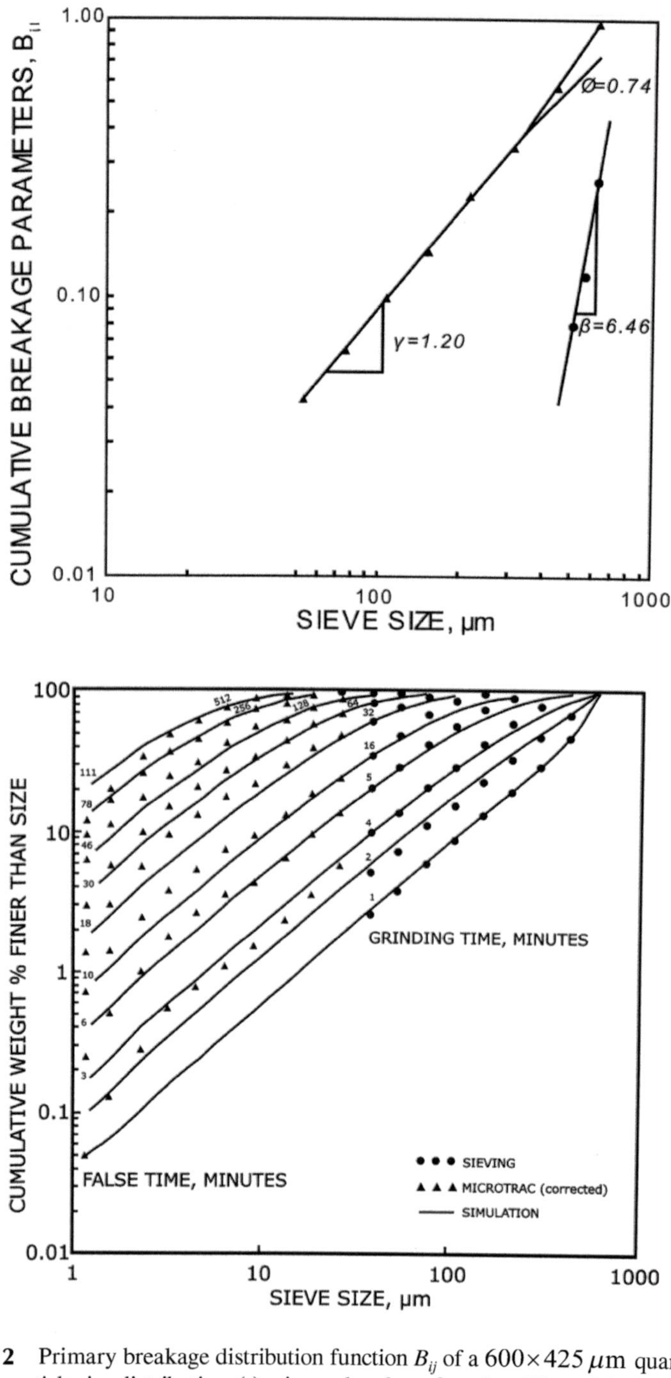

FIGURE 7.12 Primary breakage distribution function B_{ij} of a $600 \times 425\ \mu m$ quartz feed and the resulting particle size distribution: (a) primary breakage function, (b) particle size distribution.

Source: Yekeler M., Ozkan, A., and Austin, L.G. 2001. Kinetics of fine wet grinding in a laboratory ball mill. ***Powder Technology*** **114:224–228.**

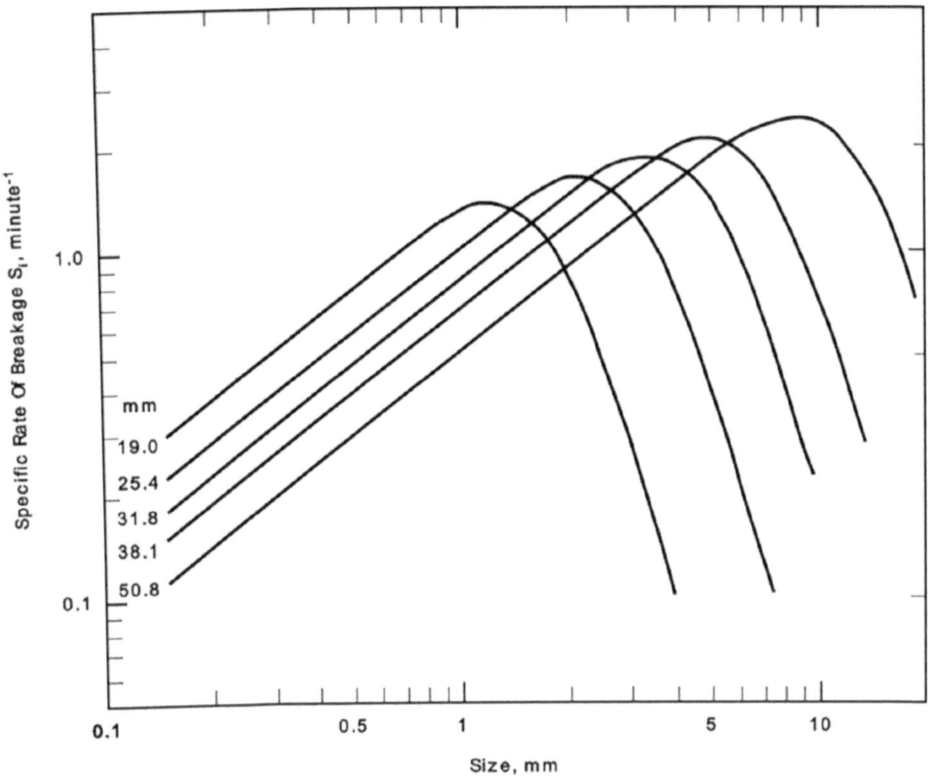

FIGURE 7.13 Effect of ball diameters on the primary rate of breakage.

Source: Austin, L.G., Klimpel, R.R., and Luckie, P.T. 1984. *The Process Engineering of Size Reduction: Ball Milling.* New York: SME-AIME.

7.2.1 Energy Consumption and Size Reduction

For a long time, the study of the relationship between energy consumed and the size reduction that produce was reduced to the expressions proposed by Kick and Rittinger related to the paradox that theoretically materials should be much harder than they are in practice, in which the former assumed that energy was associated with the volume of the rock and the second to the surface.

7.2.1.1 Kick's Postulate (1885)

Kick suggests that "the energy required to produce analogous changes in the size of geometrically similar bodies is proportional to the volume of these bodies," which means that certain amounts of energy should produce equal geometric changes in the size of a solid. Kick's assumption is backed by the theory of solid mechanics.

$$E_K = -\int_{x_0}^{x_c} F(x)dx \tag{7.37}$$

Where E is the work done by the applied force F to a cube of volume x^3 to break a rock to size x_c. Since the force is σx^2 and $\sigma = Y\varepsilon$ where σ is the applied stress and ε the deformation at rupture:

$$E_K = -\int_{x_0}^{x_c} Y\varepsilon x^2 dx = \frac{1}{2} Y\varepsilon \left(x_0^3 - x_c^3\right) \text{ and } \widehat{E}_K = -\underbrace{Y\varepsilon}_{K_K} \int_{x_0}^{x_c} \frac{x^2}{x^3} dx = K_K \ln\left(\frac{x_0}{x_c}\right) \quad (7.38)$$

7.2.1.2 Rittinger's Postulate (1867)

Rittinger assumes that "the energy consumed in the size reduction is proportional to the newly created surface area":

$$\widehat{E}_R = -\underbrace{6C}_{K_R}\left(\left.\frac{x^2}{x^3}\right|_0 - \left.\frac{x^2}{x^3}\right|_c\right) = K_R\left(\frac{1}{x_c} - \frac{1}{x_0}\right) \quad (7.39)$$

7.2.1.3 Bonds Postulate (1952)

As the postulates of Kick and Rittinger did not satisfy all the cases found in practice and, as some rules were needed in industry to classify materials according to their responses to comminution, Bond postulated an empirical law, which says "using the value of/and/for $x_{80}[\mu m]$ to characterize the size of a particle, the specific energy consumed to reduce the size of a material is inversely proportional to the root of that size" (see section 7.2.1.4).

$$\widehat{E}_B = W_i\left(\frac{10}{\sqrt{P}} - \frac{10}{\sqrt{F}}\right) \quad (7.40)$$

Where the parameter W_i is called the Work Index, function of the material that must be determined in the laboratory with the formula

$$W_i = \frac{44.5}{P_1^{0.23} \times Gbp^{0.82} \times \left(\frac{10}{\sqrt{P}} - \frac{10}{\sqrt{F}}\right)} \quad (7.41)$$

In this expression Gbp are the grams of mpol34jwqaterial produced by grinding per moll revolution [g/rev].

7.2.1.4 Empirical Differential Equation

Several authors led by Walker (Walker et al. 1937) realized that Rittinger, Kicks, and Bond postulates could be obtained as a solution of a simple differential equation with values for the value of the parameter $n = 1, 2$ or 3 respectively.

$$\widehat{E} = -C\frac{dx}{dx^n} \quad (7.42)$$

7.2.2 Operating Conditions

Bond (1960) presented several empirical variables of a mill including the power P[kW], specific energy $\hat{E}$(kWh/ton), and capacity F[tph] of a ball mill with diameter $D > 2$[m] and a mass ball charge W_B (Figure 7.14):

$$P[\text{kW}] = 7.33J\phi_c(1-0.937J)\left(1-\frac{0.1}{2^{9-10\phi_c}}\right)\rho_B L D^{2.3}$$

$$\hat{E}[\text{kWh/ton}] = \frac{P}{F} = Wi\left(\frac{10}{\sqrt{P_{80}}}-\frac{10}{\sqrt{F_{80}}}\right),[\text{kWh/ton}]$$

$$F[\text{tph}] = \frac{7.33J\phi_c(1-0.937J)\left(1-\frac{0.1}{2^{9-10\phi_c}}\right)\rho_B L D^{2.3}}{Wi\left(\frac{10}{\sqrt{P_{80}}}-\frac{10}{\sqrt{F_{80}}}\right)} \tag{7.43}$$

$$d_B[\text{mm}] = 0.2393\times F_{80}^{0.5}\times\left(\frac{\rho_s W_i}{\phi_c D^{0.5}}\right)^{1/3}$$

$$\rho_B\left[\text{ton/m}^3\right] = \rho_B(1-J)$$

$$W_B[\text{ton}] = \left(\frac{\pi}{4}D^2L\right)\times J\times\rho_B$$

with Units: $F_{80}, P_{80}[\mu\text{m}], \phi_c[-], J[-], L, D[\text{m}], \rho_B\left[\text{ton/m}^3\right]$

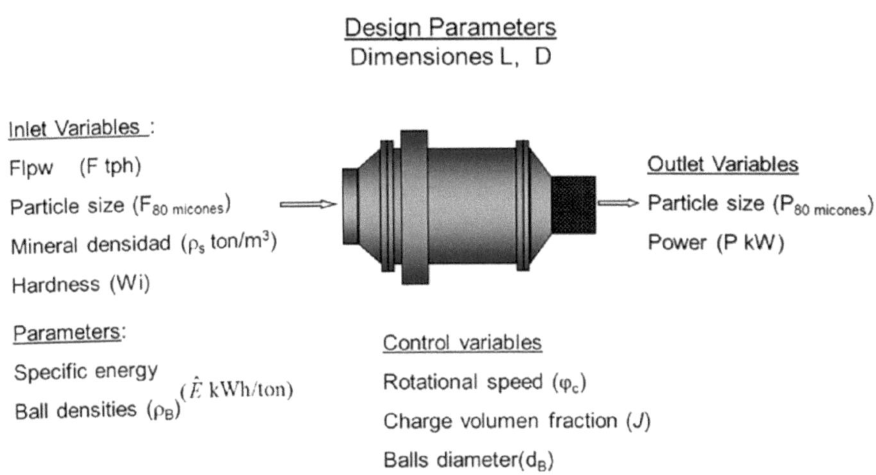

FIGURE 7.14 Variables in a rotary grinding mill.

7.2.2.1 Example

1. Determine the capacity of a ball mill of $D = 5.0[\text{m}]$ in diameter and $L = 6.5[\text{m}]$ in length treating copper ore of density $\rho_s\left[ton/\text{m}^3\right]$ with particle size of $F_{80} = 5000[\text{mm}]$ having a Work Index of $Wi = 13.2[\text{kWh}/\text{ton}]$ and delivering a product of $P_{80} = 200[\text{mm}]$ with a ball volume fraction charge of $J = 0.38$ and $\rho_B = 7.5\left[ton/\text{m}^3\right]$ in density.
2. Consider the following restrictions: $J < 0.4$; $\phi_c < 0.78$; $d_B \geq 1[\text{in}]$, that crushing can deliver a product of 4.500[μm] and that the flotation feed admits well sized up to 280[μm], calculate the optimum operating conditions. See Table 7.1 for the results calculations.

7.2.3 Population Balance Model of a Continuous Ball Mill

Consider a cylindrical continuous mill as a well-mixed reactor containing a mass W of coarse ore that will experience fracture for a time t. If the feed to the mill is restricted to one particle if size i, experience has shown that the rate of disappearance by fracture of the mass of the fraction of this size is governed by a law of the first order (see Figure 7.11).

A population balance in a continuous mill at steady state can be represented by the equation

TABLE 7.1
Bond Power Calculation Estimates Example

DATA		OPTIMIZED VALUES	
L[m]	6.50	L[m]	6.50
D[m]	5.00	D[m]	5.00
Wi [kWh/ton]	13.20	Wi [kWh/ton]	13.20
F_{80} [Rm]	5,000.00	F_{80} [Rm]	5,000.00
P_{80} [Rm]	200.00	P_{80} [Rm]	200.00
ϕ_c[–]	0.72	ϕ_c[–]	0.78
J [–]	0.38	J [–]	0.40
ρ_s [ton/m^3]	2.65	ρ_s [ton/m^3]	2.65
ρ_B [ton/m^3]	7.50	ρ_B [ton/m^3]	7.50
RESULTS		**RESULTS**	
L/D	1.30	L/D	1.30
P [kw]	2,477.52	P [kw]	2,701.19
Wi [ton]	7.47	Wi [ton]	7.47
F_F [tph]	331.79	F_F [tph]	361.75
E [kWh/ton]	7.47	E [kWh/ton]	7.47
ρ_{ap} [ton/m3]	4.65	ρ_{ap} [ton/m3]	4.50
d_B [mm]	47.22	d_B [mm]	45.97
d_B [in]	1.93	d_B [in]	1.88

$$\underbrace{\frac{d}{dt}\left(w_i(t)\right)}_{\text{Rate of change of mass of particles of size } x_i \text{ in } W(t).} = \underbrace{\frac{F_i(t)}{W}\left(f_i(t)-(t)w_i(t)\right)}_{\text{Net rate of flow of particles with property of size } x_i \text{ in and out from } W(t).} + \underbrace{w_i(t)}_{\text{Mass of particles of size } x_i \text{ in the hodlup } W(t)}$$

$$- \underbrace{S_i(t)w_i + \sum_{j=1}^{l-1} b_{ij} S_j(t)}_{\text{Net ratew of generation of particles of size } x_i \text{ in } W(t).} \tag{7.44}$$

where W, F, and f_i are respectively the total mass of particles in the mill, the feed rate, and in the particle size in the feed (Herbst 1992; Monov et al. 2012).

7.2.3.1 Residence Time Distribution

Another way of considering a continuous grinding is to open the batch mill and add to the process a transport of the pulp to the discharge through a residence time distribution. The transit of the mill charge in the mill has different residence time described by the term $\tau = F / W = \int_0^\infty E(t)tdt$,

$$E(t) = \frac{1}{\tau}\exp\left(-\frac{1}{t}\right) \tag{7.45}$$

where $E(t)$ is the residence time distribution function (RTD) for a well-mixed pulp is represented in Figure 7.15.

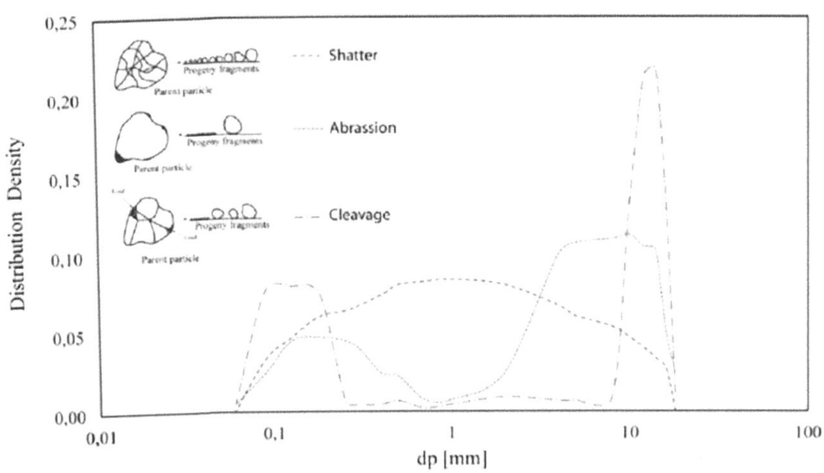

FIGURE 7.15 Residence time distribution function for a well-mixed pulp.

Then, to the solution $w_{ibatch}(t)$ of Equation 7.35, we can add the pulp transport in the mill:

$$w_i(t) = \int_0^{\infty} w_{ibatch}(t)\frac{1}{\tau}\exp\left(-\frac{t}{\tau}\right)\mathrm{dt} \tag{7.46}$$

Power P [kW], specific energy $\hat{E}$ [kWh/ton], and capacity F [tph] of a ball mill with diameter $D > 2$ [m] and a mass ball charge W_B is given in the expressions in the next section (Bond).

7.2.4 Semiautogenous Grinding

Semiautogenous grinding (SAG) represents an attempt to lower the costs of the operation of size reduction by eliminating the various stages of breakage, secondary crushers, primary and secondary mills, replacing them by a single-stage SAG mill (Figure 7.16). They use a mixture of mineral rock and a small proportion of steel balls ($\approx 12\%$) as grinding media. Semiautogenous mills are cylindrical mills of diameter twice the length with their walls covered with steel plates (liners). The liners serve the function of preventing the slippage of the ore and balls, creating a cascading effect favored by the large diameter of the mill. The dimensions of the plates and their arrangement play a fundamental role in the operation.

The size reduction mechanisms in SAG mills are the crushing, due to the impact of rocks with the mill walls and grinding by the impact of the balls and attrition between

FIGURE 7.16 Semiautogenous mill.

the grinding media. To perform this task, the mill must crush the coarse ore to smaller pieces and then grind these pieces to finer sizes with an appropriate balance between the crushing and grinding speeds. When this balance is obtained, the load reaches its equilibrium state.

A balanced size reduction media is just under 30% of the mill volume. On the contrary to ball mills, an increase in production does not translate into an increase in energy consumption since around the equilibrium state the energy consumption is relatively constant. SAG mills accumulate in their interior rock fragments too small to produce crushing on other rocks, but too large to be fragmented by other larger rocks, "critical size." Steel balls help with this problem, but close circuit with external classification is also used.

A SAG mill is different from a ball mill in three aspects, its geometry with $D/L \geq 2$, the use of a combination of coarse mineral with steel balls as grinding media, and its behavior as having an internal classifier. See Figure 7.17 and 7.18.

Assuming a perfect mixing in the mill, the mass balance for an SAG mill at steady state is

$$\underbrace{\frac{dw_i(t)}{dt}}_{\substack{\text{Rate of change of}\\ \text{particle of size i}}} = \underbrace{\frac{F}{W} f_i}_{\substack{\text{Rate of inflow}\\ \text{of particle of size}\, i}} - \underbrace{\frac{P}{W} p_i}_{\substack{\text{Rate of outtlow}\\ \text{of particle of size}}} \underbrace{- S_i(t)w_i + \sum_{j=1}^{i-1} b_{ij} S_j(t) w_j(t)}_{\substack{\text{Net rate of generation}\\ \text{of particles of size}\, i}} \tag{7.47}$$

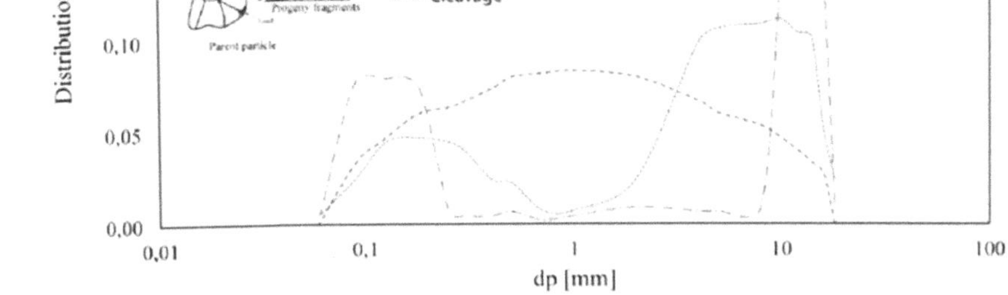

FIGURE 7.17 Particle fracture mechanism and the frequency functions of the progeny.

Source: Monov, V., Sokolov, B., and Stoenchev, S. 2012. Grinding in ball mills: Modelling and process control. *Cybernetics and Information Technologies* 12(2):51–68.

where f_i and p_i are the particle size at the mill feed and discharge

$$\frac{dw_i(t)}{dt} = \frac{F}{W} f_i - \left(\frac{P}{W} + S_i(t) \right) p_i + \sum_{j=1}^{i-1} b_{ij} S_j(t) w_j(t) \tag{7.48}$$

Denominating $C(x_i)$ the classification function, the feed plus the circulating load is $(F/W) f_i + C_i ((P/W) w_i)$, then Equation 7.48 becomes

$$\frac{dw_i(t)}{dt} = \frac{F}{W} f_i - \left(\frac{P}{W} (1 - C_i) + S_i(t) \right) w_i + \sum_{j=1}^{I-1} b_{ij} S_j(t) w_i(t) \tag{7.49}$$

For an SAG mill, the first order kinetic function S_i and the internal classification functions have the form shown in Figure 7.19.

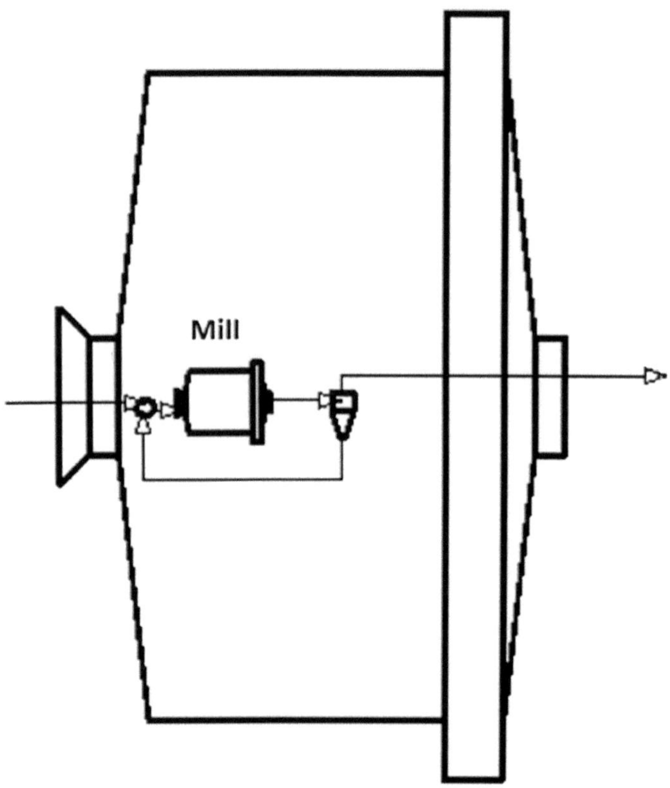

FIGURE 7.18 Semiautogenous mill with its classification action.

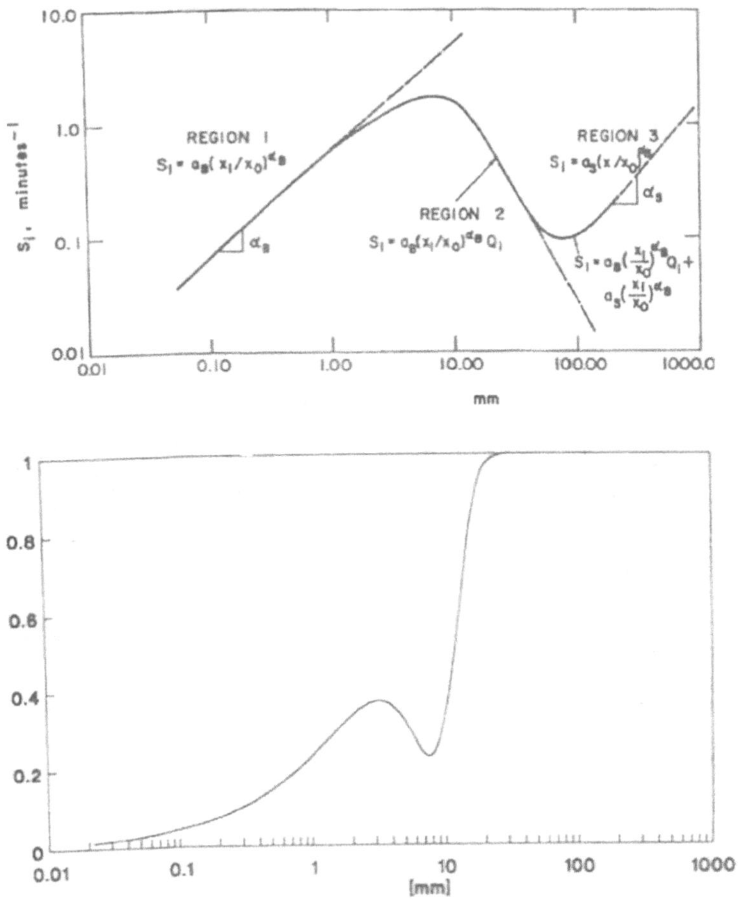

FIGURE 7.19 Characteristic curves for an SAG mill: (a) rate of breakage function, (b) internal classification curve.

Source: Austin, L.G., Klimpel, R.R., and Luckie, P.T. 1984. *The Process Engineering of Size Reduction: Ball Milling.* New York: SME-AIME.

7.4 HIGH-PRESSURE GRINDING ROLLS

7.4.1 Empirical Correlations

High-pressure grinding roll (HPGR) technology was introduced by Schönert in 1985 to be applied in the cement industry. Today, its application has become of great importance in the mineral industry of harder ores. Fuerstenau et al. (1991) and Kapur and Fuerstenau (1992) showed that, on a uniform basis of comparison, the grinding efficiency in HPGR was up to five times greater than in ball milling. See Figure 7.20.

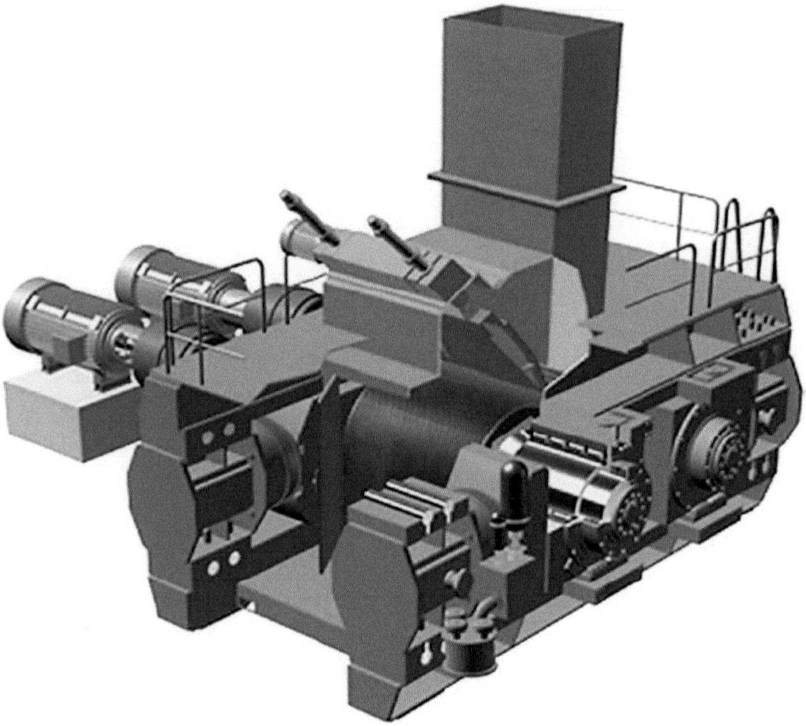

FIGURE 7.20 High-pressure grinding roll (Morley 2006).

The HPGR consists of two counter-rotating rolls enclosed in a frame and separated by a small gap that prevent their direct contact, as shown in Figure 7.21. The rolls are driven by two separate motors connected to the roll shafts with the use of gear reducers. The ore is choke-fed into the nip of the rollers, and as the distance between the rolls is decreasing, the pressing force causing the failure of the particles which produces micro-cracks, but comminution takes place in the nip zone, where the maximum pressing force is obtained. The product is a flake with thickness that depends on the particle properties and HPGR operating conditions (Saramak 2011; Anticoi 2019).

Unlike in tumbling mills, there is no explicit grinding time in HPGR, only a fixed-time duration for one pass of solids through the rolls; therefore, the *primary* variable defining the size of the product is the *energy* (pressure during the grinding time) delivered by the movement of pistons pressing the rolls. The *second* variable determining the throughput of HPGR is the *roll speed*.

The applied pressure to the HPGR is given by

$$F = \left(\pi D_4^2 / 4\right) \times n \times P_h \tag{7.50}$$

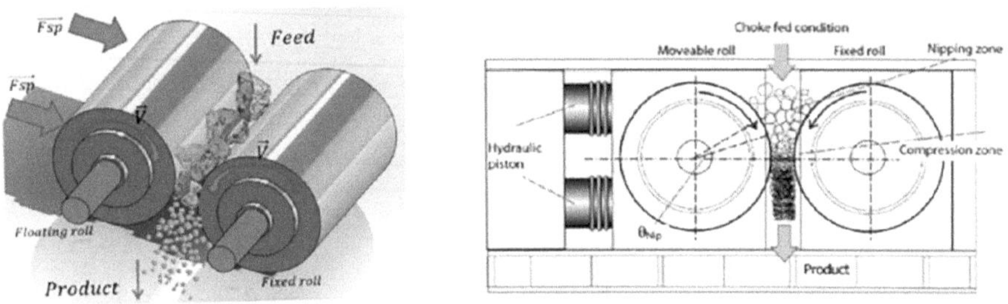

FIGURE 7.21 Schematic drawing of a choke-fed high-compression-roll mill.

Source: Anticoi, H.F. 2019. Strategic minerals milling modelling of high-pressure grind rolls and process parameters dependency. PhD thesis, Polytechnic University of Catalonia.

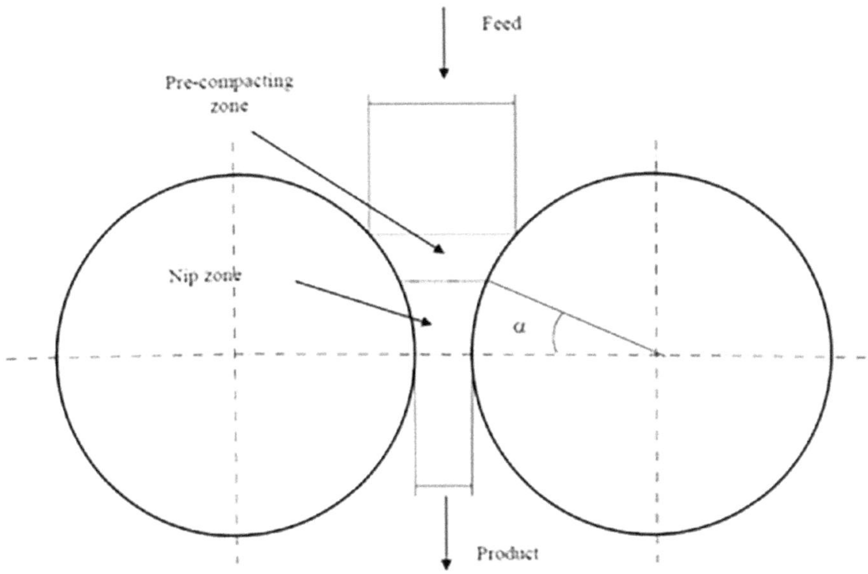

FIGURE 7.22 Schematic view of a roller mill.

Source: Saramak, D. 2011. Technological issues of high-pressure grinding rolls operation in ore comminution processes. *Archives of Mining Sciences* 56:517–526.

where F[kN] is the pressing force, D[m] is the roll's diameter, d[m] is the piston diameters, n is the number of piston (4), and P_h[kPa] is the hydraulic pressure.

The energy is given by

$$E_{sp} = \frac{F \times D \times L \times v_c}{Q} [\text{kW / tph}] \tag{7.51}$$

Saramak (2011) in Figure 7.23 gives experimental values of the increasing pressure suffered by the feed during its pass through the rolls. In the figure,

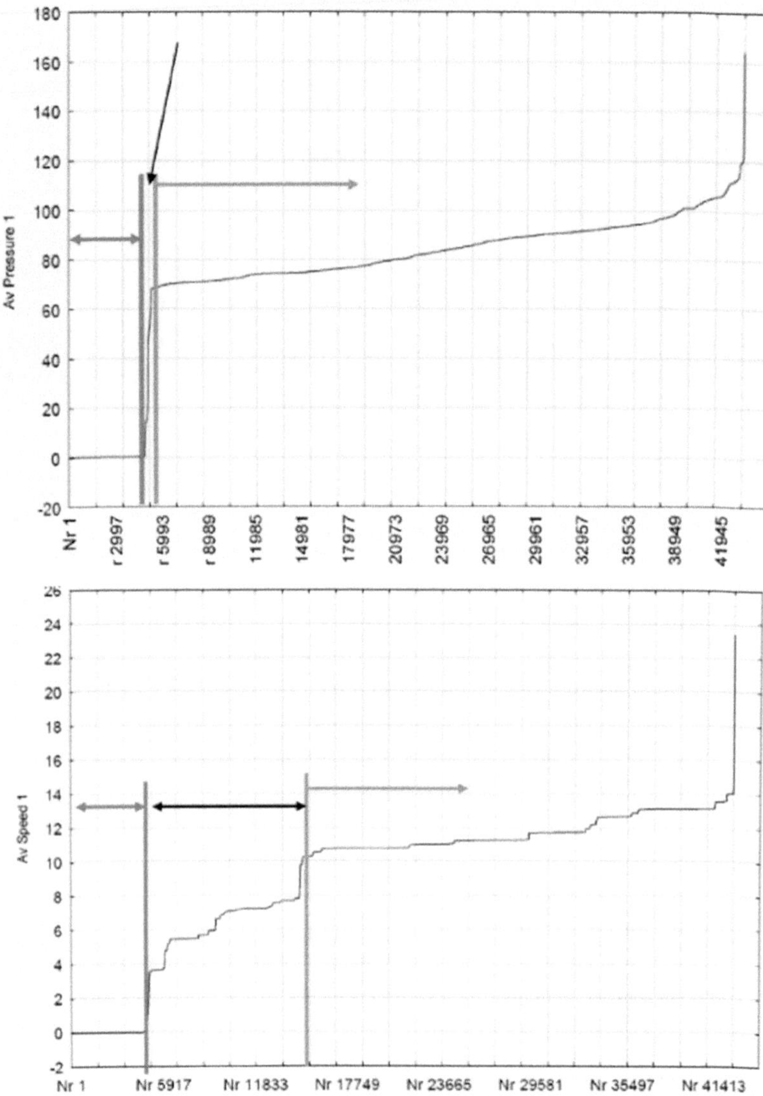

FIGURE 7.23 Average pressure and rolls speed grouped in cumulative order: (a) pressure, (b) speed.

vertical axes represent the pressure in bars in (a) and the speed rev/min in (b), while all data analyzed are in the horizontal (over 40,000 cases). The red arrow represents inactive HPGR, black arrows show the HPGR in unstable operating conditions, and the horizontal green arrow shows stable working conditions (86%).

The data considered for the analysis are those when operation was stable, discarding the very last values when pressure elevated. There might be, however, some data at the very end of such an organized sequence, which we should take into consideration.

Saramak (2011), following the results of Naziemiec and Saramak (2009), present the following empirical expressions for the specific consumption of specific energy $E_{sp}[\text{kW / ton}]$ and the throughput $Q_{[\text{tph}]}$, both in terms of the hydraulic pressure p_h and the peripherical roll speed, the first with a $R^2 = 0.821$ and the second with $R^2 = 0.725$.

$$E_{sp}(p_h) = (-0.537) + 0.026 \times p_h - 0.072 \times v + \left(10^{-5} P_h^2 + 10^{-2} v^2\right) - 10^{-3} \times p_h \times v \tag{7.52}$$

$$Q = -2500.01 + 21.38 \times p_h + 398.57 \times v - 0.05 \times p_h^2 - 10.19 \times v^2 - 1.32 \times p_h \times v \tag{7.53}$$

Figure 7.24 shows these relationships in graphical form. Table 7.2 shows the roller mill data and results.

TABLE 7.2
Roller Mill Data and Results

			ROLL SIZES 2 X L/D		
DATA	UNITS	120/120	140/140	140/160	170/180
Roll Diameter, D	mm	1200	1400	1400	1700
Roll length, L	mm	1200	1400	1600	1800
Total force, F	kN	10000	12500	15000	20000
Specific force, F_{sp}	N/mm^2	6.9	6.4	6.7	6.5
Capacity, Q	t/h	480	770	880	1470
Power, P	kW	2000	3000	4000	6000
Tangential velocity, v	m/s	1.55	1.8	1.8	2.2
gap dg	m	0.0116	0.0137	0.0137	0.0167
Constant C	-	14000	14000	14000	14000
Mineral density ρ_S	ton/m^3	2.65	2.65	2.65	2.65
Aparent density ρ_{ap}	ton/m^3	1.59	1.59	1.59	1.59
Porosity φ	-	0.4	0.4	0.4	0.4
Roll Diameter, D	m	1.2	1.4	1.4	1.7
Roll length, L	m	1.2	1.4	1.6	1.8
Total force, F	N	1.00E+07	1.25E+07	1.50E+07	2.00E+07
RESULTS					
Hydraulic pressure h_p	N/m^2	6.94E+06	6.38E+06	6.70E+06	6.54E+06
Specific energy Esp	kWh/ton	4.17	3.90	4.55	4.08
Capacity Q	tph	480	769	878	1472
Power (for two motors)	kW	2001	2994	3992	6009

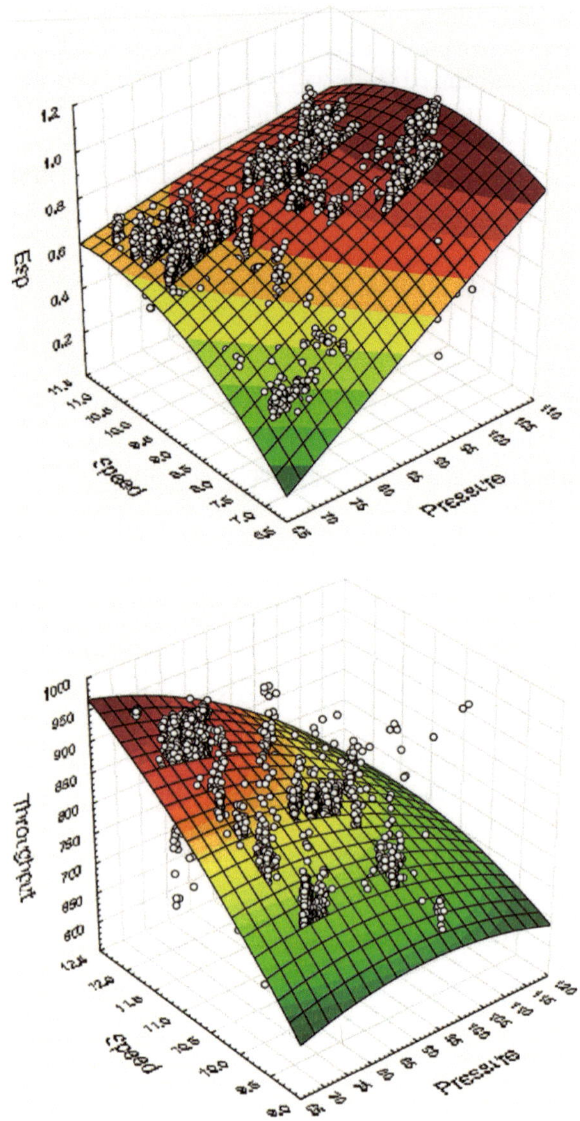

FIGURE 7.24 Consumption of energy and the throughput: (a) energy, (b) throughput.

A different empirical equation for the capacity, specific force, and specific energy of a HPGR is

$$Q = C \times \left(\rho_a L d_g\right) \times v[\text{tph}], \quad p_h = \frac{F}{DL},\left[\text{N}/\text{m}^2\right], \quad E_{sp} = \frac{P}{Q},[\text{kwh}/\text{ton}] \quad (7.54)$$

Where C is a constant; ρ_a is the apparent density of the feed; L, D are the roll's length and diameter; p_h is the hydraulic pressure; d_g is the gap between the rolls; and v is their tangential velocity.

7.4.2 Population Balance Model for HPGR

Fuerstenau et al. (1991, 1993) and Kapur and Fuerstenau (1992) presented a model of grinding in roll mills by modifying the energy dissipation in the standard population balance grinding equation shown by Equation 7.55:

$$\underbrace{\frac{\partial w_i}{\partial t}}_{\substack{\text{Rate of change of}\\ \text{particle of size}\, i}} = \underbrace{-S_i w_i(t)}_{\substack{\text{Rate of dissapearance}\\ \text{of particle of size}\, i}} + \underbrace{\sum_{j=1}^{i-1} b_{ij} S_j w_j(t)}_{\substack{\text{Rate of generation of size}\, i\\ \text{from break age of size}\, j}} =, \text{ for } n \geq i \geq j \geq 1 \tag{7.55}$$

The modification of this model consists in changing the time/with the *relevant* energy E_i', because the material in the rolls gap becomes increasingly compressed, resulting in a large energy dissipation, so that the real energy used in breaking the particles was only a fraction of the overall energy consumed. The model related the *relevant energy* expended to *feed size*, two of the important process variables.

$$\underbrace{\frac{\partial w_i(E')}{\partial E'}}_{\substack{\text{Rate of change of}\\ \text{particle of size}\, i\\ \text{with the energy}\, E'}} = \underbrace{-S_i' w_i(E')}_{\substack{\text{Rate of dissapearance}\\ \text{of particle of size}\, i\, \text{with}\\ \text{consumtion of energy}\, E'}} + \underbrace{\sum_{j=1}^{i-1} b_{ij} S_i' w_j(E)}_{\substack{\text{Rate of generation of size}\, i\\ \text{from breakage of size}\, j\, \text{with}\\ \text{consumtion of energy}\, E'}} , \text{ for } n \geq i \geq j \geq 1 \tag{7.56}$$

Where E and S_i are the relevant energy and the normalized breakage function with respect to energy.

Figure 7.25 shows the comparison of experiments and simulation with this model.

7.4.2.1 Example

With the following data: total applied force of = 10000 [kN], tangential speed of = 1.55 [m/s], gap = 11.6 [mm], density = 2650 [kg/m3] with a porosity of the particle bed of 0.4 [fraction] and a constant in expression (1.54) of K = 14000, determine for HPGR, the capacity [tph], the applied hydraulic pressure [N/m^2], and the specific energy [kWh/ton]. Table 7.3 shows the grinding parameters and results.

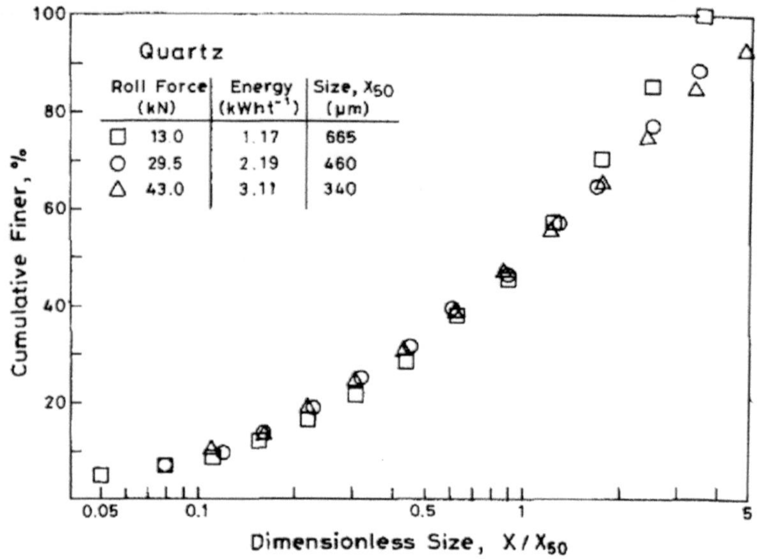

FIGURE 7.25 Self-similar size distributions of limestone.

Source: Fuerstenau, D.W, Shuklab, A., and Kapur, P.C. 1991. Energy consumption and product size distributions in choke-fed, high-compression roll mills. *International Journal of Mineral Processing* 32:59–79.

TABLE 7.3
Grinding Parameters and Results

DATA		OPTIMIZED VALUES	
L[m]	6.50	L[m]	6.50
D[m]	5.00	D[m]	5.00
Wi [kWh/ton]	13.20	Wi [kWh/ton]	13.20
F_{80} [^m]	5,000.00	F_{80} [μm]	5,000.00
P_{80} [μm]	200.00	P_{80} [μm]	200.00
ϕ_c[–]	0.72	ϕ_c[–]	0.78
J [–]	0.38	J [–]	0.40
ρ_s [ton/m^3]	2.65	ρ_s [ton/m^3]	2.65
ρ_B[ton/m^3]	7.50	ρ_B [ton/m^3]	7.50
RESULTS		**RESULTS**	
L/D	1.30	L/D	1.30
P [kw]	2,477.52	P [kw]	2,701.19
Wi [ton]	7.47	Wi [ton]	7.47
F_F [tph]	331.79	F_F [tph]	361.75
E [kWh/ton]	7.47	E [kWh/ton]	7.47
ρ_{ap} [ton/m3]	4.65	ρ_{ap} [ton/m3]	4.50
d_B [mm]	47.22	d_B [mm]	45.97
d_B [in]	1.93	d_B [in]	1.88

7.5 CLOSED-GRINDING CLASSIFICATION CIRCUIT

In most cases the grinding actions of the mill will result in an appreciable amount of non-desirable number of fines in the product. For this reason, it is necessary to remove these fines from the discharge before the material continues to the next stage and return them to the feed of the mill. See Figure 7.26.

7.5.1 Direct-Grinding Classification

The grinding-classification process in this case is called *direct grinding circuit* (Figure 7.26).

7.5.1.1 Solid Mass Balance in a Direct Grinding Circuit

For solid, pulp and water, and the fractions of solid of size x in terms of solid in the feed F, solid fraction and particle size in the feed $w_H, f_H(x)$, underflow $w_U, f_U(x)$ and overflow $w_0, f_0(x)$ of the hydrocyclone, and the circulation ratio $C = D/N$ and the fraction of water going to the underflow λW_H, as shown in Table 7.4 and Table 7.5.

The grinding task is to obtain a particle size $f_R(x)$ from a feed with particles size $f_N(x)$. The outlet of the mill contains the particle size $f_D(x) \equiv f_H(x)$ from particle size $f_F(x)$ obtained from solving the population balance. At the discharge of the hydrocyclone, the amount of fines should be proportional, $\lambda \times f_H(x)$. The fines should be obtained in terms of them $\left(f_N(x), f_H, \lambda, C\right)$, with C given by

$$C = \frac{U}{N} = \frac{\left(w_H - w_R\right) w_U}{\left(w_U - w_H\right) w_R} = \frac{f_H(x) - f_R(x)}{f_U(x) - f_H(x)} \tag{7.57}$$

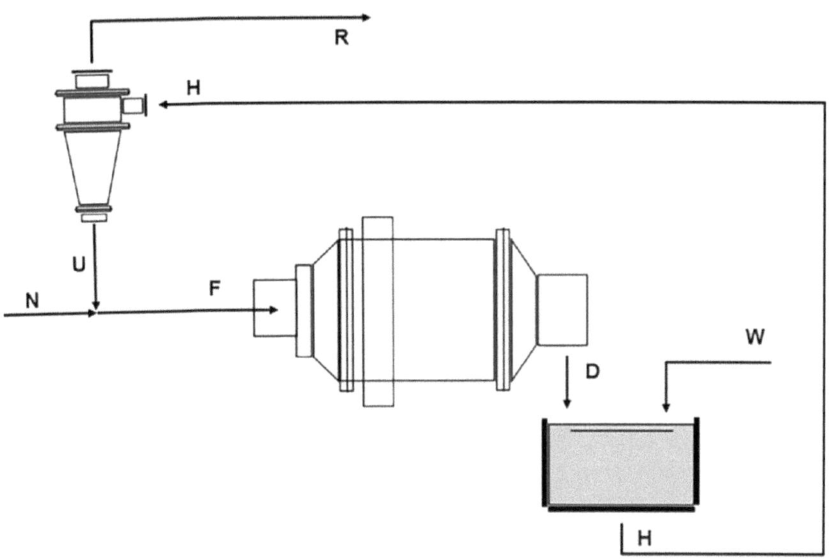

FIGURE 7.26 Direct closed-circuit grinding.

TABLE 7.4
Direct Grinding Circuit Process Streams Variables

Solids	Pulp	Water
$N = N$	$N^* = N(1/w_N)$	$W_N = N(1/w_N)\times(1-w_N)$
$F = N\times(1+C)$	$F^* = N(1/w_N + C/w_U)$	$W_F = N(1/w_N)\times(1-w_N)+(C/w_U)\times(1-w_U)$
$D = N\times(1+C)$	$D^* = N(1/w_N + C/w_U)$	$W_D = N(1/w_N + C/U(1-w_U))$
$W = 0$	$W^* = 0$	$W = W$
$H = N\times(1+C)$	$H^* = N(1/w_R + C/w_U)$	$W_H = N((C/w_U)\times(1-w_U)+1/w_R)(1-w_R)$
$U = N\times(1+C)$	$U^* = N(C/w_U)$	$W_U = N(C/\lambda w_U)(1-\lambda w_U)$
$R = N$	$R^* = N(1/w_R)$	$W_R = N(1/w_R)(1-w_R)$

TABLE 7.5
Mass Balance for Particle Size x

$$f_N(x) = f_N(x)$$
$$f_F(x) = (\lambda M(x)(1+C)+1)/(1+C)f_N(x)$$
$$f_D(x) = M(x)f_N(x)$$
$$f_W(x) = 0$$
$$f_H(x) = M(x)f_N(x)$$
$$f_U(x) = \lambda M(x)f_N(x)$$
$$f_R(x) = (1+C)M(x)f_N(x)$$

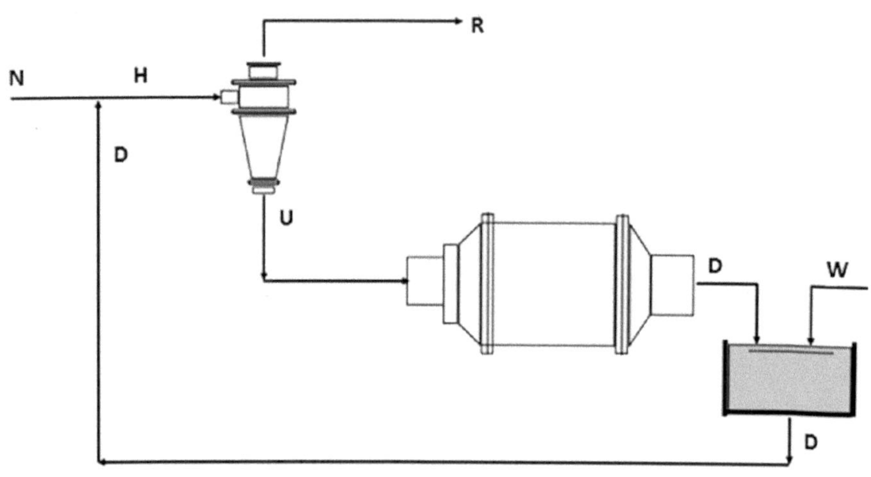

FIGURE 7.27 Reverse grinding classification circuit.

7.5.2 Reverse-Grinding Classification Circuit

Alternatively, it is possible that the ore already contains an appreciable amount of fine material, more than the one desired size in the discharge. In this case it is necessary to remove these fines before entering the mill, which will give it a greater capacity. This scheme is called the *reverse-grinding classification circuit* (Figure 7.27).

7.5.3 Grinding-Classification Simulation at Steady State

The development of the ball mill model is based on theoretical population balance considerations by Herbst et al. (1973). In this case, the model equations have been simplified especially for use in a portable computer device (Herbst and Bascur 1978). This simplification has been proven to be a good approximation in a number of experimental studies by Rajamani (1979), Hinde and King (1977), and Bascur et al. (1986). The hydrocyclone model is an empirical model of the form proposed by Rao-Lynch by Lynch (1977).

The first step in developing a simulator is to define the process variables to use for the mass balance equations of the process flow diagram, as shown in Figure 7.28. The conceptual process flow diagram with the list of simulation variables to be solved are shown in Figure 7.29.

7.5.3.1 Simplified Linear Kinetic Model

In recent years considerable progress has been made in the development of detailed models to describe grinding circuit behavior. The study of process behavior in the context of these detailed models has resulted in many important simplifications in the model equations (Rajamani 1979; Herbst et al. 1978; Pate and Herbst 1983).

As discussed elsewhere, the size discretized batch grinding model appears to be the most flexible and useful kinetic model for tumbling mill grinding developed to date. This model is obtained by dividing the particulate assembly being ground into n narrow size intervals, $d_i, d_{i+1})'$ $i = 1, 2 \ldots n$. Then, a mass balance for the material in the *i*-the size interval at time *t* yields, for $i = 1, 2 \ldots n$:

TABLE 7.6
Reverse Closed Grinding Circuit Process Variables

Solids	Pulp	Water
$N = N$	$N^* = N(1/w_N)$	$W_N = N(1/w_N)(1-w_N)$
$H = N(1+C)$	$H^* = N(1/w_N)(1+C(1/w_D))$	$W_H = N(1/w_N)(1+C/w_D)(1-w_H)$
$D = U = CN$	$D^* = N(1/w_N)C(1/w_D)$	$W_D = N(1/w_N)C(1/w_D)(1-w_D)$
$R = N$	$U^* = N(1/w_N)C(1/w_U)$	$W_U = N(1/w_N)C(1/w_U)(1-w_U)$
$W = 0$	$R^* = N(1/w_R)$	$W = W$
	$W^* = 0$	$W_R = N(1/w_R)(1-w_N)$

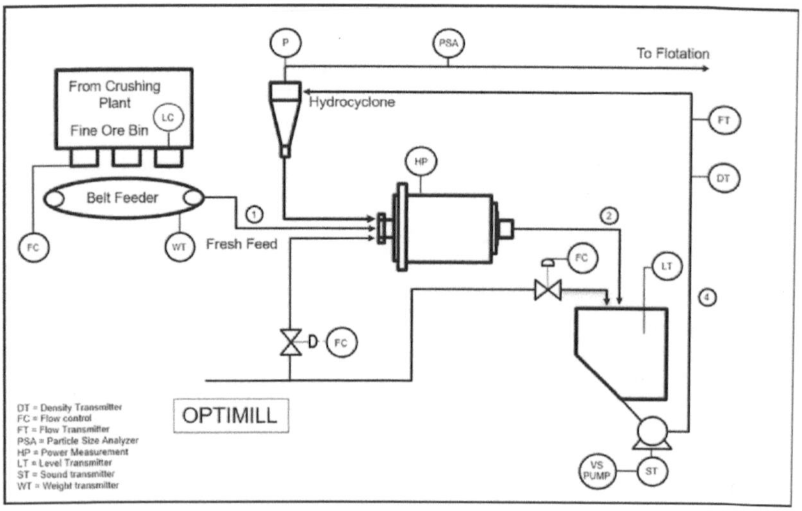

FIGURE 7.28 OPTIMILL™ process flow grinding classification diagram.

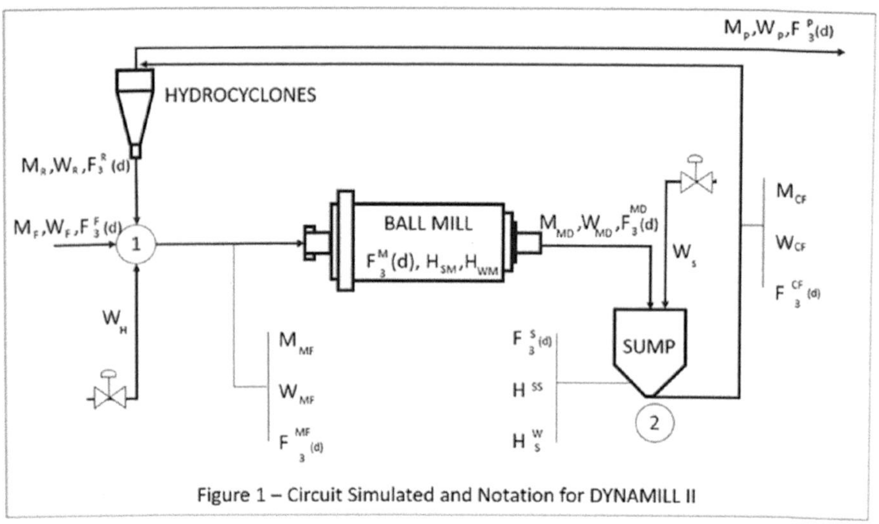

FIGURE 7.29 Typical process flow grinding classification variables.

$$\frac{d\left[Hm_i(t)\right]}{dt} = -S_i Hm_i(t) + \sum_{j=1}^{i-1} b_{ij} S_j Hm_j(t) \tag{7.58}$$

where $m_i(t)$ is the mass fraction of material in the i-th size interval, and H is the total mass of material being ground. In Equation 7.58, S_i, the size-discretized selection function for the i-th size interval denotes the fractional rate at which material is broken out of the i-th size interval and b_{ij} the size discretized breakage function, represents the fraction of the primary breakage product of material in the j-th size interval which appears in the i-th size interval. When the selection and breakage function parameters are independent of both the size and time, the kinetic model is linear with constant coefficients.

In order to demonstrate the simplification of Equation 7.59, it is convenient to express it in terms of the cumulative fraction finer than size:

$$d_i, i.e. F_3(d_i, t) = \sum_{j=n}^{i} m_j(t). \tag{7.59}$$

This transformation results in the equation:

$$\frac{dHF_3(d_i,t)}{dt} = \sum_{j=1}^{i-1} B_{ij} S_j H\left[F_3(d_{j+1},t) - F_3(d_j,t)\right] \tag{7.60}$$

In this case $B_{ij} = \sum_{k=n}^{i} bk_j$ is the cumulative breakage function—it gives the cumulative fraction of primary breakage fragments produced from the breakage of particles in the size interval d_{j+1} to d_j which are finer than size d_i.

A simple batch model is obtained for the particular case in which the fractional rate at which particles finer than a given size are produced, is independent of the parent frequent size, that is, $B_{ij} S_j = k_i \; j = 1, 2 \ldots, i-1$ and $i = 1, 2, \ldots, n$.

In this case, the kinetic model given by Equation 7.60 becomes

$$\frac{d}{dt}\left[H\left(1 - F_3(d_i)\right]\right] = -K_i H\left[1 - F_3(d_i)\right] \tag{7.61}$$

Under these conditions the integration of Equation 7.61 for batch grinding subject to the initial condition $F_3(d_i, 0) = F_3^{IN}(d_i)$ yields

$$[1 - F_3(d_I, t)_{BATCH} = \left[1 - F_3^{IN}(d_i)\right]\exp\left[-K_i t\right] \tag{7.62}$$

Although the $B_{ij} S_j = K_i$ assumption appears to be quite restrictive (Austin et al. 1972), the previous solution can be used to "derive" many of the empirical energy size distribution relationships and size distribution equations which have been reported in the literature.

7.5.3.2 Scale-Up of Kinetic Parameters

Since the general kinetic grinding models derived from the population balance point of view are phenomenological in nature, there is nothing inherent in their structure which allows the a priori prediction of the dependence of selection and breakage functions on mill dimensions and operating variables. Such relationships are required for scale-up design and certain types of simulation and automatic control.

Several research papers (Herbst et al. 1985) have shown that the following approximations are valid for ball mill grinding:

1. $S_i = S_i^E \left(\frac{P}{H} \right)$ where P is the net power draft of the mill and the S_i^E, specific selection function for the i-th size interval is a constant independent of mill dimensions and operating variables within the normal operating range.
2. b_{ij} is independent of mill dimensions and operating variables.

Given that condition 1 is valid, the S values can be obtained explicitly in terms of mill dimensions and operating variables by using standard correlations for mill power draft. The commercial mill power draft equation used in the program is the Bond's Equation 7.63 (Bond 1962):

$$P_{NET} = 2.2\rho_{BALLS} \left(\frac{L}{D} \right) D^{3.4} M_B^* \left(3.2 - 3M_B^* \right) N^* \left(1 - 0.1 \big/ 2^{9-1ON*} \right) \quad (7.63)$$

where ρ_{BALLS} is bulk density of ball charge (tons/ft3), D is the mill diameter (ft), N* fraction of critical speed and MB* is fraction of mill volume occupied by balls.

Referring back to the simplified model $B_{ij} S_j = k_i j = 1, 2 \ldots, i-1$ and $i = 1, 2, \ldots, n$, since $K_i = B_{ij} S_j$ we can substitute condition 1 and obtain

$$K_i = K_i^E \left(P/H \right) \quad (7.64)$$

where

$$K_i^E = B_{ij} S_J^E \quad (7.65)$$

It has been found that the breakage rate function dependence on particle size can often be approximated by

$$K_i^E = K_1^E \left(d_i^* \right)^{\propto} \quad (7.66)$$

In this expression d_i^* is a normalized particle size,

$$d_i^* = \sqrt{d_i d_{i+1}} \Big/ \sqrt{d_1 d_2} \quad (7.67)$$

7.5.3.3 Residence Time Distribution

A description of continuous tumbling mill requires not only a description of the breakage kinetics but also a mathematical description of material transport through the mill. In general, a distributed parameter model, in which variations down the length of the mill are accounted for is required in order to obtain the desired relationships between the input and output mass flow rates and size distributions. However, if linear breakage kinetics prevail and all particle sizes are characterized by a single residence time distribution, a general relationship between input and output can be obtained without requiring a detailed model for particle transport within the mill (Herbst et al. 1985).

Linear systems have the properties of additivity and homogeneity. Due to the additivity property, the response of a linear system to an arbitrary input can be represented as the sum of responses of the system to a series of appropriately weighted impulses.

Specifically, at steady state, the product size distribution from an open circuit mill is determined by the sum of the responses of the mill to the feed size distribution, treated as a sequence of impulses. Thus, the product size distribution in term of the cumulative size distribution is given by

$$\left[1-F_{3_{MD}}\left(d_i\right)\right]_{SS} = \int_0^\infty \left[1-F_3\left(d_i,t\right)_{BATCH}\right]E\left(t\right)dt \tag{7.68}$$

An average of batch responses weighted with respect the distribution of residence times of material in the mill. In Equation 7.68, $\left[1-F_3(d_i,t\right]_{BATCH}$ is the response of the mill operated in a batch fashion with an initial size distribution, $\left[1-F3\left(d_i\right)\right]IN$, equivalent to the steady state size $\left[1-F_{3_{MF}}\left(d_i\right)\right].E\left(t\right)dt$ distribution of the mill feed. The exit size (residence time) distribution of the material is an experimentally determinable function which denotes the mass fraction of material in the mill discharge which has resided in the mill for a time interval t tot+ dt.

In many instances, the residence time distribution, E(t), can be represented with a highly flexible mixers-in-series model (Gamma distribution function), that is,

$$E\left(t\right)=\frac{N^N\left(t/\tau\right)^{N-1}}{\tau\Gamma\left(N\right)}\exp\left(-\frac{Nt}{\tau}\right) \tag{7.69}$$

Where

$$\tau=\frac{mass\ hold-up\ in\ the\ mill}{solid\ mass\ flowrate}=\frac{H}{M_{MF}}=mass\ residence\ time \tag{7.70}$$

Substitution of Equations 7.63 and 7.69 into Equation 7.68 and integrating yields the simple expression

$$\left[1-F_{3_{MD}}\left(d_i\right)\right]_{SS} \cong \left[1-F_{3_{MF}}\left(d_i\right)\right]\left[1+\frac{K_i^E P}{NM_{MF}}\right]^{-N} \tag{7.71}$$

which is the final form of the approximate steady state grinding model.

7.5.3.4 Cyclone Model

The hydrocyclone model used here is a modified form of the empirical model developed by Rao-Lynch (Lynch 1977). The revised cyclone model consists of a set of three equations, namely, the water split equation, the d50 equation, and the reduced efficiency curve. Since the behavior of a cyclone is highly dependent on the inlet size distribution, Lynch has proposed the use of different water split and ct50 equations depending upon the type of feed. For the program described here, the equations used corresponding to a feed of intermediate fineness:

$$\text{Water split equation}: \; R_f = \frac{A1(SPIG)}{W_{CF}} + \left(\frac{A2}{W_{CF}}\right) + K1 \tag{7.72}$$

$$\text{d50 equation}: \; \log d_{50_{corr}} = A3(SPIG) + A4(VF) + AS(FS) + A6(Q) + K2 \tag{7.73}$$

Rather than the Lynch equation for the corrected classifier efficiency curve, the functional form proposed by Plitt (1976) is used:

$$\text{Corrected efficiency Plitt equation}: \; E_{corr}(d_i) = 1 - exp\left[-A7\left(\frac{d_i}{d_{50}}\right)^m\right] \tag{7.74}$$

Where:

A1 = empirical constant = 281.1
A2 = empirical constant = 137.2
A3 = empirical constant = 0.0325
A4 = empirical constant = 0.1738
A5 = empirical constant = 1.73
A6 = empirical constant = 1.82×10^{-4}
A7 = empirical constant = 0.693
W_{CF} = water mass flowrate in feed (TPH)
W_{OF} = water mas flowrate overflow (TPH)
R_f = fraction of feedwater in underflow
SPIG = spigot diameter cm
VF = vortex finder diameter cm
FS = fraction of solids in feed
Q_{CF} = slurry volumetric flowrate lt/min
$E_{corr}(d_i)$ = corrected classification efficiency = weight fraction of feed size d undergoing classification which appear in the cyclone underflow, corrected for short circuiting.

According to Lynch, the complete specification of the model for a given ore-cyclone system consists of the evaluation of the three parameters Kl, K2, and m, while the values of all other constants remain the same for any cyclone (within the range of variables studied). The values of adjustable parameters chosen for this simulation are as follows:

m = parameter = 2
Kl = parameter = 7.5
K2 = parameter = 1.3

In addition to the equations mentioned previously there are additional relationships required to simulate the cyclone separation process. Using the notation given in Figure 7.29 these relationships are as follows:

The overall classifier efficiency *E(di)* is defined by the equation:

$$E(d_i) = \frac{M_R\left[F_3^R(d_i) - F_3^R(d_{i+1})\right]}{M_{MD}\left[F_3^{MD}(d_i) - F_3^{MD}(d_{i+1})\right]} \tag{7.75}$$

Overall efficiency can be expressed in terms of corrected efficiency as follows:

$$E(d_i) = (1 - RF)E_{(corr)}(d_i) + RF \tag{7.76}$$

Where RF is the fraction of material that short-circuits the classification unit. RF can be calculated from the water split equation with the assumption being that fine (usually −400 mesh) short circuit to the underflow in proportion to the amount of water which goes to the spigot product, that is, $RF = Rf$.

The equation, which relates the volumetric flow with the mass flow of solids and water in the cyclone feed, is

$$Q_{CF} = \frac{M_{CF}}{\rho_S} + \frac{W_{CF}}{\rho L} \tag{7.77}$$

where ρ_S is density of solids and ρL the density of the liquid (H_2O).

Additional equations that result from solid and water mass balance around the classifier are as follows:

$$\text{Water Balance}: W_{CF} = W_R + W_p \tag{7.78}$$

$$\text{Solids Balance}: \dot{M}_{CF} = \dot{M}_R + \dot{M}_P \tag{7.79}$$

$$\text{Solids Balance for Each Size}: \dot{M}_{CF}F_3^{CF}(d_i) = \dot{M}_R F_3^R(d_i) + \dot{M}_P F_3^P(d_i) \tag{7.80}$$

With Equations 7.72 to 7.80 one can represent the behavior of a set of hydrocyclones of specified dimensions, that is, given the input variables $FS = \frac{\dot{M}_{CF}}{\dot{M}_{CF+}W_{CF}}$, $\dot{W}_{CF}$, $F_3^{CF}(d_i)$ one can calculate the output variables:

$$FR = \frac{\dot{M}_R}{\dot{M}_R + \dot{W}_R}, FP = \frac{\dot{M}_P}{\dot{M}_P + \dot{W}_P}, \dot{W}_R, \dot{W}_P, F_3^R(d_i) \text{ and } F_3^P(d_i) \tag{7.81}$$

The final relationships required for circuit simulation are those that specify the mill input in terms of the fresh feed and recycle and the cyclone feed in terms of the mill discharge and sump water addition. These relationships arise from steady state mass balance around the node (or mixing) points designated as (1) and (2) in Figure 7.28. These relationships for node 1 are as follows:

$$\text{Water Balance}: W_{MF} = W_F + W_R + W_H \tag{7.82}$$

$$\text{Solids Balance}: \mathrm{M_{MF}} = \mathrm{M_F} + \mathrm{M_R} \tag{7.83}$$

$$\text{Solids Balance for Each Size}: \dot{M}_{MF} F_3^{MF}(d_i) = \dot{M}_F F_3^F(d_i) + \dot{M}_R F_3^R(d_i) \tag{7.84}$$

And for node 2, they are as follows:

$$\text{Water Balance}: \mathrm{W}_{CF} = \mathrm{W_{MD}} + \mathrm{W_S} \tag{7.85}$$

$$\text{Solids Balance}: \dot{M}_{CF} = M_{MD} \tag{7.86}$$

$$\text{Solids Balance for Each Size}: \dot{M}_{CF} F_3^{CF}(d_i) = \dot{M}_{MD} F_3^{MD}(d_i) \tag{7.87}$$

7.5.3.5 Solution

A rough flowchart of OPTIMILL™ is given in Figure 7.30. The program consists of four modules: (1) data input, (2) initiation, (3) circuit calculations, and (4) output (mass balance, particle size distributions, and process flow diagram results).

All runs begin with data input, including the specification of the ball mill and hydrocyclone dimensions, process feed characteristics, and particle size distribution. Since the steady state values for the recirculated product are not known at the outset, an iterative solution (involving several passes through the circuit calculation module) is required. The procedure used in the program follows a fixed-point iteration method in which an initial guess is made for the recycle stream variables. These values are combined with the fresh feed values to produce a "trial" value for the mill feed. Using these trial mill feed values, a mill product and subsequently cyclone products are calculated during a pass through the circuit calculation module. In turn, the new recycle values are combined with the fresh feed to produce updated trial mill feed values for another pass through the calculated module. This procedure is repeated until the trial values stop changing.

A typical example of the results for grinding circuits is shown in Figure 7.31 a and b.

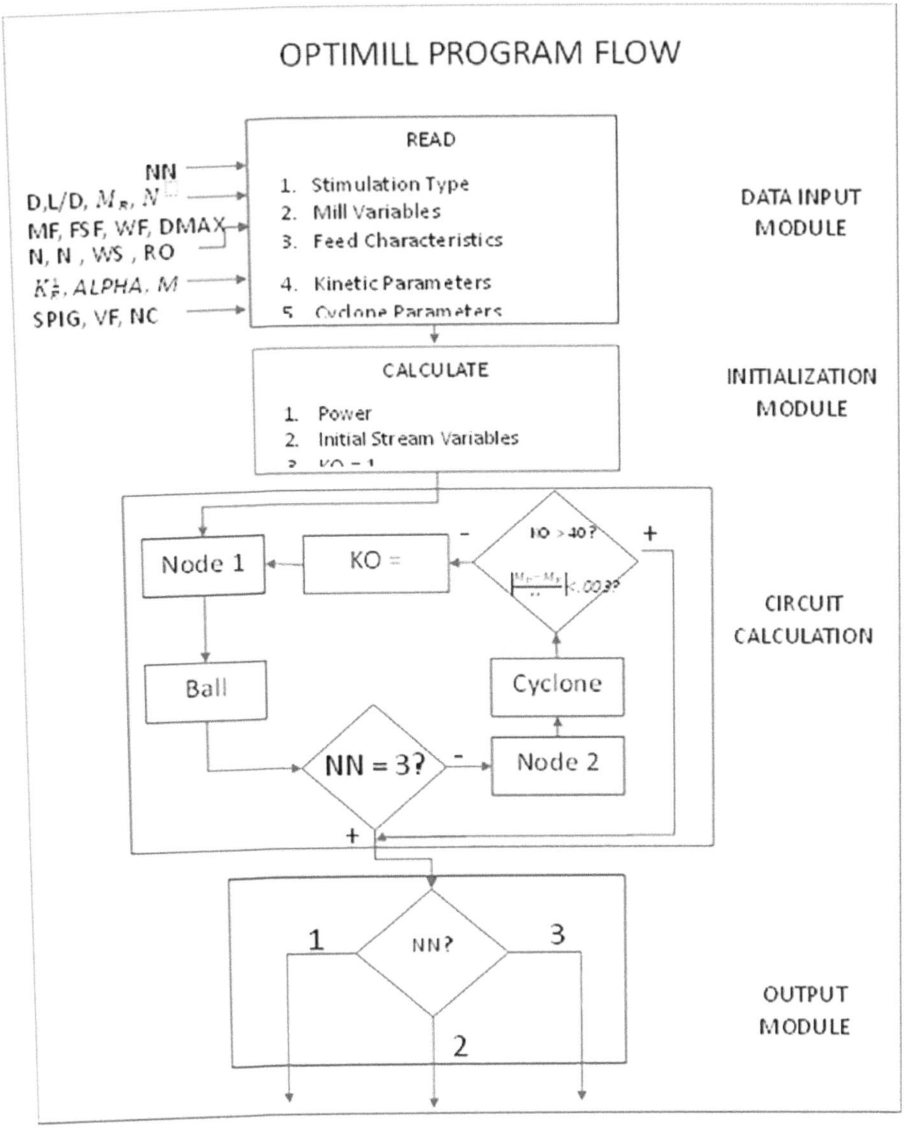

FIGURE 7.30 OPTIMILL™ program algorithm.

7.5.4 Dynamic Grinding-Classification Simulation for Online Optimization

This simplified digital simulator for a grinding-classification circuit is obtained from a series of mass balances for solids, water, and the amount of material that is finer than a certain size is then around the mill, the sump, and cyclone(s) (Herbst et al. 1978; Bascur et al. 1986).

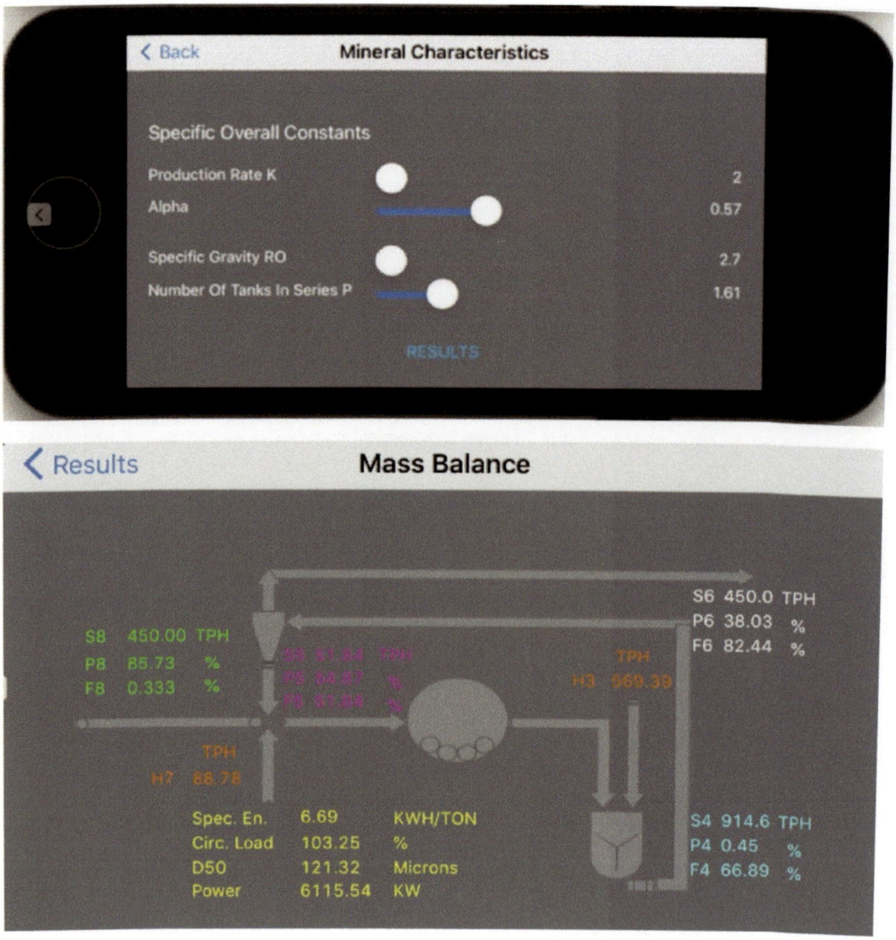

FIGURE 7.31 OPTIMILL™ settings, tabular and graphical results.

Source: Bascur, O.A., and Junge, P.A. 2011. Dynamill, DynaFlote and Optimill [iPad software apps]. www.apple.com/app-store/.

Instead of considering a large number of size fractions as is done with more detailed simulators, one variable characterizes the entire size distribution, that is, cumulative fraction finer than a specified size.

The mathematical model of the ball mill is based on the assumption that the mill acts as a perfectly mixed vessel with constant volume (overflow mill). With these assumptions, the dynamics of the ball mill are represented by three differential equations—the total mass balance, the balance of the cumulative fraction finer than a certain size, and for the fraction of solids in the mill pulp.

The model of the sump assumes constant volume (perfect level control) and perfect mixing within its volume. The dynamics of this subsystem are followed through

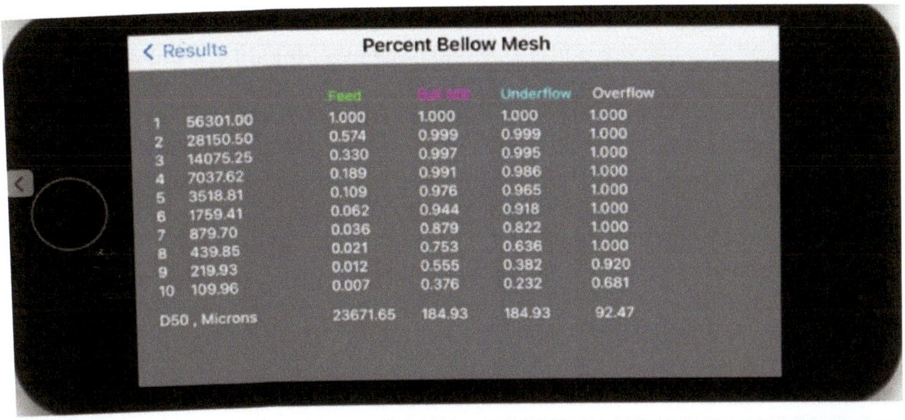

< Results Percent Bellow Mesh

		Feed	[illegible]	Underflow	Overflow
1	56301.00	1.000	1.000	1.000	1.000
2	28150.50	0.574	0.999	0.999	1.000
3	14075.25	0.330	0.997	0.995	1.000
4	7037.62	0.189	0.991	0.986	1.000
5	3518.81	0.109	0.976	0.965	1.000
6	1759.41	0.062	0.944	0.918	1.000
7	879.70	0.036	0.879	0.822	1.000
8	439.85	0.021	0.753	0.636	1.000
9	219.93	0.012	0.555	0.382	0.920
10	109.96	0.007	0.376	0.232	0.681
D50, Microns		23671.65	184.93	184.93	92.47

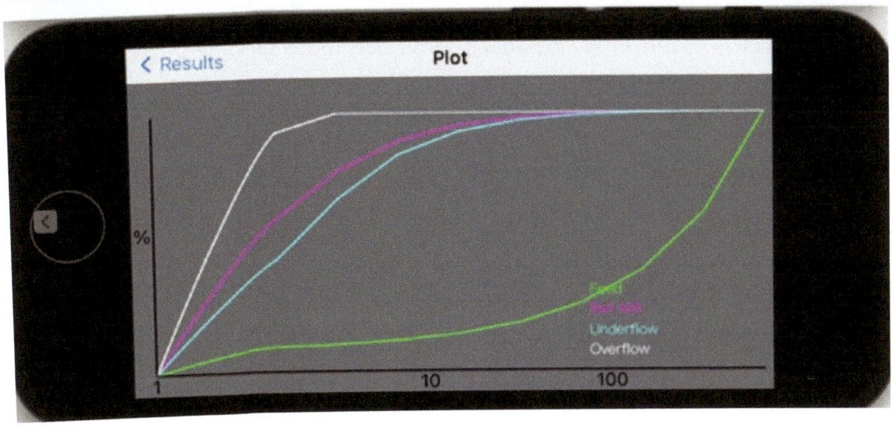

FIGURE 7.31 Continued

variations in the pulp density, mass of solids, and the size distribution in the sump (three differential equations).

In the modeling of the cyclone, empirical models of Rao and Lynch are used along with the Plitt equation for the corrected efficiency. This latter relationship is used to obtain an appropriate approximation for the classification of a feed consisting of two size fractions.

The dynamics model equations are solved simultaneously using a fourth order Runge-Kutta routine.

Automatic control of the circuit is simulated by introducing the digital equivalent of two PI (proportional and integral action) controllers. Also, the program allows for different sampling rates of the two controlled variables, that is, product size fraction and recycle flow rate. The manipulated variables in this case are fresh feed rate and water to the sump.

The program can be used for obtaining the dynamic behavior of the systems for different step changes in the perturbated variables: fresh feed rate and cumulative fraction finer than a certain size in the feed with control and without control.

The output gives the metallurgical balance updated with time.

The single stage grinding circuit which is being modeled here is shown in Figure 7.28.

After identification of the important subsystems in the circuit (ball mill, sump, hydrocyclone, pump, and piping), it is necessary to specify an appropriate model for each of them.

Only subsystems with long response times need to be modeled as dynamic elements. In this regard, the hydrocyclone, which has a very rapid response (retention time of a few seconds), requires only steady state model for its description. Since the objective is to develop a simple model which represents the main dynamics of the system, the influence of variations in sump level are considered negligible. Since sump level control is quite standard, the variations themselves are expected to be small. Another assumption is that no significant lags are introduced by the piping.

Only subsystems with long response times need to be modeled as dynamic elements. In this regard, the hydrocyclone, which has a very rapid response (retention time of a few seconds), requires only steady state model for its description. Since the objective is to develop a simple model which represents the main dynamics of the system, the influence of variations in sump level are considered negligible. Since sump level control is quite standard, the variations themselves are expected to be small. Another assumption is that no significant lags are introduced by the piping.

7.5.4.1 Ball Mill Model

The ball mill model is based on a set of dynamic mass balances written for particles of a given size, the total mass, and the fraction of solids in the mill. The mill is assumed to be perfectly mixed with regard to water/solids and size distribution. The active volume of the mill is assumed to be fixed at a value V_m independent of flowrate. The kinetics of breakage are assumed to follow the simplified first order rate law described in reference.

The net rate of grinding of material $(NRGM)$ coarser than size d_i is calculated according to Equation 7.88.

$$NRGM\ coarser\ than\ d_i = K_i^E * \left(\frac{P}{H_M}\right) * \left[1 - F_3^M\left(d_i\right)\right] \tag{7.88}$$

In this case, the application of the general mass conservation law (Equation 7.89) results in the following equations: Equation 7.90, total pulp mass balance; Equation 7.91, solid's mass balance; and Equation 7.92, solids coarser than d_i mass balance.

$$\left[Accumulation\right] = \left[Input\right] - \left[Output\right] + \left[Generation\right] \tag{7.89}$$

$$\frac{dM_m\left(t\right)}{dt} = W_{feedwater} + W_e\left(t\right) - W_B\left(t\right) \tag{7.90}$$

$$\frac{d[M_m(t)*ps_m(t)]}{dt} = ps_e(t)*W_e(t) - ps_B(t)*W_B(t) \tag{7.91}$$

$$\begin{aligned}\frac{d[M_m(t)*ps_m(t)*R_{3m}(d_i,t)]}{dt} &= ps_e(t)*W_e(t)*R_{3e}(d_i,t)\\ &- ps_B(t)*W_B(t)*R_{3B}(d_i,t)\\ &- ps_B(t)*M_m(t)*K(d_i)*R_{3m}(d_i,t)\end{aligned} \tag{7.92}$$

Where the cumulative weight fraction of solids coarser than size d_i is represented for Equation 7.93.

$$R_3(d_i,t) = 1 - F_3(d_i,t) \tag{7.93}$$

Assuming that an overflow mill behaves like a perfect mixer with constant volume, we have the following relations represented in the Equations 7.94, 7.95, 7.96.

$$ps_m(t) = ps_B(t) \tag{7.94}$$

$$R_{3m}(d_i,t) = R_{3B}(d_i,t) \tag{7.95}$$

$$V = Constant \rightarrow V_m \neq V_m(t) \tag{7.96}$$

Another important relationship is the pulp density in terms of the fraction of solids represented in Equation 7.97.

$$\rho_{pm} = \frac{\rho_s * \rho_F}{\rho_s * \left[1 - ps_m(t)\right] + \rho_F * ps_m(t)} \tag{7.97}$$

Taking the specific gravity of solids of the fluid (water) equals one we have Equation 7.98.

$$\rho_{pm} = \frac{\rho_s}{\rho_s * \left[1 - ps_m(t)\right] + ps_m(t)} \tag{7.98}$$

Using the relationship represented in Equation 7.99 and replacing in Equation 7.90, we have the Equation 7.100.

$$M_m(t) = \rho_{pm}(t) * V_m \tag{7.99}$$

$$W_B(t) = W_{feedwater} + W_e(t) - \frac{dM_m(t)}{dt} = W_{feedwater} + W_e(t) - V_m * \frac{d\rho_{pm}(t)}{dt} \quad (7.100)$$

Now using Equation 7.98 after differentiation respect to time, we have Equation 7.101.

$$W_B(t) = W_{feedwater} + W_e(t) - \frac{V_m * \left[\rho_s * (1-\rho_s)\right] * \left(\frac{dps_m(t)}{dt}\right)}{\left[\rho_s * \left(1 - ps_m(t)\right) + ps_m(t)\right]^2} \quad (7.101)$$

Differentiation of Equation 7.91 and using Equation 7.94, we have Equation 7.102.

$$\frac{dps_m(t)}{dt} = \frac{W_e(t)}{M_m(t)} * \left[ps_e(t) - ps_m(t)\right] \quad (7.102)$$

Using Equation 7.90 and rearranging, then we have Equation 7.102. This last equation is the dynamic equation for the fraction of solids in the mill.

Then, using the relationship represented in Equation 7.103, then we have Equation 7.104.

$$K(d_i) = \frac{K_i^E * P}{ps_m(t) * M_m(t)} \quad (7.103)$$

$$\frac{dR_{3m}(d_i,t)}{dt} = \frac{ps_e(t) * W_e(t)}{ps_m(t) * M_m(t)} * \left[R_{3e}(d_i,t) - R_{3m}(d_i,t)\right] - K(d_i) * R_{3m}(d_i,t) \quad (7.104)$$

Where P is the power supplied to the mill and Equation 7.105 the solids holdup of the mill.

$$ps_m(t) * M_m(t) = H_m(t) \quad (7.105)$$

Finally, the dynamic equation for the cumulative size fraction course then d_i is represented in Equation 7.106. The dynamic of the ball mill is modeled using Equations 7.101, 7.102, and 7.106.

$$\frac{dR_{3m}(d_i,t)}{dt} = \frac{ps_e(t) * W_e(t) * \left[R_{3e}(d_i,t) - R_{3m}(d_i,t)\right] - K_i^E * P * R_{3m}(d_i,t)}{ps_m(t) * M_m(t)} \quad (7.106)$$

7.5.4.2 Sump-Pump Model

For the steady state simulation of the grinding circuit performance, a model of the sump and pump is not required since at steady state the solid/liquid ratio in the inlet and outlet of the sump are the same, and since no grinding or classification occurs in the sump, the steady state size distribution in the inlet and the outlet are identical. The sump is, however, a very important dynamic element of a grinding circuit and therefore must be properly modeled. Its dynamic importance arises from the fact that a disturbance in the feed to a typical ball mill sump will persist for several minutes. A persistent disturbance in the sump has a significant dynamic impact on the entire circuit through the cyclone and resulting recycle.

The sump-pump model used in Dynamill II (Bascur 1979) assumes (1) perfect mixing in the sump with respect to solid/liquid and size distribution (2) constant volume of slurry in the sump. Assumption 2 is premised on the availability of a variable speed pump plus a fast control loop so that increases or decreases in slurry flow to sump are compensated for immediately by an equivalent change in pumping rate.

In this case, the application of the general mass conservation law Equation 7.89 results in the following equations: Equation 7.107, total slurry mass balance; Equation 7.108, solid's mass balance; and Equation 7.109, solids coarser than d_i mass balance.

$$\frac{dM_s(t)}{dt} = W_{water}(t) + W_B(t) - W_A(t) \tag{7.107}$$

$$\frac{d[M_s(t) * ps_s(t)]}{dt} = ps_B(t) * W_B(t) - ps_A(t) * W_A(t) \tag{7.108}$$

$$\begin{aligned} \frac{d[M_s(t) * ps_s(t) * R_{3s}(d_i,t)]}{dt} &= ps_B(t) * W_B(t) * R_{3B}(d_i,t) \\ &\quad - ps_A(t) * W_A(t) * R_{3A}(d_i,t) \end{aligned} \tag{7.109}$$

The assumptions already stated gives the following relationships represented in Equations 7.110, 7.111, and 7.112.

$$V_s \neq V_s(t) \tag{7.110}$$

$$ps_s(t) = ps_A(t) \tag{7.111}$$

$$R_{3s}(d_i,t) = R_{3A}(d_i,t) \tag{7.112}$$

Using a similar procedure as for the ball mill, we obtain Equations 7.113, 7.114, and 7.115. These equations represent the dynamic model for the sump.

$$W_A(t) = W_B(t) + W_{water}(t) - \frac{V_p * [\rho_s * (1-\rho_s] * \left(\frac{dps_s(t)}{dt}\right)}{\left[\rho_s * (1 - ps_s(t)] + ps_s(t)\right]^2} \tag{7.113}$$

$$\frac{dps_s(t)}{dt} = \frac{ps_m(t) * W_B(t) - ps_s(t) * \left[W_{water}(t) + W_B(t)\right]}{M_s(t)} \tag{7.114}$$

$$\frac{dR_s(t)}{dt} = \frac{ps_m(t) * W_B(t) * \left[R_{3m}(d_i,t) - R_{3s}(d_i,t)\right]}{M_s(t) * ps_s(t)} \tag{7.115}$$

7.5.4.3 Cyclone Model

Due to the very fast dynamic response of a hydrocyclone, it is appropriate to use a steady state model for circuit simulation. The most suitable model available at the present time is the empirical Rao-Lynch Model. The most recent form of this model consists of the following empirical equations: Equation 7.116 water split and Equation 7.117 D50.

$$R_f = A_1 * \left(\frac{SPIG}{W_{water}}\right) + \left(\frac{A_2}{W_{water}}\right) + K_1 \tag{7.116}$$

$$Log\, d_{50_{corr}} = A_3 * SPIG + A_4 * VF + A_5 * FS + A_6 * Q + K_2 \tag{7.117}$$

Rather than the Lynch equation for the corrected classifier efficiency curve, the functional form proposed by Plitt is used, represented in the Equation 7.118. The meaning and values of the constants utilized are resumed in Table 7.7.

$$E_{corr_{(d_i)}} = 1 - \exp\left[-A_7 * \left(\frac{d_i}{d_{50}}\right)^m\right] \tag{7.118}$$

Since this simulator is being developed in terms of the cumulative fraction of mineral coarser than a certain size, it is necessary to deal with the global efficiency classifies rather than the differential efficiency.

The global efficiency is the fraction of material in a load (not differential) size range, for example, *dl* to *du* in the cyclone feed which is returned to the mill, represented in Equation 7.119.

$$E_T(dl, du) = \frac{W_R * \int_{dl}^{du} F_3^R(d)d(d)}{W_{CF} * \int_{dl}^{au} F_3^{CF}(d)d(d)} \tag{7.119}$$

TABLE 7.7
Meaning and Values of the Constants

Constant	Meaning	Valor
A_1	*Empirical Constant*	281,6
A_2	*Empirical Constant*	137,2
A_3	*Empirical Constant*	0,00325
A_4	*Empirical Constant*	0,1738
A_5	*Empirical Constant*	1,73
A_6	*Empirical Constant*	$1,82*10^{-4}$
A_7	*Empirical Constant*	0,693
W_{water_F}	*Water mass flowrate in the feed (TPH)*	
W_{water}	*Water mass flowrate in the overflow (TPH)*	
R_f	*Fraction of feed water going to cyclone underflow*	$1-\frac{W_{watero_o}}{W_{water_F}}$
SPIG	*Spigot Diameter (cm)*	-
VF	*Vortex Finder Diameter (cm)*	-
FS	*Fraction of Solids in Cyclone Feed*	-
Q	*Slurry Volumetric Flowrate to Cyclone (1t / min)*	-
$E_{corr_{(d)}}$	*Corrected Efficiency of Classification Weight fraction of feed sized undergoing classification which appears in the cyclone underflow, corrected for short circuiting*	-

Where the fraction of material in stream K with size between d and $d^+ d(d)$ is represented for Equation 7.120 and the density function for size in stream K is given by Equation 7.121.

$$F_3^K(d)d(d) \tag{7.120}$$

$$F_3^K(d) = \frac{dF_3^K(d)}{d(d)} \tag{7.121}$$

By introducing the definition of differential efficiency, represented in Equation 7.122, Equation 7.119 becomes Equation 7.123.

$$E(d) = \frac{W_R * F_3^R(d)}{W_{CF} * F_3^{CF}(d)} \tag{7.122}$$

$$E_T(dl, du) = \frac{\int_{di}^{du} E(d) * F_3^{CF}(d)d(d)}{\int_{d}^{di} F_3^{CF}(d)d(d)} \tag{7.123}$$

Since we have the corrected differential efficiency in Equation 7.118, the latter can be corrected to the overall differential efficiency, *E*(*d*), by the relationship in Equation 7.124.

$$E_{coT[d_i)} = \frac{E(d_i) - R_f}{1 - R_f} \tag{7.124}$$

Where R_f is the fraction of fine material which short-circuits to the cyclone underflow according to Equation 7.125.

$$R_f = \frac{Water\ Underflow}{Water\ Feed\ to\ Cyclone} \tag{7.125}$$

For this simulator we are interested in the material coarser and the material finer than a specific size d^+, thus we must obtain the global efficiency for these two size fractions, $dl = 0$, $du = d^+ : F_3^{CF}\left(d^+\right)$, and $dl = d^+$, $du = d_{\max} : 1 - F_3^{CF}\left(d^+\right)$. In general, we need the entire density function to evaluate Equation 7.123 for the range of interest, however, this information is not available.

Approximate expressions for $E_T\left(0, d^+\right)$ and $E_T\left(d^+, d_{\max}\right)$ can be obtained by assuming the density function is approximately constant over the two separate intervals, in which case Equation 7.123 reduces to Equations 7.126 and 7.127, respectively.

$$E_T\left(0, d^+\right) = \frac{\int_0^{d^+} E(d)d(d)}{\int_0^{d^+} d(d)} \cong C_1 \tag{7.126}$$

$$E_T\left(d^+, d_{\max}\right) = \frac{\int_{a^+}^{d_{\max}} E(d)d(d)}{\int_{d^+}^{d_{\max}} d(d)} \cong C_2 \tag{7.127}$$

These equations state that the global efficiency can be approximated by the average differential efficiency for the intervals 0 to $d^+ : C_1$ and d^+ to $d_{\max} : C_2$. Expressions for C_1 and C_2 were obtained in terms of the cyclone separation size by substituting Equations 7.124 and 7.118 with $m = 2$ into Equations 7.126 and 7.127, respectively. In this case, the integration can be done easily to obtain C_1 and C_2 according to Equations 7.128 and 7.129, where $erf(x)$ is the standard error function represented in Equation 7.130, which can be readily evaluated with a polynomial expansion.

$$C_1 = 1 - (1 - RF) * \left(\frac{d_{50}}{dt}\right) * \sqrt{\frac{\pi}{0{,}693}} * \left[erf(0) - erf\left(d^+\right)\right] \tag{7.128}$$

$$C_2 = 1 - (1 - RF) * \left(\frac{d_{50}}{dt}\right) * \sqrt{\frac{\pi}{0{,}693}} * \left[erf\left(d^+\right) - erf\left(d_{\max}\right)\right] \tag{7.129}$$

$$erf(x) = \frac{1}{2\pi} * \int_x^{\infty} e^{\left(\frac{t^2}{2}\right)} dt \tag{7.130}$$

The algorithm used here to simulate cyclone performance can be summarized as follows:

1. Calculate fraction of solids and amount of water in the feed.
2. Calculate Q_{CF} in $\left(\frac{lt}{\min}\right)$
3. Calculate $R_f = RF$ from Equation 7.116.
4. Calculate d_{50} from Equation 7.117.
5. Evaluate the Global Efficiency Integrals for C_1 and C_2 using Equations 7.128 and 7.129.
6. Calculate cyclone products according to the Equations 7.131 and 7.132.

$$W_R * \rho_R * R_3^R\left(d^+\right) = C_2 * W_{CF} * \rho_{CF} * R_3^{CF}\left(d^+\right) \tag{7.131}$$

$$W_P * \rho_P * R_3^P\left(d^+\right) = \left(1 - C_2\right) * W_{CF} * R_3^{CF}\left(d^+\right) \tag{7.132}$$

While the total solids recirculation is given by Equation 7.133 and the recycle water is given by Equation 7.134.

$$W_R * P_R = C_2 * W_{CF} * P_{CF} * R_3^{CF}\left(d^+\right) + C_1 * W_{CF} * P_{CF} * \left[1 - R_3^{CF}\left(d^+\right)\right] \tag{7.133}$$

$$W_{WR} = R_f * W_{WCF} \tag{7.134}$$

7.5.4.4 Controller Models

Dynamill II permits the analysis of system dynamics with and without automatic control loops to stabilize particle size product and/or circulating load. In designing a control system, it is of interest to know how the system will behave without controller for a variety of circuit variables changes. This program is capable of producing step changes in the (1) fresh feed water, (2) fresh feed size distribution, or (3) sump water addition rate by a series of options which are specified during the running of the program. Any other change can be affected by restarting the program (after achieving steady state) with a different value for any input.

7.5.4.5 Multivariable Control System

The ball mill circuit simulated here can be controlled using a series of easily implemented alternatives. The control of the circuit product size distribution can be accomplished by manipulating the amount of water to the sump and/or manipulating the fresh feed rate. The control of the circulating load can be achieved with either or both of these manipulated variables as well. In both cases control action is based on a feedback controller with proportional-integral (PI) action components. Also, two different sampling rates have been introduced in order to analyze those effects in the dynamic of the systems. It is well-known that the online particle size measurement takes about three minutes and in the other part the flow rate measurement of the circulating load takes only a few seconds.

A classical single input-single output Type II controller manipulates sump water addition u, to control the fraction finer than d^+ in the cyclone overflow $F_3^P(d^+)$ in a feedback fashion according to magnitude error. (The difference between the set point $F_{3_e}^P$ and F_3^6) and the integral error according to the relationship in Equation 7.135.

$$u_1(water\ to\ sump) = K_p * \left(e + K_I * \int_0^t edt \right) \tag{7.135}$$

Where, in this case, we have the expressions in Equations 7.136 through 7.140.

$$u_1 = W_{water}(t) - W_{\mathrm{waty}}^{SS} \tag{7.136}$$

$$e = F_3^P(d^+) - F_{3sp}^P(d^+) = R_{3sp}^P(d^+) - R_3(d^+) \tag{7.137}$$

$$K_p = feedback\ controller\ gain \tag{7.138}$$

$$K_I = \frac{1}{\delta_I} = feedback\ reset\ constant\left(\frac{1}{min}\right) \tag{7.139}$$

$$\delta_I = feedback\ reset\ time\,(min) \tag{7.140}$$

Control actions of this type has been generalized to allow the manipulation of variables to be determined by a combination of errors according to Equations 7.141 and 7.142.

$$u_1(water\ to\ sump) = K_{p_{w,ss}} * \left(e_1 + K_{I_{w,Ds}} * \int_0^t e_1 dt \right) + K_{p_{w,a}} * \left(e_2 + K_{I_{w_l,I}} * \int_0^t e_2 dt \right) \tag{7.141}$$

$$u_2(fresh\ feed) = K_{p_{F,},ss} * \left(e_1 + K_{I_{FF,ss}} * \int_0^1 e_1 dt \right) + K_{pF,CL} * \left(e_2 + K_{I_{FF,CL}} * \int_0^1 e_2 dt \right) \tag{7.142}$$

Where, in this case, we have the expressions in Equations 7.143 through 7.146.

$$e_1 = \left[F_3^P \left(d^+ \right) - F_{3_v}^P \left(d^+ \right) \right] \tag{7.143}$$

$$u_1 = W_{\text{water}} - W_{\text{water}}^{SS} \tag{7.144}$$

$$e_2 = \left[W_{CL} - W_{CL_{sv}} \right] \tag{7.145}$$

$$u_2 = \left[W_{FF} - W_{FF}^{SS} \right] \tag{7.146}$$

The subindex are resumed in Table 7.8.

In order to evaluate equations for the simulation of control actions for the simulation of control actions, we need to integrate the integral error term numerically according to Equation 7.147. The derivative of ERINT is just the error, represented in Equation 7.148.

$$\int_0^t edt = ERINT \tag{7.147}$$

$$\frac{d \int edt}{dt} = e \tag{7.148}$$

Therefore, a Euler type of algorithm for generating ERINT is represented in Equation 7.149, where *DELTA* is the time step size used for the numerical integration of the mill and sump equations using the Runge-Kutta algorithm.

TABLE 7.8
Subindexes and Meanings

Subindex	Meaning
W, PS	*Manipulation of sump water based on particle sizeerrors*
W, CL	*Manipulation of sump water based oncirculating load errors*
FF, PS	*Manipulation of fresh feed based on particle sizeerrors*
FF, CL	*Manipulation of fresh feed based oncirculating load errors*

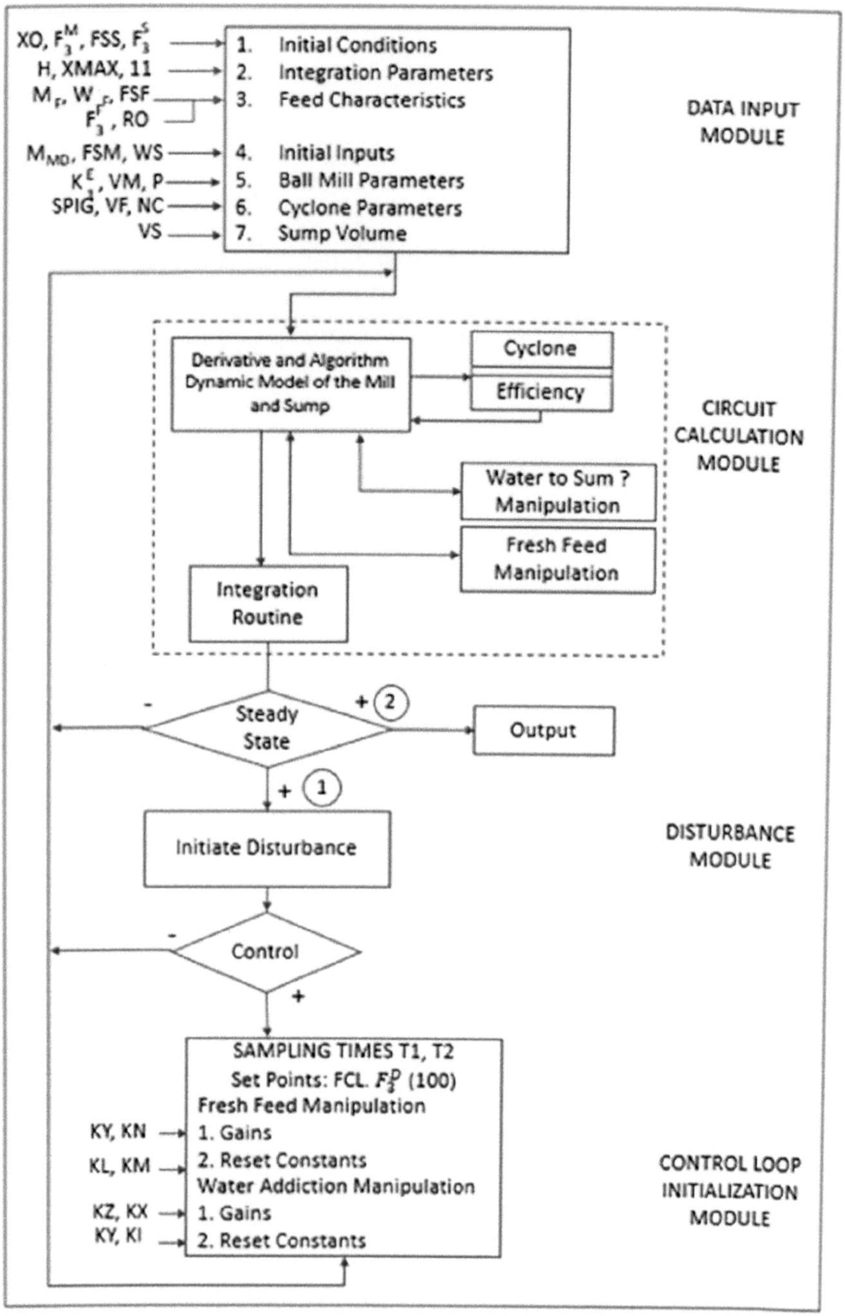

FIGURE 7.32 Dynamill II program flowchart.

$$(ERINT)_{n+1} = (ERINT)_n + DELTA * (ERROR)_n \tag{7.149}$$

As in practice the sump water addition and fresh feed flowrates must satisfy certain constraints. The program requires that these values must be positive and less than a specified maximum value.

7.5.4.6 Node Point Relationships

The final relationships required for dynamic simulation are those which specify the mill input in terms of the fresh feed and recycle. The relationships arise from a mass balance around the node in rapid mixing point which corresponds to the mill feed box designated as 1 in Figure 7.28. The relationships are Equation 7.150, total pulp mass balance; Equation 7.151, solid's mass balance; and Equation 7.152, solid's balance for size fraction.

$$W_{MF} = W_{FF} + W_R + W_{\text{Water to Mill}} \tag{7.150}$$

$$ps_{MF} * W_{MF} = p_{FF} * W_{FF} + p_R * W_R \tag{7.151}$$

$$ps_{MF} * W_{MF} * R_3^{MF}\left(d^+\right) = p_{FF} * W_{FF} * R_3^{FF}\left(d^+\right) + p_R^* W_R * R_3^R\left(d^+\right) \tag{7.152}$$

7.6 USE OF THE PROGRAM

7.6.1 Program Structure

A rough flowchart of Dynamill II program is given in Figure 7.32. The program consists of five modules: (1) data input, (2) circuit calculation, (3) disturbance initiation, (4) control initialization, and (5) output. A run is initiated in the data input module with a series of "reads" to establish circuit conditions prior to the disturbance. The program takes theses initial conditions and establishes a steady state with them in the circuit calculation module. This module involves a mode I, mixing calculation, and a sump product subroutine calculation which uses a fourth order Runge-Kutta integration subroutine to evaluate the system differential equations, cyclone subroutine calculation global efficiency subroutine calculation, erf function subroutine calculation. Then time is advanced H minutes, and the sequence is repeated until steady state values for all dependent variables is achieved for the initial input variables. This flow chart has been adapted to run in a modern environment such as an iPad using Xcode program. This type of object-oriented code enables you to touch the screen with your finger to manipulate the object while the program is running and change the equipment characteristics, mill, sump and hydrocyclones, feed characteristics, controllers, initiate disturbance is the feed, water addition to the mill or to the sump.

The objective is to start with the current mill setup and understand the relationship between the controlled (particle size and circulating load) and the manipulated variables (water addition to the sump, and process feed rate). While maintaining the right sump level controller and water addition to the mill to avoid violating pulp transport constraints, mill overflow, increase or hydrocyclone roping. Once you understand the limitation of the current circuit initiate step response to define your best control strategy to maximize throughput while maintain the right particle size shape by adding the right amount of water to the sump and to the mill.

We have created a more advanced dynamic simulator including a SAG mill and two ball mills with hydrocyclone classification. Here an inverse grinding circuit is modeled using the same approach as described earlier. Dynamill Pro™ was developed to run in the Google cloud for testing online models and developing integrated control strategies and production planning strategies including a flotation subsystem and thickening subsystem dynamic models based on hybrid industrial models.

After the initial steady state has been achieved, the disturbance initiation module is used to identify the type and magnitude of the disturbance which response is of interest. Step changes in water, feed rate, and feed size distribution are all option. In all cases the values of W_s, W_{FF}^{solid}, $F_3^F\left(d^+\right)$ (sump water, fresh feed, and fresh feed size distribution) immediately following the disturbance are entered whether they have changed or not. Figure 7.33 shows a screenshot of the process flow diagram of Dynamill dashboard to perform the simulation and manipulated the parameters setting by touching with your finger the objects. With two figures you get to see the dynamic trends showing the dynamic behavior of the controlled variables subject to the manipulated variables and process equipment parameters.

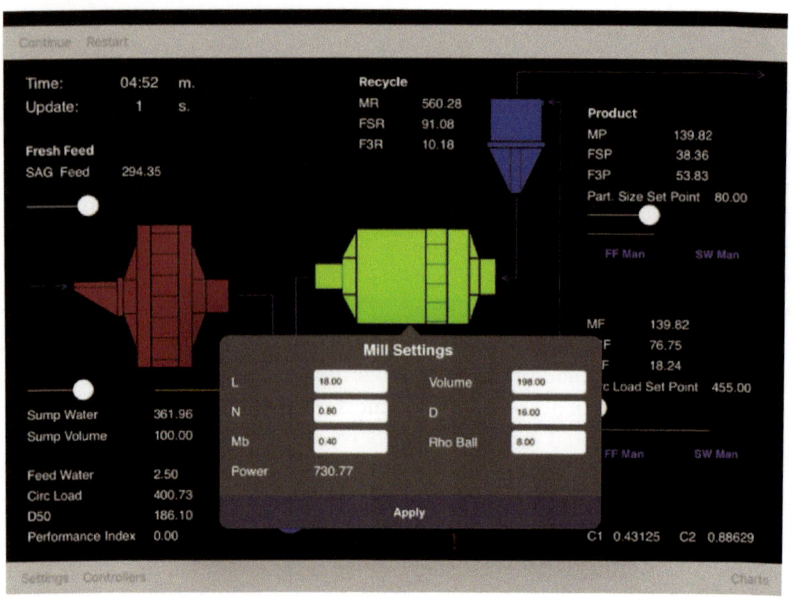

FIGURE 7.33 Dynamill hydrocyclone settings module obtained by touching the cyclone object.

Source: Dynamill, DynaFlote and Optimill [iPad software apps]. www.apple.com/app-store/.

Once the disturbance is defined, a decision must be made as to whether a controlled response or an uncontrolled response is desired. If the uncontrolled response is desired, the program launches back to the circuit circulation module and proceeds to fluid a new steady state. Intermediate results are printed out after each one minute of elapsed time after the disturbance.

If control options are to be exercised once the disturbance is defined, the program transfers to the control initialization module. Here a total of four gains and four reset tunes are entered to establish various linkages between controlled variables, particle, size and recycle mass flow (circulating load) and manipulated variables, sump water, and fresh feed rate. Once the tuning constants have been selected, program control is transferred to the circuit calculation module which computes the controlled approach to steady state. Once again intermediate results are printed out after each one minute of elapsed circuit time after the disturbance.

At the completion of either controlled or uncontrolled calculation program control is transferred to the output for final display the stop criteria are the steady-state condition, represented in Equation 7.153.

$$\frac{Mf - Mp}{Mp}\left\langle 0{,}004\,or\big/_{and}\;a\,max.time\,T\right\rangle TMAX \tag{7.153}$$

Note: the first checking of the steady state in this problem is very important since it is a recycle problem. At the first run this one has to be guessed, and then by itself the program finds a steady state by successive approximations (fixed point method).

7.6.2 Specification of Control Actions

The circuit simulated here is assumed to have an external level control look which maintains the level of sump constant. In addition, one or more of three measurement systems—a particle size analyzer for the cyclone overflow, solids mass flow measurement for the fresh feed, a volume flowmeter density gauge for the cyclone feed, volume flowmeters for feed water and sump water—are assumed available. With this instrumentation two principal variables can be controlled: the fraction finer than a specified size d^+ in the cyclone overflow, $F_3^P\left(d^+\right)$ and the recycle mass flowrate M_R. Possible manipulated variables are the sump water addition rate and the fresh feed rate. The control strategy option available within the program structure follow the multivariable feedback control laws given by Equations 7.141 and 7.142.

If no control action is required, the decision block following the disturbance module (Figure 7.32) and program control will be transferred to the circuit calculation module, where the uncontrolled response will be calculated. If control is required, set $NN = 1$ at which time the program will request the time intervals for sampling the controlled variables and eight controllers' gains requires for Equations 7.141 and 7.142. Selection of the "best" tunning constants is a very difficult task. If only one control loop (single input-single output) is required, several empirical techniques are available for tuning, including Ziegler Nichols and ultimate sensitivity.

When both the particle size and circulating load loops are considered simultaneous, serious controller interactions can occur which severely limit the use fulness of the single loop tunning techniques. Some techniques for decoupling the system are available like the relative gain. Another possibility is to use some experimental design search strategy for the best tuning constants.

Finally, in some instance it may be desirable to determine "optimal control trajectories" analytically by applying the latest optimization tools available.

7.6.3 Program Testing

The program has been tested using industrial data from many sites. A test run is scheduled to perform a set of experiments in the field to obtain the grinding characteristics parameters and hydrocyclone model parameters.

Figure 7.34 shows a typical grinding circuit predicting P80 and Gaudin size module M > 0.37. This is at the edge of having too many fines for flotation and flocculation of the rougher tailings with the consequence of losing metal and water to the tailings due to high percent solids in the output of the mill.

The program has been tested for various cases over the years and implemented many controls strategies by exploring the alternatives with the dynamic simulator (Bascur et al. 1986, 2022; Bascur and Herbst 1985; Bascur 1991; Bascur and Soudek 2019).

The seminal paper by Bascur et al. (1986) shows the importance of the pump in grinding circuit and the great investment it represents in augmenting the throughput in a typical operation. It gave in practice an increase of 35% more throughput with an increase of the metal production as reported by Bascur (1990) and generalized in Bascur (1991).

Here are a few examples. After a typical steady state for typical conditions and some data adjustment of the circuit parameters, the following test was found:

1. Figure 7.35 shows a 10% step change in the Particle Size Set Point. The dynamic response of product size distribution is shown, and the product flowrate is plotted since it is the stopping criteria for finding the new steady state. It can be seen that this system is highly sensitive to a sudden change of water to sump and that this step change produces a perturbation that last for about 15 min. This is very similar to real practice.
2. Control of the particle size distribution by adding water to the sump. Tuning.

Figure 7.36 presents the dynamic response of the particle size distribution to a set point change for various tuning constants. It can be seen that the best controlling action is obtained using $K_p = 450$ and $K_i = 180$. Also, in Figure 7.36, the manipulated variable (water to sump) is plotted together with the controlled variable P80.

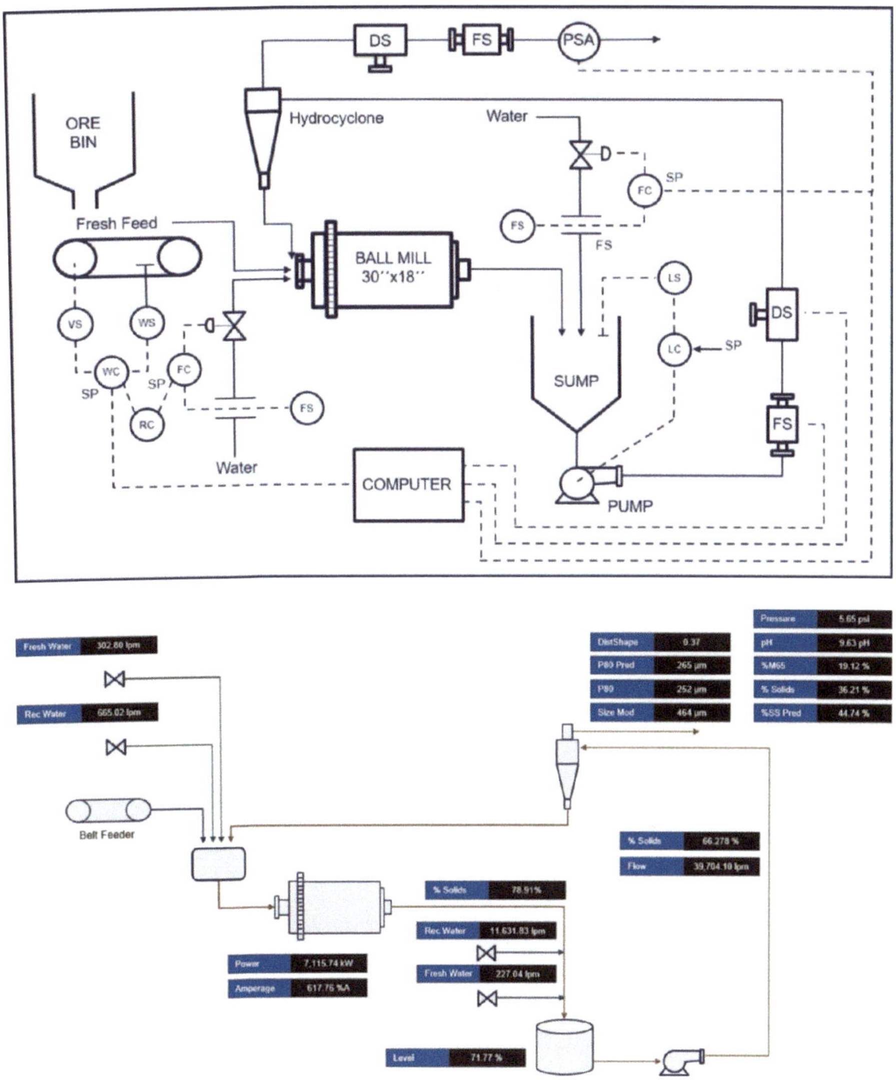

FIGURE 7.34 Dynamill predictions of the P80 and Gaudin size shape module at 0.37.

It can be seen that the product size distribution arrives to the steady state in about 17 min while the water addition continues regulating the system till 38 min after the set point change was produced.

Figure 7.36 shows the tuning of the particle size controller using the sump water control tuning using a variable speed pump to control the sump level of the grinding circuit. The sump/pump plays an important role in a grinding circuit, as it has been discussed by Bascur et al. (1986). A variable speed pump enables to increase the throughput of the grinding circuits by allowing to adapt the circulating load of the

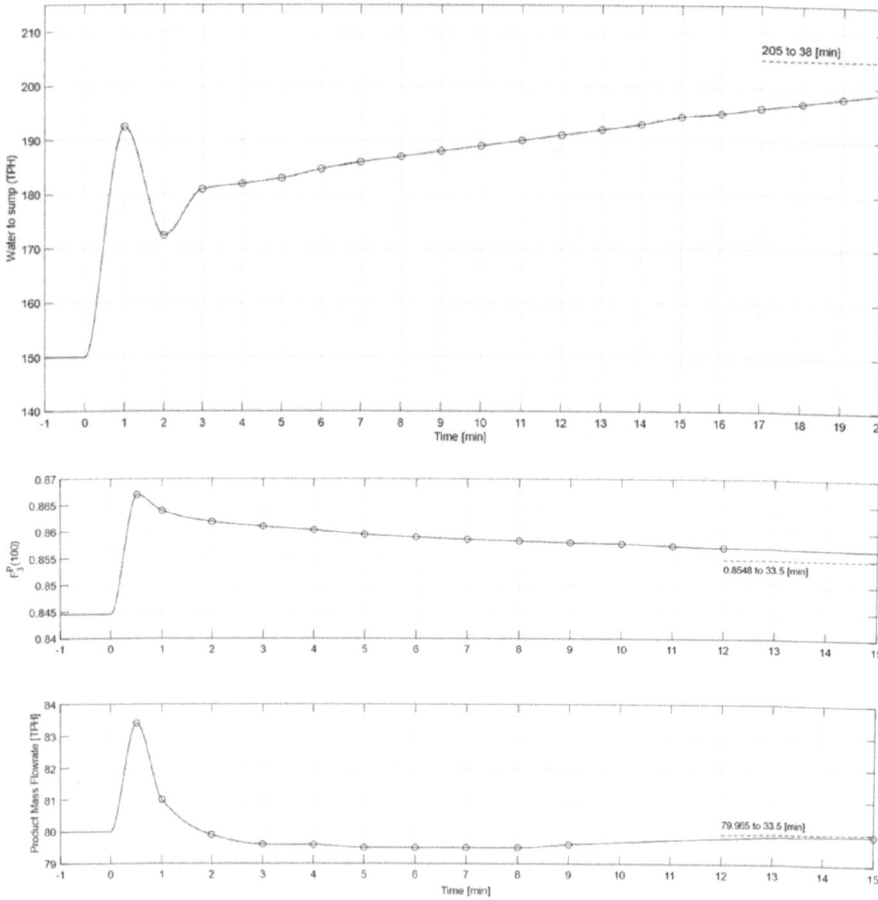

FIGURE 7.35 Dynamic trend showing effect of Particle Size under a Particle Size Set Point step change.

circuit and maintaining the right flow to the hydrocyclone bank. It is important to note that the simulator enables you to change the number of hydrocyclones if you need to increase the circulating load to achieve the throughput and desired particle size distribution. It is important to respect the rheological characters of the pulp as the feed to the cyclone, underflow discharge, and ball mill % solids by adjusting the water to mill accordingly. The underflow of the cyclone is about 75% solids so need to compensate by adding the right amount of water in sump to achieve a good particle size separation. The hydrocyclone spigot diameter plays an important role in a grinding circuit since it defines the short circuit in general of water and fines toward the ball mill in a reverse or direct circuit.

Another important aspect of including the pump model in the dynamic simulator in a grinding circuit is to see the effect of the size of the sump which is almost ignored in traditional steady state models since it is not required. The SAG mill

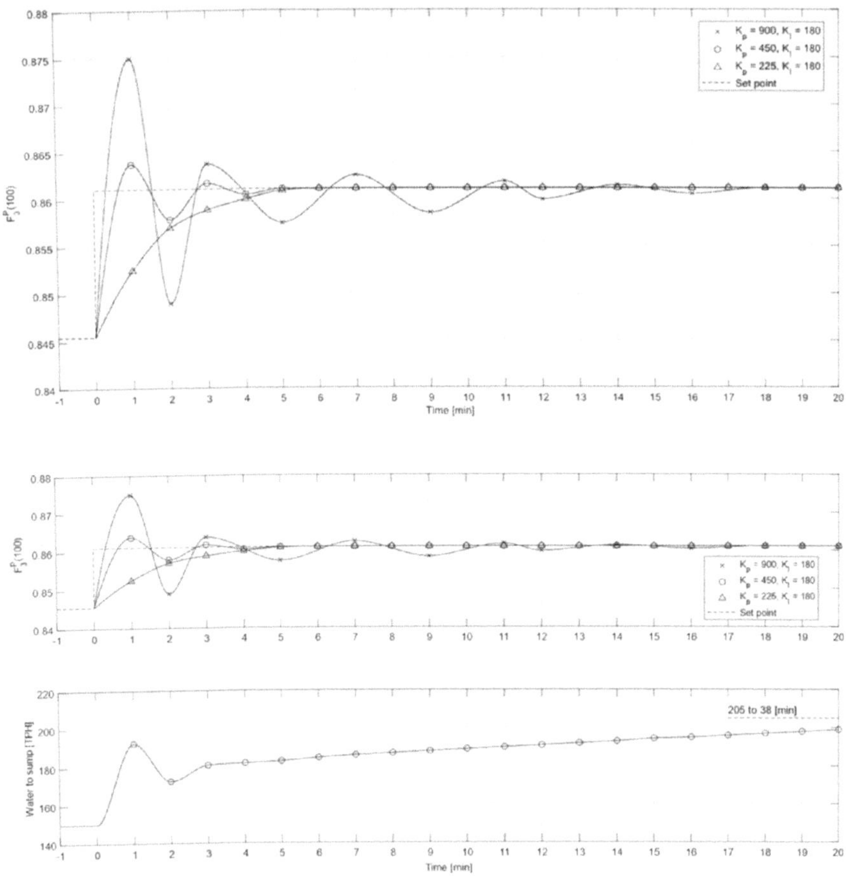

FIGURE 7.36 Particle size versus sump water control loop tuning.

holdup, two ball mills holdups, and the sump form an integrated dynamic system where the SUMP becomes the shock absorber of the disturbances of the SAG mill and the cyclone underflow to the ball mills. In a grinding circuit, it absorbs the ore hardness, particle size distributions coming from the stockpile, or the bin feeding the mill via a belt conveyor or series of belt conveyors. These actuators are also modeled in the Dynamill Pro version of the problem since lags are required to be modeled for a good fidelity of the dynamic characteristics of a comminution circuit system. More on this subject is presented in Chapter 12 where we simulate using operational data the whole plant using a digital twin plant integrating the comminution circuit with the rougher flotation and the water recovery from the processing of the rougher flotation tailings in the thickeners. The water recovered from the thickeners is vital for providing the right particle size distribution for flotation of mineral particles and flocculation of the rock (waste) particles.

Figure 7.37 shows a screenshot of Dynamill running under particle size control with water addition to sump while pushing throughput and not having the right sump level. As you can see immediately, instability of the grinding circuit system appears in the circuit. As such, the dynamic grinding simulator is a great tool to design the right size of the sump depending on the size of the SAG, ball mills, sump, pumps, and hydrocyclone bank. This type of behavior never appears using the traditional steady state models such as Optimill for example. Reason, that having a dynamic simulator for grinding classification become very important when drastic changes in the type of ore containing variable amounts of rocks containing soft and hard species changing the grindability and thus, the rheology of the pulp. Caution has to be given to maintain a pulp suspension in the mill. For typical copper ores with a density of 2.7 when the percent solids by weight in the mill goes over 75–77% solids, a drastic change in pulp transport become present. This produces the generating of extra fines which will be detrimental for the flotation process and most importantly for the flocculation of the ground particles which are the rougher flotation tailing from where we need to recover as much water as possible to control the rheology in the mill not to create fines. The fines have difficulty floating and they reduce the concentrate quality grade by entrainment of undesired fines by the water in flotation.

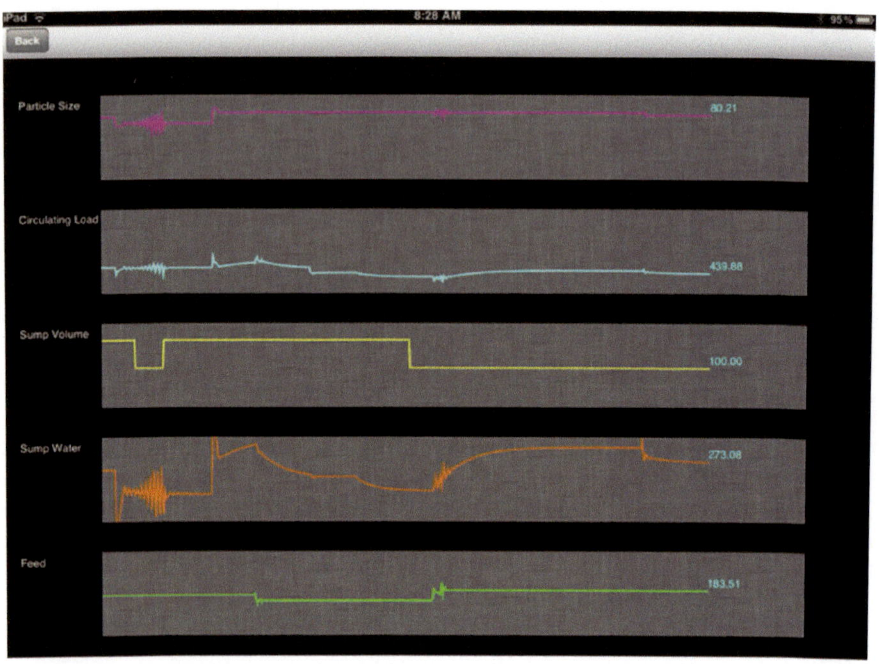

FIGURE 7.37 Dynamill results when controlling P80 with water and having the right pumping capabilities in the sump.

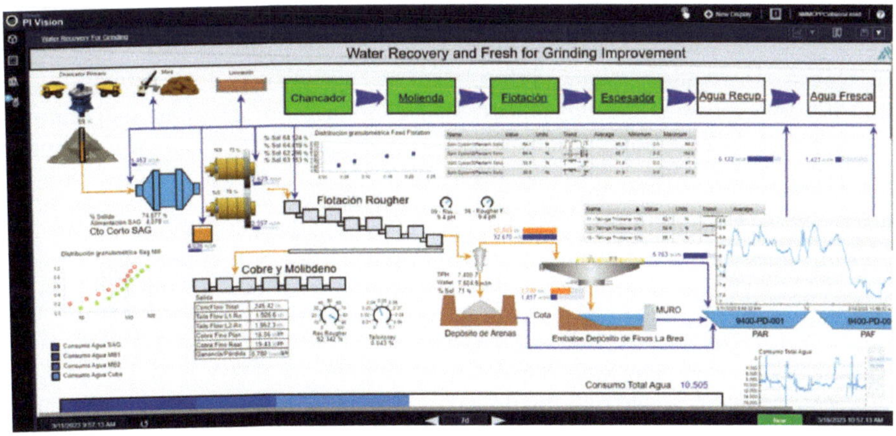

FIGURE 7.38 Grinding/classification/flotation/water separation OSB digital twin.

Source: Bascur, O.A. 2023. Digital twins in mining and mineral processing. Presented at the 26th World Mining Conference: Resourcing Tomorrow, Creating Value for Society, Brisbane.

7.7 CONCLUSION

Dynamill is a program for the dynamic simulation of a grinding classification circuit in a good qualitative and quantitative manner. It can be used for testing different control strategies such as fresh feed manipulating for controlling the circulating load and particle size of the product and/or water to sump for controlling the circulating load and particle size of the product. This program can be used in a course in process control for metallurgical engineers.

The main objective to present a steady state grinding circuit and a dynamic grinding circuit is to show the imperative importance to include in the simulation the pump effects, the sump volume effect, to understand the effect of the disturbances in the designing of the mill for designing best practices. These best practices cannot be identified using a traditional steady state simulation. Today, with the high variability of the low-grade ores to produce the necessary metals for humanity, we need to capture the knowledge in dynamic simulators to learn and to be able to design adequate digital twins for implementation in industrial plants. Today the size of a grinding line is in the order of 5000 TPH. Before, a grinding line was designed to run between 250–700 TPH. Today an unscheduled downtime in grinding can cost millions of dollars and there is NO redundancy of the grinding lines.

By running a dynamic model in an industrial system enables us to avoid costly violation of process and equipment constraints by predicting in real time if a critical variable might trigger an event based on the Digital Twin Plant Model (Bascur 2023). The amount of data generated by the process control system is too high for an operator or an engineer to comprehend fast enough. As such, the implementation of digital twins for optimal planning and operational support is vital (Bascur and Retamal 2023).

Figure 7.38 shows an example of a typical SAG mill, pebble crusher, sump, pumps, cyclones, and ball mills to grind and liberate the minerals from the rocks. Having the computer to process the data into information and generating predictive conditions ahead of them happening is a way to increase the efficiency of the mineral operations.

REFERENCES

Anticoi, H.F. 2019. Strategic minerals milling modelling of high-pressure grind rolls and process parameters dependency. PhD thesis, Polytechnic University of Catalonia.

Austin, L.G., and Concha, F. 1994. *Grinding-classification circuit design [in Spanish]*. Chiguayante: CETTEM.

Austin, L.G., Klimpel, R.R., and Luckie, P.T. 1984. *The Process Engineering of Size Reduction: Ball Milling*. New York: SME-AIME.

Austin, L.G., Luckie, P.T., and Klimpel, R.R. 1972. Solutions of the batch grinding equations leading to Rosin-Rammler distributions. *AIME Transactions* 252:87.

Bascur, O.A. 1979. *Dynamill PRO: Chemical Engineering Process Dynamics, CHEM E MI System HP2100 Dynamic Simulator*. Unpublished paper. Utah: University of Utah.

Bascur, O.A. 1990. Profit-based grinding controls. *Minerals and Metallurgical Processing* (February): 9–15.

Bascur, O.A. 1991. Integrated grinding/flotation controls and management. Presented at the Copper 91/Cobre 91 International Symposium, Ottawa, Ontario.

Bascur, O.A. 2023. Digital twins in mining and mineral processing. Presented at the 26th World Mining Conference: Resourcing Tomorrow, Creating Value for Society, Brisbane.

Bascur, O.A., Freeh, E.J., and Herbst, J.A. 1986. Dynamic simulation and estimation of industrial grinding circuit control characteristics. *ISA Transactions* 15(2):77–84.

Bascur, O.A., and Herbst, J.A. 1985. Dynamic simulators for training personnel in the control of grinding/flotation systems. Presented at IFAC MMM: Automation for Mineral Resource Development, Brisbane.

Bascur, O.A., and Retamal, F. 2023. Increasing water recovery to maximize copper production with low grade ores. Presented at the 62nd Conference of Metallurgists: MET SOC COM 2023, Toronto, Ontario, August 21–24.

Bascur, O.A., Retamal, F., and Espinosa, S. 2022. Treating low grades ores using a digital in mining and mineral processing plants. In *Proceedings of the Copper 2022 International Conference: IIMChile. Vol. 7: Process Optimization through Innovation, Technology, Automation and Design*. Santiago: Gecamin.

Bascur, O.A., and Soudek, A. 2019. Grinding and flotation optimization using operational intelligence. *Mining, Metallurgy and Exploration* 36:139–149. https://doi.org/10.1007/s42461-018-0036-4.

Bond, F.C. 1960. Crushing and grinding calculations. *British Chemical Engineering* 6:378–391, 543–548.

Bond, F.C. 1962. *Crushing and Grinding Calculations*. Milwaukee, WI: Allis Chalmers.

Bürger, R., Bustamante, O., Fullac, M.R., and Rivera, L.E. 2018. A population balance model of ball wear in grinding mills: An experimental case study. *Minerals Engineering* 128:288–293.

Concha, F., Magne, L., and Austin, J.G. 1992. Optimization of the make-up ball charge in a grinding mill. *International Journal of Mineral Processing* 34:231–241.

Fuerstenau, D.W., Kapur, P.C., and Gutsche, O. 1993. Comminution of single particles in a rigidly mounted roll mill. Part 1: Mill torque model and energy investment. *Powder Technology* 76:253–262.

Fuerstenau, D.W., Shuklab, A., and Kapur, P.C. 1991. Energy consumption and product size distributions in choke-fed, high-compression roll mills. *International Journal of Mineral Processing* 32:59–79.

Herbst, J.A. 1992. Microscale comminution studies for ball mill modeling. In *Comminution: Theory and Practice*. Edited by K. Kawatra. Littleton, CO: SME.

Herbst, J.A., and Bascur, O.A. 1978. *STEMILL: A Program for the Simulation of Steady State Grinding Circuits Behavior Using a Small Computer*. Salt Lake City, UT: University of Utah.

Herbst, J.A., Bascur, O.A., and Rajamani, R. 1978. *Dynamill II: A Program for the Simulation of Grinding Circuits Dynamics and Control Using a Small Computer*. Salt Lake City, UT: University of Utah.

Herbst, J.A., Grandy, G.A., and Fuerstenau, D.W. 1973. Population balance models for the design of continuous grinding mills. In *Proceedings of the 10th International Mineral Processing Congress*. Edited by M.J. Jones. London: Institution of Mining and Metallurgy.

Herbst, J.A., Lo, Y.C., and Rajamani, K. 1985. Population balance model predictions of the performance of large-diameter mills. *Mining, Metallurgy & Exploration* 2:114–120.

Hinde, A.L., and King, R.P. 1977. Minimal variance control strategies for wet milling circuits. Paper Presented at AIME Annual Meeting, Atlanta, March 1977.

Kapur, G.S., and Fuerstenau, D.W. 1992. Grinding of heterogeneous mixtures in a high-pressure roll mill. In *Comminution: Theory and Practice*. Edited by K. Kawatra. Littleton, CO: SME.

Lynch, A.J. 1977. Mineral crushing and grinding circuits, their simulation, optimization, design and control. In *Developments in Mineral Processing*. New York: Elsevier Scientific.

Menacho, J., and Concha, F. 1986. Mathematical model of ball wear in grinding mills. *Powder Technology* 47:87–96.

Monov, V., Sokolov, B., and Stoenchev, S. 2012. Grinding in ball mills: Modelling and process control. *Cybernetics and Information Technologies* 12(2):51–68.

Morley, C. 2006. *High Pressure Grinding Rolls—A Technology Review*. SME Preprint No. 15–39. Littleton, CO: SME.

Naziemiec, Z., and Saramak, D. 2009. Analiza zmian obciążenia materiału w strefi e zgniotu pras walcowych. *Górnictwo i Geoinżynieria* 4.

Pate, W.T., and Herbst, J.A. 1983. *Dynamic Estimation of Unmeasured Variables in an Industrial Grinding Circuit Using a Kalman Filter*. SME Preprint No. 83–124. Littleton, CO: SME.

Plitt, L.R. 1976. A mathematical model of the hydrocyclone classifier. *CIM Bulletin* (December): 114–123.

Rajamani, K. 1979. Optimal control of closed grinding circuit ball mill grinding. PhD dissertation, University of Utah.

Saramak, D. 2011. Technological issues of high-pressure grinding rolls operation in ore comminution processes. *Archives of Mining Sciences* 56:517–526.

Sastry, K.V., and Concha, F. 1985. Proc. Int. Symp. In Mineral Processing, 5. Population Balance with Applications in Mineral Processing, University of Conception, Chile. p. 5.

SolidsWiki Contributors. 2013. Rod mills. *SolidsWiki*. www.solidswiki.com/index.php?title=Rod_Mills&oldid=15378. Accessed July 2023.

Walker, W.H., et al. 1937. *Principles of Chemical Engineering* (Chemical engineering series) Hardcover – January 1, 1937. NY: McGraw-Hill. p. 255. https://www.amazon.com/Principles-chemical-engineering-Chemical/dp/B00085BF6U

FURTHER READING

Austin, L.G., Menacho, J.M., and Pearcy, F. 1987. A general model for semi-autogenous and autogenous milling. In *APCOM 87: Proceedings of the Twentieth International Symposium on the Application of Computers and Mathematics in the Mineral Industries. Volume 2: Metallurgy*, Southern African Institute of Mining and Metallurgy, Johannesburg. pp. 107–126.

Bascur, O.A., and Junge, P.A. 2011. Dynamill, DynaFlote and Optimill [iPad software apps]. www.apple.com/app-store/.

Jankovic, A. 2015. Chapter 8: Developments in iron ore comminution and classification technologies. In *Mineralogy, Processing and Environmental Sustainability*. Edited by B. Lu. New York: Woodhead.

Morrell, S. 1996. Power draw of wet tumbling mills and its relationship to charge dynamics. Part 1: A continuum approach to mathematical modelling of mill power draw. *Transactions of the Institution of Mining and Metallurgy, Sect. C*: 43–54.

Mulenga, F.K. 2008. Effect of ball size distribution on milling parameters. Master of Science in Engineering thesis, University of the Witwatersrand.

Schönert, K., and Sander, U. 2002. Shear stresses and material slip in high pressure roller mills. *Powder Technology* 122(2–3):136–144.

Whiten, W.J., Walter, G.W., and White, M.E. 1979. A breakage function suitable for crusher models. *Fracture at Work* :19.1–19.3.

Yekeler, M., Ozkan, A., and Austin, L.G. 2001. Kinetics of fine wet grinding in a laboratory ball mill. *Powder Technology* 114:224–228.

8 Particle Size Classification

8.1 INTRODUCTION

Classification in mineral processing is the operation of separating a mixture of particle into two or more fractions according to size, or other physical property, with each resulting group resulting more uniform in the defined property than the original mixture. There are two types of classifiers: mechanical and hydraulic classifiers. During screening, the classical mechanical classifier, the material is subjected to the action of a series of screens through which small particles pass, while coarser particles greater than screen openings are retained. In screening, particle shape also has an important effect. The mechanism involved in separating particle sizes in a hydraulic classifier is sedimentation. Particles are subjected to body forces such as gravity or centrifugal forces, and in this case, the factors affecting the classifications are the solid particle properties of size, density, and shape and the fluid temperature, density, and viscosity.

In most cases the product of classification comprises two fractions relative to the separation size, the coarser and finer fractions. In a perfect operation the coarse fraction is called the "underflow," which contains all particles that are greater than the separation size, and the fine fraction called the "overflow." During hydraulic classification, the presence of different particle concentration, random dispersion, and other particles properties such as mineralogical composition and particle shape hinders a perfect classification. When hydraulic classification is performed on a heterogeneous material with particles of different densities, which often happens in practice, it takes the name sorting and becomes a pre-concentration process.

8.1.1 DEFINITIONS

There are several ways of measuring solid concentration in a suspension, solid and fluid densities, weight or volume fractions of solids and dilution, therefore it is essential to know how to convert one into another. If we designate by ρ_s and ρ_f the solid and fluid density, $\Delta\rho$ their difference, w and φ the mass and volume fractions of solids, ρ the suspension density, also known as pulp density, and D the dilution, that is, the ratio of mass of water to mass of solid, from a mass balance of the total mass and the mass of the solid and the fluid, is shown in Equation 8.1.

$$\begin{aligned} \underbrace{\rho V}_{\text{Solid mass}} &= \underbrace{\rho_s \varphi}_{\text{solid mass}} + \underbrace{\rho_f (1-\varphi)}_{\text{Fluid mass}} \\ \underbrace{D}_{\text{Dilution}} &= \frac{\rho_f (1-\varphi)}{\rho_s \varphi} \end{aligned}, \tag{8.1}$$

DOI: 10.1201/9781032614212-8

From which the conversion between these variables are given in the following equation:

$$\text{Pulp density: } \rho = \frac{\rho_s \rho_f}{\rho_s(1-X)+\rho_f X} = \rho_f + \Delta\rho\varphi = \frac{\rho_x \rho_f (1-D)}{\rho_s D + \rho_f}$$

$$\text{Volume fraction of solids: } \varphi = \frac{\rho - \rho_f}{\rho_s - \rho_f} = \frac{\rho_f w}{\rho_s(1-w)+\rho_f w} = \frac{\rho_f}{\rho_f + \rho_s D}$$

$$\text{Mass fraction of solids: } w = \frac{\rho_s \varphi}{\rho_f + \varphi\Delta\rho} = \frac{\rho_s\left(\rho - \rho_f\right)}{\rho\Delta\rho} = \frac{1}{1+D} \tag{8.2}$$

$$\text{Percent solid (\%) } \% = 100 \times \frac{\rho_s \varphi}{\rho_f + \varphi\Delta\rho} = 100 \times \frac{\rho_s\left(\rho - \rho_f\right)}{\rho\Delta\rho} = 100 \times \frac{1}{1+D}$$

$$\text{Dilution } D = \frac{\rho_f(1-\varphi)}{\rho_s \varphi} = \frac{1-w}{w} = \frac{\rho_f\left(\rho_s - \rho\right)}{\rho_s\left(\rho - \rho_f\right)}$$

8.1.2 Mass Balances

In a classifier at steady state a macroscopic balance may be expressed in terms of volume flow of pulp, flow of solids, of particles of size x, and flow of particles of sizes smaller or larger than size x. Figure 8.1 and Equation 8.3 show those balances, where the asterisk symbol represents "of the pulp."

$$\text{Pulp balance} \qquad F^* = O^* + U^*$$

$$\text{Solid balance} \quad \begin{cases} F = O + U \\ F^* w_F = O_O^* w_O + U_U^* w_U \end{cases}$$

$$\text{Fluid balance} \quad \begin{cases} F^*\left(1-w_F\right) = O^*\left(1-w_O\right) + U^*\left(1-w_U\right) \\ F\dfrac{1-w_F}{w_F}\left(1-w_F\right) = O\dfrac{\left(1-w_O\right)}{w_O} + U\dfrac{\left(1-w_U\right)}{w_U} \end{cases} \tag{8.3}$$

$$f_F(x) = Of_0(x) + Uf_U(x)$$
$$F_F(x) = O_O(x) + UR_U(x)$$
$$R_F(x) = OR_O(x) + UR_U(x)$$

$$F^*, F, w_F, x, F_F(x), R_F(x) \rightarrow \boxed{\text{CLASSIFIER}} \rightarrow O^*, O, w_O, F_O(x), R_O(x)$$

Overflow

$$\downarrow$$

$$U^*, U, w_U, x, F_U(x), R_U(x)$$

Underflow

FIGURE 8.1 Variables of a classifier.

Combining Equation 8.3 yields

$$\frac{U^*}{F^*} = \frac{(w_F - w_O)}{(w_U - w_O)}$$

$$\frac{U}{F} = \frac{(w_F - w_O)}{(w_U - w_O)}\frac{w_F}{w_U} = \frac{f_F(x) - f_O(x)}{f_U(x) - f_O(x)} \tag{8.4}$$

$$= \frac{F_F(x) - O_O(x)}{U_U(x) - O_O(x)} = \frac{R_F(x) - R_O(x)}{R_U(x) - R_O(x)}$$

Equation 8.4 shows that the fraction of feed that appears in the underflow can be calculated from samples taken from the feed, overflow and underflow of the classifier, and the determination of the solid concentration and particle size analysis.

8.1.3 Classification Efficiency

The quality of the separation in a classifier is quantified by the *efficiency of classification* whose mathematical expression is known as the *selectivity function* and as *Tromp curve*. The efficiency defines the mass of particles of size x in the feed $Ff_F(x)$ that goes to the underflow $Uf_U(x)$:

$$S(x) = \frac{U_U f_U(x)}{F_F f_F(x)} \tag{8.5}$$

Using one Equation 8.4 yields

$$S(x) = \frac{(f_F(x) - f_0(x)) f_U(x)}{(f_U(x) - f_0(x)) f_F(x)} \tag{8.6}$$

To visualize the form of a selectivity function, we will use experimental data and mass balance in a classifier (Table 8.1).

A plot of the selectivity function with data from Table 8.1 is given in Figure 8.2.

Figure 8.2 shows that particles finer than 30μm go to the underflow together with the oversize particles. These misallocated particles are called the *short circuit* or *bypass of fines*, since in effect, they are not classified. Their amount is designated by α. The short circuit in this case is 20%.

In some cases, even smaller fine particles appear in the underflow beyond the amount that would be expected from the short circuit, especially in hydrocyclones. This effect is known as the *fishhook*. Kraipech et al. (2002) showed that operational conditions in small hydrocyclones affect the shape of the selectivity curve and the corresponding fishhook (Figure 8.3).

TABLE 8.1
Experimental Data of the Size Distribution around a Classifier

Size *x*		Average size *x*	Feed		Undererflow		Overflow		Selectivity
Mesh #	µm	µm	F_F ton/h	f_F (X)	U_U ton/h	*f* u(x)	O_O ton/h	*f* u(*x*)	S (*x*)
28	600	505	0.000	0.00	0.000	0.00	0.00	0.00	1.000
35	425	357	2.122	3.49	2.122	8.60	0.00	0.00	1.000
48	300	252	1.678	2.76	1.678	6.80	0.00	0.00	1.000
65	212	178	3.028	4.99	2.812	11.40	0.22	0.60	0.929
100	150	126	5.904	9.72	3.923	15.90	1.98	5.49	0.664
150	106	89	10.266	16.91	4.860	19.70	5.41	14.99	0.473
200	75	63	12.414	20.44	3.627	14.70	8.79	24.37	0.292
270	53	45	8.752	14.41	2.270	9.20	6.48	17.98	0.259
-270	27	14	16.560	27.27	3.379	13.70	13.18	36.56	0.204
		Total	**60.724**	**100.00**	**24.671**	**100.00**	**36.05**	**100.00**	
			W_F=	**48.60**	W_u=	**65.5**	W_o=	**41.30**	

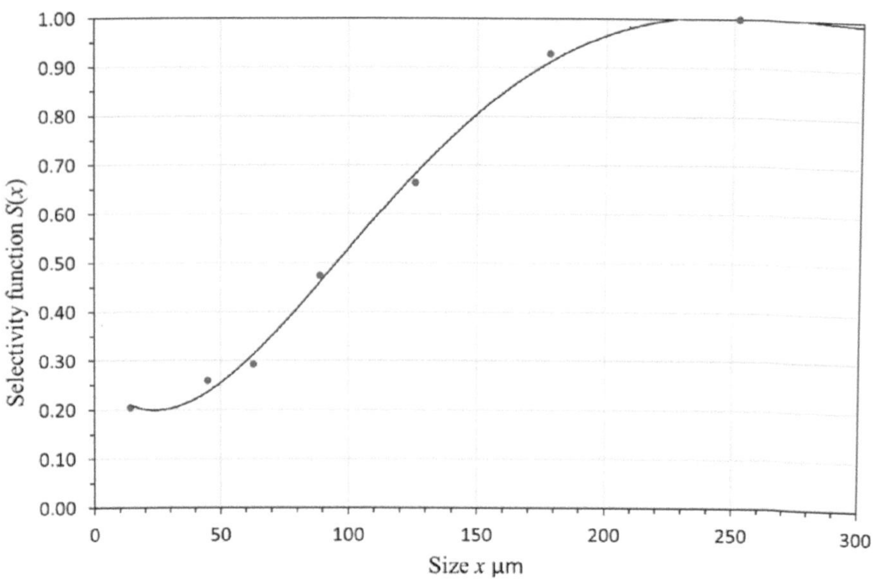

FIGURE 8.2 Selectivity function.

Of the several aspects causing fishhook in the selectivity function of a suspension, the most plausible are particles with different densities (Finch 1983; Rouse et al. 1987), agglomeration of fines to coarse solids (Hoffmann and Stein 2008), flocculation of the feed material (Gerhard 2001), entrapment of fin particles in the wake of larger particles (Kraipech et al. 2002), and inlet velocity of the pulp (Zhu and Liow 2014).

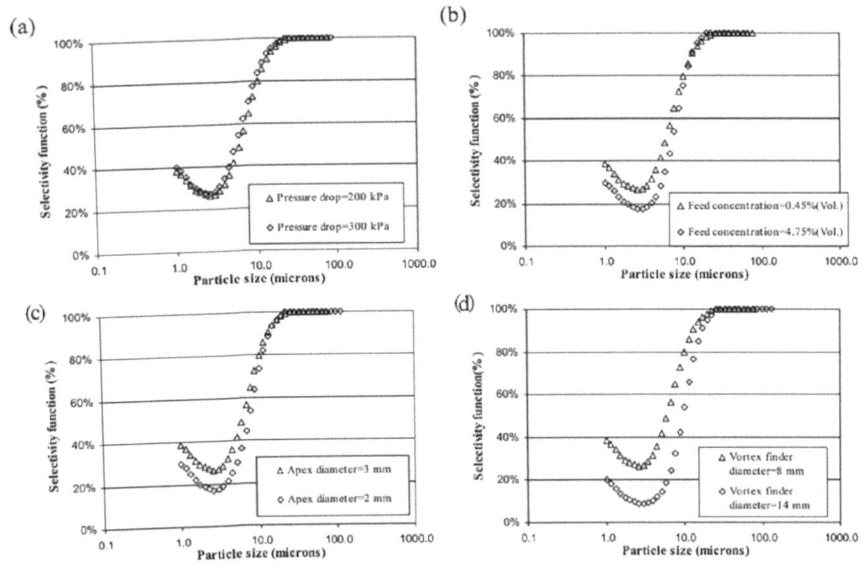

FIGURE 8.3 The effect of pressure, feed concentration, apex, and vortex finder sizes on the effect on selectivity curve in certain materials.

Source: Kraipech, W., Chen, W., Parma, F.J., and Dyakowski, T. 2002. Modelling the fish-hook effect of the flow within hydrocyclones. *International Journal of Mineral Processing* 66:49–65.

Dueck et al. (2007, 2013) explain with the aid of a kinematic model of entrainment of small particle by large one the presence of a fishhook in the selectivity curve, as shown in Figure 8.4.

8.1.4 Classification Function

For a given material, the amount of the short circuit depends on the operating condition of the grinding-classification circuit, which we will study in the next chapter, therefore, the selectivity function does not represent the real material that suffered classification. The correct *classification function* $C(x)$ is equal to the selectivity function from which the short circuit "α" has been subtracted.

Since the short circuit is material not classified, it should have the particle size composition of the feed, then

$$C(x) = \frac{Uf_U(x) - aF_F f_F(x)}{F_F f_F(x) - aF_F f_F(x)} \tag{8.7}$$

Dividing every term by $F_F f_F(x)$ yields

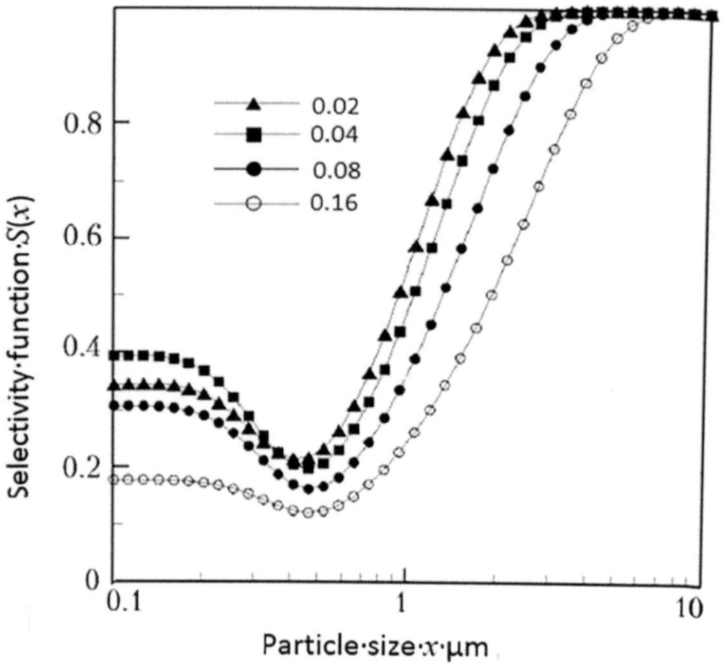

FIGURE 8.4 Influence of the solid concentration on the separation function.

Source: Dueck, J.G.T., Minkov, L.L., and Pikushchak, E.V. 2007. Modeling of the "fish-hook." *Journal of Engineering Physics and Thermophysics* 80:64–73.

$$C(x) = \frac{S(x) - a}{1 - a} \tag{8.8}$$

To Table 8.1 we can add the calculation of classification shown in Table 8.2. The results are presented in a form of sigmoidal curves in Figure 8.5. The corrected Efficiency curve is also called the Classification Function. This curve is corrected by the short circuit "alpha" or "Rf" (in the term of the water bypass or short circuit in an industrial classifier).

Neesse et al. (2004) demonstrated that the entrapment of small particles in the wake of larger particles is due to the turbulent diffusion of particles smaller than a threshold size. Figure 8.6 shows the effect of the feed concentration.

8.1.5 Reduced Classification

We define the *reduced separation sizes* x / x_{50}, x / x_{75}, and x / x_{25} as the particle size values at which the 50%, 75%, and 25% of the material goes to the underflow in a *reduced classification function* with x / x_{50} as the abscissa. Figure 8.7 shows the reduced classification function versus the reduced separation size x / x_{50}.

TABLE 8.2
Hydrocyclone Classification and Selectivity Analysis by Size

Size *x*		Average size x	Feed		Underflow		Overflow		Selectivity	Classification
Mesh #	μm	pm	F_F ton/h	*f* F(X)	U_U ton/h	f_U (*x*)	O_o ton/h	$f_u(x)$	S (x)	C(x)
28	600	505	0.000	0.00	0.000	0.00	0.00	0.00	1.00	1.00
35	425	357	2.122	3.49	2.122	8.60	0.00	0.00	1.00	1.00
48	300	252	1.678	2.76	1.678	6.80	0.00	0.00	1.00	1.00
65	212	178	3.028	4.99	2.812	11.40	0.22	0.60	0.93	0.91
100	150	126	5.904	9.72	3.923	15.90	1.98	5.49	0.66	0.58
150	106	89	10.266	16.91	4.860	19.70	5.41	14.99	0.47	0.34
200	75	63	12.414	20.44	3.627	14.70	8.79	24.37	0.29	0.11
270	53	45	8.752	14.41	2.270	9.20	6.48	17.98	0.26	0.07
-270	27	14	16.560	27.27	3.379	13.70	13.18	36.56	0.20	0.00
		Total	**60.724**	**100.00**	**24.671**	**100.00**	**36.05**	**100.00**		
			W_F=	**48.60**	**Wu=**	**65.5**	**a=**	**0.204**	**d50=**	**113.83**

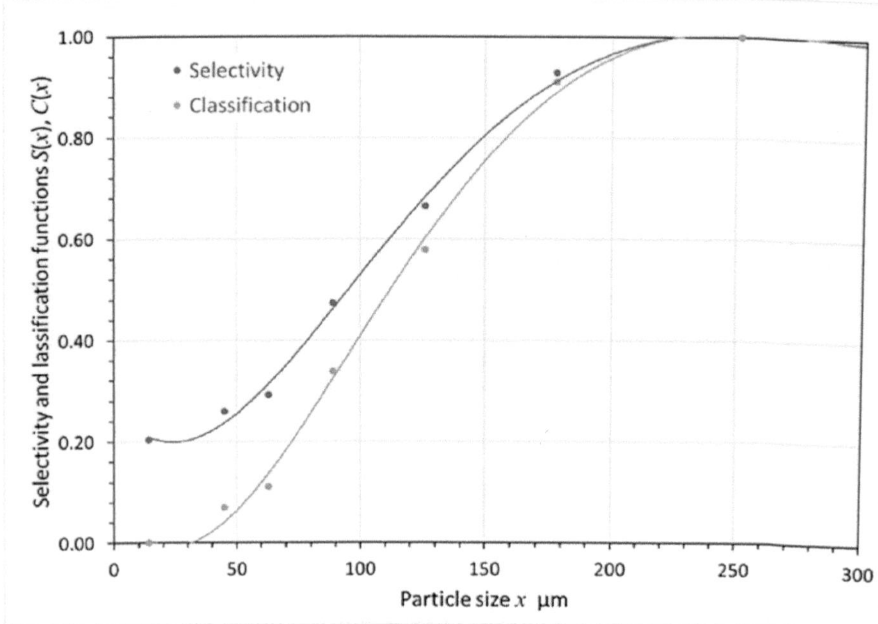

FIGURE 8.5 Plot of the selectivity and classification functions.

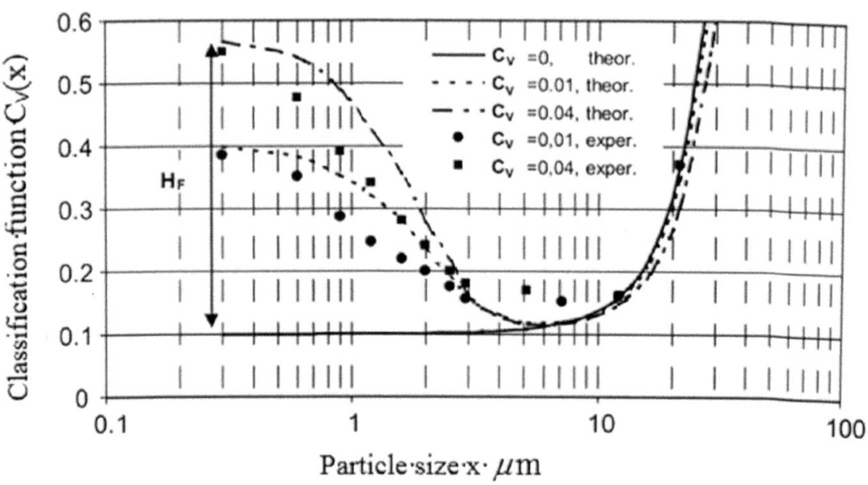

FIGURE 8.6 Classification function with a fishhook for ground quartz of less than 50 µm at different solids concentrations.

Source: Neesse, T., Schneider, M., Dueck, J., Golyk, V., Buntenbach, S., Tiefel, H. 2004. Hydrocyclone operation at the transition point rope/spray discharge. *Minerals Engineering* 17:733–737.

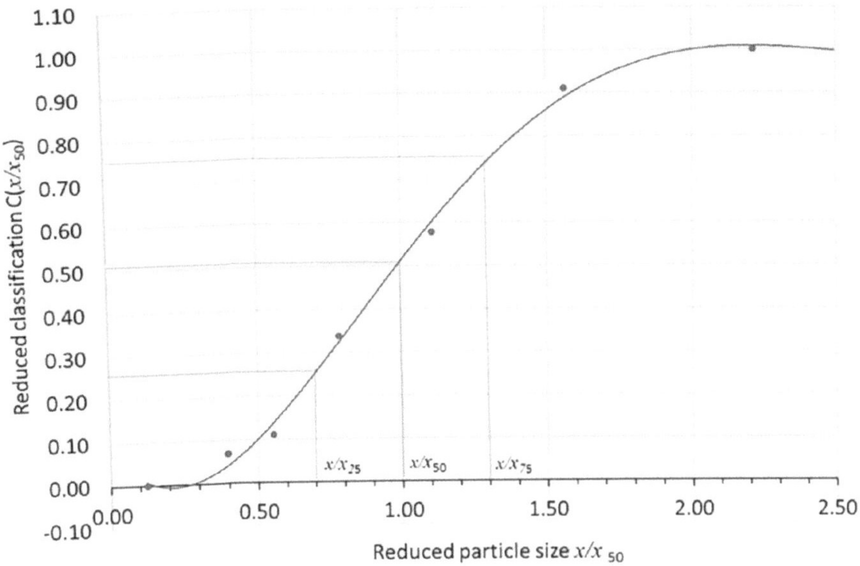

FIGURE 8.7 Reduced classification function.

From Table 8.2, we can calculate the values of the reduced separation sizes.

$$\begin{aligned} x_{50} &= 126.00 - \frac{126-89}{0.58-0.34} \times (0.58-0.50) = 114\,\mu\text{m} \\ x_{25} &= 89.16 - \frac{89.16-63.05}{0.34-0.11} \times (0.34-0.25) = 78.8\,\mu\text{m} \\ x_{75} &= 212.00 - \frac{178-126}{0.91-0.58} \times (0.91-0.75) = 153.0\,\mu\text{m} \end{aligned} \tag{8.9}$$

As discussed in Chapter 7, there are many ways to model this empirical relationship between efficiency, size, d50, and short circuit.

$$E_{corr_{(d_i)}} = 1 - \exp\left[-A_7 * \left(\frac{d_i}{d_{50}}\right)^m\right] \tag{8.10}$$

We use the Plitt equation to fit this curve from industrial data. Where d_i is the size of a particle, d_{50} is the size at 50% efficiency, A7 is Ln (2) = 0.693, m is the Plitt parameter obtained from reconciled operational data. The industrial data taken from industrial test runs contain errors so the first step in to run a data reconciliation method to transform the raw data into information. Once we have the information, we can run a least square fitting program to obtain the parameter "m" in the equation. This parameter "m" is a measure of the sharpness of the separation. In general, for typical hydrocyclone operation it is in order of 2. This number 2 is very important since, as you saw in Chapter 7, we can calculate the global efficiency by integration of the differential efficiency curve using the erf function. As such, we can have an online estimator of hydrocyclone banks in industrial to predict in real time the quality of the separation and most importantly to be able to run mass balances estimation around a grinding circuit or a separation circuit before going to further coarse or fine flotation or in some case coarse tailings and fine tailings separation going to high-capacity thickeners after proper flocculation of the particles.

We can use the tools developed by Sepulveda (1984, 2022) called *Moly Cop tools*. As described in Chapter 7, we have been using this Cyclobal tools since 1984. Figure 8.8 shows a screenshot of the results after reconciling the data and providing the separation parameters, as shown.

For practical purposes, the Rao-Lynch empirical models are a good starting point to estimate the d_{50} and the a (short circuit or Rf) are obtained from empirical relationships as presented in Chapter 7. In practice, we see that the d_{50} is influenced by the pulp volumetric flow, the cyclone feed percent solids, rheology of the suspension, represented by the cyclone feed percent solids, the amount of −325 mesh material, the density of ore, the size of the vortex finder, and most importantly the size of the apex, which is the major effect of the short circuits effect. This is related to two phase flow continuity equation in separation systems as the flux or tails flow rate controlling the shock wave between the pulp phases formation due to the gravity or in this case the centrifugal force creating the air core as we will try to understand phenomenologically in the next sections of this chapter (Bascur 1991; Bascur and

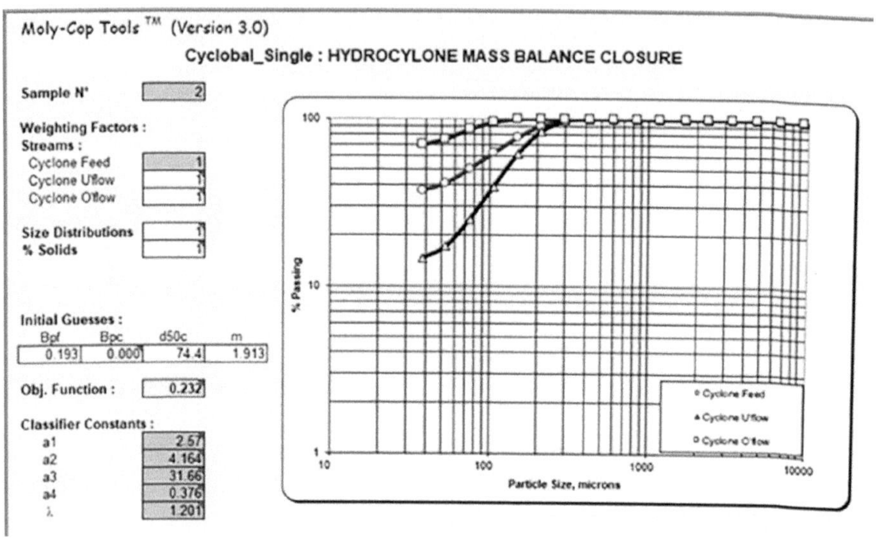

FIGURE 8.8 Cyclobal: hydrocyclone mass balance closure results.

Barrientos 1977). In all these suspensions systems, there are nonlinearities caused by the concentration of solids, particle size distributions, and density of the particles. As a practical advice for rock of density 2.7, the cyclone feed maximum is 65% solids by weight, the cyclone overflow is in the range of 34%–45% solids by weight, cyclone underflow is 70%–80% solids, in a ball mill the maximum percent solids are in the order of 75%–77% solids by weight, and a thickener underflow is in the order of 60%–65% solids by weight. In aerated pulps such as the flotation cells, everything changes to a typical percent solid by weight of 35% solids with an air holdup of about 7%–30% range depending on the scale of the turbulence available, as we will discuss in the particulate flotation chapter. The change is very minor to go from a fluid suspension to a very difficult to move paste. These are non-Newtonian fluids, as we have described in Chapter 6.

The most important parameter is the Plitt "m" parameter representing the sharpness of the separation under the current industrial conditions. In this case the M is equal to 1.913.

Recently, they have been innovations in the area of separation methods as reported by Starrett and Galvin (2023). They use an inclined fluidized bed to separate coarse and fine particle with less short circuit than a traditional hydrocyclone. The REFLUX™ classifier, with water as the fluidizing medium, to classify feed suspensions based on particle size. This work built on an earlier approach, known as *split fluidization*, to achieve sharp separations, focusing on the potential to control the separation size, and to deliver efficient separations at high throughputs. The partition curves were remarkably sharp, with the Imperfection, $I = Ep/D50$, typically less than 0.14 for separation sizes covering a range from 182 μm down to 44 μm, with virtually no ultrafine entrainment with the coarse underflow, and virtually no oversize particles misplaced in the overflow. The solids throughputs ranged from 39 $t/m^2/h$ down to 8 $t/m^2/h$. Importantly, the study achieved a deeper understanding of how to control the separation size, with the fluidization velocity identified as the key control variable. The underflow rate is then increased to a level sufficient to prevent the bed level from rising, while ensuring the solids throughput is below the limit applicable for a given separation size.

They are also additional option such as the double effect Cavex (Lopez and Cepeda 2015) double headed hydrocyclone with wash water inlet used in large mineral complexes to separate the rougher flotation particle size before going to coarse tailings, and enhanced water recovery thickeners for the finer particles. The modeling of these particle separation processes is vital to increase the water recovery requirements in lower grade ore (Bascur, Retamal, Espinosa 2022). The water recovery is used for the right grinding particle size distribution required for the type of flotation machines with a constraint of (20–220 microns) particle sizes.

8.2 HYDROCYCLONE DYNAMICS

8.2.1 Introduction

The hydrocyclone is the most widely used classification devise in grinding-classification circuits in mineral processing and in several other industries especially industrial minerals. It is a relatively small and simple devise with no moving parts and high capacities, which makes it ideal for the mining industry. Hydrocyclones has a cylindrical-conical shape with a pulp feed input tangential to the upper cylindrical section and two axial outlets, an upper and a bottom cylindrical tube called vortex and apex. The shape of the device induces a three-dimensional fluid motion, called vortex motion, with tangential, axial, and radial velocities. The vortex motion produces a low-pressure zone in the center of the device that sucks air from outside, forming an air core. The axial and tangential velocities are highly turbulent and have order of magnitudes of several meters per second, while the radial velocity is one order of magnitude smaller.

The measurement of the velocity distribution in a hydrocyclone is the first step in predicting the flow pattern from a fluid mechanics perspective. The first detailed measurements of these velocities were made by Kelsall (1952, 1963) and are shown in Figure 8.9.

The development of laser Doppler velocimetry (LDV) years after Kelsall's measurements significantly improved these measurements. Very accurate measurements of the tangential and axial velocity distributions were made by Hsieh and Rajamani (1988a, 1988b) shown in Figure 8.10.

Chiné and Concha (2000) used a two-beam, 300-mW laser Doppler velocimeter to determine the tangential and axial velocity fields for the water flow in a 102-mm modular hydrocyclone. Figure 8.11 and 8.12 show results for the tangential velocity in cylinder-conical and flat-bottom hydrocyclones.

Oure approach will be to use the experimental values of the velocity distributions and the mass and linear momentum equations solved for several flow conditions, inviscid approach combining a free vortex with a sink, rotational inviscid models, models based on the exact solution of the Reynolds equation for highly idealized conditions and solutions using different mathematical techniques such as asymptotic expansions, similarity solutions, and boundary layer flows.

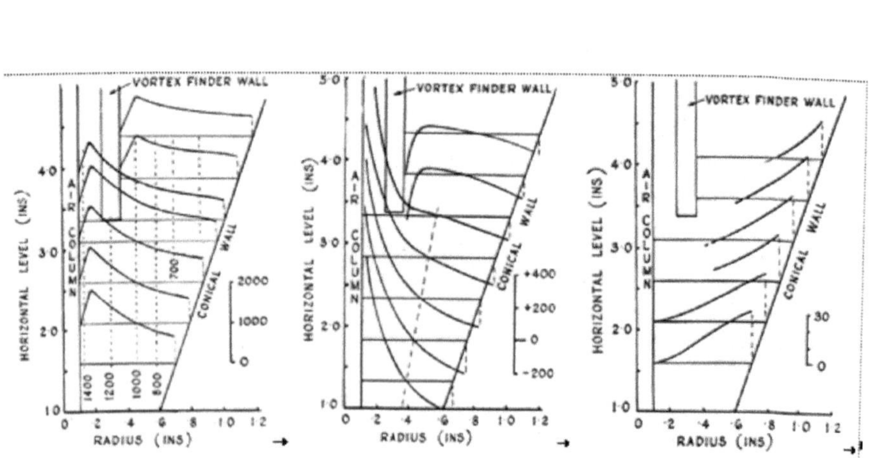

FIGURE 8.9 Velocity components according to Kelsall (1963).

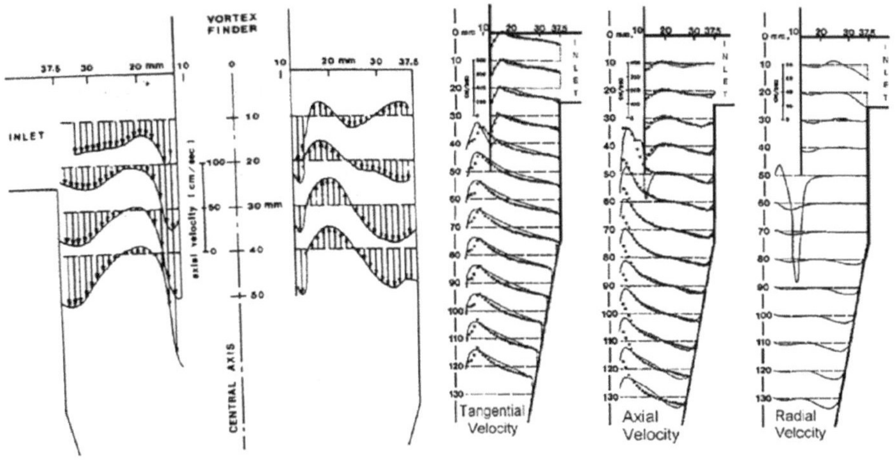

FIGURE 8.10 (a) Axial velocity profile in the inlet region, (b) tangential, axial, and radial velocity components (Hsieh and Rajamani 1988b).

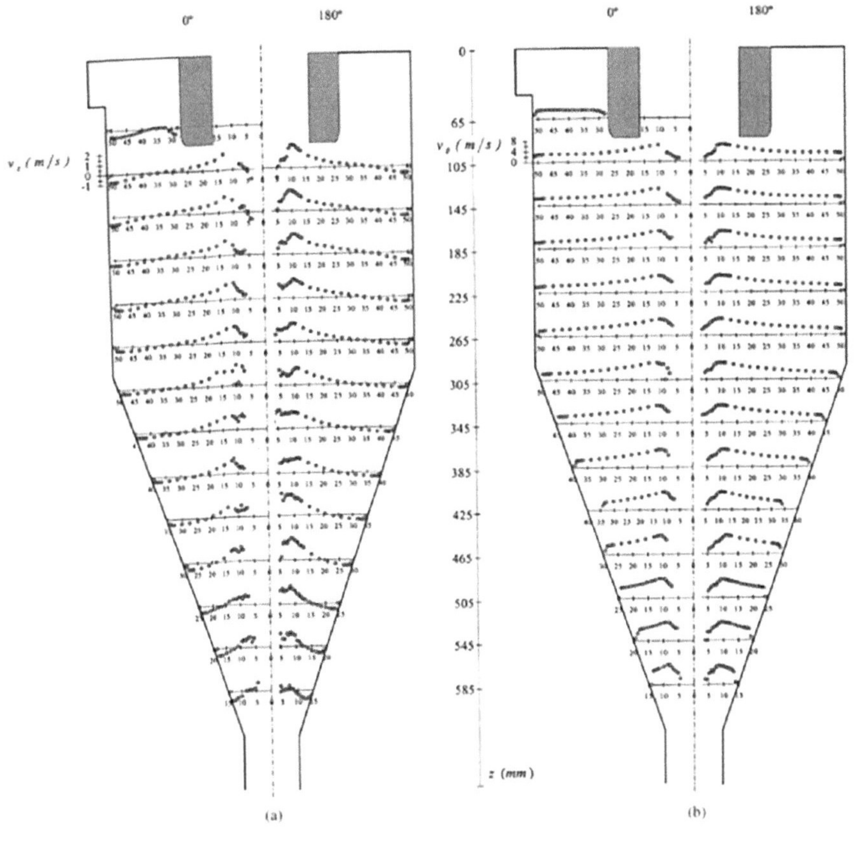

FIGURE 8.11 (a) Axial and (b) tangential velocity profiles for cylinder-conical hydrocyclones.

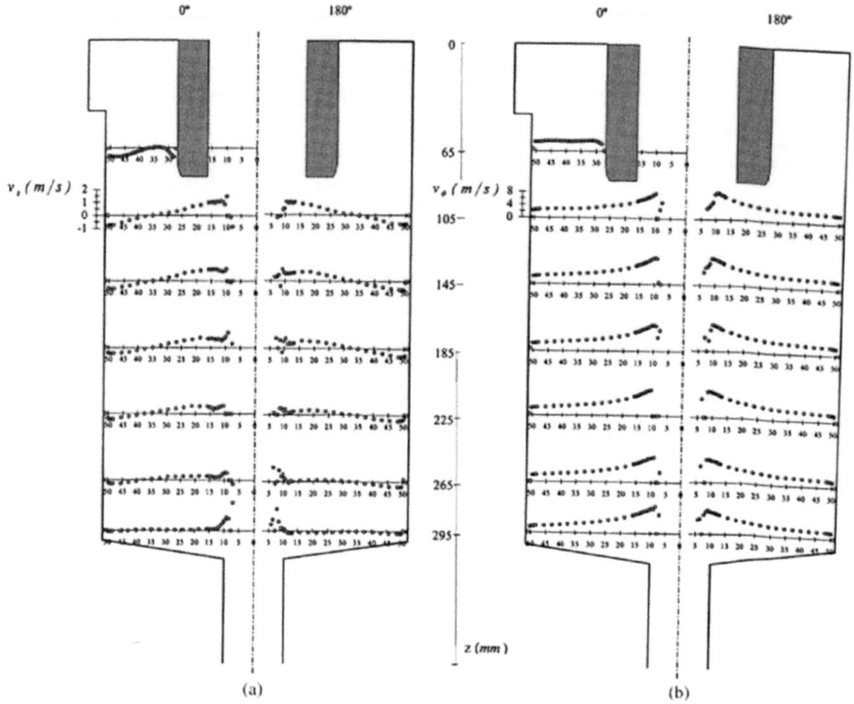

FIGURE 8.12 (a) Axial and (b) tangential velocity profiles for flat-bottom hydrocyclones.

The Navier-Stokes equations have been for many years the starting point to solve fluid dynamic problems involving simple fluids such as liquids and gases at moderate temperature and pressure. For laminar incompressible flows, they may be represented by the equations:

$$\frac{\partial \rho \boldsymbol{v}}{\partial t} + \nabla \cdot \rho \boldsymbol{v}\boldsymbol{v} = \nabla \cdot \boldsymbol{T} + \boldsymbol{b}$$
$$\nabla \cdot \boldsymbol{v} = 0 \tag{8.11}$$

where ρ and $\boldsymbol{v}$ are the density and velocity of the fluid and $\boldsymbol{T}$ and $\boldsymbol{b}$ are the stress tensor and the external force field, respectively. To study specific problems, it is necessary to set a boundary value problem stipulating boundary condition.

For the case of turbulent flow, the Newtonian fluid is modeled with the instantaneous and time-averaged velocity, stress, and external force fields of the Navier-Stokes equations, $\boldsymbol{v} = \overline{\boldsymbol{v}} + \boldsymbol{v}'$, $\boldsymbol{T} = \overline{\boldsymbol{T}} + \boldsymbol{T}'$, and $\boldsymbol{b} = \overline{\boldsymbol{b}} + \boldsymbol{b}'$. Reynolds equations are obtained by integrating the resulting equations for a short time:

$$\frac{\partial \rho \overline{\boldsymbol{v}}}{\partial t} + \nabla \cdot \rho \overline{\boldsymbol{v}} = \nabla \cdot \left(\overline{\boldsymbol{T}} - \rho \overline{\boldsymbol{v}'\boldsymbol{v}} \right) + \overline{\boldsymbol{b}}$$
$$\nabla \cdot \overline{\boldsymbol{v}} = 0 \tag{8.12}$$

where $\overline{v}$, $\overline{T}$, and $\overline{b}$ are the mean velocity, mean stress tensor, and mean body force fields, while v' is the fluctuating velocity field, ρ is the constant fluid density, and $T_t = -\rho\overline{v'v'}$ are the Reynolds stresses. In most cases, researchers have used Reynolds equations to model hydrocyclones.

Unfortunately, the Reynolds equations are nonlinear and therefore it is difficult to solve the boundary-value problems. Several approaches can be used to solve them: *exact solutions, extreme solutions,* and *similarity* and *asymptotic models.*

8.2.2 Exact Solutions of Reynolds Equation

Exact solutions for highly idealized situations occur when the nonlinear terms of the Reynolds equations vanish due to the geometry or symmetry of the flow. These solutions are rarely valid for real configurations, but often give an indication of the flow and pressure distribution to be expected in the real problem.

In general, the turbulent velocity distribution in a hydrocyclone is three dimensional, as we have already discussed, but the flow can be approximated, except for the region close to the entrance, as an axial-symmetric motion. Kelsall work (1952) showed that the radial and tangential velocities depend only on the r coordinate. Then, neglecting the body forces, Reynolds equations in cylindrical coordinates, with the turbulence represented by a constant eddy viscosity ν_ε, much greater than the kinematical shear viscosity, can be written in the form:

$$\frac{\partial}{\partial r}(rv_r)+\frac{\partial}{\partial z}(rv_z)=0$$

$$\frac{1}{r}\frac{\partial}{\partial r}(rv_\theta)=\nu_\varepsilon\frac{\partial}{\partial r}\left(\frac{1}{r}\frac{\partial}{\partial r}(rv_\theta)\right) \tag{8.13}$$

$$v_r\frac{\partial v_z}{\partial r}+v_z\frac{\partial v_z}{\partial z}=\nu_\varepsilon\frac{1}{r}\frac{\partial}{\partial r}\left(\frac{1}{r}\frac{\partial}{\partial r}\left(r\frac{\partial v_z}{\partial r}\right)+\right)\frac{\partial^2 v_z}{\partial z^2}$$

Concha (2007) organized the solution of this model based on the following assumptions.

8.2.2.1 Models with $v_r = v_r(r)$, $v_\theta = v_\theta(r)$, and $v_z = 0$

Consider a cylindrical hydrocyclone with the feed entering through the entire cylindrical mantle at $r = R$ and having a radial velocity v_R and a tangential velocity $v_{\varepsilon R}$ at the entrance (Figure 8.13).

Consider a flow rate into the hydrocyclone given by $Q = 2\pi RLv_R$ with a circulation at $r = R$ of $\Gamma(R) = 2\pi Rv_{\theta R}$, where L is the height of the cylinder. Rietema and Krajenbrink (1958) analyzed two cases, a hydrocyclone with and without an air core, but since in the mining industry all hydrocyclones have an air core, we will consider only this case. Then the boundary conditions is

$$v_\theta(R)=v_{\theta R}, v_r(R)=v_R, v_r(R_a=R_o)=0 \tag{8.14}$$

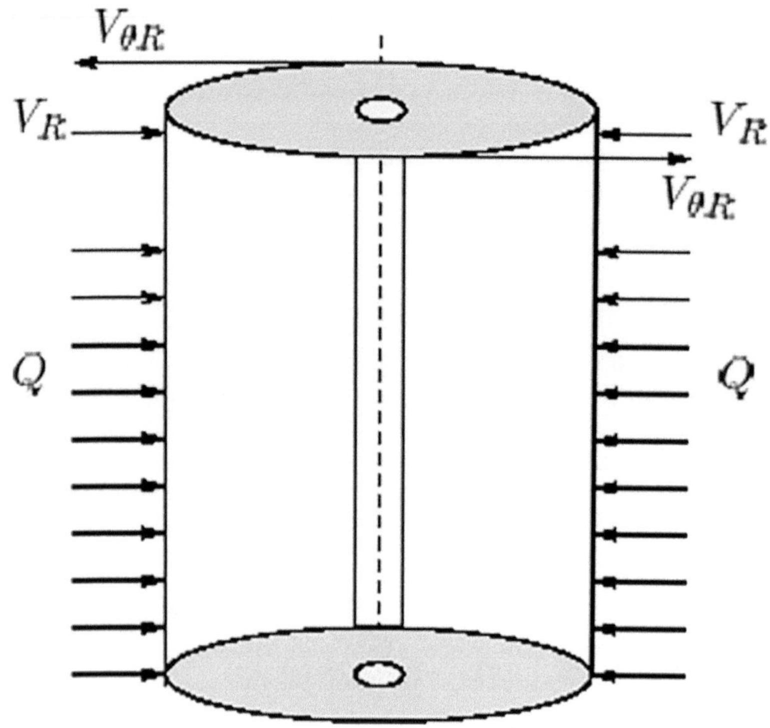

FIGURE 8.13 Physical model of a cylindrical hydrocyclone.

where R_a is the radius of the air core and R_0 is the radius of the vortex-finder. Since the hydrocyclone feed entered through the entire cylindrical mantle, they assumed no axial velocity. This is equivalent to assuming that a *linear sink* evacuates the fluid at the cylinder axis. Thus, from the continuity Equation 8.13 and boundary condition (Equation 8.14) with the radial velocity $r = R$ becomes:

$$v_r = Rv_R / r \tag{8.15}$$

The solutions to Equation 8.13 is given in Equation 8.16, and Figure 8.14 shows the plot of the tangential velocity versus the dimensionless radius.

$$v_\theta^* = \begin{cases} \dfrac{\mathrm{Re}\, R_o^{*(2-\mathrm{Re})} - 2r^{*(2-\mathrm{Re})}}{\mathrm{Re}\, R_0^{*(2-\mathrm{Re})} - 2} & \text{for Re} \neq 2 > 0 \\ \dfrac{1 - 2\ln\left(R_0^* / r^*\right)}{1 - 2\ln\left(R_0^*\right)} & \text{for Re} = 2 \\ 1 & \text{for Re} = \infty \end{cases} \tag{8.16}$$

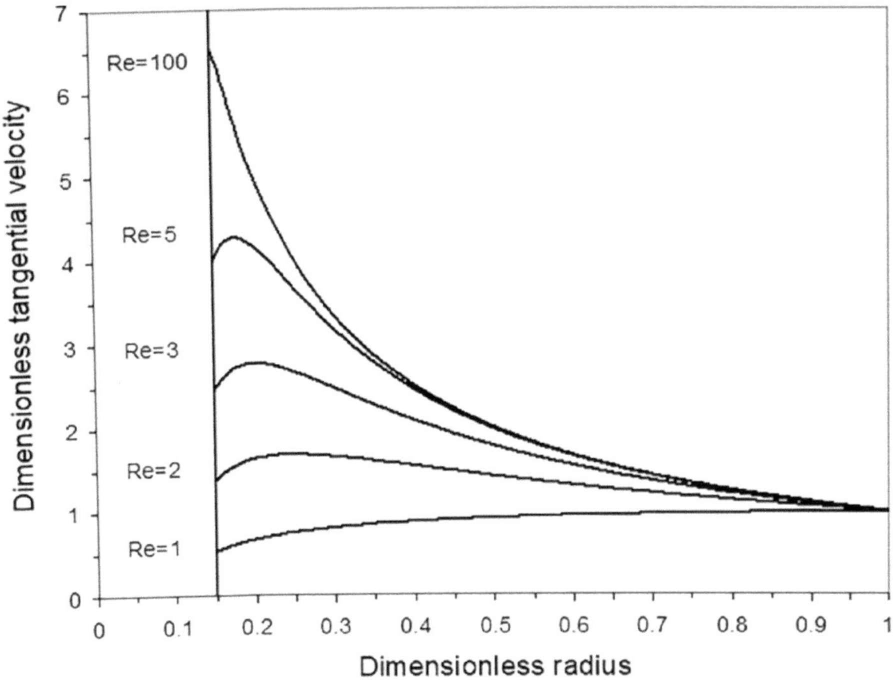

FIGURE 8.14 Tangential velocity for a cylindrical hydrocyclone with an air core the size equal to the overflow, $R_a^* = R_a^* = 0.15$.

Source: Rietema, K., and Krajenbrink, H.J. 1959. Theoretical derivation of tangential velocity profiles in a flat vortex chamber. Influence of turbulence and wall friction. *Applied Scientific Research, Section A* 8:177–197. https://doi.org/10.1007/BF00411748.

Rietema and Krajenbrink compared their result with experimental information concluding that the correct radial Reynolds number in this case was Re≈3. In other words, the turbulent eddy viscosity should be a constant on the order of $v_\varepsilon \approx 3\times10^{-4}\,\mathrm{m/s}$. Rietema (1961a, 1962) obtained the results shown in Figure 8.15 and 8.16.

The problem with these models is that they do not explicitly consider the axial velocity. Furthermore, their solutions serve only to calculate the centrifugal force but give no clues as to how the particles are separated and how the particles and fluid are evacuated from the device. Furthermore, the authors were not concerned about whether their models satisfied the continuity equation. The conclusion is that it is essential to incorporate the axial component of the Reynolds equation into the dynamic process.

8.2.2.2 Models with $v_r = v_r(r)$, $v_\theta = v_\theta(r)$, and $v_z = v_z(r, z)$

Considering that the velocity distribution depends in this case from the following variables, $v_r = v_r(r)$, $v_\theta = v_\theta(r)$, and $v_z = v_z(r,z)$, integrating the continuity Equation 8.13 yields

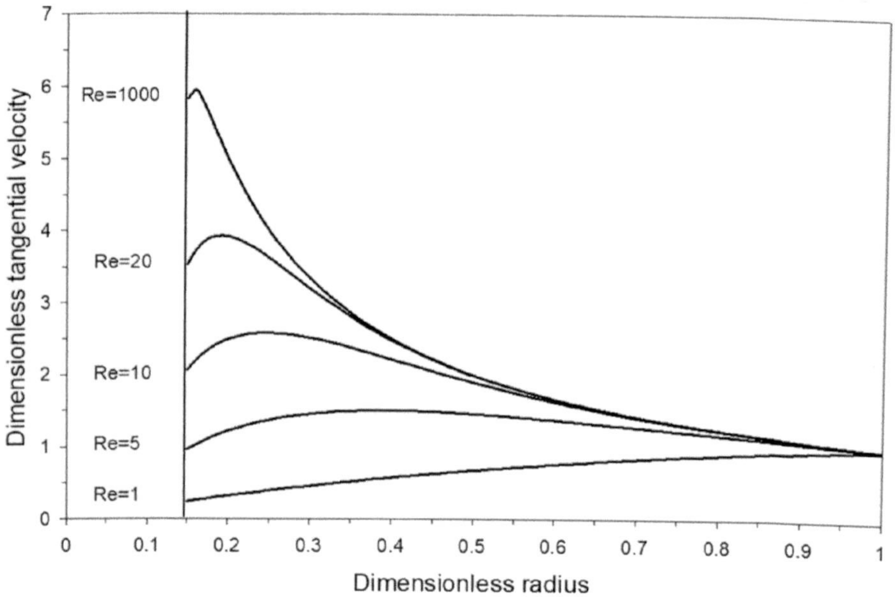

FIGURE 8.15 Tangential velocity (Rietema 1961b).

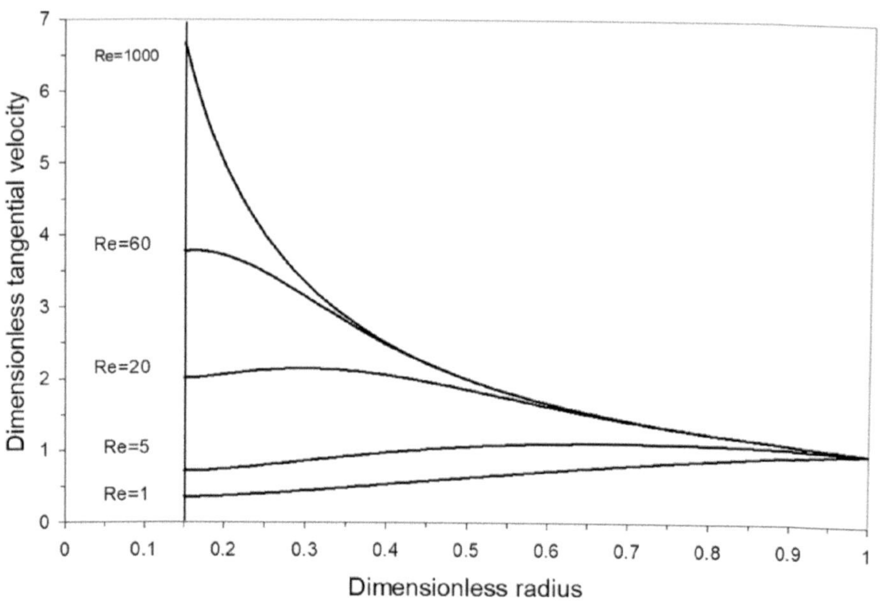

FIGURE 8.16 Tangential velocity.

Source: Rietema, K. 1962. Liquid-Solids Separation in a Cyclone: The Effect of Turbulence on Separation. Symposium on the Interaction between Fluids and Particles. London: Institution of Chemical Engineers. pp. 275–281.

$$rv_r(r) = -\int \frac{\partial v_z}{\partial z} r dr + K \tag{8.17}$$

Since, by assumption, the radial component of the velocity is a function or r only, it is clear from Equation 8.17 that the axial velocity gradient $\partial v_z / \partial z$ must be independent of z, that is, the most general form of the axial velocity should be according to Donaldson and Sullivan (1960):

$$v_z = f(r)z + g(r) \tag{8.18}$$

8.2.2.2.1 *Case 1:* f = 0; v_z = g(r)

First consider the case in which the axial velocity is an arbitrary function of r independent of z; that is:

$$v_z = g(r), \frac{\partial v_z}{\partial z} = 0 \text{ and } v_r = \frac{Rv_R}{r} \tag{8.19}$$

Equation 8.19 implies that radial flow becomes infinite at the cylinder axis; therefore, a singularity must be introduced into the equations if the axis of symmetry is included in the domain. In that case, no solution exists for which the tangential velocity diminishes as the radius diminishes, reaching a finite value at the axis. On the other hand, for a hydrocyclone with an air core of radius R_a, the tangential velocity distribution has the following form:

$$v_\theta^* = \frac{1}{r^*} \frac{(2/\text{Re})(r/R_a)^{-(R_a-2)} - 1}{(2/\text{Re})(R_a/R)^{-(R_a-2)} - 1} \tag{8.20}$$

Since the function $g(r)$ is arbitrary, assume the following form for the dimensionless axial velocity, which compares favorably with experimental data:

$$v_z^* = 1 - 1.8 \sin^6((r - 0.5)/R) \tag{8.21}$$

Figure 8.17 shows the tangential and axial velocity distributions given by Equations 8.20 and 8.21.

8.2.2.2.2 *Case 2:* V_z = f(r)z + g(r)

Consider the case of an axial velocity as a linear function of z and an arbitrary function of r. The problems analyzed by Rietema (1961b) and Upadrashta and Venkateswarlu (1983) correspond to this case. Assuming a constant radial velocity $v_r = v_R$, $g(r)$ can be an arbitrary function of r and was chosen, giving an axial velocity distribution as given in Figure 8.18.

$$v_z = \frac{v_R}{r} z - 1.8 \sin^6 \left(\frac{r - 0.5}{R} \right) \tag{8.22}$$

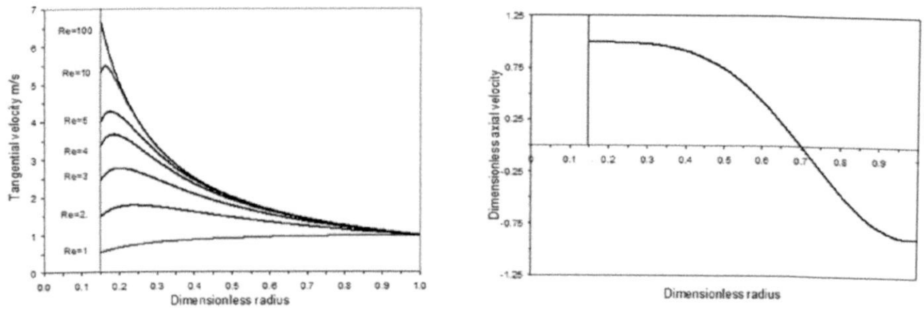

(a) Tangential velocity (b) Axial velocity

FIGURE 8.17 Tangential and axial velocity profiles.

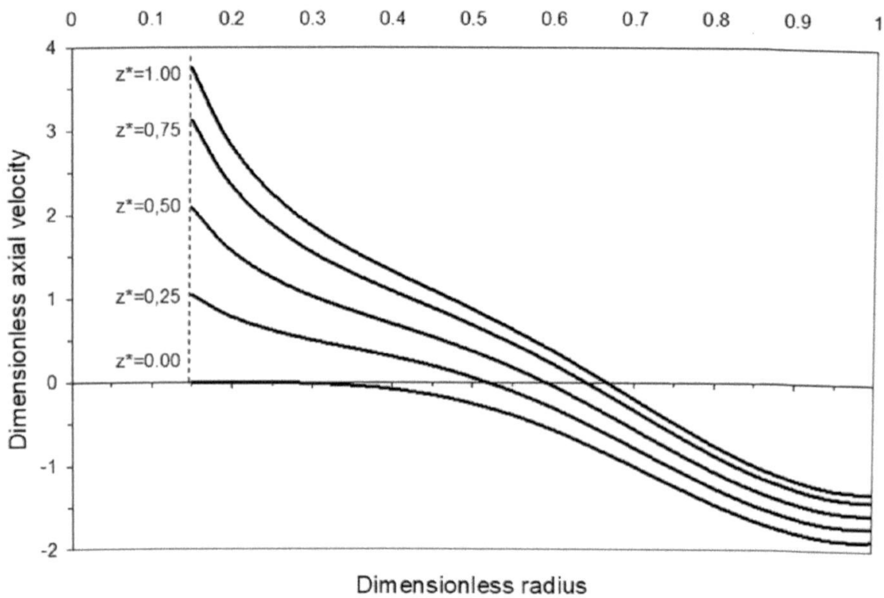

FIGURE 8.18 Axial velocity profile according to equation.

Collantes et al. (2000) applied this case to a flat-bottom hydrocyclone, with the functions f and g as arbitrary functions of r to be determined. Introducing the general form of the axial velocity into the field equations, they wrote the following equations for continuity and the radial, tangential, and axial components of the velocity:

$$\text{Continuity}: \ \frac{1}{r}\frac{d}{dr}\left(rv_r(r)\right)+f(r)=0$$

$$\text{Radial component: } v_r(r)\frac{dv_r(r)}{dr}-\frac{v_\theta^2}{r}-v_z\frac{d}{dr}\left(\frac{1}{r}\frac{d}{dr}\left(rv_r(r)\right)\right)=-\frac{1}{\rho}\frac{\partial p(r,z)}{\partial r}$$

$$\text{Tangential component: } \frac{v_r(r)}{r}\frac{d}{dr}\left(rv_\theta(r)\right) = v_z\frac{d}{dr}\left(\frac{1}{r}\frac{d}{dr}\left(rv_\theta(r)\right)\right)$$

$$\begin{aligned}&\text{Axial component: } \left(v_r(r)\frac{df(r)}{dr} + f^2(r) - \frac{v_z}{r}\frac{d}{dr}\left(r\frac{df(r)}{dr}\right)\right)z \\ &+\left(v_r(r)\frac{dg(r)}{dr} + f(r)g(r) - \frac{v_z}{r}\frac{d}{dr}\left(r\frac{dg(r)}{dr}\right)\right) = -\frac{1}{r}\frac{\partial p(r,z)}{\partial r}\end{aligned} \tag{8.23}$$

Eliminating the pressure from the radial component and axial component in Equation 8.23, the following set of equations is obtained for $f(r)$ and $g(r)$:

$$\begin{aligned}&\frac{d}{dr}\left(v_r(r)\frac{df(r)}{dr} + f^2(r) - v_z\frac{1}{r}\frac{d}{dr}\left(r\frac{df}{dr}\right)\right) = 0 \\ &\frac{d}{dr}\left(v_r(r)\frac{dg(r)}{dr} + f(r)g(r) - v_z\frac{1}{r}\frac{d}{dr}\left(r\frac{dg}{dr}\right)\right) = 0\end{aligned} \tag{8.24}$$

To satisfy the continuity equation, a stream function ψ is defined, such that

$$v_r(r) = \frac{1}{r}\frac{\partial \psi}{\partial z} \equiv \frac{q(r)}{r} \quad \text{and} \quad v_z = -\frac{1}{r}\frac{\partial \psi}{\partial r} \equiv -\frac{1}{r}\frac{dq}{dr}z + g(r) \tag{8.25}$$

Defining the dimensionless variable $\eta = (r/R)^2$, introducing these expressions into Equation 8.24, and neglecting the viscous terms in Equation 8.24, we obtain for $q(\eta)$ and $g(\eta)$

$$\begin{aligned}&\frac{dq(\eta)}{d\eta}\frac{d^2q(\eta)}{d\eta^2} - q(\eta)\frac{d^3q(\eta)}{d\eta^3} = 0 \\ &\frac{d}{d\eta}\left(q(\eta)\frac{dg(\eta)}{d\eta}\right) - \frac{d}{d\eta}\left(\frac{dq(\eta)}{d\eta}g(\eta)\right) = 0\end{aligned} \tag{8.26}$$

A solution to this ordinary differential equation system, compatible with the boundary conditions for a flat bottom hydrocyclone, where the feed enters the equipment through a ring of height L_A in the upper part of the cylinder is as follows (Figure 8.19):

8.2.2.2.3 Zone I

$$\begin{aligned}&v_r(r) = -Q/2\pi L_A r \equiv v_R R/r\left(\frac{Q_0}{Q}\left(1-\exp\left(-5.6\left(\frac{r}{R_0}\right)^2\right)\right) - \sin\left(\beta\left(\frac{r}{R}\right)^2\right)\right) \\ &v_z(r,z) = \frac{Q}{\pi R^2}\left(\beta\cos\left(\beta\left(\frac{r}{R}\right)^2\right) + \left(5.6\frac{Q_0}{Q}\left(\frac{R}{R_0}\right)^2\exp\left(-5.6\left(\frac{r}{R}\right)^2\right) - \beta\left(\frac{r}{R}\right)^2\right)\frac{z}{L_A}\right)\end{aligned} \tag{8.27}$$

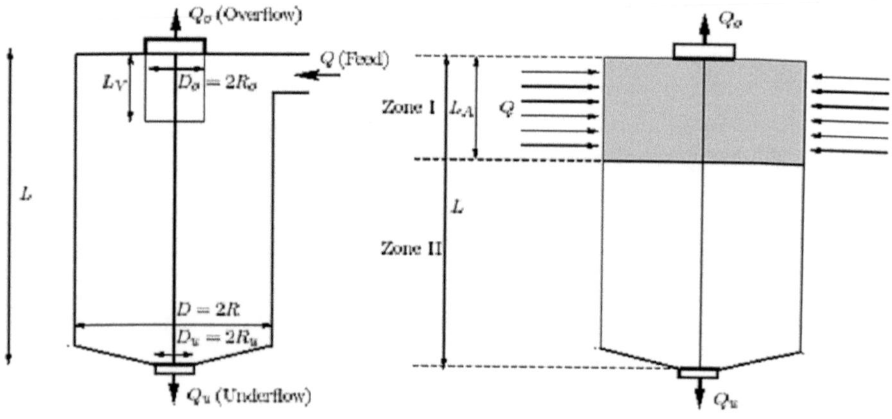

FIGURE 8.19 Schematic representation of the flat-bottom hydrocyclone.

$$v_\theta(r) = \frac{\Gamma_R}{2\pi r}\left(\frac{1-\exp\left(-\frac{1}{2}\text{Re}\,\beta\left(\frac{r}{R}\right)^2\right)}{1-\exp\left(-\frac{1}{2}\text{Re}\,\beta\right)}\right)$$

Where $\beta = \pi + \arcsin(Q_u / Q)$, and where Q, Q_u, and Q_0 are the volume feed and the overflow rate respectively, and R is the hydrocyclone radius. All the volume flowrates are taken positive. Note that in Equation 8.27, the factors are $-Q / 2\pi L_A r \equiv v_R R / r$, $Q / \pi R^2 \equiv 2 v_R L_A / R$, and $\Gamma_R / 2\pi r = v_{\theta R} R / r$.

8.2.2.2.4 Zone II

$$v_r(r) = -\frac{Q}{2\pi(L-L_A)r}\left(\sin\left(\beta\left(\frac{r}{R}\right)^2\right) + \frac{Q_u}{Q}\left(1-\exp\left(-5.6\left(\frac{r}{R}\right)^2\right)\right)\right)$$

$$v_z(r,z) = -\frac{Q}{\pi R^2}\left(\begin{array}{c} 5.6\frac{Q_u}{Q}\left(\frac{R}{R_u}\right)^2 \exp\left(-5.6\left(\frac{r}{R}\right)^2\right) - \left(\beta\cos\left(\frac{r}{R}\right)^2\right) \\ +5.6\frac{Q_u}{Q}\left(\frac{R}{R_u}\right)^2 \exp\left(-5.6\left(\frac{r}{R}\right)^2\right)\frac{z}{L-L_A} \end{array}\right) \quad (8.28)$$

$$v_\theta(r) = \frac{\Gamma_R}{2\pi r}\left(\frac{1-\exp\left(\frac{\text{Re}}{2}\beta\frac{L_A}{L-L_A}\left(\frac{r}{R}\right)^2\right)}{1-\exp\left(\frac{\text{Re}}{2}\beta\frac{L_A}{L-L_A}\right)}\right)$$

In Equation 8.28, $Q/2\pi\left(L-L_A\right)r = v_R R L_A / r\left(L-L_A\right)$ and (8.1.27) $Q/\pi R^2 \equiv 2v_R L_A / R$ The results are presented in Figures 8.20, 8.21, and 8.22.

In both Zones I and II the pressure distribution can be calculated from

$$p(r,z) = p_0 - \frac{\rho}{2}\left(v_r(r)f'(r) + f^2(r)\right)z^2 \\ -\rho\left(v_r(r)g'(r) + f(r)g(r)\right)z + \rho\int\frac{v_\theta^2}{r}\mathrm{d}r - \frac{1}{2}v_r^2(r) \tag{8.29}$$

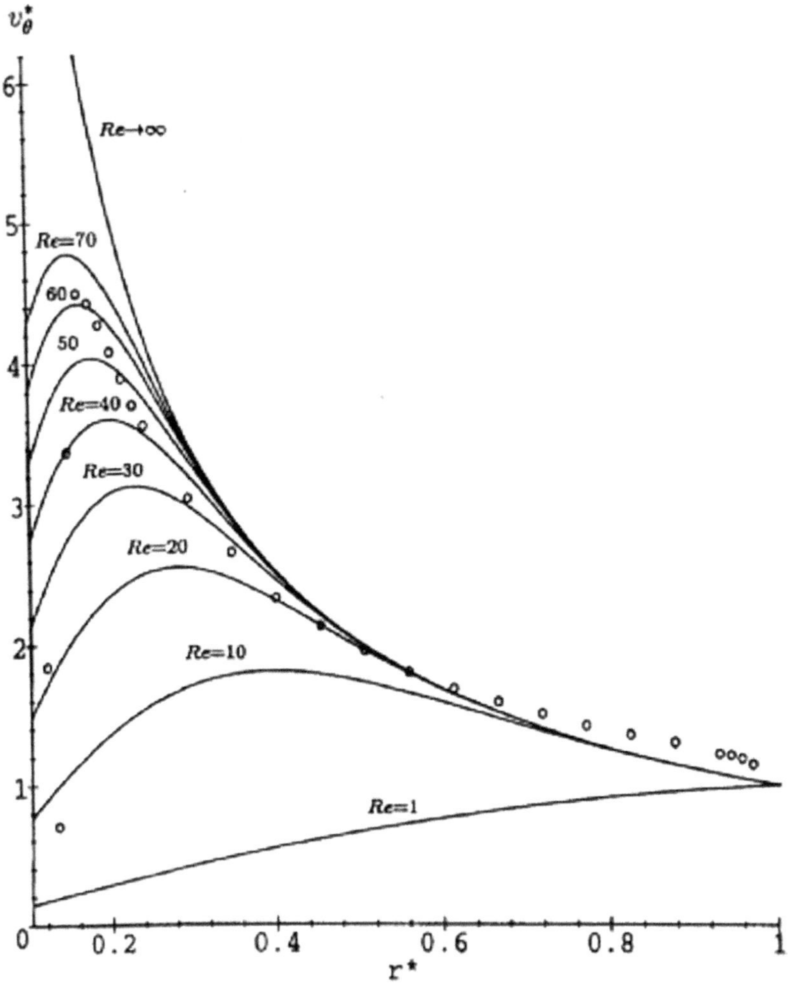

FIGURE 8.20 Simulated tangential velocity distribution according to Collantes et al. (2000). Comparison with experimental data from Chiné and Concha (2000).

Source: Collantes, J., Concha, F., and Chiné, B. 2000. Axial symmetric model for a flat bottom hydrocyclone. *Chemical Engineering Journal* 80:257–265.

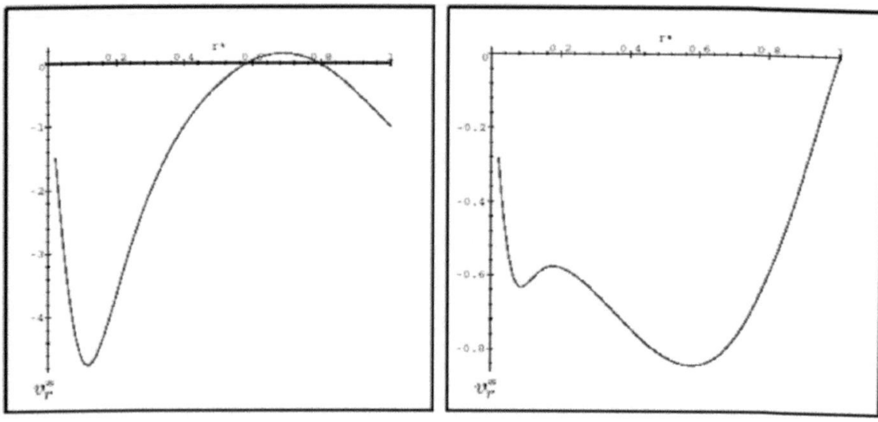

FIGURE 8.21 Radial velocity distribution for Zones I and II, according to Collantes et al. (2000): (a) Zone I, (b) Zone II.

Source: Collantes, J., Concha, F., and Chiné, B. 2000. Axial symmetric model for a flat bottom hydrocyclone. ***Chemical Engineering Journal*** **80:257–265.**

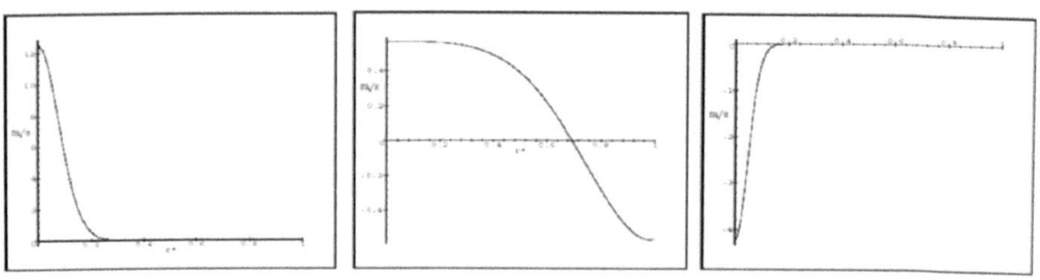

FIGURE 8.22 Simulated axial velocity distribution, according to Collantes et al. (2000): (a) At the top wall. (b) At the interface I/II. (c) At the bottom wall.

Source: Collantes, J., Concha, F., and Chiné, B. 2000. Axial symmetric model for a flat bottom hydrocyclone. ***Chemical Engineering Journal*** **80:257–265.**

From the exact solutions of the Reynolds equations, as given in this section, and from the conclusions of the various researchers who have modeled the hydrocyclone in this manner, we can conclude that the radial velocity depends highly on the shape of the axial velocity. One can obtain completely different radial velocity distributions by choosing appropriate axial flows. Nevertheless, the shape of the tangential velocity, as a combination of a free vortex and a rigid motion, is insensitive to the form of these axial and radial velocities. Only the maximum value of the tangential velocity and its position vary as the radial velocity changes.

For the axial velocity to be a consequence of the radial velocity, it must be a function of r and at least a linear function of z. Finally, we can indicate that, in modeling the hydrocyclone, selecting among the several solutions of the field equations by comparing the value of the maximum tangential velocity with experimental values is not the proper way to go. The three, or at least two, velocity components must be compared, and the axial velocity in particular must be reasonable.

8.2.2.3 Models with $v_\theta = v_\theta(r)$, $v_r = v_r(r, z)$, and $v_z = v_z(r, z)$

Experimental data of Hsieh and Rajamani (1988b) confirm that the radial and axial velocities are functions of the radial and axial coordinates.

Bloor and Ingham (1974, 1983) analyzed the inviscid isochoric flow within a hydrocyclone, as we already saw. Assuming a free vortex for the tangential velocity, they obtained the radial and axial component as functions of the radial and axial coordinates. Since these velocities compared favorably with Kelsall's experimental information, Bloor and Ingham used the resulting axial and radial velocities to recalculate the tangential velocity from Reynolds's equations with variable eddy viscosity. Consider the tangential component of Reynolds's equation:

$$\frac{v_r}{r}\frac{\partial}{\partial r}(rv_\theta)+v_z\frac{\partial v_z}{\partial z}=\frac{1}{r^2}\frac{\partial}{\partial r}\left(v_z r^3\frac{\partial}{\partial r}\left(\frac{v_\theta}{r}\right)+\frac{\partial}{\partial z}\left(v_z\frac{\partial v_z}{\partial z}\right)\right) \tag{8.30}$$

Using dimensionless coordinates: $r^* = r/R$, $v_\theta^* = v_\theta / v_{\theta R}$, radial and axial velocities given by $v_r = -A\dfrac{r^{3/2}}{z^2}$ and $v_z = (1/2)A\left(1/r^{1/2}\right)(3\alpha - 5(r/z))$, and assuming that the eddy viscosity has the following form: $v_\varepsilon = v_t / z^2$ (necessary for the tangential velocity to be independent of z), with boundary conditions $v_\theta^*(1) = 1$ and $v_\theta^*(0) = 0$:

$$v_\theta^* = \frac{1}{r^*}\gamma\left(\frac{4}{5},\frac{2}{5},r^{*/2}\lambda_{\text{Re}}\right)/\gamma\left(\frac{4}{5},\frac{2}{5},\lambda_{\text{Re}}\right) \tag{8.31}$$

Where λ_{Re} is a type of Reynolds number defined by $\lambda_{\text{Re}} = AR^{5/2} / v_t$, where γ is the incomplete gamma function, and v_t is fluid kinematical viscosity. The assumption that $v_\varepsilon = v_t / z^2$ implies that the length scale l of the turbulence in the Prandtl's mixing length theory, $v_\varepsilon = l^2\left|K + \dfrac{\partial v_\theta}{\partial r} - \dfrac{v_\theta}{r}\right|$, varies as $l = h(r)/z$ (Bloor and Ingham 1983). Figure 8.23 gives the new tangential velocity distribution.

Bloor and Ingham (1987) show that their solution compared well with experimental values of Kelsall (1952) and of Knowles et al. (1973). They indicate that the way the entry conditions are modeled has a profound effect on the nature of the flow. Hwang et al. (1993) extended the work of Bloor and Ingham (1987) to include underflow through the apex. To do this, they use a stream function for the sink in the form:

$$\psi_1 = \frac{Q\cos\phi\cos\alpha}{2\pi\left(1+Q_o/Q_u\right)(1-\cos\alpha)} \tag{8.32}$$

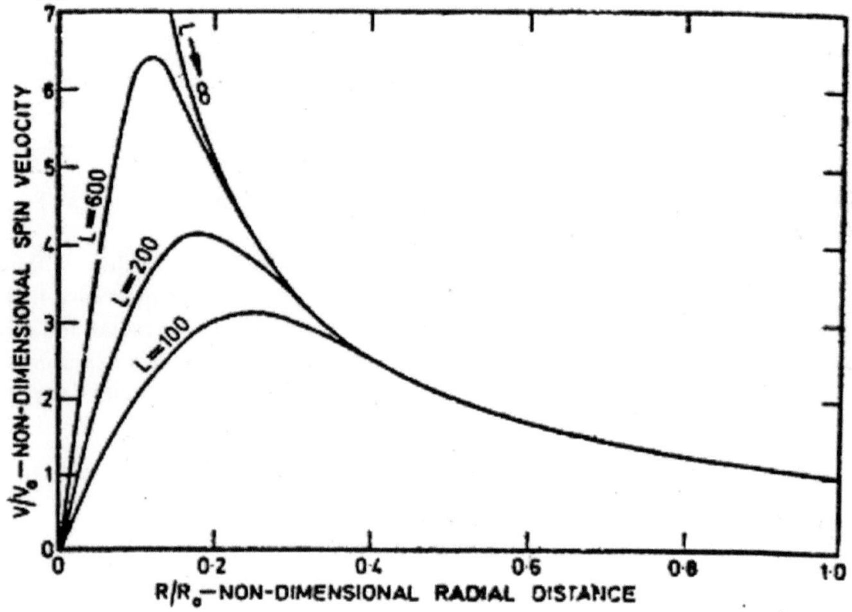

FIGURE 8.23 Tangential velocity distribution according to Bloor and Ingham (1976).

Source: Bloor, M.I.G., and Ingham, D.B. 1976. Boundary layer flow on the side walls of conical hydrocyclones. ***Transactions of the Institution of Chemical Engineers*** **54:276.**

With this modification the velocity distributions become

$$v_\theta = \frac{R}{r} v_{\theta R} \left(1 - \frac{2\psi}{R^2 v_z}\right)^{1/2} \tag{8.33}$$

$$v_r = -\frac{Q\sin\phi}{2\pi R^2}\left(\frac{1+\wedge}{\eta^2}\frac{1}{1-\cos\alpha} + 2\sigma\left(\begin{array}{l}\dfrac{D(\alpha)}{\sin^2\alpha}\cos\phi - 1 - \cos\phi\ln(\tan(\phi/2)) \\ +\dfrac{Q\cos\phi}{2\pi R^2}\left(\dfrac{D(\alpha)}{\sin^2\alpha}\sin\phi - \dfrac{D(\alpha)}{\sin\phi}\right)\end{array}\right)\right) \tag{8.34}$$

$$\begin{aligned} v_z = &-\frac{Q\cos\phi}{2\pi R^2}\left(\frac{1+\wedge}{\eta^2}\frac{1}{1-\cos\alpha} + 2\sigma\left(\frac{D(\alpha)}{\sin^2\alpha}\cos\phi - 1 - \cos\phi\ln(\tan(\phi/2))\right)\right) \\ &-\frac{Q\sin\phi}{2\pi R^2}\left(2\sigma\left(\frac{D(\alpha)}{\sin^2\alpha}\sin\phi - \frac{D(\alpha)}{\sin\phi}\right)\right) \end{aligned} \tag{8.35}$$

where Q is the feed volume flowrate, R is the radius of the cylindrical part of the hydrocyclone, where $\sigma = -\pi R^2 v_{\theta R}^2 / Q v_z \sigma$, $\wedge = Q_o / Q_u$ is the flow partition between

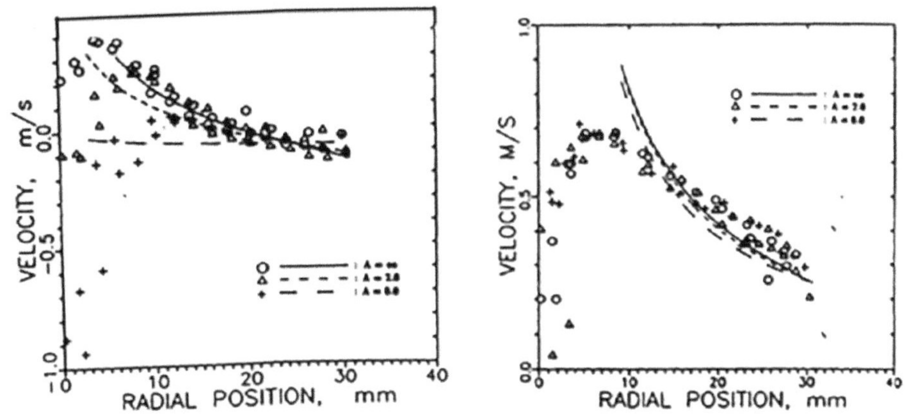

(a) Tangential velocity distribution (b) Axial velocity distribution

FIGURE 8.24 Simulated velocity distribution compared with measurements.

Source: Hwang, C.C., Shen, H.Q., Zhu, G., and Khonsari, M.M. 1993. On the main flow pattern in hydrocyclones. *Journal of Fluids Engineering* 155:21–25.

overflow and underflow streams as shown in Figure 8.24, and where the operator is $D(\circ) = 1 - \cos(\circ) + \sin^2(\circ)\ln(\tan(\circ)/2)$.

8.2.3 Extreme Models

8.2.3.1 Inviscid Isochoric Models

Since for a hydrocyclone the tangential velocity is very close to a free vortex, several research workers have used inviscid isochoric flow, also known as *potential flow*, for the hydrocyclone. Within ideal flows, we can distinguish rotational and irrotational flows. The field equations for potential flows are as follows (Concha 2007):

$$\nabla \cdot \boldsymbol{v} = 0 \quad \nabla B = \boldsymbol{v x w} \quad B = \frac{p}{\rho} + \frac{1}{2} v^2 \tag{8.36}$$

where $\boldsymbol{v}$, $\boldsymbol{w}$, and p are the velocity, vorticity and pressure fields and the Bernoulli body forces have been neglected.

8.2.3.2 Potential Flow Model (Irrotational Inviscid Isochoric Flow)

For potential flow, the field equations become

$$\nabla \cdot \boldsymbol{v} = 0 \qquad \nabla \times \boldsymbol{v} = 0 \qquad \nabla B = 0 \tag{8.37}$$

Combining these equations, we get

$$\boldsymbol{v} = -\nabla \phi, \quad \nabla^2 \phi = 0, \quad \frac{1}{2} v^2 + \frac{p}{\rho} = \text{constant} \tag{8.38}$$

Where ϕ is the velocity potential. For cylindrical plane flows, the Laplace equation is valid:

$$\frac{\partial^2 \phi}{\partial r^2}+\frac{1}{r}\frac{\partial \phi}{\partial r}+\frac{1}{r^2}\frac{\partial^2 \phi}{\partial \theta^2}=0, \quad p(r)=p(R)-\frac{1}{2}\rho\left(v_R^2+v_{\theta R}^2\right)\left(1-\frac{v_r^2+v_\theta^2}{v_R^2+v_{\theta R}^2}\right) \tag{8.39}$$

$$v_r=-\frac{\partial \phi}{\partial r}, \quad v_\theta=-\frac{1}{r}\frac{\partial \phi}{\partial \theta} \tag{8.40}$$

The measurements of the flow patterns in hydrocyclones (Kelsall 1952; Hsieh and Rajamani 1988a; Quian et al. 1989) have shown that the tangential velocity has the form of a free vortex from the cyclone wall in the direction of the axis. Near the axis, the tangential velocity resembles that of a solid body rotation. The best assumption that can be made, from Hsieh and Rajamani's measurement, is that the radial flow is similar to a sink close to the cyclone wall but is proportional to the radius closer to the cyclone axis. Taking these observations into consideration, the simplest model for the hydrocyclone is a combination of a free vortex and a rigid motion with a sink. This combination is also called a *forced vortex with a sink.*

Consider the flow in a hydrocyclone, working without an air core, divided into two regions, the vortex core for $0 \le r \le R_c$, where the flow is a rigid body motion with constant angular velocity $\omega = v_\theta\left(R_c\right)/R_c$ and the exterior region $R_c \le r \le R$, where the flow is a free vortex with a constant circulation $\Gamma = 2\pi R v_{\theta R}$. The total flow rate into the cyclone per unit length is $\widehat{Q} = 2\pi R v_R$.

8.2.3.2.1 *Region* Rc ≤ r ≤ R

The solution of Equations 8.39 and 8.40 in this region yields a free vortex and a sink for the tangential and radial velocity respectively:

$$v_\theta=\frac{\Gamma}{2\pi r}, \quad v_r=-\frac{\widehat{Q}}{2\pi r} \tag{8.41}$$

From Bernoulli's equation:

$$p(r)-p(R)=\frac{1}{2}\rho\left(v^2(R)-v^2(r)\right)$$

Since $v^2 = v_\theta^2 + v_r^2$, substituting the values of (10) at r and R, yields

$$p(r)=p(R)-\frac{\rho}{2}\left(\frac{\Gamma}{2\pi R}\right)^2\left(1+\left(\frac{\widehat{Q}}{\Gamma}\right)^2\right)\left(1-\left(\frac{R}{r}\right)^2\right) \tag{8.42}$$

8.2.3.2.2 *Region* 0 ≤ r ≤ R$_c$

In this region a forced vortex and a sink represent the tangential and the radial velocity respectively:

$$v_\theta(r) = \omega r \quad v_r = -\frac{\hat{Q}}{2\pi r} = -\left(\frac{\hat{Q}}{\Gamma}\right)\left(\frac{R}{r}\right) \tag{8.43}$$

Where ω is the angular velocity of the rigid motion. At $r = R_c$, the velocity fields of both regions intersect, that is, $\omega R_c = \Gamma / 2\pi R_c$, from which R_c can be calculated as follows:

$$R_c = \left(\frac{\Gamma}{2\pi\omega}\right)^{1/2} \tag{8.44}$$

Figure 8.25 shows the tangential and radial velocity profiles for the *forced vortex with a sink* model.

For a rigid body motion, the pressure can be calculated from $\nabla B = 2\rho\omega^2 r\boldsymbol{e}_r$. Integrating from R_c to r yields:

$$p(\xi) + \frac{1}{2}\rho v^2(\xi)\Big|_{\xi=R_c}^{\xi=r} = \rho\omega^2\eta^2\Big|_{\eta=R_c}^{\eta=r}$$

$$p(r) = p(R_c) + \frac{1}{2}\rho v^2(R_c) - \frac{1}{2}\rho v^2(r) + \rho\omega^2 r^2 - \rho\omega^2 R_c^2$$

Replacing v from 8.43 gives the following:

$$p(r) = p(R_c) + \frac{1}{2}\rho\omega^2\left(r^2 - R_c^2\right) + \frac{1}{2}\rho\left(\frac{\hat{Q}}{\Gamma}\right)^2\left(\left(\frac{R}{R_c}\right)^2 - \left(\frac{R}{r}\right)^2\right) \text{for } 0 \le r \le R_c \tag{8.45}$$

Figure 8.26 shows the pressure distribution.

Several authors, for example, Fontein and Dijksman (1952), suggested models of this type.

8.2.3.3 Rotational Inviscid Isochoric Model

For steady flow $\dot{B} = \nabla B \cdot v$, therefore, from Equation 8.37, $\dot{B} = \boldsymbol{v} \times \boldsymbol{w} \cdot \boldsymbol{v} \equiv 0$, which indicates that Bernoulli is constant along streamlines. Then, in *spherical coordinates* (ξ, ϕ, θ) we can write the following:

$$v_\xi \frac{\partial}{\partial\xi}(\xi \sin\phi v_\theta) + \frac{v_\phi}{\xi}\frac{\partial}{\partial\phi}(\xi \sin\phi v_\theta) = 0 \tag{8.46}$$

Therefore, $\xi \sin\phi v_\theta$ is constant along a streamline:

$$\xi \sin\phi v_\theta = f(\psi) \tag{8.47}$$

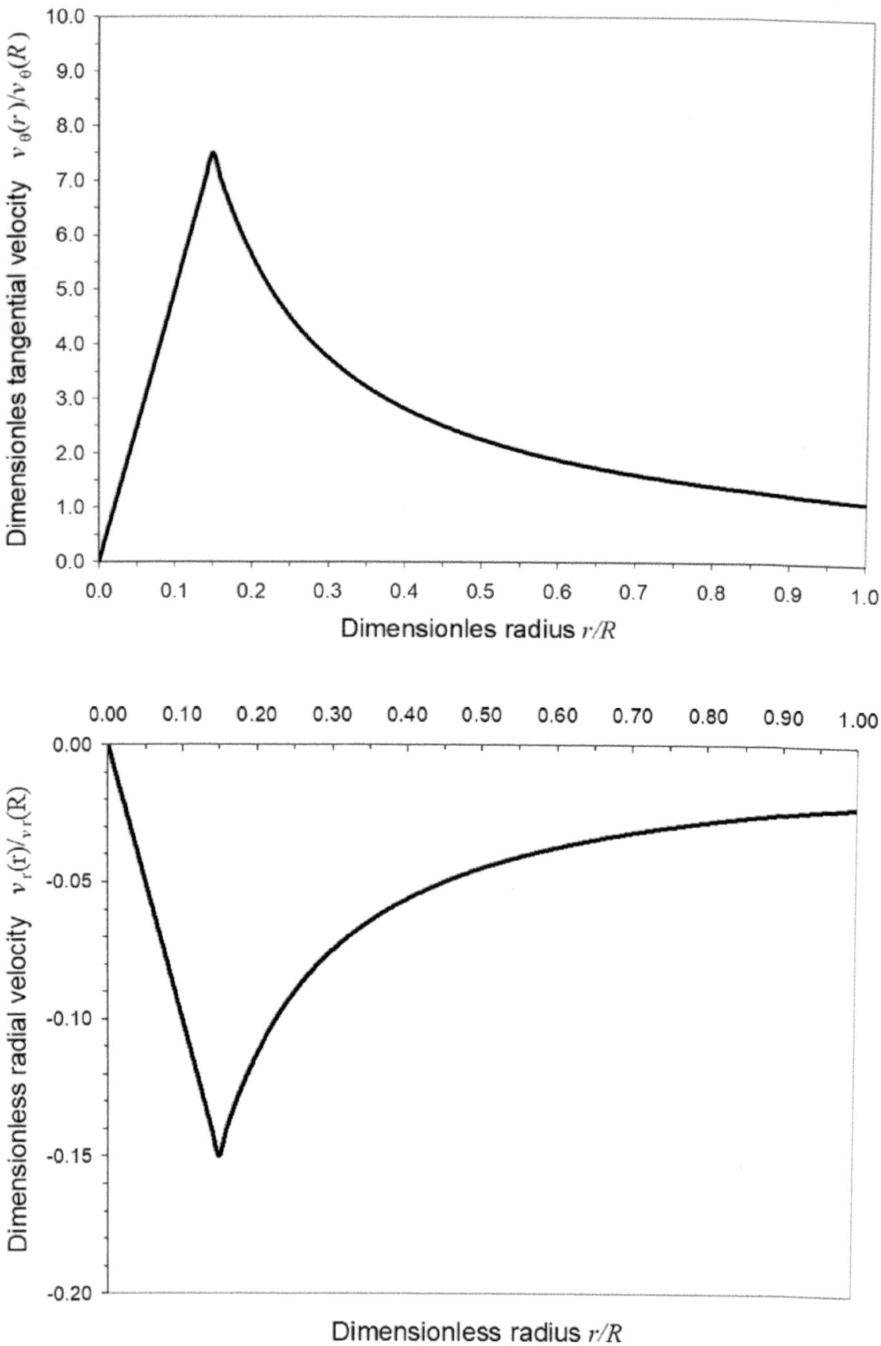

FIGURE 8.25 Tangential and radial velocity profiles for a potential flow model of a hydrocyclone for $\Gamma = 7.065$, $Q = 0.1413$, and $\omega = 50$.

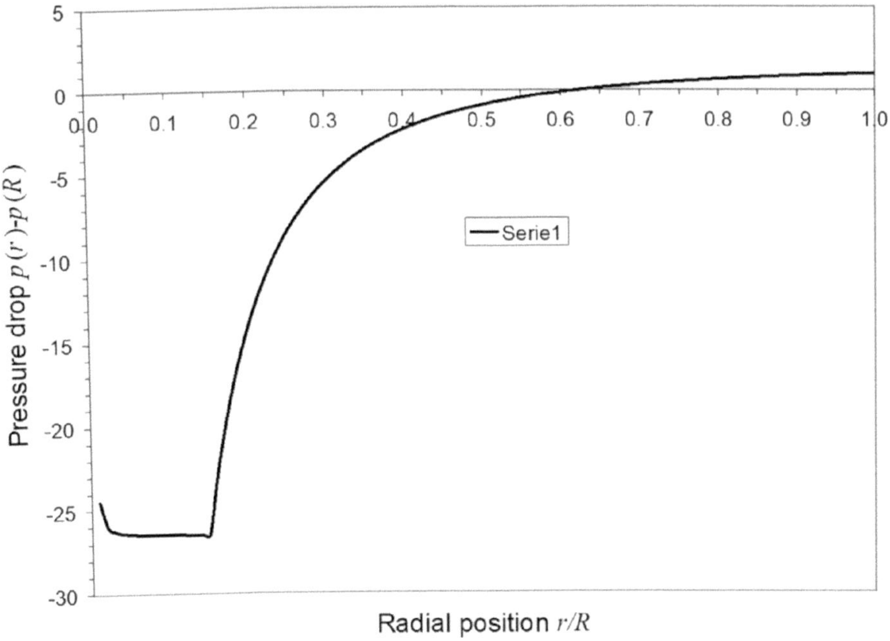

FIGURE 8.26 Pressure distribution for a potential flow model of a hydrocyclone for $\Gamma = 7.065$, $Q = 0.1413$, and $\omega = 50$.

Where $f(\psi)$ is the vorticity function. At the wall, the streamline must be a constant, $f(\psi)|_{\phi=\alpha} = k$. Bloor and Ingham (1974, 1984a) assumed that the tangential velocity was a free vortex everywhere, that $v_\theta = f(\psi)/\xi \sin\phi$. Using the stream function to satisfy the continuity equation, the velocity components may be written in the form:

$$v_\xi = \frac{1}{\xi^2 \sin\phi}\frac{\partial\psi}{\partial\phi} \text{ and } v_\phi = -\frac{1}{\xi \sin\phi}\frac{\partial\psi}{\partial\xi} \tag{8.48}$$

Substituting Equations 8.47 and 8.48 into 8.46, neglecting the first term and using the fact that for small ϕ, sen$\phi \approx \phi$, yields:

$$\frac{\partial}{\partial\phi}\left(\frac{1}{\phi}\frac{\partial\psi}{\partial\phi}\right) = -\xi^4 f(\psi) \tag{8.49}$$

Assuming the following form for the vorticity function,

$$f(\psi) = F\psi^{-5/3} \tag{8.50}$$

the solution of Equation 8.49 by Bloor and Ingham (1973a,b) yields $\psi(\xi,\phi) = K\xi^{3/2}\phi^{3/2}(\alpha-\phi)$, so that

$$v_\xi = \frac{3}{2}\frac{A\phi^{1/2}}{\xi\phi}(3\alpha - 5\phi) \quad v_\phi = -\frac{3}{2}\frac{A\phi^{1/2}}{\xi}(\alpha - \phi) \tag{8.51}$$

and the radial, tangential, and axial velocities in cylindrical coordinates become (Bloor and Ingham 1975a, 1975b)

$$v_r = -A\frac{r^{3/2}}{z^2} \qquad v_\phi = \frac{k}{r} \qquad v_z = \frac{1}{2}Ar^{-1/2}(3\alpha - r(r/z)) \tag{8.52}$$

$$\text{where } k = \frac{0.8QR}{A_i} \text{ and } A = \frac{Q}{2\pi\left(d_v/2\right)^{3/2}\left(\alpha - d_v/2l\right)} \tag{8.53}$$

and Q is the volume flow rate to the hydrocyclone, A_i is the feed inlet area, d_v is the external diameter of the vortex finder tube, and l is the distance from the apex to the lower part of the vortex finder tube. Figure 8.27 shows the tangential and radial velocity, while Figure 8.28 shows the axial velocity distribution, according to Bloor and Ingham (1974).

Brayshaw (1978a, 1978b) explored other vorticity functions for Equation 8.49. Using *cylindrical coordinates*, they derived the following boundary conditions for the stream function $f(\psi)$ at the entry section, at the underflow, and at the overflow respectively:

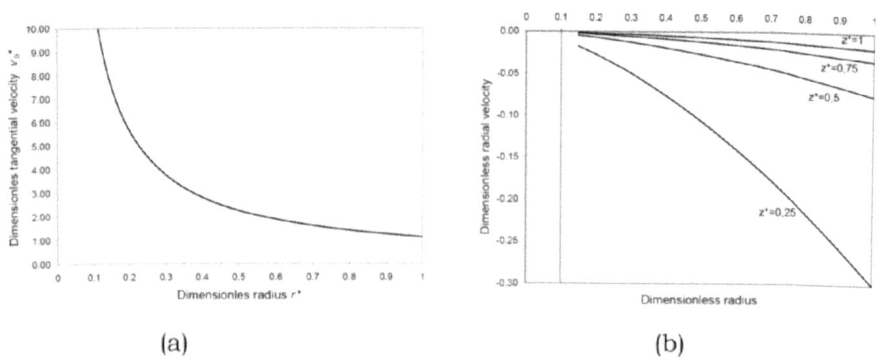

FIGURE 8.27 Velocity distribution for the isochoric rotational flow model of a hydrocyclone according to Bloor and Ingham (1974): (a) Dimensionless tangential velocity. (b) Dimensionless radial velocity with A = 0.01875.

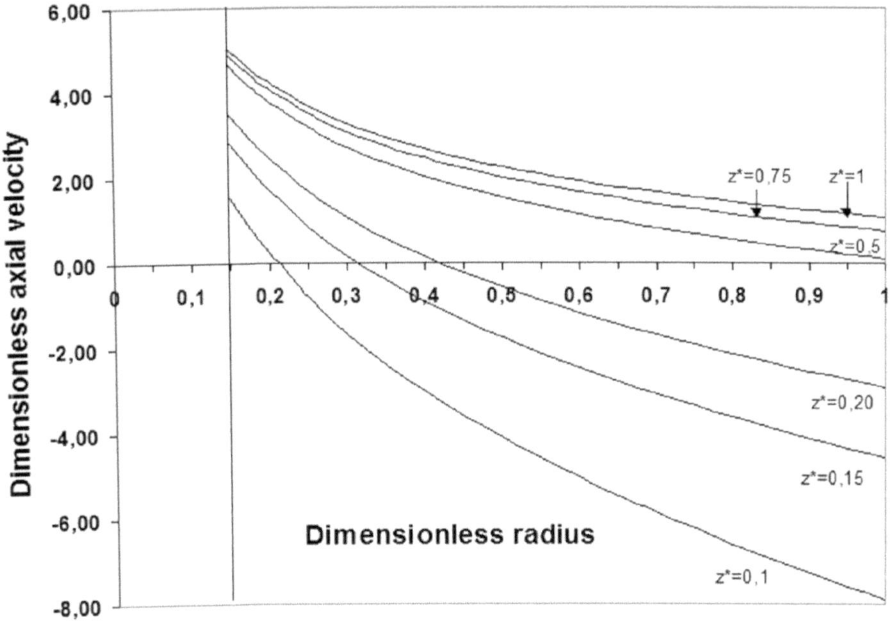

FIGURE 8.28 Dimensionless axial velocity distributions for the isochoric rotational flow model of a hydrocyclone.

Source: Bloor, M.I.G., and Ingham, D.B. 1974. Theoretical analysis of the conical cyclone. Paper E-6. In *Proceedings of the First European Conference on Mixing and Centrifugal Separation*. pp. 95–106.

$$\frac{1}{R^2}\frac{d^2\psi}{dz^2}\;\frac{1}{r^2}\frac{d^2\psi}{dr^2}\bigg|_{r=R_u} -\frac{1}{r^3}\frac{d\psi}{dr}\bigg|_{r=R_u}, -\frac{1}{r^2}\frac{d^2\psi}{dr^2}\bigg|_{r=R_o} +\frac{1}{r^3}\frac{d\psi}{dr}\bigg|_{r=R_u} \tag{8.54}$$

He found that a gamma function and an exponential function were suitable for $f(\psi)$ and gave rise to different particle selectivity curves, with the gamma function showing a sharper separation. Figure 8.29 shows a plot of different vorticity function as given by Brayshaw (1990).

Davidson (1988) indicated that Boor and Ingham's solution of Equation 8.49 could be generalized in the form:

$$\psi = (\xi\phi)^m\left(\alpha^\beta - \phi^\beta\right);\ \text{with}\ \beta = 2(2-\mathrm{m}) \tag{8.55}$$

Unfortunately, no analytical solution of this form can be obtained. Instead, Davidson (1988) proposes a similar solution, in cylindrical coordinates with the tangential velocity independent of the axial coordinate, of the form

$$v_z = z^n F(r/z),\quad v_r = z^n G(r/z),\quad p = P(r) + \rho z^{2n} p_0(r/z) \tag{8.56}$$

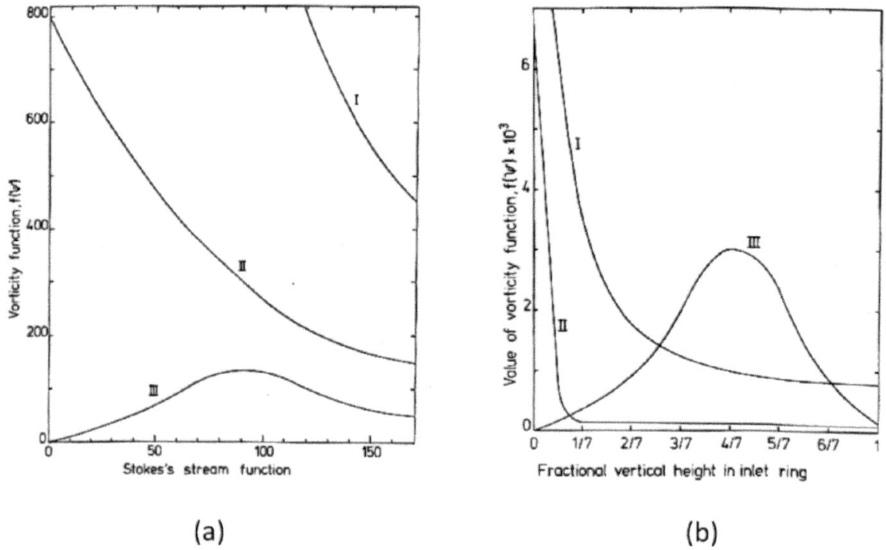

FIGURE 8.29 Three different vorticity functions: (a) Vorticity versus stream function. (b) Vorticity versus height in inlet ring.

Source: Brayshaw, M.D. 1990. A numerical model for inviscid flow of fluid in a hydrocyclone. *International Journal of Mineral Processing* 29:51–75.

$$\psi = z^{n+2}H(r/z), \quad \frac{dH}{d(r/z)} = F(r/z), \quad \mathrm{G}(\mathrm{r/z}) = (r/z)^2 F - (n+1)H \tag{8.57}$$

$$H = \gamma^{1/2}(r/z)_c^2 \times \left(d^2 - \phi^2\right), \quad F = a\gamma^{1/2}\phi\left(d^2 - \phi^2\right)^{-1/2} \tag{8.58}$$

where ϕ is a function obtained by solving equation

$$\sum_{i=0}^{k} a_i \phi^{2k-2i+1} = \frac{1}{4}(-1)^{k+1}\left[\left(\frac{r/z}{r/z_c}\right) - 0.5\right] \tag{8.59}$$

Davidson solved these equations numerically obtaining excellent agreement for the velocity distributions with $n = 0.5$ (corresponding to $m = 3/2$) compared to the experimental values of Knowles et al. (1973) and Kelsall (1952). In this case the axial velocity is

$$v_z = Ar^{-1/2}[1.5\alpha - 2.5(r/z)] \tag{8.60}$$

where A is a constant. Figure 8.30 shows this axial velocity:

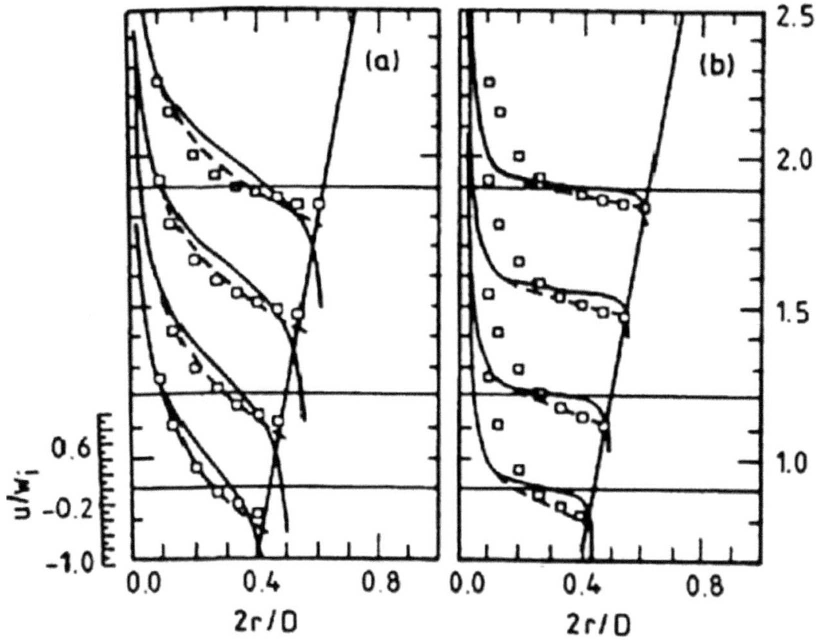

FIGURE 8.30 Axial velocity for the isochoric rotational flow model of a hydrocyclone according to Davidson (1988), compared with experimental data of Kelsall (1952). (a) n = −0.5; (b) n = −1.8.

Source: Davidson, M.R. 1988. Similarity solutions for the flow in hydrocyclones. *Chemical Engineering Science* 43(7):1499–1505.

8.2.4 Boundary Layer Models

In previous sections we have learned that the tangential and axial velocities have an order of magnitude of 1 m/s and that the tangential velocity is insensitive to the axial coordinate, while the axial velocity depends on z. The radial velocity is two orders of magnitude smaller. Since the axial velocity is of the order of 1 m/s, the length of the hydrocyclone is of the order of 1 m, and the kinematical viscosity is of the order of 10^{-4} m^2/s, the axial Reynolds number is of the order of 10^4. With this high Reynolds number, it is only natural to think of the flow inside the hydrocyclone as divided into a viscous boundary layer close to the walls and an inviscid flow in the central region. The inviscid flow was thoroughly analyzed by Bloor and Ingham in several papers (1973a, 1973b, 1974, 1975a, 1975b, 1984a, 1984b) yielding the velocity distribution. In later papers, Bloor and Ingham (1975b, 1976) studied the boundary layer along the lateral walls and at the top wall of a conical hydrocyclone giving the velocity distribution and volume flow within the boundary layer and the boundary layer thickness.

Consider the Reynolds equations for isotropic turbulent swirling flow in a cone, characterized with a turbulent eddy viscosity v_ε (Concha 2007):

$$\frac{\partial}{\partial r}(rv_r)+\frac{\partial}{\partial z}(rv_z)=0 \tag{8.61}$$

$$v_r\frac{\partial v_r}{\partial r}+v_z\frac{\partial v_r}{\partial z}-\frac{v_\theta^2}{r}=-\frac{1}{\rho}\frac{\partial p}{\partial r}+v_z\left(\frac{\partial}{\partial r}\left(\frac{1}{r}\right)\frac{\partial}{\partial r}(rv_r)+\frac{\partial^2 v_r}{\partial z^2}\right) \tag{8.62}$$

$$\frac{v_r}{r}\frac{\partial}{\partial r}(rv_\theta)+v_z\frac{\partial v_\theta}{\partial z}=v_\varepsilon\left(\frac{\partial}{\partial r}\left(\frac{1}{r}\frac{\partial}{\partial r}(rv_\theta)\right)+\frac{\partial^2 v_\theta}{\partial z^2}\right) \tag{8.63}$$

$$v_r\frac{\partial v_z}{\partial r}+v_z\frac{\partial v_z}{\partial z}=-\frac{1}{\rho}\frac{\partial p}{\partial z}+v_\varepsilon\left(\frac{1}{r}\frac{\partial}{\partial r}\left(r\frac{\partial v_z}{\partial r}\right)+\frac{\partial^2 v_z}{\partial z^2}\right) \tag{8.64}$$

8.2.4.1 Boundary Layer at the Lateral Walls of a Hydrocyclone

Since the length scale in the direction "y" perpendicular to the lateral wall is much smaller than in the direction "x" along the wall, it is convenient to make a coordinate transformation from (z,r) to (x,y) by means of

$$\begin{aligned} z &= x\cos\alpha + y\sin\alpha \\ r &= x\sin\alpha - y\cos\alpha \\ \theta &= \theta \end{aligned} \tag{8.65}$$

where α is the semi angle of the hydrocyclone. In addition, it is convenient to define dimensionless variables:

$$\begin{aligned} x^* &= \frac{x}{L_x}, y^* = \frac{y}{L_y}, v_x^* = \frac{v_x}{u_x}, v_y^* = \frac{v_y}{u_y}, v_\theta^* = \frac{v_\theta}{u_x}, p^* = \frac{p}{p_0}, \frac{\partial}{\partial x^*}(\circ) \\ &= L_x\frac{\partial}{\partial x}(\circ), \frac{\partial}{\partial y^*}(\circ) = L_y\frac{\partial}{\partial y}(\circ) \end{aligned} \tag{8.66}$$

With these considerations the field equations become the following:
Continuity equation:

$$\begin{aligned} &\frac{\partial}{\partial x^*}\left(x^*v_x^*\right)+\left\{\frac{L_x}{L_y}\frac{u_y}{u_x}\right\}\frac{\partial}{\partial y^*}\left(x^*v_y^*\right)-\cot\alpha\left\{\frac{u_y}{u_x}\right\}\frac{\partial}{\partial y^*}\left(y^*v_y^*\right) \\ &+\left\{\frac{L_y}{L_x}\right\}\frac{\partial}{\partial x^*}\left(y^*v_x^*\right)=0 \end{aligned} \tag{8.67}$$

x^* component:

$$v_x^* \frac{\partial v_x^*}{\partial x^*} + \left\{ \frac{L_x}{L_y} \frac{u_y}{u_x} \right\} v_y^* \frac{\partial v_x^*}{\partial y^*} - \frac{\sin\alpha\, v_\theta^{*2}}{x^* \sin\alpha - \left\{ \frac{L_y}{L_x} \right\} y^* \cos\alpha} = -\left\{ \frac{p_0}{\rho u_x^{*2}} \right\} \frac{\partial p}{\partial x^*}$$

$$+ \left\{ \frac{L_x}{L_y} \frac{u_y}{u_x} \right\} \frac{\partial^2 v_x^*}{\partial x^{*2}} + \frac{1}{\text{Re}} \left\{ \frac{L_y}{L_x} \right\} \frac{\partial^2 v_x^*}{\partial y^{*2}} + \frac{1}{\text{Re}} \frac{\sin\alpha \frac{\partial v_x^*}{\partial x^*} - \cos\alpha \left\{ \frac{L_y}{L_x} \right\} \frac{\partial v_x^*}{\partial y^*}}{x^* \sin\alpha - \left\{ \frac{L_y}{L_x} \right\} y^* \cos\alpha} \quad (8.68)$$

$$+ \frac{1}{\text{Re}} \frac{\sin\alpha \cos\alpha \frac{u_x}{u_y} v_y^*}{\left(x^* \sin\alpha - \left\{ \frac{L_y}{L_x} \right\} y^* \cos\alpha \right)^2} - \frac{1}{\text{Re}} \frac{\sin^2 \alpha v_x^*}{\left(x^* \sin\alpha - \left\{ \frac{L_y}{L_x} \right\} y^* \cos\alpha \right)^2}$$

θ component:

$$v_x^* \frac{\partial v_\theta^*}{\partial x^*} + \left\{ \frac{L_x}{L_y} \frac{u_y}{u_x} \right\} v_y^* \frac{\partial v_\theta^*}{\partial y^*} - \frac{\sin\alpha\, v_\theta^* - \left\{ \frac{u_y}{u_x} \right\} \cos\alpha\, v_y^*}{x^* \sin\alpha - \left\{ \frac{L_y}{L_x} \right\} y^* \cos\alpha} = \frac{1}{\text{Re}} \frac{\partial^2 v_\theta^*}{\partial x^{*2}}$$

$$+ \frac{1}{\text{Re}} \left\{ \frac{L_x}{L_y} \right\} \frac{\partial^2 v_\theta^*}{\partial y^*} + \frac{1}{\text{Re}} \frac{\cos\alpha \left\{ \frac{L_x}{L_y} \right\} v_\theta^*}{\left(x^* \sin\alpha - \left\{ \frac{L_y}{L_x} \right\} y^* \cos\alpha \right)^2} \quad (8.69)$$

$$+ \frac{1}{\text{Re}} \frac{\sin\alpha \frac{\partial v_\theta^*}{\partial x^*} - \cos\alpha \left\{ \frac{L_x}{L_y} \right\} \frac{\partial v_\theta^*}{\partial y^*}}{\left(x^* \sin\alpha - \left\{ \frac{L_y}{L_x} \right\} y^* \cos\alpha \right)^2}$$

y^* component:

$$v_x^* \frac{\partial v_y^*}{\partial x^*} + \left\{ \frac{L_x}{L_y} \frac{u_y}{u_x} \right\} v_y^* \frac{\partial v_y^*}{\partial y^*} - \frac{\left\{ \frac{u_y}{u_x} \right\} \cos\alpha\, v_\theta^*}{x^* \sin\alpha - \left\{ \frac{L_y}{L_x} \right\} y^* \cos\alpha} = -\left\{ \frac{p_0}{\rho u_x^2} \right\} \left\{ \frac{L_x}{L_y} \frac{u_y}{u_x} \right\} \frac{\partial p^*}{\partial y^*} \quad (8.70)$$

$$\frac{1}{\text{Re}}\frac{\partial^2 v_y^*}{\partial x^{*2}}+\frac{1}{\text{Re}}\left\{\frac{L_x}{L_y}\right\}\frac{\partial^2 v_y^*}{\partial y^{*2}}+\frac{1}{\text{Re}}\frac{\sin\alpha\frac{\partial v_y^*}{\partial x^*}-\cos\alpha\left\{\frac{L_x}{L_y}\right\}\frac{\partial v_y^*}{\partial y^*}}{x^*\sin\alpha-\left\{\frac{L_y}{L_x}\right\}y^*\cos\alpha}$$

$$+\frac{1}{\text{Re}}\frac{\sin\alpha\cos\alpha\left\{\frac{u_x}{u_y}\right\}v_x^*}{\left(x^*\sin\alpha-\left\{\frac{L_y}{L_x}\right\}y^*\cos\alpha\right)^2}$$

A study of order of magnitude of the terms shows that

$$\left\{\frac{L_x}{L_y}\right\}^2=\text{Re},\left\{\frac{u_x}{u_y}\right\}^2=\text{Re},\left\{\frac{L_x}{L_y}\frac{u_y}{u_x}\right\}=1 \tag{8.71}$$

Substituting into the previous equations and neglecting all terms with powers to the Re^{-1}, the boundary layers equations in dimensional form for the swirling flow in a cone are

$$\frac{\partial}{\partial x}(xv_x)+\frac{\partial}{\partial y}(xv_y)=0 \tag{8.72}$$

$$v_x\frac{\partial v_\theta}{\partial x}+v_y\frac{\partial v_\theta}{\partial y}=v_\varepsilon\frac{\partial^2 v_\theta}{\partial y^2} \tag{8.73}$$

$$v_x\frac{\partial v_x}{\partial x}+v_y\frac{\partial v_x}{\partial y}=-\frac{1}{\rho}\frac{\partial p}{\partial x}+v_\varepsilon\frac{\partial^2 v_x}{\partial y^2} \tag{8.74}$$

$$\frac{\partial p}{\partial y}=0 \tag{8.75}$$

Since the pressure is independent of y, the pressure gradient in Equation 8.74 can be written in terms of the velocity outside the boundary layer, according to Bernoulli's equation.

Bloor and Ingham used the velocities obtained for the inviscid flow (see previous analysis) as boundary conditions at the boundary layer limit, then:

$$\frac{\partial}{\partial r}(rv_r)+\frac{\partial}{\partial z}(rv_z)=0 \quad v_x=v_x(R)\left(\frac{x}{x_0}\right)^{-1/2}, \quad v_\theta=\frac{Rv_\theta(R)}{r}\frac{\gamma\left(\frac{4}{5},\frac{2}{5}\frac{Lr}{R^{5/2}}\right)}{\gamma\left(\frac{4}{5},\frac{2}{5}L\right)} \tag{8.76}$$

Where $x_0 = l/\tan\alpha$ and R is the radius of the cylindrical portion of the hydrocyclone. Bloor and Ingham (1976) obtained a solution to these equations using the approximate integral method of von Polhausen. Defining the new similar variable $\eta = y/\delta(x)$, where $\delta(x)$ is the thickness of the boundary layer, they described the velocity distributions in this form:

$$\frac{v_x}{v_x(R)} = 2\eta - \eta^2 + a(x)\eta(1-\eta)^2, \quad \frac{v_\theta}{v_\theta(R)} = 2\eta - \eta^2 \tag{8.77}$$

Substituting these expressions into the von Karman integrals of the field equations results in two ordinary differential equations, the result of which indicate that the dimensionless thickness of the boundary layer $\delta^*\left(x^*\right)$ increases from $x^* = 0$ to $x^* = 0.5$, then stays constant and finally diminishes to 0 at $x^* = 1$. At the points of maximum thickness, the boundary layer has the form:

$$\delta_m = \frac{1.88x_0}{\text{Re}^{1/2}} \tag{8.78}$$

where $x_0 = l/\tan\alpha$.

8.2.4.2 Boundary Layer at the Top Wall of a Hydrocyclone

A dimensional analysis to equations *equation reference goes here* yields the following set of field equations for the boundary layer at the top wall of the hydrocyclone:

$$\frac{\partial}{\partial r}(rv_r) + \frac{\partial}{\partial z}(rv_z) = 0 \tag{8.79}$$

$$v_r \frac{\partial v_r}{\partial r} + v_z \frac{\partial v_r}{\partial z} - \frac{v_\theta^2}{r} = -\frac{1}{\rho}\frac{\partial p}{\partial r} + \frac{1}{\rho}\frac{\partial T_{rz}^E}{\partial z} \tag{8.80}$$

$$\frac{v_r}{r}\frac{\partial}{\partial r}(rv_\theta) + v_z \frac{\partial v_\theta}{\partial z} = \frac{1}{\rho}\frac{\partial T_{\theta z}^E}{\partial z} \tag{8.81}$$

$$0 = \frac{1}{\rho}\frac{\partial p}{\partial z} \tag{8.82}$$

with boundary conditions:

$$\begin{aligned} v_r(r,0) &= 0 \qquad & v_\theta(r,0) &= 0 \\ v_r(r,\delta(r)) &= V_r \qquad & v_\theta(r,\delta(r)) &= V_\theta \end{aligned} \tag{8.83}$$

Since the pressure is independent of z, the pressure gradient in Equation 8.82 can be written in terms of the velocity outside the boundary layer, according to Bernoulli's equation: $V_r \partial V_r / \partial r - V_\theta^2 / r = -(1/\rho)\partial p / \partial r$. Using the method of von Karman integrals, Bloor and Ingham (1975b) approximated the velocity distribution with the following functions of a similar variable ξ:

$$v_r = V_r \xi^{1/7}\left(\gamma\left(\frac{15}{7} - 2\gamma\right)\xi + \left(\gamma - \frac{8}{7}\right)\xi^2\right); \quad v_\theta = \frac{1}{7} V_\theta \xi^{1/7}(8 - \xi) \qquad (8.84)$$

where $\xi = z / \delta(r)$

Smith (1962) analyzed the stability of the boundary layer in the vortex flow of a pneumatic cyclone. He explains that the flow is naturally unstable close to the wall, since in this region the angular momentum diminishes with increasing radius. The instability manifest itself with the intermittent burst of fluid from the boundary layer moving radially inwards where it diffuses and loses identity. These bursts make the flow in the cyclone very turbulent even if there is no turbulence convection into the equipment by the incoming fluid or produced by the interaction of the incoming fluid with the flow in the cyclone. This turbulence, resulting from the unstable boundary layer, is inherent in any device employing a rotating flow within a stationary container. On a conical wall, the secondary flows and the generation of turbulence occur in combination. During a burst, the heavy particles are left behind forming the streaks that are observed commonly in dust separation cyclones. The lighter particles are dragged by the fluid and remix into the main flow. This is a mechanism of re-dispersion of small particles trapped in the boundary layer. Although these mechanisms were observed in pneumatic cyclones, they can also explain the high turbulent motion in hydrocyclones and the re-dispersion of small particles.

Kumari and Nath (1985) studied the unsteady laminar motion in the boundary layer of nozzles, diffusers, and hydrocyclones. They concluded that the tangential velocity at the edge of the boundary layer is of a non-free vortex type, in contrast to the steady flow. In the flow through hydrocyclones the swirl flow dominates the axial flow, which increases approaching the apex.

8.2.5 The Air Core

The role of the air core in the performance of a hydrocyclone was not recognized until 1987, when Plitt and co-workers commented that it relates to the flow transition from a spray discharge to roping. That same year Davidson et al. (1987) indicated that as the slurry viscosity increased, the angular velocity dropped in the hydrocyclone and the air core contracts. Castro (1990) found that the water split is mainly controlled by the air core size for an apex/vortex ratio and, therefore, under such conditions the water split is governed by those operating conditions and variables which affect the air core. Bustamante (1991) found that working a hydrocyclone with water only might be used to predict roping conditions. He performed tests in which, for a fixed apex/vortex ratio, he measured the underflow rate for increasing values of the feed rate. Those conditions when no underflow was obtained, because the air core obstructed completely the apex, coincided with the onset of roping when the hydrocyclone was operated with pulp (Barrientos et al. 1993; Concha et al. 1996b).

This late recognition of the importance of the air core in the operation of a hydrocyclone is strange, since research workers had developed models for the air core size in vortex motions long before.

Binnie et al. (1948) analyzed the drainage of water from an open cylindrical tank with a controlled axial and tangential inflow and a free discharge at the bottom and center of the tank. They observed a relationship between the total head and the discharge rate that depended on the ratio of tangential to radial inflow. They also observed a critical pressure head that could maintain an air core. Figure 8.31 shows such curves.

When a combination of tangential and axial inflow of constant ratio was used, they observed an interesting phenomenon of surging. This effect was related to the forming and breakdown of the air core. Consider the tank full up to a total head of 0.16 m with a tangential inflow rate of 0.24×10^{-3} m^3/s and an axial inflow rate of 0.66×10^{-3} m^3/s. The total inflow was maintained at $0.9 \times 10^{-3} m^3/s$. The curve for the total head versus the discharge flow rate is also given in Figure 8.31. At $H = 0.16$m, the ratio of tangential to axial inflow rates of 0.37 is not sufficient to maintain an

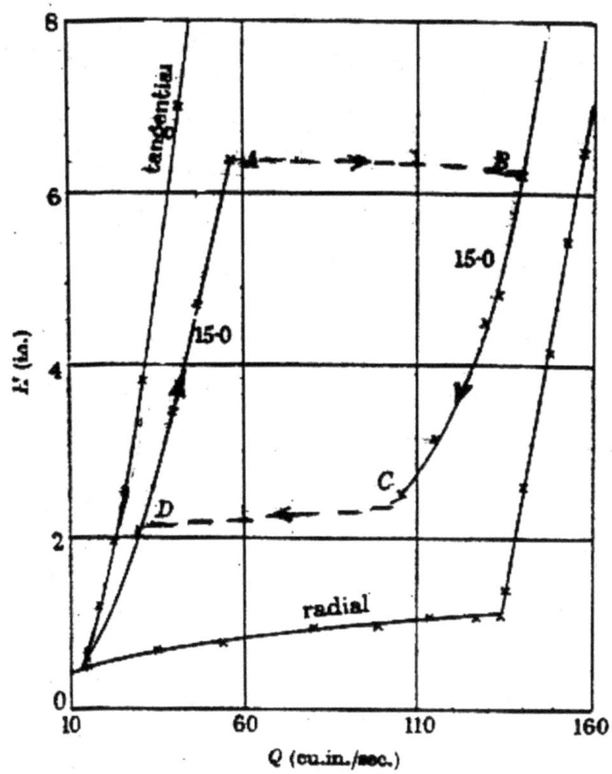

FIGURE 8.31 Discharge volume flow rate Q_D versus total head H for axial or tangential inlet of water of 0.9×10^{-3} m^3/s. Also shown is the curve for a tangential inflow of 0.24×10^{-3} m^3/s and axial inflow of 0.66×10^{-3} m^3/s.

air core, and the entire discharge leaves through the exit pipe with a flow rate of 2.3×10^{-3} m³/s. Since this outflow is larger than the inflow of 0.90×10^{-3} m³/s, the container begins to loose water and the total head, and discharge rate diminish until H becomes $H = 0.06$m. This is the lowest total head at which a coreless discharge is possible with the given inflow ratio of 0.37, and once again, an air core forms, suddenly diminishing abruptly the discharge rate to 0.57×10^{-3} m³/s. This outflow is now smaller than the inflow of 0.90×10^{-3} m³/s, and the water level will rise until the discharge rates equates the inflow rate. At this point, with a total head is $H = 0.16$m, the air core again becomes unstable, collapsing to a coreless outflow of 2.3×10^{-3} m³/s. Then, a new cycle begins.

Binnie et al. (1948) obtained an approximate expression for the discharge rate in the previous problem, by considering a potential flow. With the Bernoulli given by $B = p/\rho + (1/2)\left(v_\theta^2 + v_z^2\right)$, the total head is $H = B/g$, and the axial velocity at the interface, where $p = 0$, is given by $v_z = \left(2gH - v_\theta^2\right)^{1/2}$. If the underflow and the air core radii are denoted by R_u and R_a respectively, the outflow volume is given by

$$Q_u = \pi\left(R_u^2 - R_a^2\right)v_z \tag{8.85}$$

and therefore

$$Q_u = \pi\left(R_u^2 - R_a^2\right)\left(2gH - V_\theta^2\right)^{1/2} \tag{8.86}$$

For an inviscid flow, the tangential velocity is a free vortex of the form $v_\theta = V_{\theta u}/r$, where $V_{\theta u}$ is the tangential velocity at $r = R_u$, and therefore, for $r = R_a$:

$$Q_u = Q_u = \pi\left(R_u^2 - R_a^2\right)\left(2gH - \left(V_{\theta u}/R_a\right)^2\right)^{1/2} \tag{8.87}$$

Binnie and Hookings argued that, for the stability of the flow, R_a should adjust itself so that the discharge is maximum, which means that $\partial Q_u/\partial R = 0$, and leads to the following condition:

$$4gHR_a^4 - V_{\theta_u}^2 R_a^2 - V_{\theta_u}^2 R_u^2 = 0 \tag{8.88}$$

from which the radius of the air core is

$$R_a = \left(\frac{V_{\theta u}^4 + \left(V_{\theta u}^4 + 16gHV_{\theta u}^2 R_u^2\right)^{1/2}}{8gH}\right)^{1/2} \tag{8.89}$$

We can also be written Equation 8.89 in the form:

$$4gHR_a^4 - 2V_{\theta u}^2 R_a^2 = V_{\theta u}^2 R_a^2 + V_{\theta u}^2 R_u^2 - 2V_{\theta u}^2 R_a^2 = 0$$

$$2gH - V_{\theta u}^2 = \frac{V_{\theta u}^2}{2R_a^2}\left(\frac{R_u^2}{R_a^2} - 1\right) \equiv v_z^2\left(R_a\right) \tag{8.90}$$

This is the value of the axial velocity in the discharge tube. Introducing this value into Equation 8.85, the flow rate through the discharge can be found:

$$Q_u^2 = \frac{1}{2}\pi R_a^2 V_{\theta u}^2 \left(\left(\frac{R_u}{R_a}\right)^2 - 1\right)^3 \quad (8.91)$$

or

$$\left(\frac{Q_u}{\Gamma_u}\right)^2 = \frac{1}{8}\left(\frac{R_a}{R_u}\right)^2 \left(\left(\frac{R_u}{R_a}\right)^2 - 1\right)^3 \quad (8.92)$$

where $\Gamma_u = 2\pi R_u V_{\theta u}$ is the circulation at $r = R$. This equation was derived by Smith (1962) for the outflow of a hydrocyclone.

Davidson (1994) followed Binnie and Hookings approach to calculate the air core size, for one outlet at the underflow and for two outlets at the underflow and overflow. For one outlet he described the tangential velocity with the equation $v_\theta = A / r^n$. The Bernoulli can now be written in the form $B = p/\rho + v_z^2 + A^2 / r^{2n}$. At the air core interface, the pressure is zero, $r = R_a$ and $A = V_{\theta u}$, then as follows:

$$v_z = \left(2B - V_{\theta u}^2 / R_a^{2n}\right)^{1/2} \quad (8.93)$$

So that the volume flow results as follows:

$$Q_u = \pi\left(R_u^2 - R_a^2\right)\left(2B - V_{\theta u}^2 / R_a^{2n}\right)^{1/2} \quad (8.94)$$

Differentiating this expression with respect to R_a and equating the result to zero yields the following:

$$\frac{R_a}{R_u} = \left(\frac{nV_{\theta u}^2}{nV_{\theta u}^2 + 2V_z^2}\right)^{1/2} \quad (8.95)$$

For $n = 1$, Equation 8.92, derived by Smith (1962), is recovered:

For two outlets Davidson (1994) described the axial velocities with the following equations: $V_{zu} = 2B - V_{\theta u}^2 / R_a^{2n}$ and $V_{zo} = 2B - V_{\theta o}^2 / R_a^{2m}$. The total flowrate is given by the following:

$$Q = \pi\left(R_u^2 - R_a^2\right)\left(2B - V_{\theta u}^2 / R_a^{2n}\right)^{1/2} + \pi\left(R_o^2 - R_a^2\right)\left(2B - V_{\theta o}^2 / R_a^{2m}\right)^{1/2} \quad (8.96)$$

After differentiating with respect to R_a, the radius of the air core becomes as follows:

$$R_a = \left(\frac{nR_u^2 V_{\theta u}^2 V_{zu} + mR_o^2 V_{\theta o}^2 V_{zo}}{\left(2V_{zu}^2 + nV_{\theta u}^2\right)V_{zo} + \left(2V_{zo}^2 + mV_{\theta o}^2\right)V_{zu}} \right)^{1/2} \tag{8.97}$$

For large $V_{\theta u}, R_a \rightarrow R_u$ and for large $V_{\theta 0}, R_a \rightarrow R_o$. Since R_u is smaller than R_o, the apex will be blocked first.

Davidson (1995) indicates that, using inviscid flow to calculate the air core size, Equation 8.97 gives correct results when the fluid is water. For a suspension he assumes that a correction factor proportional to the ratio of the kinematical viscosity of the suspension v_s to the kinematical viscosity of the water v_w can be used for the air core radius as follows:

$$R_a\Big|_{\text{suspension}} \approx R_a\Big|_{\text{water}} \left(\frac{v_s}{v_w} \right)^{0.3} \tag{8.98}$$

He further indicates that his equation for the air core size compares well with the experiments of Petersen (1993) and those of Kelsall (1952), Hsieh and Rajamani (1991), and Svarovsky and Marasinghe (1980). The agreement with the experiments of Steffens et al. (1993) is only fair.

Mikaylov and Romensky (1974) proposed the following equation for the air core in a conical hydrocyclone:

$$R_a = k_a d_o (\tan \alpha)^{0.30} \tag{8.99}$$

Where 2α is the angle of the conical section of the hydrocyclone and k_a is a constant.

Nepomnyashchii and Pavlobskii (1977) derived the following equation for the air core radius

$$R_a = \frac{R_F}{R} \left(\frac{\rho v_F^2}{3\Delta p} \right)^{1/3} \tag{8.100}$$

where R_F and R are the eccentricity of the hydrocyclone feed and the hydrocyclone radius respectively, v_F is the inlet velocity of the feed.

Svarovsky (1984) presented this equation in dimensionless form:

$$\frac{d_a}{d_c} = \frac{d_F}{d_c} \left(\frac{1}{3Eu} \left(\frac{d_c}{d_F} \right)^4 \right)^{1/3} \tag{8.101}$$

Where d_a, d_F, and d_c are the air core diameter, twice the inlet eccentricity and the hydrocyclone diameter respectively, and $\text{Eu} = \Delta p / (1/2)\rho v^2$ is Euler's number.

Barrientos et al. (1993) calculated the air core radius using the momentum jump balance for the normal stresses. Across the interface, a jump of the normal stresses exists and is given by the Young-Laplace equation:

$$[\boldsymbol{Tn}\cdot\boldsymbol{n}] = -2H\gamma \tag{8.102}$$

where the square brackets represent the jump of a property across the interface, $\boldsymbol{T}$ is the stress tensor, $\boldsymbol{n}$ is the normal unit vector to the interface, H is the main curvature of the air core, and g is the liquid surface tension. Assuming the air behaves as an ideal fluid rotating as a rigid body with density ρ_a and angular velocity ω_a and that the liquid is a Newtonian fluid with viscosity m, equation can be written in the form Equation 8.102:

$$\frac{1}{2}\rho_a\omega_a^2R_a^3 + \left(2\mu\frac{\partial v_r}{\partial r}\bigg|_{r=R_a} - \Delta p_a\right)R_a - \gamma = 0 \tag{8.103}$$

Introducing the dimensionless variables and making a study of order of magnitude of the resulting equation yields the following equation for the air core:

$$R_a \approx \frac{\gamma}{2\mu v - \Delta p_a} \tag{8.104}$$

Based on this equation, Concha et al. (1996b) and Castro et al. (1996) developed empirical equations for the air core diameter in terms of the operating variables in a laboratory and industrial hydrocyclone, respectively. Concha et al. (1996b) obtained the following equation for the air core:

$$d_a = \frac{3.03\gamma d_o}{270.3\mu_w\left(1+\dfrac{1.18\times10^5}{\Delta P^{2.61}}\right)\left(\dfrac{d_u}{d_o}\right)^{-0.20} - 1.38\times10^{-3}} \tag{8.105}$$

In this equation, γ is measured in N/m, μ_w is in kg/ms, ΔP is in Pa and d_a, d_o, and d_u are in m. In this case ΔP is the total pressure drop in the hydrocyclone.

Castro et al. (1996) obtained the following relationship:

$$d_a = \frac{103.6(d_u/d_o)^{0.89}}{6.3(\mu/\mu_w)\left(1+\dfrac{0.98}{\Delta P^{1.15}} - \dfrac{1.41}{\Delta P^{1.57}}\right)} \tag{8.106}$$

where μ and μ_w are the viscosities of the pulp and the water, respectively.

Ovalle and Concha (2005) studied the generation and propagation of waves at the air core-water interface of a conical hydrocyclone. They showed that inside the hydrocyclone the flow is subcritical (Froude number less than 1) and therefore in a favorable condition for the existence of undulatory motion in the fluid. They used a hydraulic analogy consisting of a launder with a ramp, where the water enters through the bottom at the left side and overflows both at the left and the right of the upper part of the physical model (Figure 8.32). The centrifugal force of the hydrocyclone is substituted by the gravitational force in the hydraulic model.

The flow in this hydraulic analogy was modeled as inviscid and irrotational, with the form of the free surface obtained as follows:

$$\eta(x,y,t) = 2A\frac{k}{\omega}\sinh(k(y+h))\cos(kx-\omega t) \tag{8.107}$$

where A is a constant and $\omega(t)$ and $k(x)$ represent the temporal and spatial behavior, respectively. Figure 8.33 shows some isobaric lines resulting from the calculations of the temporal evolution of the waves imposed over the free surface.

The authors concluded that the form of the free surface depends mainly on the global geometrical characteristics of the flow and possibly to a minor extent on physicochemical properties of the surface. They showed that the waves propagating through the water-air interface of the air core will increase their amplitude as the hydrocyclone diameter becomes smaller, so that the air core may become of the size of the apex, in which case instability is induced in the underflow. The apex will intermittently become blocked by the air core, inducing roping. Unfortunately, the presence of waves in a conical hydrocyclone is inherent to the physics of the flow.

Although most of the experimental evidence on the flow pattern in a hydrocyclone is based on studies with water, one can say that the flow pattern in a

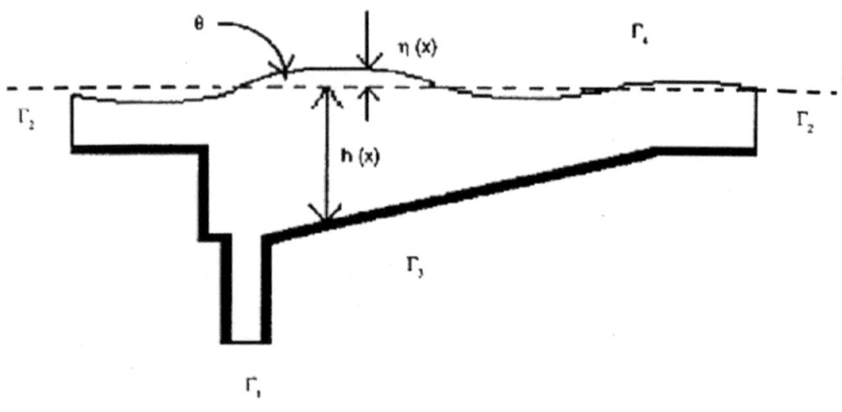

FIGURE 8.32 Hydraulic analogy of a conical hydrocyclone.

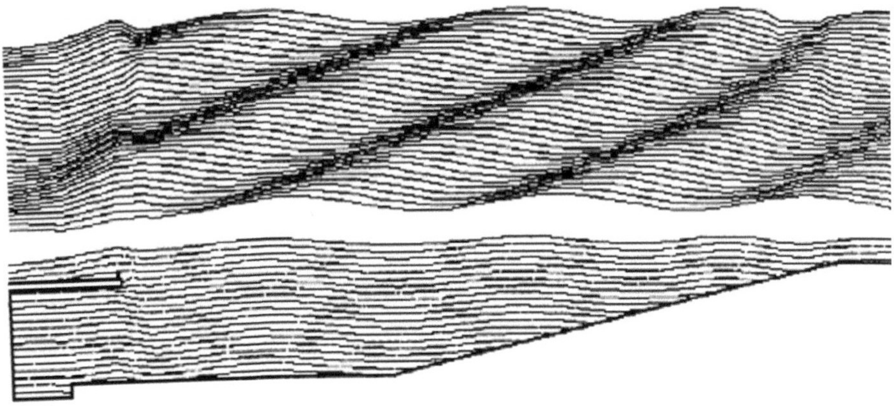

FIGURE 8.33 Isobaric lines in the hydraulic model of a conical hydrocyclone.

hydrocyclone treating mineral suspensions is well understood. What is not yet sufficiently clear is a model of the air core and how it affects the flow pattern, the fluctuating velocity component and hydrocyclone operation. Nevertheless, our knowledge of the flow pattern in a hydrocyclone is sufficient to develop classification models of the device.

8.2.6 Discharge and Roping

Although roping in hydrocyclones is a frequent problem in industries and has been studied by many researchers, we do not yet have a theory that relates all the variables involved (Pérez et al. 2018). Several experimental works with different approaches such as mechanical energy balance, work on hydrocyclones with water only, and hydrocyclone air core measurements with different instruments in the laboratory and plant have attempted to explain the roping phenomenon. They have addressed design variables such as apex-to-vortex diameters and different cone angles, as well as operating variables such as inlet solid concentration, particle size, pressure and overflow and underflow flow rates and concentrations in order to understand their effect on the roping condition.

Bustamante (1991), operating a 6-inch Krebs hydrocyclone with water only measured the air core diameter increase as the inlet pressure was increased. He showed that at a certain point the air core equaled the apex diameter, and no water leaves through the underflow. He established that when operating the hydrocyclone with pulp at the same operating conditions, the transition from spray to roping occurred when the apex was blocked and no air came in to form the air core. Using the Young-Laplace equation, Concha et al. (1996a) showed for that this hydrocyclone the apex to vortex diameter roping always exists and for there is no roping under any operating conditions. In the intermediate range, spray and roping can coexist depending on other variables such as pressure, feed concentration, particle feed size, and size distribution. The conclusion of this work was that the precondition for roping is the absence of the air core in the apex and the formation of a bed of particles in the bottom of the hydrocyclone.

This same year (Neesse et al. 1991) pointed out that a hydrocyclone can operate with a spray, semi-rope, and rope discharge depending on several variables but mainly on the particle concentration of the feed. Figure 8.34 shows the representation of these phenomena. In dense flow separation, more solids are stored in the conical part of the hydrocyclone and slowly moved upward reducing solids recovery to the underflow. The air core does not extend over the entire length of the hydrocyclone and oscillates intensively. The discharge assumes the shape of a rope and is characterized by high solids content and fewer fines, which is the advantage of this variant. The transition state between spray and rope discharge presents the advantages of both high solid recovery and high solids content in the underflow.

Using CFD Pérez et al. (2018) confirmed that the incremental increase in inlet pressure produces an expansion of the air core diameter leading to roping, which is characterized by a change in the frequency spectrum of the pressure oscillations at the hydrocyclone wall with high amplitudes and low frequencies for spray discharge and noisy signals in all spectrums and damped low frequencies for roping. They also

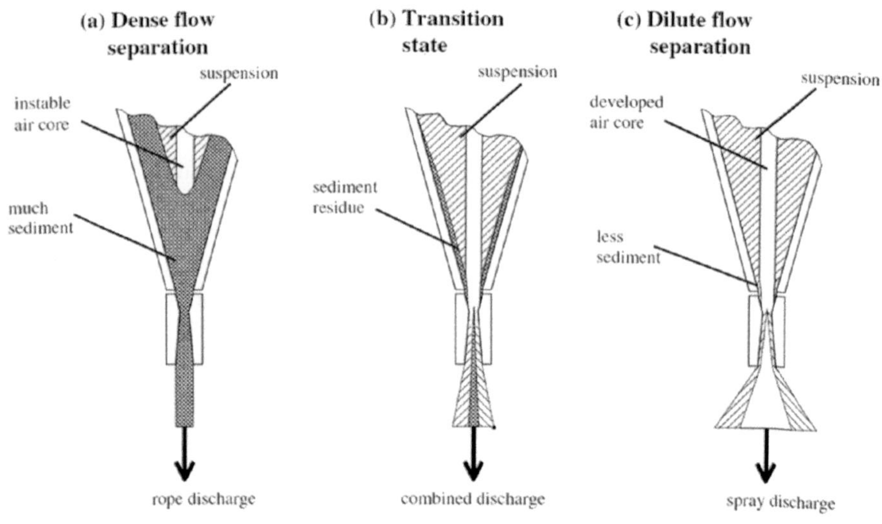

FIGURE 8.34 Operating state of a hydrocyclone.

TABLE 8.3
Operating hydrocyclone discharge states

	Case I:	Case II	Case III
Particle size (µm)	20	34	70
Pressure (psi)	8.6, 19.2, 33.2, 50.2, 70.5	7.4,19.1, 25.8, 50.2, 58.8	4.7, 4.8, 8.6, 14

indicated that transition from spray to roping is generated by an increase in particle diameter at a constant solid concentration in the feed flow. With 20-μm-diameter particles in the feed, an increase in pressure results in a decrease in the discharge angle, although it always maintains a spray shape, while 34-μm-diameter particles generate a semi-rope discharge and 70-μm-diameter particles always roping condition, as presented in Table 8.3. Figure 8.35 shows slurry-air contours and discharge angles as the inlet pressure increases for the following three cases.

In a recent work, Daza et al. (2018) examined the air core and material distribution inside the hydrocyclone, the underflow shape and spray angle, the flow and solid concentrations, the efficiency curve and separation size, and the particle size distribution in the underflow sizes. The study confirmed that in a feed with particle size distribution, the feed has a direct impact on the underflow behavior. When the device operates with coarser material it tends to rope with a decrease in the spray angle. The same happens with an increase in the feed pressure inlet. The roping condition generated by the coarse material directly affects the efficiency of the hydrocyclone, triggering an increase in the separation size. The underflow particle size distribution tends to be coarser with an increase in the inlet pressure. Figure 8.35 shows the CFD simulation.

8.2.7 Phenomenological Models

Phenomenological hydrocyclone models are based on the study of the physical phenomena occurring during the flow in the equipment. The first such studies were conducted in the 1950s (Rietema and Krajenbrink 1958) and have progressed very slowly due to the formidable task of solving sets of nonlinear differential equations that describe the flow in the device (Concha and Barrientos 1991).

The tangential velocity must be modeled as a function of the radial coordinate in the hydrocyclone in order to ascertain the centrifugal force generated in the device by the vortex motion. This force is responsible for the distribution of the particles by size along the radius. Based on this idea, Rietema (1961a) proposed the theory of the radius of stability, according to which the equilibrium between the centrifugal and radial drag and buoyancy force on a particle is obtained at different radii of the hydrocyclone (radius of stability) for different particle sizes. Particles whose position is smaller than the radius of stability will leave through the overflow, while all the other particles will leave the device through the underflow. This theory considers the radial balance of forces only and assumes that the radius of stability is always reached by all particles.

8.2.7.1 Equilibrium Orbit Model

The axial and tangential velocities in a hydrocyclone are of the order of m/s, while the radial velocity is one order of magnitude smaller; therefore, the velocity of a small particle in the radial direction can be considered in Stokes regime. Consider the fluid flow in the hydrocyclone with a radial velocity v_r. If a particle of radius r is dragged through a fluid by the centrifugal force with a velocity u_r in the radial direction of the hydrocyclone, the drag force F_D exerted by the fluid on the particle is as follows:

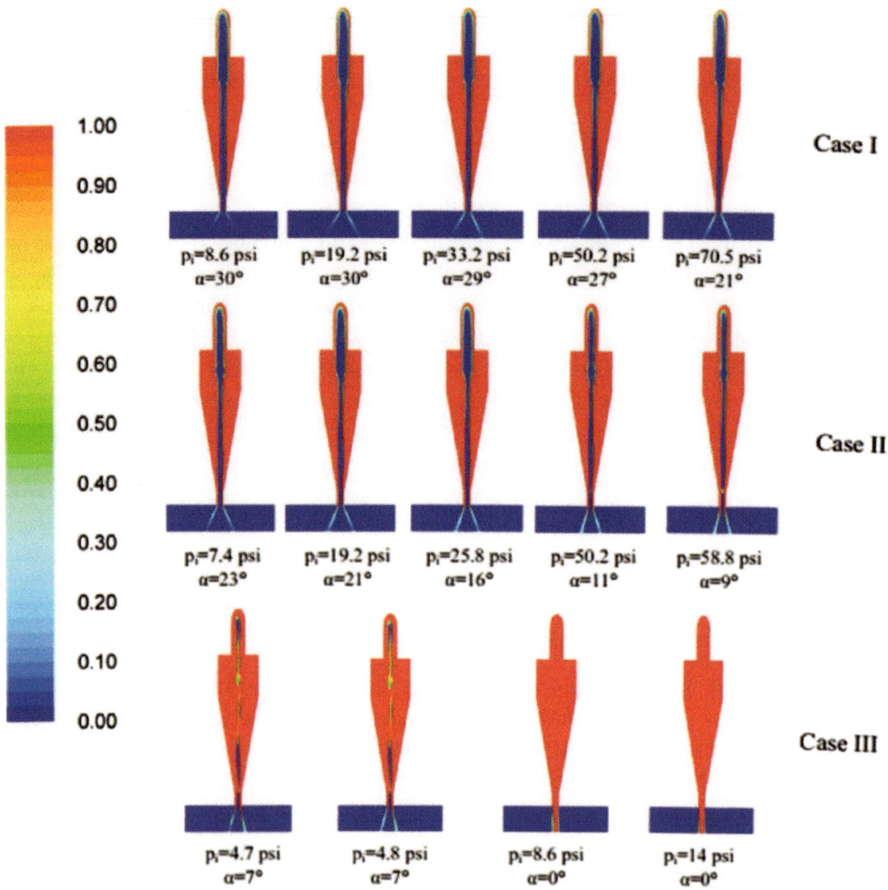

FIGURE 8.35 Slurry-air contours and discharge angles as the inlet pressure increases.

$$F_D = -6\pi ur\left(u_r - v_r\right) \tag{8.108}$$

while the centrifugal force is as follows:

$$F_C = \left(\rho_s - \rho_f\right)\frac{v_\theta^2}{r}\left(\frac{4}{3}\pi r^3\right)$$

where ρ_s and ρ_f are the solid and the fluid velocity, respectively, and v_θ is the tangential velocity of the particle. The balance of these forces gives the following:

$$r = -\frac{9}{2}\frac{\left(u_r - v_r\right)}{\left(\rho_s - \rho_f\right)v_\theta^2} \tag{8.109}$$

The particle with a 50% chance of passing to the overflow will be called d_{50}, and we will assume that its position corresponds to the locus of zero particle velocity; thus:

$$d_{50} = \frac{9}{2}\frac{\mu}{\left(\rho_s - \rho_f\right)}\frac{v_r}{v_\theta^2} \tag{8.110}$$

Equation 8.110 is called the equilibrium orbit hypothesis.

8.2.7.2 Phenomenological Model

Concha and Barrientos (1991) and Barrientos and Concha (1992) presented a phenomenological model of a hydrocyclone based on the experimental observations of Hsieh and Rajamani (1988a, 1988b) and the works of Bloor and Ingham (1975b, 1976).

They expressed that the tangential fluid entry produces a three-dimensional vortex in the interior of the hydrocyclone. The high tangential and axial velocities, which have values of several meters per second, and the fact that all velocities must vanish at the hydrocyclone walls, enable the flow near the walls of the equipment to be regarded as a viscous boundary layer and an inviscid flow far from it (Bloor and Ingham 1975b, 1976). On the other hand, the high velocities within the device produce low pressures near the axis, sucking air from the outside through the apex and forming an air core.

They divided the equipment into six zones, as shown in Figure 8.36.

8.2.7.2.1 Zone I

Zone I is the upper cylindrical part of the hydrocyclone, from the top of the device to the bottom of the vortex finder tube. Due to the presence of a toroidal vortex in this zone, it can be assumed to be well-mixed and therefore homogeneous, with the same composition as the feed to the hydrocyclone.

8.2.7.2.2 Zone II

Zone II is the lower cylindrical part of the hydrocyclone. It has most of the properties of the conical section below and serves as a transition region between the homogeneous Zone I and the conical Zone III. We assume the existence of a sink at the interface of Zones I and II, close to the vortex finder tube, and more precisely from point p, where the axial velocity is zero, to the border of the vortex finder tube (see Figure 8.361), such that all particles reaching this region from Zone II will flow out through the vortex. In Zone II we assume that Bloor and Ingham's solution (1973a, 1973b) for a viscid flow is valid with an empirical correction $f(z)$ for the radial velocity. With this correction the radial velocity $v_r(r,z)$ is set to zero at the I/II interface, giving the following equations for the radial axial and tangential velocity:

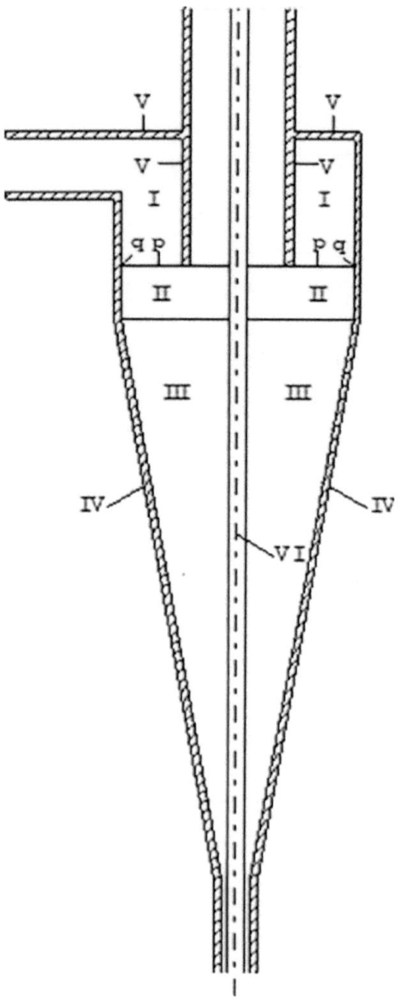

FIGURE 8.36 Zones in a conical hydrocyclone.

$$v_r = -Bz^{-2}r^{3/2}(1 - f(z)) \tag{8.111}$$

$$v_z = \frac{1}{2}Br^{-1/2}\left(3\alpha - \frac{5r}{z + l_{pc}}\right) \tag{8.112}$$

$$v_\phi = \frac{2k}{d_c}\frac{\left(\frac{4}{5}, \frac{2}{5}Lr^{*5/2}\right)}{r^*\gamma\left(\frac{4}{5}, \frac{2}{5}L\right)} \tag{8.113}$$

The tangential velocity v_ϕ expressed by Equation 8.113 gives an excellent approximation of the available experimental information of Hsieh (1988). Nevertheless, in order to make the calculations more rapidly, an empirical fitting to the prediction from the incomplete gamma function γ was made, by the combination of a free and a forced vortex with a common boundary at

$$r^* = 0.367 - 0.00111L \tag{8.114}$$

In the previous equations B, k, L, and r^* are parameters that depend on the operational conditions in the device and are expressed by (Concha and Barrientos 1991)

$$k = -\frac{c_f Q_a}{2A_a}, \quad L = \frac{\rho B(d_c/2)}{\mu}, \quad r^* = \frac{2r}{d_c} \tag{8.115}$$

$$B = \frac{\varepsilon Q_F}{2\pi\alpha\left[\left(\frac{d_c}{2}\right)^{5/2} - \left(\frac{d_r}{2}\right)^{5/2}\right]\left[\frac{2}{\alpha\ell_{pc} + d_d} - 1\right]} \tag{8.116}$$

where c_f in Equation 8.115 is a number that takes into account the fact that not all the flow enters the hydrocyclone tangentially, d_c is the hydrocyclone diameter, A_f is the cross section of the feed opening, Q_f is the feed pulp volume flow rate, z_0 is the coordinate specifying the position of the apex, l_{pc} is the length of the conical section of the hydrocyclone, d_o and d_U are the internal diameters of the vortex finder and apex tubes, respectively, and ε is a parameter.

The correction f_z for the radial velocity is based on the experimental observations of Hsieh (1988):

$$f(z) = \left(\frac{z - z_o - l_{pc}}{l - l_{pc}}\right)^4 \tag{8.117}$$

Where l is the distance from the apex to the bottom part of the vortex-finder tube.

8.2.7.2.3 *Zone III*

Zone III is the region in the conical section of the hydrocyclone of inviscid flow outside the boundary layer. We assume that in this region Bloor and Ingham's solution for the velocity profiles is valid (1973a, 1973b, 1975a, 1983, 1987). Point q in the boundary between Zones I and II denotes the border of the lateral wall boundary layer. The velocity distribution in this region III is the same given for Zone II, except that here $f(z) = 0$.

8.2.7.2.4 *Zone IV*

Zone IV is the boundary layer on the lateral wall of the hydrocyclone. The boundary layer starts at the feed and develops its maximum thickness when reaching the interface of Zones I and II. This thickness is relatively constant over the rest of the wall but diminishes when approaching the apex (Bloor and Ingham 1976). The maximum thickness of this boundary layer is given by

$$\delta_m = \frac{\delta_o}{(v_o x_o / \nu)^{1/2}}, \quad \delta_o = \beta x_0 \tag{8.118}$$

where x_0 is the distance along the lateral wall from the bottom of the vortex finder to the apex; v_0 is the exterior flow velocity, parallel to the lateral wall at the I/II interface just outside the boundary layer; ν is the fluid kinematical viscosity and $\delta_0 = \beta x_0$, where β is a number. Bloor and Ingham's model predicted $\beta = 1.88$, but we use this parameter for upscaling purposes. By integrating the axial velocity within the boundary layer, the total flow in the boundary layer may be calculated.

8.2.7.2.5 Zone V

Zone V corresponds to the boundary layer over the top wall of the hydrocyclone. Bloor and Ingham (1975b) calculated the flow in this layer that descends through the outer wall of the vortex finder tube and discharges directly into the overflow. They called this flow leakage and gave the following value for the volume flow within the boundary layer:

$$Q_k = \frac{\pi d_c \,|k|\, Q^*(d_v / d_c)}{\mathrm{Re}^{1/5}} \tag{8.119}$$

with the Reynolds number given by $\mathrm{Re} = \frac{(|k|/d_c)d_c}{U}$ and where Q_l is the leakage volume flow; $|k|$ is the parameter, with dimensions of [L^2/T], given in (5); d_c is the hydrocyclone diameter and the dimensionless flow rate is $Q^*(d_v^*) = 0.153 - 0.0897 d_v^* - 0.0621 d_v^{*2}$, where $d_v^* = d_v / d_c$ and d_v is the external diameter of the vortex finder tube.

8.2.7.2.6 Zone VI

Zone VI is the air core, which is assumed to have a cylindrical shape. The air core is modeled according to the work of Concha et al. (1994):

$$d_n = \frac{0.303\sigma d_r}{2703\mu\left(1 + \frac{0.678}{\Delta p^{2.61}}\right)\left(\frac{d_d}{d_r}\right)^{-0.20} - 0.138} \tag{8.120}$$

where d_n is the air core diameter; d_o and d_U are the vortex and apex diameters, respectively; Δp is the pressure drop; and μ and σ are the suspension viscosity and surface tension at the overflow concentration.

8.2.7.3 Simulation Model

Concha et al. (1996b) used the phenomenological model of the hydrocyclone, described in the previous section, describing the fluid velocity field within the hydrocyclone as the particle trajectories inside the classifier. From these equations it is possible to determine the short circuit, the classification function, and the size distribution of the overflow and underflow. The simulation model is based on the following assumptions:

- Zone I is fully mixed and contains a homogeneous suspension, such that all particles in this zone have the same opportunity to enter the boundary layer or pass to Zone II. Therefore, we selected the hydrocyclone feed at the interface between Zones II and III between points p and q (see Figure 8.36).
- Particles entering the lateral boundary layer will never leave it because there is no force capable of forcing them out; therefore, they are discharged from the hydrocyclone through the apex. From Zones II and III particles will enter the boundary layer by classification and join the particles coming from the feed. Therefore, we postulate that the flow in the lateral boundary layer transfers a certain number of particles directly from the feed (Zone I) to the underflow without classification. These particles are the short circuit to the underflow and must have the composition of the well-mixed Zone I (the feed). During the transit of the short-circuited stream, other particles join by classification, as we have already said, and are discharged together with the non-classified particles.
- Early works accepted the hypothesis that a well-designed and operated hydrocyclone suffers no leakage of particles to the overflow. The basis for this belief is that selectivity curves normally reach the value of 100 percent at a certain particle size, and therefore, particles of this size and larger do not leak to the overflow. We think that there is leakage in all cases, and we assume that the flow in the upper boundary layer constitutes a direct transfer of particles from the feed to the overflow. But not all particles entering the boundary layer leave through the overflow since some particles, especially coarser particles during their movement in the boundary layer from the feed to the overflow, have the chance to escape the boundary layer due to the effect of the drag of the fluid flow and the effect of gravity.
- Classification occurs in Zones II and III and is obtained by solving the momentum equations for individual particles and tracing their trajectories.
- Integrating the axial velocity component at the apex, between the boundary layer and the air core, and adding the boundary layer flow yields the underflow volume rate and the underflow concentration.
- The air core is the main cause of roping phenomena. For a fixed geometry, the air core diameter depends on the operating conditions and fluid properties. In some conditions the air core blocks most of the apex, leading to roping.

From experiences with empirical models, we know that five equations are needed to simulate the operation of a hydrocyclone: (1) the classification function, (2) the selectivity function, (3) the short circuit to the underflow, (4) the characteristic separation size, and (5) the capacity-pressure drop relationship. We will see in the following section that a sixth equation is needed to allow for the leakage of particles to the overflow.

8.2.7.3.1 Classification Function

The classification function $C(x)$ represents the fraction of particles of size x in the well-mixed zone of the hydrocyclone (Zone I) that reports to the underflow by classification. In our model this definition requires that the particles undergo classification following trajectories that passes through Zones II and III. By calculating the trajectories of enough particles, initially evenly distributed at the I/II interface, it is possible to estimate the classification function. To calculate it the momentum equations for the individual particles are solved, and their trajectories in Zones II and III must be traced. If the relative solid-fluid flow occurs in the Stokes regime, that is, if the particle Reynolds number $\mathrm{Re} = \rho\,|\boldsymbol{u}|\,x/\mu << 1$, where x is the particle diameter, ρ and μ are the suspension density and viscosity, respectively, and $\boldsymbol{u}$ is the solid-fluid relative velocity, Euler's equation reduces to

$$3\pi\mu x\boldsymbol{u} + \Delta\rho(\pi/6)x^3\left(\boldsymbol{g} + \left(v_\phi/r\right)\boldsymbol{e}_r\right) \tag{8.121}$$

where $\Delta\rho = \rho_s - \rho_f$ is the solid-suspension density difference, $\boldsymbol{g}$ is the acceleration of gravity vector, v_ϕ is the tangential velocity component, and $\boldsymbol{e}_r$ is the radial unit vector. Thus

$$\boldsymbol{u} = -\frac{\Delta\rho x^2}{18\mu}\left(\boldsymbol{g} - \frac{v_\phi^2}{r}\boldsymbol{e}_\mathrm{r}\right) \tag{8.122}$$

where the particle velocity components are

$$v_{pr} = v_r + \frac{\Delta\rho x^2}{18\mu}\frac{v_\phi^2}{r} \tag{8.123}$$

$$v_{p\phi} = v_\phi \tag{8.124}$$

$$v_{pz} = v_z + \frac{\Delta\rho x^2 g}{18\mu} \tag{8.125}$$

In these equations v_r, v_ϕ and v_z are given by Equations 8.122 to 125 and $\boldsymbol{g} = -g\boldsymbol{k}$, where $\boldsymbol{k}$ is the unit vector pointing in the upward direction. The particle trajectories are obtained by integration of Equation 8.122.

When the particle Reynolds number for the solid-fluid motion is greater than $\mathrm{Re}_p > 1$, it is necessary to use a more general equation for the hydrodynamic force. In this case we use the concept of drag coefficient and Equation 8.121 must be replaced by:

$$-3\rho\,|u|\,\mathrm{C_d}(\mathrm{Re})\boldsymbol{u} + 4\Delta\rho x\left(\boldsymbol{g} + \frac{v_\phi^2}{r}\boldsymbol{e}_r\right) \tag{8.126}$$

where $|\boldsymbol{u}|$ is the magnitude of the relative solid-fluid velocity and C_D is the drag coefficient, which can be calculated by Concha and Almendra's equation (1979) or by Concha and Christiansen's equation (1986) if the particles are not spherical. The relative solid-fluid velocity components are now

$$r\,\text{direction}\ \left|v_p - \boldsymbol{v}\right| C_d(\text{Re})\left(v_{pr} - v_r\right) = -\frac{4}{3}\frac{\Delta\rho v_\phi^2}{\rho r} \tag{8.127}$$

$$z\ \text{direction}\left|v_p - \boldsymbol{v}\right| C_d(\text{Re})\left(v_{pz} - v_z\right) = -\frac{4}{3}\frac{\Delta\rho g}{\rho r} \tag{8.128}$$

$$\phi\ \text{direction}\ \ v_{p\phi} = v_\phi \tag{8.129}$$

where

$$\left|v_\text{p} - \boldsymbol{v}\right| = \left\{\left(v_{pr} - v_r\right)^2 - \left(v_{pz} - v_z\right)^2\right\}^{1/2} \tag{8.130}$$

is the magnitude of the relative particle-fluid velocity. The particle trajectories are obtained from Equations 8.127 to 8.130.

The plot of the classification values versus size is the *classification function*, which can be parameterized through the *sharpness index* (*S.I.*) and the characteristic separation size x_{50}, defined by

$$S.I. = x_{75} / x_{25} \tag{8.131}$$

where x_{25}, x_{50}, and x_{75} are defined by $C(25) = 0.25$, $C(50) = 0.50$, and $C(75) = 0.75$. An ideal classification yields $S.I. = 1$.

8.2.7.3.2 *Short Circuit to the Under Flow*

By integrating the velocity field, at the I/II interface $r = R$, from the wall to the border of the boundary layer δ_m.

$$a = \frac{\left\{\left(\frac{d_c}{2}\right)^{5/2} - \left(\frac{d_c}{2} - \delta_m\right)^{5/2}\right\} - \left\{z_o + l_{pc}\right\}\left\{\left(\frac{d_c}{2}\right)^{3/2} - \left(\frac{d_c}{2} - \delta_m\right)^{3/2}\right\}}{\alpha\left\{\left(\frac{d_c}{2}\right)^{5/2} - \left(\frac{3\alpha\left(z_o + l_{pc}\right)}{5}\right)^{5/2}\right\} - \alpha\left\{z_o + l_{pc}\right\}\left\{\left(\frac{d_c}{2}\right)^{3/2} - \left(\frac{3\alpha\left(z_o + l_{pc}\right)}{5}\right)^{3/2}\right\}} \tag{8.132}$$

Based on experimental observations, Barrientos and Concha (1992) showed that, when a hydrocyclone is working properly, the air core has a diameter close to that of the apex, and in general, only a small amount of water passes from the classification zone (III) to the underflow. This implies that almost all the water appearing at the underflow comes from the boundary layer flow. According to this observation, our hypothesis on the origin of the short circuit produces exactly the observed result that the short circuit is proportional to the water split.

8.2.7.3.3 Leakage to the Overflow

Barrientos et al. (1995) obtained the leakage of particles to the overflow by calculating the trajectories of particles of different sizes within the boundary layer. The particles that crossed the boundary layer edge before reaching the vortex finder wall returned to the well-mixed Zone I without going to the overflow. All the rest constituted the leakage. It was shown that the fraction of particles that leak to the overflow can be expressed in the form

$$b\left(x,x^*\right)=b_l f\left(x,x^*\right) \tag{8.133}$$

Where $b_l = Q_l / Q_F$ is the fraction of the total volume flow represented by the leakage Q_l (given by Equation 8.132), Q_F is the feed volume flow rate of pulp into the hydrocyclone, and x^* is the greatest size that leaks to the overflow through the upper boundary layer. By definition $f(x)$ is given by (Barrientos et al. 1995)

$$f(x)=\frac{a_1}{1+\exp\left(a_2\left(x-a_3\right)\right)} \tag{8.134}$$

Where a_1 is the maximum value in the leakage curve, a_2 is the slope of the curve, and a_3 is the particle size for which the value of $f(x)=a_1/2$. These parameters are a function of the geometry of the hydrocyclone:

$$a_1 = 230.7\left(1-\exp\left(-7.420\left(d_v / d_c\right)\right)\right)-180.3 \tag{8.135}$$

TABLE 8.4
Geometrical and Operational Data

Geometrical parameters	ERAL 150 mm	Operating conditions	Parameter	150 mm ERAL
Hydrocyclone diameter	150	Feed volume flow	$QF \times 103$ [m3/s]	5.3
Feed diameter	55	Feed concentration	X_F [% solid]	13.9
Internal apex diameter	30	Underflow volume flow	$Q_U \times 103$ [m3/s]	1.62
Internal vortex diameter	45	Underflow concentration	φ_U [% solid]	0.558
External vortex diameter	96	Overflow volume flow	$Q_O \times 103$ [m3/s]	3.71
Length cylindrical part	190	Overflow	X_O: [% solid]	34.8
Length conical part	500	Pressure drop	Δp [psi]	10
Length vortex finder	190	Solid density	ρ_s [kg/m3]	2,650
		Fluid density	ρ_f [kg/m3]	1,000

$$a_2 = 0.476\exp\left(-4.251\left(d_v / d_c\right)\right) + 0.226(1 - \exp(-0.0114\alpha)) - 0.0251 \quad (8.136)$$

$$a_3 = 1.371\exp\left(8.090\left(d_v / d_c\right)\right) - 207.9\exp(-0.0283\alpha) + 284.7 \quad (8.137)$$

where d_v, d_c, and α are the external diameter of the vortex finder tube, the hydrocyclone diameter, and the semi-angle of the hydrocyclone body.

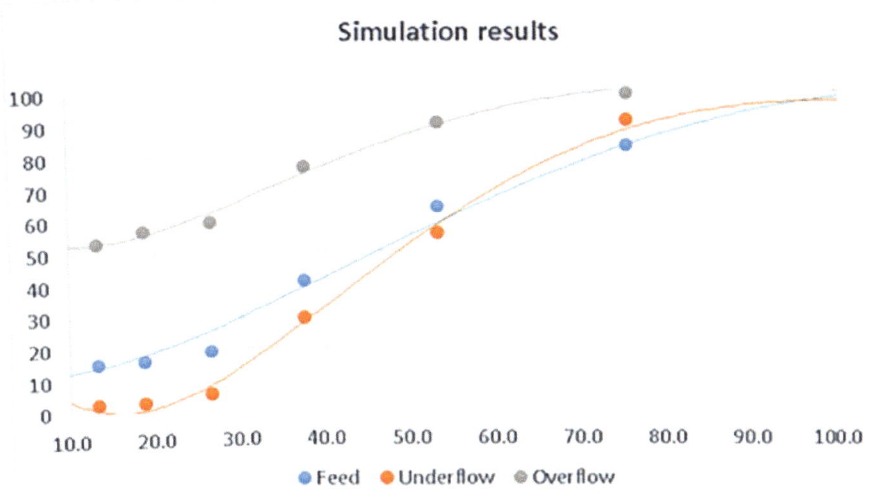

FIGURE 8.37 Particle size distribution results for a 150-mm-diameter hydrocyclone.

TABLE 8.5
Experimental Selectivity

ERAL 150 mm	
Particle size *x*	**S(*x*)**
μm	**%**
300.0	100.0
212.0	100.0
150.0	98.6
106.0	96.7
75.0	93.0
53.0	63.9
37.5	40.6
26.5	18.9
18.8	146.3
13.3	14.1
9.4	15.3

8.2.7.3.4 Example

The phenomenological model presented previously was applied to a 150 mm ERAL hydrocyclone treating a copper ore. Table 8.4 gives the geometrical and operational data.

Using the particle size analysis of a quartz suspension, the experimental selectivity function was calculated, and the short circuit established at $a = 15$ µm, as shown in Figure 8.37. Numerous particle trajectories were calculated, and the hydrocyclone parameters adjusted to the experimental selectivity values.

From the geometric and experimental data, as shown in Table 8.5, the selectivity function was determined shown in Figure 8.38.

The hydrocyclone parameters were adjusted to the experimental selectivity values.

With this information presented in Table 8.6 and the equations presented previously, numerous particle trajectories were calculated, the velocity distribution and the theoretical selective function was determined, as shown in Figures 8.39 and 8.40.

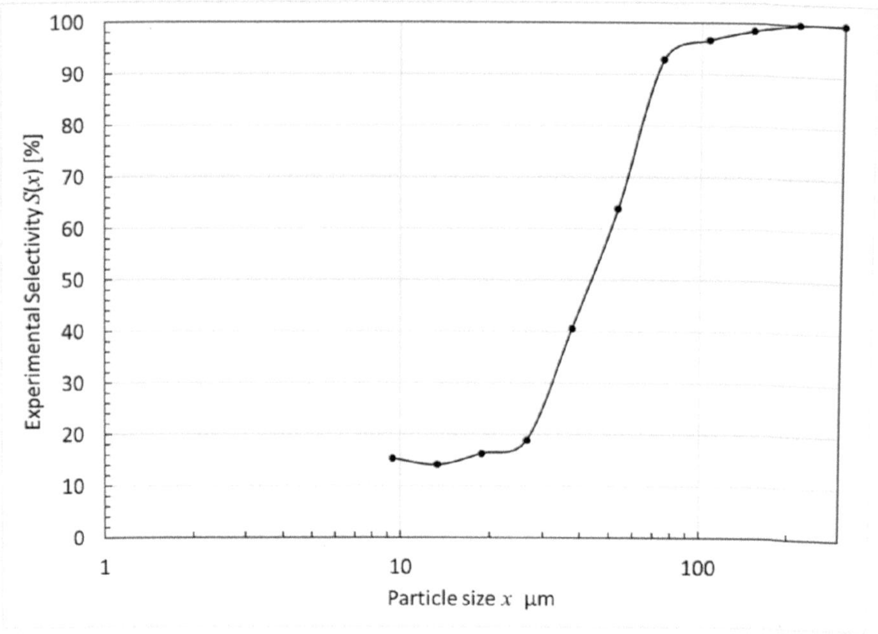

FIGURE 8.38 Experimental selectivity.

TABLE 8.6
Hydrocyclone Parameters

a	b	x_{50}	CF	ε	β
14.33%	11.67%	43.37	0.8	0.7124	1.7227

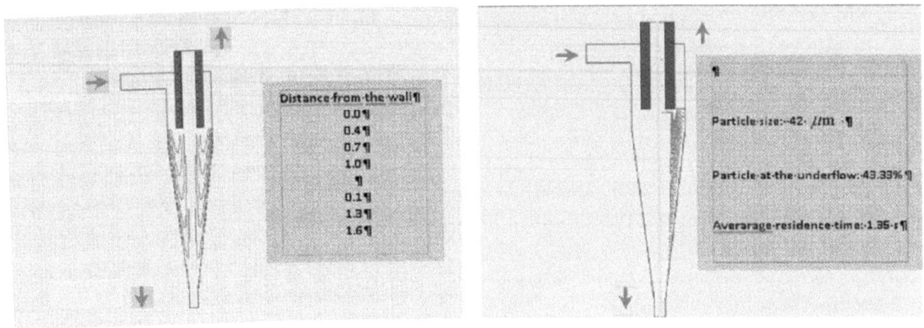

FIGURE 8.39 Simulation of the streamlines (a) and particle trajectories (b) in the hydrocyclone.

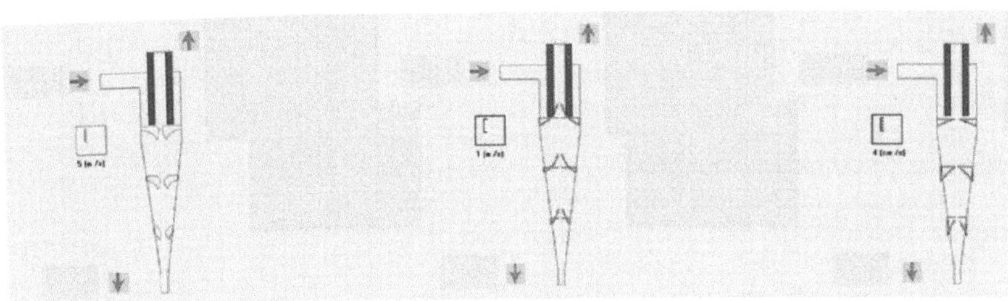

Tangential velocity   Axial velocity   Radial velocity

FIGURE 8.40 Simulation of the velocity patterns in a hydrocyclone.

REFERENCES

Barrientos, A., and Concha F. 1992. Chapter 21: Phenomenological model of classification in conventional hydrocyclones. In *Comminution: Theory and Practice*. Edited by K. Kawatra. New York: AIME. pp. 288–305.

Bascur, O.A. 1991. A unified solid liquid separation framework. *Fluid Particle Separation Journal* 4:117–122.

Bascur, O.A., and Barrientos, A. 1977. Phenomenological model of sedimentation: Experimental parameters. Presented at the IV Chilean Chemical Engineering Congress, University de Concepcion.

Bascur, O.A., Retamal, F., and Espinosa, S. 2022. Treating low grades ores using a digital in mining and mineral processing plants. In *Proceedings of the Copper 2022 International Conference*, IIMChile, Santiago.

Binnie, A.M., Hookings, G.A., and Kamel, M.Y.M. 1948. The flow of swirling water through a convergent-divergent nozzle. *Proceedings of the Royal Society A* 194:398–415.

Bloor, M.I.G., and Ingham, D.B. 1973a. On the efficiency of the industrial cyclone. *Transactions of the Institution of Chemical Engineers* 51:173.

Bloor, M.I.G., and Ingham, D.B. 1973b. Theoretical investigation of the flow in a conical hydrocyclone. *Transactions of the Institution of Chemical Engineers* 51(1):36–41.

Bloor, M.I.G., and Ingham, D.B. 1975a. The leakage effect in the industrial cyclone. *Transactions of the Institution of Chemical Engineers* 53:7–11.

Bloor, M.I.G., and Ingham, D.B. 1975b. Turbulent spin in a cyclone. *Transactions of the Institution of Chemical Engineers* 53:1–6.

Bloor, M.I.G., and Ingham, D.B. 1976. Boundary layer flow on the side walls of conical hydrocyclones. *Transactions of the Institution of Chemical Engineers* 54:276.

Bloor, M.I.G., and Ingham, D.B. 1983. Theoretical aspects of hydrocyclone flow. In *Progress in Filtration and Separation* 3. Edited by R.J. Wakeman. New York: Elsevier. pp. 57–148.

Bloor, M.I.G., and Ingham, D.B. 1984a. The influence of vorticity on the efficiency of the hydrocyclone, *Proceedings of the 2nd International Conference on Hydrocyclones*, BATH, Cranfield, Paper B2. pp. 51–66.

Bloor, M.I.G., and Ingham, D.B. 1984b. A theoretical investigation of the fluid mechanics of the hydrocyclone. *Filtration*: 266–269.

Bloor, M.I.G., and Ingham, D.B. 1987. The flow in industrial cyclones. *Journal of Fluid Mechanics* 178:507–519.

Brayshaw, M.D. 1978a. The effect of changes in the vorticity function of the flow field of a hydrocyclone on the sharpness of particle separation. Technical Report. Vanderbijlparck, South Africa: Department of Metallurgical Engineering, University of Potchestroom.

Brayshaw, M.D. 1978b. A numerical model for inviscid flow of fluid in a hydrocyclone. Technical Report. Vanderbijlparck, South Africa: Department of Metallurgical Engineering, University of Potchestroom.

Brayshaw, M.D. 1990. A numerical model for inviscid flow of fluid in a hydrocyclone. *International Journal of Mineral Processing* 29:51–75.

Bustamante, M.O. 1991. Effect of the hydrocyclone geometry on normal operation condition [in Spanish]. MSc thesis, University of Concepción.

Castro, O. 1990. Pulp rheology effects for hydrocyclone models. MSc thesis, University of Queensland.

Castro, O., Concha, F., Montero, J., Miranda, J., Castro, J., and Urizar, D. 1996. Air core modelling for an industrial hydrocyclone. In *Hydrocyclones 96*. Edited by D. Claxton, L. Svarovsky, and M.T. Thew. London: MEP.

Chiné, B., and Concha, F. 2000. Flow patterns in conical and flat bottom hydrocyclones. *Chemical Engineering Journal* 80:267–273.

Collantes, J., Concha, F., and Chiné, B. 2000. Axial symmetric model for a flat bottom hydrocyclone. *Chemical Engineering Journal* 80:257–265.

Concha F. 2007. Flow pattern in hydrocyclones. *KONA Powder and Particle* 25:97–132.

Concha, F., and Almendra, E.R. 1979. Settling velocities of particulate systems. 1. Settling velocities of individual spherical particles. *International Journal of Mineral Processing* 5:349–367.

Concha, F., and Barrientos, A. 1991. Modelación Fenomenológica de Hidrociclones. Monografía, Proyecto PNUD/CHI/88/011, Agosto.

Concha, F., Barrientos A., Montero J., and Sampaio, R. 1994. Air core and roping in hydrocyclones. In *8th European symposium of comminution*, Stockholm, Sweden, vol. II. pp. 814–823.

Concha, F., Barrientos, A., Montero, J., and Sampaio, R. 1996a. Air core and roping in hydrocyclones. *International Journal of Mineral Processing* 44–45:743–749.

Concha, F., Barrientos, A., Muñoz, L., Bustamante, O., and Castro, O. 1996b. A phenomenological model of a hydrocyclone. In *Hydrocyclones 96*. Edited by D. Claxton, L. Svarovsky, and M.T. Thew. London: MEP. pp. 63–82.

Concha, F., and Christiansen, A. 1986. Settling velocities of particulate systems. 5. Settling velocities of suspensions of particles of arbitrary shape, *International Journal of Mineral Processing* 5:309–322.

Davidson, L., Kittel, S., and Curtis, A. 1987. The improvement of cyclone performance with a special spigot. In *Proceedings of the 3rd International Conference of Hydrocyclones, Oxford, BHRA*. Paper K1. Edited by P. Wood. Amsterdam: Elsevier. pp. 259–62.

Davidson, M.R. 1988. Similarity solutions for the flow in hydrocyclones. *Chemical Engineering Science* 43(7):1499–1505.

Davidson, M.R. 1994. An adaptive method of predicting the air core diameter for numerical models of hydrocyclone flow. *International Journal of Mineral Processing* 1021.

Davidson, M.R. 1995. An adaptive method of predicting the air core diameter for numerical models of hydrocyclone flow. *International Journal of Mineral Processing* 43(3–4):167–177.

Daza, J., Cornejo, P., Rodriguez, C., Betancourt, F., and Concha, F. 2018. Transition from spray to roping in hydrocyclones: Influence of the feed particle size distribution. *Minerals Engineering* 123:71–78.

Donaldson, C., and Sullivan, R.D. 1960. Behavior of solutions of the Navier-Stokes equations for a complete class of three-dimensional viscous vortices. *Proceedings of the Heat Transfer and Fluid Mechanics Institute* 16–30.

Dueck, J., Farghaly, M., and Neesse, T. 2013. The theoretical partition curve of the hydrocyclone. *Mineral Engineering*. http://doi.org/10.1016/j.mineng.2013.10.004

Dueck, J.G.T., Minkov, L.L., and Pikushchak, E.V. 2007. Modeling of the "fish-hook." *Journal of Engineering Physics and Thermophysics* 80:64–73.

Finch, J.A. 1983. Modelling a fishhook in hydrocyclone selectivity curves. *Powder Technology* 36:127–129. Fontein, F.J., Dijksman, I.C. 1952. The hydrocyclone, its application and its explanation. In *Proceedings., 1st International Mineral Processing Congress, Recent Developments in Mineral Dressing*. London: Institution of Mining and Metallurgy. pp. 229–246.

Gerhard, C. 2001. Untrersuchungen zum trennverhalten von hydrozyclones niedriger Trennkorngrösse. Theses, Universität Erlangen-Nürenberg.

Hoffmann, A.C., and Stein, L.E. 2008. *Gas Cyclones and Swirl Turbines*. Berlin: Springer-Verlag. https://doi.org/10.1007/978-3-540-74696-6

Hsieh, K.T. 1988. Phenomenological model of hydrocyclone. PhD thesis, University of Utah, Salt Lake City, UT.

Hsieh, K.T., and Rajamani, K. 1988a. Phenomenological model of hydrocyclone: Model development and verification for single phase flow. *International Journal of Mineral Processing* 22:223–310.

Hsieh, K.T., and Rajamani, K. 1988b. Phenomenological model of hydrocyclone: Numerical solution and verification of particle movements. In *Proceedings of the XVI International Mineral Processing Congress*. Edited by E. Forssberg. Amsterdam: Elsevier. pp. 377–388.

Hsieh, K.T., and Rajamani, R.K. 1991. Mathematical model of the hydrocyclone based on physics of fluid flow. *AIChE Journal* 37:735.

Hwang, C.C., Shen, H.Q., Zhu, G., and Khonsari, M.M. 1993. On the main flow pattern in hydrocyclones. *Journal of Fluids Engineering* 155:21–25.

Kelsall, D.F. 1952. A study of the motion of solid particles in a hydraulic cyclone. *Transactions of the Institution of Chemical Engineers* 30:87.

Kelsall, D.F. 1963. Some applications of hydraulic cyclones in hydrometallurgical processes. *Transactions of the AIME* 226:225–231.

Knowles, S.R., Woods, D.R., and Fuerstein, I.A. 1973. The velocity distribution within a hydrocyclone operating without an air core. *Canadian Journal of Chemical Engineering* 51:263.

Kraipech, W., Chen, W., Parma, F.J., and Dyakowski, T. 2002. Modelling the fish-hook effect of the flow within hydrocyclones. *International Journal of Mineral Processing* 66: 49–65.

Kumari, K., and Nath, G. 1985. Unsteady swirling boundary layer flow with heat and mass transfer in nozzles, diffusers and hydrocyclones. *International Journal of Heat and Mass Transfer* 28(7):1377–1383.

Lopez, J., and Cepeda, E. 2015. Double Effect CAVEX: The end solution in sand classification and water saving for tailings dam, Tailing 2015. In *3rd International Seminar on Tailings Management*, Gecamin, Santiago.

Mikaylov, P.M., and Romensky, A.A. 1974. On the calculation of the flow dynamics in liquid cyclones. *Fluid Mechanics, Soviet Research* 3(1):154–159.

Neesse, T., Schneider, M., Dueck, J., Golyk, V., Buntenbach, S., and Tiefel, H. 2004. Hydrocyclone operation at the transition point rope/spray discharge. *Minerals Engineering* 17:733–737.

Neesse, T., Schubert, H., and Graichen, K. 1991. Practical and theoretical aspects of dense-flow separation. *Aufbereitungstechnik* 32(9):459–472.

Nepomnyashchii, E.A., and Pavlobskii, V.V. 1977. Hydrodynamic design of a hydrocyclone. *Theoretical Foundations of Chemical Engineering* (English Translation) 11(1):83–88.

Ovalle, E., and Concha, F. 2005. Role of wave propagation in hydrocyclone operations II. Wave propagation in the air-water interface of a conical hydrocyclone. *Chemical Engineering Journal* 111:213–223.

Pérez, D., Cornejo, P., Rodríguez, C., and Concha, F. 2018. Transition from spray to roping. *Minerals Engineering* 123:71–84.

Petersen, K.R.P. 1993. Vortex chamber modelling for hydrocyclone simulation. MSc thesis, JKMRC, University of Queensland.

Quian, L., Changlie, D., Jirun, X., Lixin, Y., and Guangai, X. 1989. Comparison of the performance of water-sealed and commercial hydrocyclone. *International Journal of Mineral Processing* 25:297–310.

Rietema, K. 1961a. Performance and design of hydrocyclones. Part 1–4. *Chemical Engineering Science* 15:298–325.

Rietema, K. 1961b. The mechanism of separation of finely dispersed solids in cyclones. In *Cyclones in Industry*. Edited by K. Rietema dn C.G. Verver. Amsterdam: Elsevier. pp. 46–63.

Rietema, K. 1962. *Liquid-Solids Separation in a Cyclone: The Effect of Turbulence on Separation. Symposium on the Interaction between Fluids and Particles*. London: Institution of Chemical Engineers. pp. 275–281.

Rietema, K., and Krajenbrink, H.J. 1958. Theoretical derivation of tangential velocity profiles in a flat vortex chamber. Influence of turbulence and wall friction. *Applied Scientific Research, Section A* 8:177–197.

Rouse, B.D., Clayton, J.S., and Brookes, G.F. 1987. Confirmation of modeling techniques for small diameter cyclones. In *Proceedings of the 3rd International Conference of Hydrocyclones, Oxford, BHRA*. Edited by P. Wood. Amsterdam: Elsevier. pp. 7–20.

Sepulveda, J. 1984. *Moly Cop Tools*. Santiago: J-Consultants.

Sepulveda, J. 2022. *Moly Cop Tools*. Santiago: J-Consultants.

Smith, J.L. Jr. 1962. An analysis of the vortex flow in the cyclone separator. *Transactions of the American Society of Mechanical Engineers* (December): 609–618.

Starrett, J.B., and Galvin, K.P. 2023. Application of inclined channels in the hydrodynamic classification of minerals by particle size. *Minerals Engineering* 195(4):108002.

Steffens, P.R., Whiten, W.J., Appleby, S., and Hitchins, J. 1993. Prediction of air core diameters for hydrocyclones. *International Journal of Mineral Processing* 39:61–74.

Svarovsky, L. 1984. *Hydrocyclones*. London: Holt, Rinehart and Winston.

Svarovsky, L., and Marasinghe, B.S. 1980. Performance of hydrocyclones at high solid concentrations. In *International Conference Hydrocyclones*. Paper 10. Cambridge: BHRA. pp. 127–142.

FURTHER READING

Aguilera, G.F., and Arismendy, J.E. 2014. Development of a hydrodynamic model of a hydrocyclone including the simulation of air-core effect, using CFD. *Holos* 30(3):184–192.

Antunes, M., and Medronho, R.A. 1988. *Modelos que prevêm o desempenho de hidrociclones: uma avaliação crítica, Relatorio de pesquisa*. Río de Janeiro: UFRJ.

Bank, N., and Gauvin, W.H. 1977. Measurement of flow characteristics in a confined vortex flow. *Canadian Journal of Chemical Engineering* 55:397–402.

Barrientos, A., Etcheverry, J., and Concha F. 1995. The leakage of particles through the top boundary layer and its effect on the classification in a conventional hydrocyclone. *International Journal of Mineral Processing*, submitted.

Barrientos, A., Sampaio, R., and Concha, F. 1993. Effect of the air core on the performance of hydrocyclone. In *XVIII international mineral processing congress*, Sydney, pp. 23–28, 267–270.

Bhaskar, K.U., Murthy, Y.R., Raju, M.R., Tivari, S., Srivastava, J.K., and Ramakrishnan, N. 2007. CFD simulation and experimental validation studies on hydrocyclone. *Minerals Engineering* 20:60–71.

Bhattacharyya, P. 1980. Theoretical study of the flow field inside a hydrocyclone with vortex finder diameter greater than the apex opening; I. Laminar case. *Applied Scientific Research* 36:213–225.

Bond, J., Cullivan, J. C., Climpson, N., Dyakowski, T., Faulks, I., Jia, X., Kostuch, J. A., Payton, D., Wang, M., Wang, S. J., West, R. M., and Williams, R. A. 1999. Industrial monitoring of hydrocyclone operation using electrical resistance tomography. In *1st World Congress on Industrial Process Tomography* (pp. 102–107). International Society for Industrial Process Tomography.

Bradley, D., and Pulling, D.J. 1959. Flow patterns in the hydraulic cyclone and their interpretation in terms of performance. *Transactions of the Institution of Chemical Engineers* 51:263.

Brennan, M. 2006. CFD simulations of hydrocyclones with an air core. *Chemical Engineering Research and Design* 84:495–505. https://doi.org/10.1205/cherd.05111.

Chiné, B., Concha, F., and Barrientos, A. 1995. A finite difference solution of the swirling flow in the hydrocyclone. In *Proceedings of the IX International Conference on Finite Elements in Fluids, Part II*, Venezia, Italia, October 15–20, 1995. pp. 581–592.

Collantes, J., and Concha, F. 1998. Asymptotic solutions of the Navier-Stokes equation for the swirling flow in a flat bottom hydrocyclone. XXVI Congresso Brasileiro de Sistemas Particulados, Teresópolis. pp. 19–22.

Collantes, J., and Concha, F. 2000. Vortex flow in flat-bottomed hydrocyclones: Leading order steady solution. In *Proceedings of the XXI International Mineral Processing Congress: Development in Mineral Processing 13*. Edited by P. Massacci. Amsterdam: Elsevier. pp. A4–77, A4–86.

Concha, F., Barrientos, A., Montero, J., and Sampaio, R. 1996. Air core and roping in hydrocyclones. *Int J Miner Process* 44–45:743–749.

Concha, F., Castro, B., Ovalle, E., and Romero, J. 1998. Numerical simulation of the flow pattern in a hydrocyclone. In *Innovation in Physical Separation Technology. Richard Mozley Symposium, Keynote Conference*, Falmouth, England, June 4–5. pp. 35–60.

Concha, F., and Collantes, J. 1999. Solution to the Navier-Stokes equation for the flow in a flat-bottom hydrocyclone with the axial velocity dependent on the radial and axial coordinate. Presented at the Conference on Solid–Liquid Separation, Hawaii.

Cui, B., Wei, W., Gao, S., Liu, W., and Feng, T. 2014. Numerical and experimental studies of flow field in hydrocyclone with air core. *Transactions of Nonferrous Metals Society of China* 24:2642–2649.

Cullivan, J.C., Williams, R.A., and Cross, C.R. 1998. Understanding the hydrocyclone separator through computational fluid dynamics. *Transactions of the IChemE, Part A, Chemical Engineering, Research and Design* 81(A4):455–466.

Cullivan, J.C., Williams, R.A., Dyakowski, T., and Cross, C.R. 2004. New understanding of a hydrocyclone flow field and separation mechanism from computational fluid dynamics. *Mineral Engineering* 17(5):651–660.

Dai, G.Q., Li, J.M., and Chen, W.M. 1999. Numerical prediction of liquid flow within a hydrocyclone. *Chemical Engineering Journal* 74:211–216.

Davailles, A., Climent, E., and Bourgeois, F. 2012. Fundamental understanding of swirling flow pattern in hydrocyclones. *Separation and Purification Technology* 92:152–160.

Delgadillo, J.A., and Rajamani, R.K. 2005. A comparative study of three turbulence-closure models for the hydrocyclone problem. *International Journal of Mineral Processing* 77:217–230.

Delgadillo, J.A., and Rajamani, R.K. 2007. Exploration of hydrocyclone designs using computational fluid dynamics. *International Journal of Mineral Processing* 84:252–261.

Delgadillo, J.A., and Rajamani, R.K. 2009. Computational fluid dynamics prediction of the air-core in hydrocyclones. *International Journal of Computational Fluid Dynamics* 23:189–197.

Doby, M.J., Nowakowski, A.F., and Dyakowski, T. 2008. Understanding air core formation in hydrocyclones by studying pressure distribution as a function of viscosity. *International Journal of Mineral Processing* 86:18–25.

Dyakowski, T., and Williams, R.A. 1993. Modeling turbulent flow within a small-diameter hydrocyclone. *Chemical Engineering Science* 6:1143–1152.

Dyakowski, T., and Williams, R.A. 1995. Prediction of air-core size and shape in a hydrocyclone. *International Journal of Mineral Processing* 43(1–2):1–14.

Finch, J.A. 1993. Modelling a fishhook in hydrocyclone selectivity curves. *Powder Technology* 36:127–129.

Fisher, M.J., and Flack, R.D. 2002. Velocity distributions in a hydrocyclone separator. *Experiments in Fluids* 32:302–312.

Narasimha, M., Spripriya, R., and Banerjee, P.K. 2005. CFD modelling of hydrocyclone—prediction of cut size. *International Journal of Mineral Processing* 75(1–2):53–68.

Nepomnyashchii, E.A., and Pavlobskii, V.V. 1980. Calculating the velocity field in a hydrocyclone on the basis of a laminar analog of the averaged turbulent flow. *Theoretical Foundations of Chemical Engineering* (English Translation) 13(5):671–674.

Nepomnyashchii, E.A., and Pavlobskii, V.V. 1986. Hydrodynamic calculation of a pressurized hydrocyclone. *Theoretical Foundations of Chemical Engineering* (English Translation) 20(2):151–156.

Nowakowski, A.F., Cullivan, J.C., Williams, R.A., and Dyakowski, T. 2005. Application of CFD to modeling the flow in hydrocyclones. Is this a realizable option or still a research challenge? *Minerals Engineering* 17(5):661–669.

Nowakowski, A.F., Kraipech, W., Williams, R.A., and Dyakowski, T. 2000. The hydrodynamics of a hydrocyclone based on a three-dimensional multi-continuum model. *Chemical Engineering Journal* 80(1–3):275–282.

Rajamani, K., and Hsieh, K.T. 1988. *Hydrocyclone Model: A Fluid Mechanical Approach SME Preprint*. Littleton, CO: SME. pp. 88–163.

Rajamani, R.K., and Devulapalli, B. 1994. Hydrodynamic modeling of swirling flow and particle classification in large-scale hydrocyclones. *Kona* 12:95–104.

Rietema, K., and Krajenbrink, H.J. 1959. Theoretical derivation of tangential velocity profiles in a flat vortex chamber. Influence of turbulence and wall friction. *Applied Scientific Research, Section A* 8:177–197.

Schubert, H. 1985. A hydrocyclone separation model in consideration of the turbulent multiphase flow. *Particulate Science and Technology* 3:1–13.

Schubert, H., and Nesse, T. 1973. The role of turbulence in wet classification. In *Proceedings of the Institution of Mining and Metallurgy 10th International Mineral Processing Congress*. Cannes Paper 29. Edited by M.J. Jones. London: IMM. pp. 5–27.

Schubert, H., and Nesse, T. 1980. A hydrocyclone separation model in consideration of the turbulent multi-phase flow. In *Proceedings of the International Conference on Hydrocyclones*. BHRA, Fluid Engineering. pp. 23–36.

Van Duijin, G., and Rietema, K. 1983. Performance of a large-cone-angle hydrocyclone, I. Hydrodynamics. *Chemical Engineering Science* 38(10):1651–1661.

Zhu, G., and Liow, J. 2014. Experimental study of particle separation and the fish-hook effect in a mini-hydrocyclone. *Chemical Engineering Science* 111:94–105.

9 Particle Flotation

9.1 INTRODUCTION

The objective of concentration is to separate from the ore the barren rock. To perform this task, the process makes use of differences in physical or physicochemical properties between the several minerals and rocks contained in the ore. In general, these operations do not change the chemical identity of mineralogical species. In the case of flotation this property is the selective hydrophobicity of some of the mineral component of the ore. Flotation is the process of separation of finely ground minerals based on the surface properties of the minerals when they come in contact with water and air bubbles, separating the hydrophobic minerals attached to the air bubbles from hydrophilic minerals that stay in the water-solid phase. Adding chemicals reagents permit the formation of a stable foam above the pulp that keeps the particles on the surface. These flotation reagents are collectors, depressants, activators, and modifiers, whose main actions are to induce *hydrophobicity* of valuable particles constituting the concentrate and *hydrophilicity* to the tailings. Taken into consideration that the flotation process depends on the *surface chemistry* to prepare the three-phase system and on the *hydrodynamic* of the flow to produce the attachment of particles to bubbles, the science involved in this field has received the name of *Physicochemical Hydrodynamics*. See Chapters 3 and 5 in this book.

9.1.1 VARIABLES IN FLOTATION

In order to improve the flotation process in a mineral concentration plant, the manipulated variables that influence the grade and recovery of the concentrate needs to be identified and solved. Mechanisms of interface transfer are represented by kinetic equations to account for attachment/detachment and entertainment/drainage between four states (phases). In each phase the influence of important operational variables must be included in the model equations. Manipulated variables such as aeration rate, frother addition, agitation degree, pulp level, feed flowrate, percent solids in the feed, grade in the feed, tailings flowrate, and froth level should be considered.

One of the most important factors that determine the water distribution in the flotation equipment is the froth. The description of the hydraulic transport through the froth entails the assumption that bubbles leaving the pulp phase carry a sheath of water, which moves into the froth phase. The number, size of the bubbles, and the thickness of the water sheath are determined by the frother concentration and the agitation conditions in the pulp volume.

Flotation is machine and variables dependent. The manipulated variables of any flotation process can be divided into two groups: the physical variable impeller speed, aeration rate, pulp level and froth and removal rate, and the operating variable including collector, frother, depressant, and modifier dosage and conditioning

DOI: 10.1201/9781032614212-9

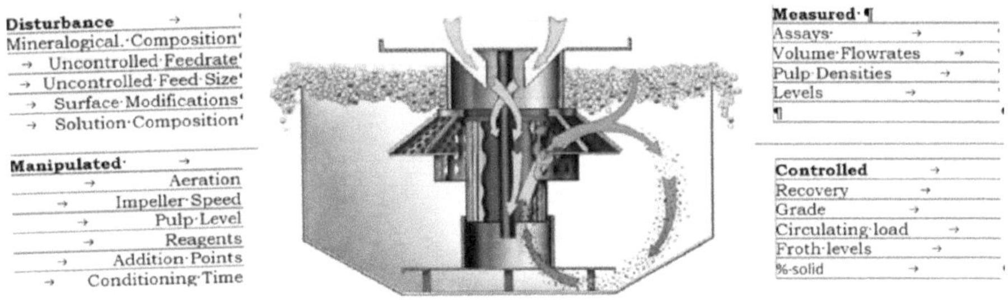

FIGURE 9.1 Flotation cell process variables for optimization.

Sources: Herbst, J.A., and Bascur, O.A. 1985. Alternative Control Strategies for Coal Flotation, Minerals and Metallurgical Processing, February, SME, Englewood, Co. pp 1–9; Smidth, F.L. 2010. *Flotation Technology.* Salt Lake City, UT: FLSmidth Minerals Processing Technology Center.

times. A complete list of manipulated and controlled variables per flotation is given in Figure 9.1.

9.1.2 Fundamental Steps in the Flotation Process

There are several steps in flotation: (a) the conditioning of the solid particles, the water, and the air to permit the formation of a three-phase contact; (b) the formation of air bubbles and foam by collision of the particles to the air bubbles; (c) attachment—the attachment, and detachment, of the mineral particles to the bubbles with ore; (d) the transport of the mineralized bubbles to and from the froth. Each of these steps are based on surface chemistry and hydrodynamics. Additional reviews are available by Jovanovic and Miljanović (2015).

9.1.2.1 Conditioning

The first step in flotation is to add the proper chemical reagent to the mixture of finely ground mineral with water contained in a stirred reactor. The fine particles have surface properties that depend on its crystalline nature, type of fracture they have been subjected, and have surface electrostatic charge that turn them naturally attracted to air (hydrophobic) or water (hydrophilic) or forced them with chemical reagent called collectors which are adsorbed at the mineral surfaces.

The most common collectors are ionic, cationic, or anionic, water-soluble heteropolar molecules with a polar group that is adsorbed (physically or chemically) on the surface of a mineral and a long non-polar part oriented towards the water. Figure 9.2 shows a particle in a liquid cover by a collector for the hydrophobicity enhancement to be able to capture the specie by an air bubble.

The air contained in the reactor has an interface with the water and tries to keep it as small as possible, achieving it by forming spherical bubbles (and froth layer) whose internal pressure p_B is greater than its static pressure p_h maintaining a balance of pressures between the inside and outside of the bubble. Responsible to these balances are the capillary forces originated by the interfacial tension γ, that is, the

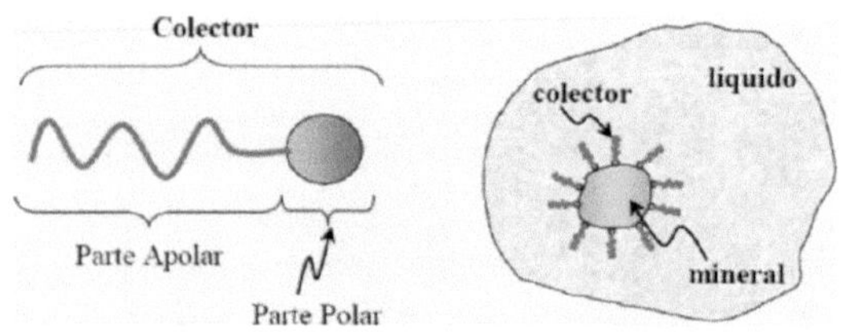

FIGURE 9.2 Mineral particle covered by the collector reagent.

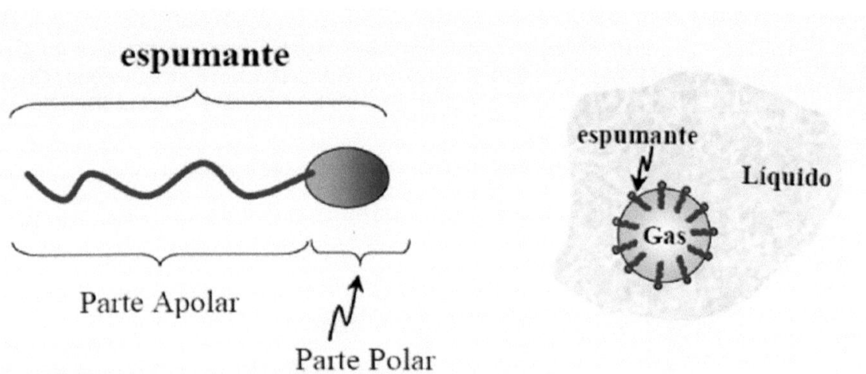

FIGURE 9.3 Gas bubble covered by the frother.

reversible work that must be done to increase the size of the interface. Figure 9.3 shows a bubble covered by a frother to enhance the creation of bubbles to capture the hydrophobic species. If R is the radius of the bubble. See Chapter 7 Section 7.1.

$$p_B - p_h = \frac{2\lambda}{R} \tag{9.1}$$

9.1.2.2 Collision

The attachment of the hydrophobic particles to the air bubbles requires bringing them together and adsorbed the collectors covering the particles to the bubble. This occurs by collision due to the strong stirring (Turbulence = Power/Pulp Mass) of the pulp in the reactor. Several factors are involved in the collision: size of the particles and bubbles, hydrodynamic regime, relative velocity between them, quality of the turbulence. Figure 9.4 shows the Bubble of Diameter dB making contact with a specie of size dP (see Chapter 7 Sections 7.1 and 7.2., Ruylov 2020 Chapter 4).

9.1.2.3 Attachment

After collision, the attachment of the particles-bubbles attachment occurs when the water film maintained by the disjoining pressure between particle and bubble

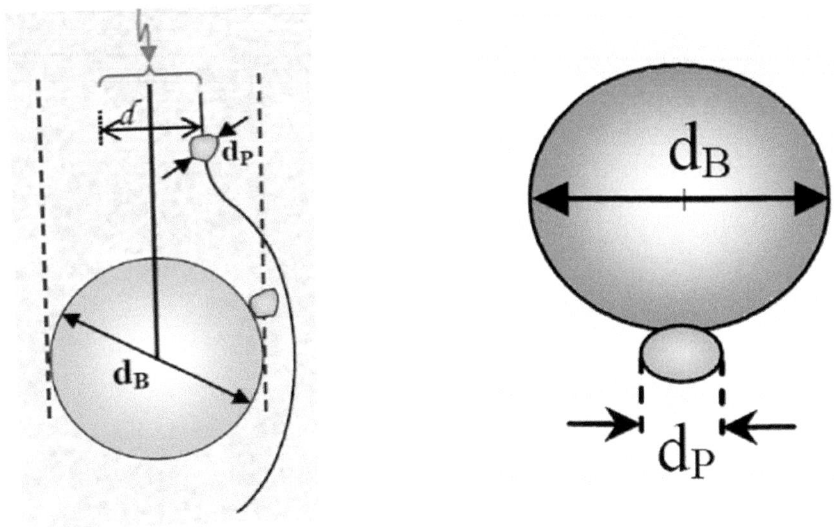

FIGURE 9.4 Particle bubble collision.

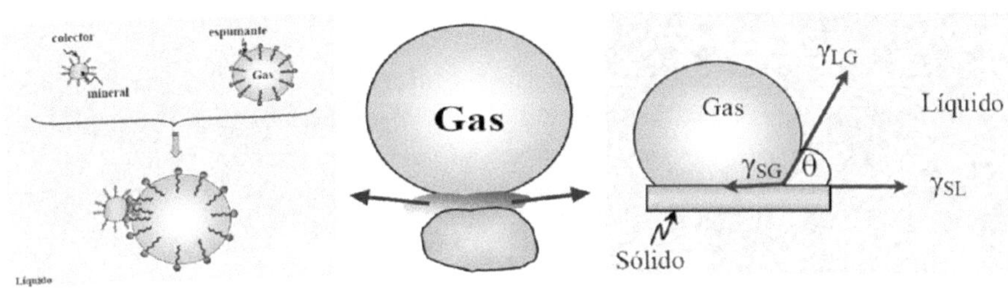

FIGURE 9.5 Mineral particle-bubble attachment: (a) attachment, (b) interfacial water film, (c) three-phase contact.

surfaces is squeezed out and the collectors and frothers tails can be adsorbed at the surface of the particles. Figure 9.5 shows a schematic of the attachment process. The adhesion can take place through chemical or electrostatic attraction (see Chapter 7, Section 7.1).

9.1.2.4 Pulp-Froth Particle Transfer

The mechanisms of mass transport of particle from the pulp into the froth are the rising of the particle-charged air bubbles and slime covering to other coarser particles. Some particles can drain back from the froth by shear forces exerted by the water flowing around the bubbles. Drainage of water returning to the pulp is modeled by analogy with the drainage through plateau borders in foams. The height of the froth is determined by the froth removal mechanism in the cell. For natural overflow, weir flow results are used. A detailed description of the flotation model can be found in

Bascur (1982), Bascur and Herbst (1982), Herbst and Bascur (1985), and an updated version in Bascur (2013).

9.2 FLOTATION DYNAMICS

Successful flotation is not only dependent on the selective interaction between solid particles and gas bubbles in an aqueous media related to the *flotation kinetics* but also in the behavior of the flotation cell as a *process reactor*.

The simplest, and most elementary form, to describe the kinetics of flotation as the rate of change of desirable particles in the flotation cell:

$$\frac{dc(t)}{dt} = -kc^n(t)dt \tag{9.2}$$

Where $c(t)$ is the concentration of desirable particles remaining in the flotation cell at time t, k is the flotation rate constant and n expresses the kinetic order. For an initial concentration of $c(0) = c_0$ for the non-floatable particles and a first order kinetics $n = 1$, integration of equation 9.2 yields

$$c(t) = c_0 \exp(-kt) \tag{9.3}$$

The behavior of the flotation cell as a process reactor is described by quality of mixing existing in the cell, which can be determined from the average of the residence time function $E(t)$. Using the simplest model of *continuous perfectly mixed reactors* (CPMR) added to the first-order flotation kinetics results in

$$c = \int_0^\infty \exp(-kt)E(t)dt \tag{9.4}$$

The integration of Equation 9.4 for different number of CPMR yields

$$c_n = c_0 \frac{1}{(1+k\tau)^n} \tag{9.5}$$

Where c_n is the outlet concentration from the nth CPMR.

9.2.1 Flotation Kinetics and Distribution of Species

Metallic and nonmetallic minerals contain several species, each with a different particle size distribution and different kinetic constants. They can be described in general with the function

$$k(x,s) = \left(f(x)\psi(s|,x)\right) \tag{9.6}$$

Where $f(x)$ is the size distribution and $\psi(s|,x)$ is the conditional size distribution of each specie.

Concha and Delgado (1977) used mineral coal as a flotation material considering that the coal has a continuous distribution of densities. They used mineral coal with particles in the range $40 \text{ to } 400[\mu m]$ and densities between 1.2 and $2.6\left[\dfrac{g}{\text{cm}^3}\right]$. Results of flotation in time and particle size analysis and density determination of a sample are shown in Figure 9.6 and Equation 9.7.

$$k(x,\rho|x) \equiv \tilde{k} = 6.5x^{0.75} \times \exp(-0.015x + 3.5\rho) \tag{9.7}$$

Usually some of these species are ignored, for example, the size distribution or the species distribution. Imaizumi and Inoue (1963) presented the concept of continuous distribution for the kinetic constant $f(k)$, with the normalization $\int_0^{\infty} f(k) = 1$, for a mono-particle size distribution, or an average constant for a particle size distribution,

$$\frac{c}{c_0} = 1 - \int_0^{K} f(k)\exp(-kt)\,dk \tag{9.8}$$

For example, for a Bu et al. (2017) presented for a Gamma, a Square and a Normal kinetic constant distribution with the parameters shown in Figure 9.7. In this figure the parameters a and c are the kinetic constants of first-order models with gamma and rectangular function, p is the exponential number of the first-order model with gamma distribution, σ and μ represent the mean rate constant of the normal distribution and the standard deviation or the width of the distribution, respectively.

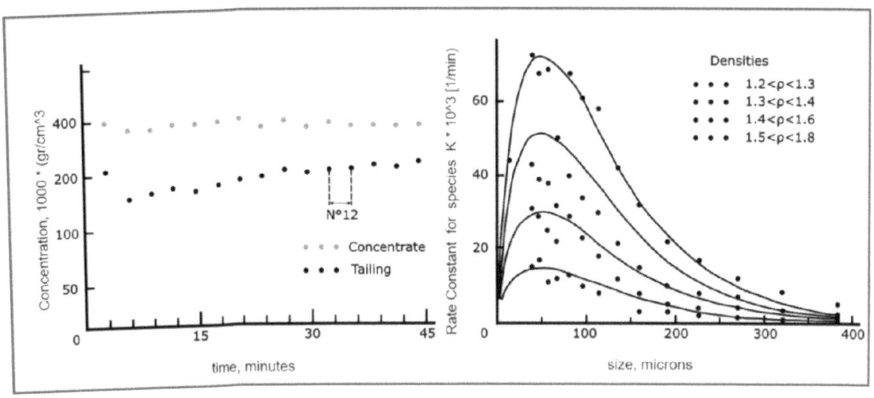

FIGURE 9.6 Results of flotation and the particle size analysis, and the determination of the densities.

Name	Form of distribution	$f(K)$	$R(t)$ (first-order)
Gamma	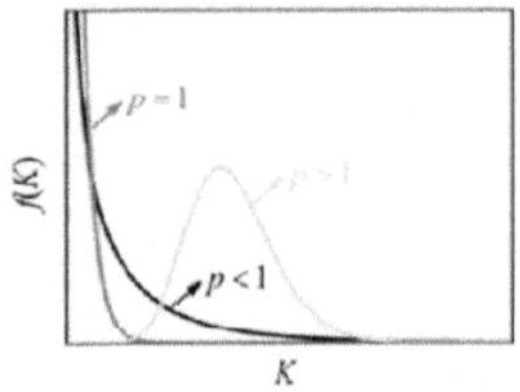	$\frac{K^{p-1}}{\Gamma(p)} a^{-p} e^{-\frac{K}{a}}$	$R_\infty\left[1-\left(\frac{1}{1+at}\right)^p\right]$
Rectangular (continuous uniform)	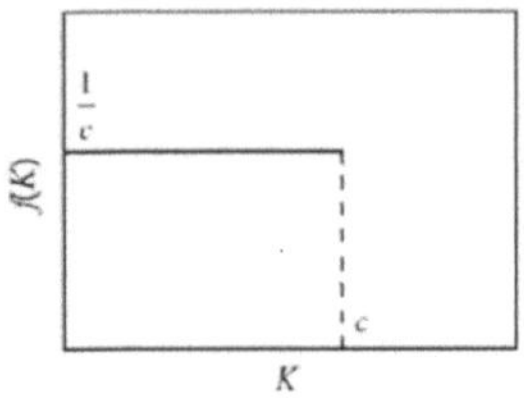	$\begin{cases} \frac{1}{c} & \text{for } 0 \le K \le c \\ 0 & \text{for } c < K < \infty \end{cases}$	$R_\infty\left[1-\frac{1-e^{-ct}}{ct}\right]$
Normal (Gaussian)	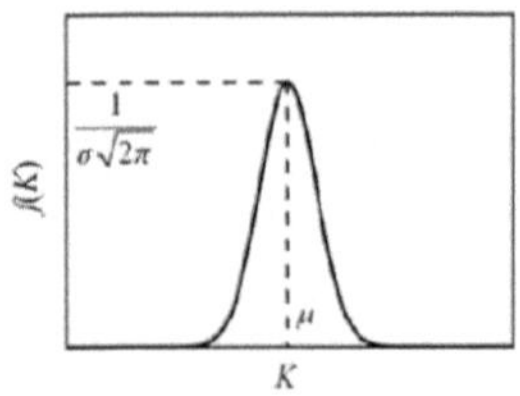	$\begin{cases} \frac{1}{\sigma\sqrt{2\pi}} e^{-\frac{1}{2}\left(\frac{K-\mu}{\sigma}\right)^2} & \text{for } 0 \le K \le 2\mu \\ 0 & \text{for } 2\mu < K < \infty \end{cases}$	$R_\infty\left[1-\frac{\text{erf}\left(\frac{\left(\sigma t+\frac{\mu}{\sigma}\right)}{\sqrt{2}}\right)-\text{erf}\left(\frac{\left(\sigma t-\frac{\mu}{\sigma}\right)}{\sqrt{2}}\right)}{2e^{\mu t-\frac{(\sigma t)^2}{2}}}\right]$

FIGURE 9.7 Distribution function for the kinetic constant.

Source: Bu, X., Xie, G-y., Peng, Y-l., Ge, L., and Ni, C. 2017. Kinetics of flotation: Order process, rate constant distribution and ultimate recovery. *Physicochemical Problems of Mineral Processing* 53(1):342–365.

9.3 FLOTATION MODELS

Several mathematical flotation models have turned into commercial simulators. The known are are JKMRC; SUPASIM; USIM-PAC 2; MODSIM. All of them require basic input data from flotation tests. Other models such as Bascur (2010), Do (2010), and Huang (2015) are based on first principles of Physicochemical Hydrodynamics.

9.3.1 JKMRC Simulator

The JKMRC flotation simulator is based on the fundamental flotation model developed by Kelley (2011). The overall flotation recovery is shown in Equation 9.9 where the parameters are obtained from flotation tests.

$$R_{i,j} = \frac{\left(P_{i,j} S_b R_{fi,j} \tau\right)\left(1-R_w\right)+ENT_i R_w}{\left(1+P_{i,j} S_b R_{fi,j} \tau\right)\left(1-R_w\right)+ENT_i R_w} \tag{9.9}$$

Where subscript i represents a particle size class, and subscript j represents a surface liberation class. P is ore floatability, R is overall recovery, τ is residence time, R_w is water recovery, S_b is the bubble surface area flux in the pulp zone, and ENT is degree of entrainment.

9.3.2 SUPASIM

The SUPASIM flotation was developed in 1980 to predict plant performance from standard laboratory flotation tests (Hay and Rule 2003):

$$R = (100 - \Theta)\left(1 - \exp(k_f t)\right) + \Theta\left(1 - \exp(k_s t)\right) \tag{9.10}$$

Where Θ is the mineral slow floating fraction; t is the flotation time; R is the percentage recovery at time; t, k_f, and k_s are the fast and slow floating rate constant respectively. The kinetic constants and the slow flotation mineral particles must be estimated from the laboratory batch flotation tests.

9.3.3 USIM PAC 2

The simulator, Villeneuve et al. 1995, is a steady-state software having several material balance flotation models and models based on flotation kinetics. The kinetic rate constants consider that the feed is composed of three "sub-populations," non-floating, fast floating, and slow floating. Assuming that each cell is represented as a perfectly mixed reactor, flotation can be described as

$$F_{fj} = F_j R_{\infty j}\left[\varphi_j\left(\frac{1}{1 + ks_j\tau}\right) + (1 - \varphi_j)\left(\frac{1}{1 + kf_j\tau}\right)\right] \tag{9.11}$$

Where F_{fj} and F_j are the flow rate of mineral j in the froth and in the feed respectively, $R_{\infty j}$ and φ_j are the maximum possible recovery of particles j in the froth, and the fraction of slow floatable material of mineral j. For details see Villeneuve et al. (1995) and Huang (2015).

9.3.4 Bascur-Herbst Model

In 1982 Bascur and Herbst developed a flotation model involving the interaction of three phases, solid, liquid, and gas governed by the physicochemical and hydrodynamic environment. The model includes physical measurements, disturbances, operating and controlled variables that influences the concentrate grade and recovery. It is necessary to identify and to promptly solve the problems referred to the frequent and inevitable mineralogical ore changes and grinding disturbances in most plants. Mechanisms of interface transfer are represented by kinetic equations to account for attachment/detachment and entertainment/drainage between four states (phases). In each phase the influence of important operational variables is represented by model equations. Manipulated variables, such as aeration rate, frother addition, agitation degree, pulp level, feed flowrate, percent solids in the feed, grade in the feed, tailings flowrate, and froth level, should be considered. One of the important factors is the water distribution between the pulp and the froth in the equipment. The description of the hydraulic transport through the froth entails the assumption that bubbles leaving the pulp phase carry water into the froth phase. The number, size of the bubbles, and the thickness of the water sheath are determined by the frother concentration and the agitation conditions in the pulp volume (Fuerstenau 1999).

A schematic flotation vessel is presented in Figure 9.8. The pulp/froth interface is characterized by the critical area which is vital in calculating the entrainment and drainage fluxes of water between the pulp and froth volumes.

9.3.5 Population Balance

The general flotation framework is based on a population balance and the hydraulic characteristics of a three-phase contacting device. The model represents the behavior of mineralogical species and any number of particle sizes. Each particle specie can exist in one of four states (phases) in the cell: (1) free in the pulp, (2) attached to bubbles in the pulp, (3) free in the froth, and (4) attached to bubbles in the froth. Water is distributed throughout the pulp and froth. The air holdup in the pulp and froth is calculated using semi-empirical equations derived from mechanistic models.

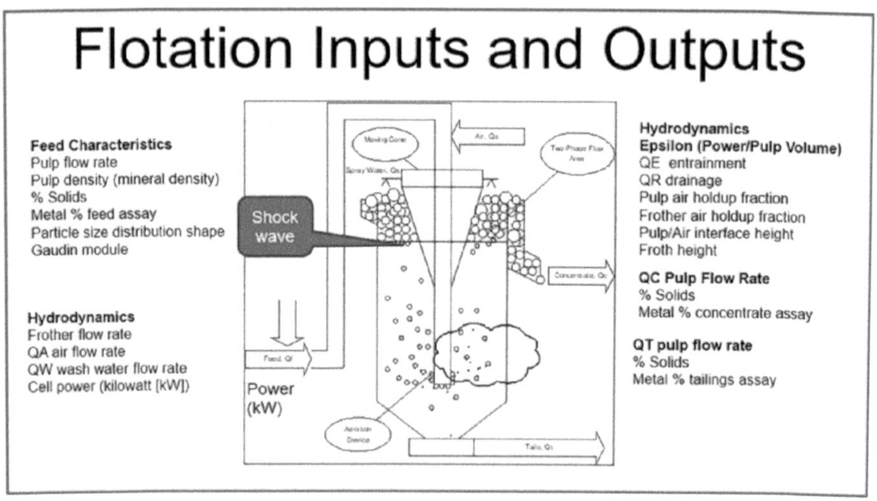

FIGURE 9.8 Flotation cell with movable froth collector.

Source: Bascur, O.A. 2013. A dynamic flotation framework for performance management. In *Separation Technologies for Minerals, Coal and Earth Resources*. Edited by C. Young and G. Luttrel. Englewood, CO: SME. pp. 40–45.

The macroscopic PBM equation can be written in its most general form as follows by Randolph and Larson (1971):

$$\frac{1}{V(t)}\frac{\partial}{\partial t}\left(V(t)\psi(t)\right)+\sum_{i=1}^{m}\frac{\partial}{\partial\xi_i}\left(v_i\psi(t)\right)+D(t)-B(t)$$
$$=\frac{1}{V(t)}\left[\sum_{p}Q_{IN}{}^{(p)}\psi_{IN}^{(p)}(t)-\Sigma Q_{OUT}^{(q)}\psi^{(q)}(t)\right] \tag{9.12}$$

In Equation 9.12 $1,\psi(t)$ represents the number density of particles in a specified state with properties values in the ranges ξ_i to $\xi_i+d\xi_i$: The first term in this equation represents the accumulation of species in any volume V(t), the second term takes into account continuous changes of particles properties within the volume; while D and B represent discrete death and birth rates of particles in the volume, and the terms on the right hand size represent the input and output flow rate of particles across the boundaries of the volume under consideration. The $\psi_{IN}^{(p)}(t)$ and $\psi^{(q)}(t)$ are the concentration of mineral particles at the inlet and outlet volume flow rate respectively. The particle properties of the greatest importance in flotation are size and composition. In this regard $\psi(t)$ in Equation 9.12 is taken to be $\psi_{ij}(t)$ where interval i is a narrow size range, and interval j is a narrow range of mineralogical compositions, there $\psi_{i,j}(t)$ gives the concentration of particles of "size i" and "mineralogical composition j" in any specified state in the flotation cell.

The flotation cell is divided into two volumes, the *pulp volume* where intimate particle-bubble contact is induced by the turbulent action of the impeller and the *froth volume* that separate and remove the valuable minerals. The pulp and the froth volumes are subdivided into a liquid pulp (VLP) and an air froth (VAP) phase. Associated with each are particles undergoing separation. The mechanisms whereby a particle can be transported from the pulp into the froth are levitation on air bubbles, water entrainment, slime covering to other particles. Some particles can drain back from the froth in the pulp volume or leave the froth volume as concentrate.

The application of Equation 9.12 to each of the four states: (1) free in the pulp, (2) attached in the pulp, (3) free in the froth, and (4) attached in the froth leads to the set of differential equations given (Equations 9.12–9.16). In each case flows into and out of a volume (Q) are considered to carry particles as depicted in Figure 9.9, while attachment and detachment processes are represented as birth and death terms characterized by rates which are proportional to the number of particles present in a given phase with attachment and detachment rate constants k^{AT} and k^{DT} determined by molecular, electrostatic, and hydrodynamic forces. It is recognized that a death event in one phase results in an immediate birth event in the complimentary phase in the same volume. Finally, the properties of a given particle in the cell are not considered to be time dependent so that term two in Equation 9.12 is zero in all cases. The set of coupled differential equations given should provide a complete description of particle dynamics in a flotation cell. In order to implement these equations, the forms of rate constants for attachment and detachment must be specified, and a hydrodynamic model for the two-phase flow must be provided.

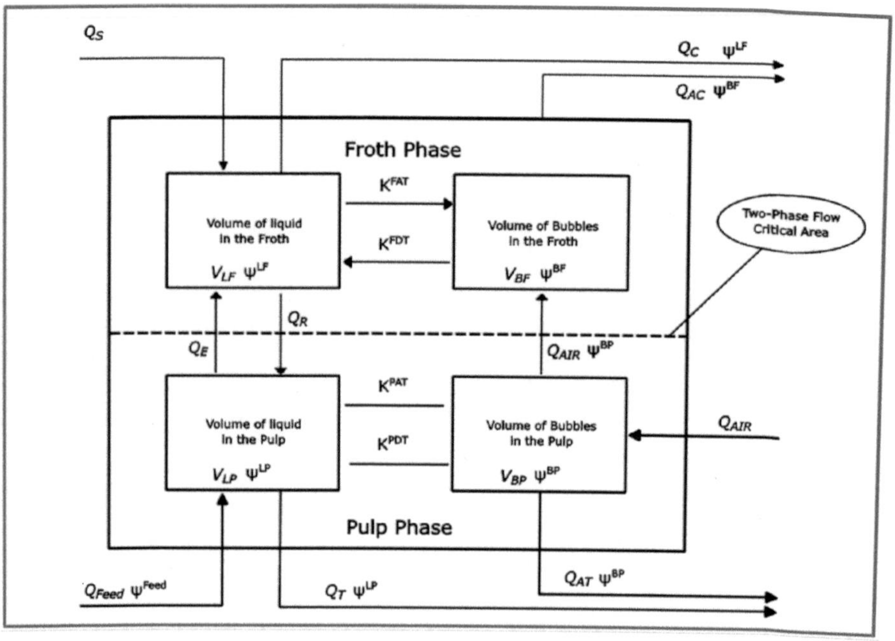

FIGURE 9.9 Schematic of the four states for particles in a flotation cell according to the population balance model.

Source: Bascur, O.A. 2013. A dynamic flotation framework for performance management. In *Separation Technologies for Minerals, Coal and Earth Resources*. Edited by C. Young and G. Luttrel. Englewood, CO: SME. pp. 40–45.

9.3.5.1 Population Balance Model for Four States in a Flotation Cell

Free in the pulp:

$$\frac{d}{dt}\left(v_{LP}\psi_{ij}^{LP}\right) + k_{ij}^{PAT} V_{LP}\psi_{ij}^{LP} - k_{ij}^{PDT} V_{AP}\psi_{ij}^{AP} = Q_{Feed}\psi^{Feed} - Q_T\psi_{ij}^{LP} + Q_R k_{ij}^{R}\psi_{ij}^{LF} - Q_E\psi_{ij}^{LP} \quad (9.13)$$

Attached in the pulp:

$$\frac{d}{dt}\left(V_{BP}\psi_{ij}^{BP}\right) + k_{ij}^{PDT}\psi_{ij}^{BP} V_{BP} - k_{ij} R_{j}^{PAT}\psi_{ij}^{LP} V_{LP} = -Q_A\psi_{i,j}^{BP} - Q_{AT}\psi_{ij}^{BP} \quad (9.14)$$

Free in the froth:

$$\frac{d}{dt}\left(V_{LF}\psi_{ij}^{LF}\right) + k_{ij}^{FAT} V_{LF}\psi_{ij}^{LF} - k_{ij}^{FDT} V_{BF}\psi_{ij}^{BF} = Q_E\psi_{ij}^{LP} - Q_R k_{ij}^{R}\psi_{ij}^{LF} - Q_C\psi_{ij}^{LF} \quad (9.15)$$

Attached in the froth:

$$\frac{d}{dt}\left(v_{BF}\psi_{ij}^{BF}\right)+k_{ij}^{FDT}V_{BF}\psi_{ij}^{BF}-k_{ij}^{FAT}V_{LF}\psi_{ij}^{LF}=Q_{A}\psi_{ij}^{BF}-Q_{AC}\psi_{ij}^{BF} \qquad (9.16)$$

9.3.5.2 Attachment Phenomena in the Pulp

Mika and Fuerstenau (1968) have critically reviewed the various viewpoints presented in the literature concerning attachment subprocesses. Other reviews have been presented by Laskowski (1974) and by Jameson et al. (1977). The formation of a particle-bubble aggregate proceeds via three important steps: (1) approach of a particle to a bubble, (2) thinning of the water film between particle and bubble, and (3) receding of the residual film to give an air-solid interface. The most relevant approaches to deriving a net attachment rate constant are those by Woodburn et al. (1971) and Schubert and Bischofberger (1981); both consider the phenomena of turbulence in their development. Bischofberger and Schubert use Abrahamson's 1975 equation for aggregate formation (Abrahamson 1975), which depends on the local energy dissipation throughout the turbulent velocity of the aggregates based on the Kolmogoroff theory of the turbulence. Kolmogorov's theory describes how energy is transferred from larger to smaller eddies, how much energy is contained by eddies of a given size, and how much energy is dissipated by eddies of each size. The number of collisions per unit of volume and time in a flotation system is given by

$$Z=5.0N_{ij}N_{B}d_{iB}^{2}U_{t} \qquad (9.17)$$

where NN_{ij}, N_B are particle and bubble number respectively available for collision, d_{iB}, the mean size of the aggregate of size i (particle/bubble). The turbulent aggregate velocity, U_t, which they use is that proposed by Liepe in Schubert and Bischofberger (1981):

$$U_t=K_t\frac{\varepsilon^{4/9}d_{iB}^{7/9}}{\left(\frac{\rho_p}{\mu}\right)^{1/3}}\left(\frac{\rho_j-\rho_p}{\rho_p}\right)^{2/3} \qquad (9.18)$$

Where ε is the average energy dissipation in a flotation given by P_g/V_{LP} in our treatment, ρ_j the specific gravity of species j, ρ_p the specific gravity of the pulp, and μ the viscosity. Kolmogorov's theory of 1941. Richardson's notion of turbulence was that a turbulent flow is composed by "eddies" of different sizes. The sizes define a characteristic length scale for the eddies, which are also characterized by flow velocity scales and time scales (turnover time) dependent on the length scale. Comparison of Equation 9.17 with the second term in Equation 9.13 yields the following:

$$k_{ij}^{PAT}=k_{j}^{PAT}S_{B}\,d_{iB}^{2}U_{t}/d_{BP}^{2} \qquad (9.19)$$

Here k_j^{PAT} needs to be determined experimentally for each mineralogical species. This equation shows the typical elephant curve described by Trahar, which defines that small and large particles are difficult to be floated (Trahar 1981; Lynch et al. 1981). The mean bubble diameter, d_{BP}, is determined with Calderbank (1958) modified Equation 9.31. S_b, defined as the bubble surface that flows across the cross-sectional area of the cell per unit of time and can be expressed in terms of J_g and d_{BF} as follows:

$$S_b = \frac{6}{d_{BP}} J_g \tag{9.20}$$

Similar experiments to those carried out by Calderbank were made by Bascur (1982). Figure 9.10 shows the relationship between Air Holdup and the Air Velocity in an Agitated Flotation Cell for several amount of frother concentrations. Experiments were run for several rotating speed and measuring the impeller torque. These resulted in the following expression for the air holdup estimate:

$$\phi_p = K3\left(\frac{P_g}{V_{LP}}\right)^{0.45} *(Q_A / A_1)^{0.67} C_F^{0.25} \frac{\Delta y}{\Delta x} \tag{9.21}$$

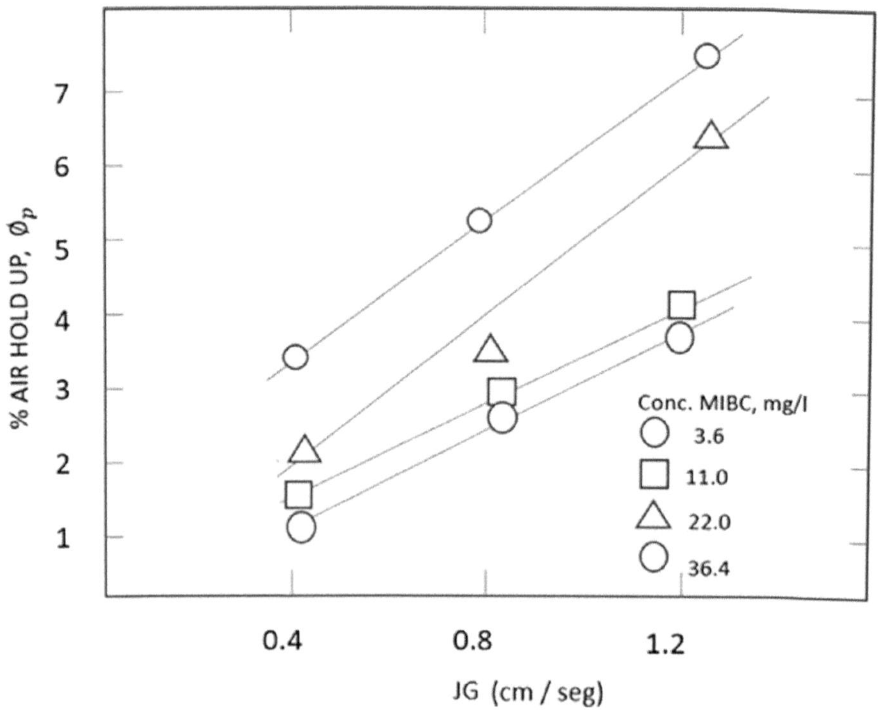

FIGURE 9.10 Pulp air holdup versus superficial air velocity for several frother additions.

QA/A1 represents the superficial gas velocity J_g where A_1 is the interfacial area between the pulp and froth. The Calderbank equation involves surface tension of the liquid, but the frother concentration term CF was introduced due to its major effect on the bubble size and the minimum effect of CF on surface tension at the typical frother addition levels in a flotation cell.

It is interesting to note that many researchers have reported the correlation between the Air Holdup and Superficial gas velocity such as Lopez-Saucedo and others (2012). They report the correlation of Metal (Zinc) Recovery has function of Bubble Surface Are Flux, Sb and gas holdup, ϕ_p. Mineral flotation is the result of the interaction of several factors as we have described in Figures 9.1 and 9.9 (pulp chemistry, hydrodynamics, mineralogy, gas particle contact, particle size distribution, etc.); nevertheless, it is clear the major importance of the gas dispersion parameter (J_g, ϕ_p, d_{BF}) addresses in the following Dynamic Flotation Model. As such, the bubble diameter and the air holdup in the pulp and in the froth are vital estimations that can be made using the shown equations in an industrial flotation system to assist in finding the right hydrodynamic characteristics.

The term Pg / V_{LP} in Equation 9.21 is average energy dissipation calculated from the power input per unit volume where Pg is the power in the presence of air and V_{LP} the volume of the liquid in the pulp. Extensive experimentation has been carried out in flotation hydrodynamics concerning the correlation between power and aeration rate by Harris (1976) and Mesa et al. (2020). In practice, the power is measured from the amperage reading of each of the cell in the flotation bank. For the derivations of the equations, a very sensible torque sensor, air flow rate, QA l/min, pulp % solids, and rotational speed, N (RPM) measurements, flotation cell pulp level, cm, and frother addition, mg/l, were used to derive all the hydrodynamic equations, which are very useful to assess the flotation hydrodynamic performance of a flotation bank.

No correlations that include the effect of frother have, however, been reported. The correlation derived is as follows:

$$\mathrm{P_g} / \mathrm{P_0} = \mathrm{C2}\left[\frac{Q_A}{\mathrm{ND_T^3}}\right]^{\mathrm{m}}\left[\frac{\mathrm{N^2D_T^3}\rho_{\mathrm{P}}}{\sigma}\right]^{n} \tag{9.22}$$

Where P_o is the power consumption without aeration, which can be correlated for Re numbers > 10^5 by $P_o = 1.64\rho_P N^3 N_T^5$ as obtained in Figure 9.11. The first term in brackets is the Aeration Number, and the second dismalness number is the Weber Number. This correlation is good enough for estimating the power transmitted to a volume of aerated pulp for froth flotation improvement practices.

QA is the air flowrate, N is the Impeller speed, RPM, DT is the impeller diameter, ρ_{P} is the specific gravity of the pulp, σ is the surface tension of the liquid, m is equal to −0.1018, and n is equal to −0.0984.

A typical experimental plot is shown for a Denver flotation cell in Figure 9.11.

It is evident that manipulating the air flowrate is very valuable strategy to maintain the right balance of air in the cell as air holdup but keeping the right power to have the right turbulence to increase the attachment of the valuable minerals.

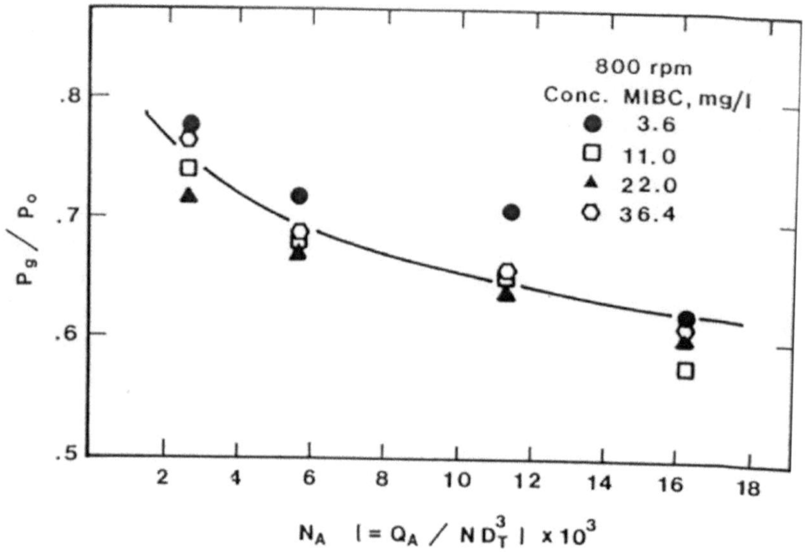

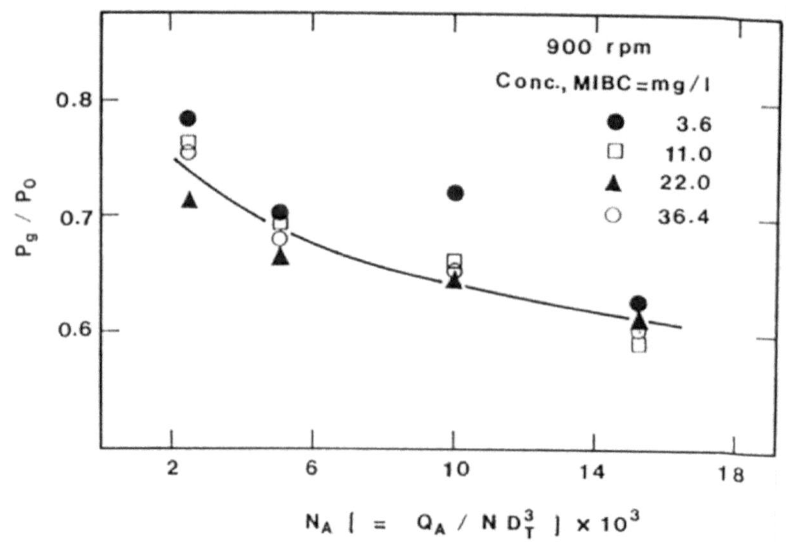

FIGURE 9.11 Effect of aeration rate and frother concentration on aerated power measurement for 800 and 900 RPM.

The higher the air holdup and the smaller the bubble diameter, the better the flotation results. As such, manipulating these variables in an industrial flotation bank is of vital importance. You can see that increasing the air flowrate decreases the effective power in the flotation cell. You can also see that increasing the surface tension assists in maintaining the right amount of power to increase the flotation rate of the floatable species. The effect of the pulp density is also understood. As such, using these equations in real time assist in improving the efficiency of the grinding and flotation systems. An increase in pulp density will decrease the power of an aerated pulp communicated to the system losing the flotation effectiveness. This is well-known that for solids of density in the order of 2.7 T/M3 the percent solids of 35% is good practice. Since we tested the models for copper sulfide and coal minerals. We can see that the flotation differs enough between the light highly floating species and the heavier copper sulfide species. For coal flotation the % solids used was 7% solids for the best removal of ash from the Illinois #6 mineral.

As such, we can see that by manipulating the impeller speed, small bubbles can be generated creating small particles and also created a higher air holdup who increase the capture of floatable species to the concentrate. These new control strategies can be simulated dynamically using Dynaflote. The same equations can be used in an industrial plant to tune the flotation bank to be able to float coarse, fine, and intermediate size particles depending on the Ball Mill Classification circuit size distribution product. As such, having the particle size distribution from the Overflow of the cyclone is very important to be able to set the Rougher Flotation bank to maximize the net metal production rate, as we will show in Chapter 12. An increase of Net Metal Production Rate of 32% can be achieved by integrating the grinding/classification and flotation data in an industrial plant. It will also be shown that the particle size distribution also affects the flocculation and, thus, settling particle velocities enhancing the capabilities to increase the water recovery to have enough water in the grinding circuit to maintain the right suspension characteristics in the grinding circuit.

9.3.5.3 Detachment Phenomena in the Pulp

A well-documented review of different approaches that have been used to understand the detachment mechanism in flotation is presented by Mika and Fuerstenau (1968).

It has been shown that the upper size of floatability in an impeller flotation machine is determined by agitation with turbulence as the characteristic parameter. Schulze (1977) points out the dominance of tensile forces both turbulent tensile and shear strengths acting in the aggregates as a function of the specific energy dissipation (Schulze 1977). Woodburn et al. (1971) derived an expression based on the maximum deformation stress set up in the bubble skin following a sudden acceleration of the bubble-particle. They obtained the following proportionality for the probability of detachment: $P_d \sim d_p^{1.5}$ Ut; where Ut is the instantaneous velocity achieved by a bubble due to its interaction with an eddy. By analogy, in general in a flotation cell the specific detachment rate as a function of size can be expressed where U_t is the turbulent velocity given by Liepe's equation and n is an adjustable parameter (1.5–4) (Schubert et al. 1979). In this model, the rate of detachment depends on the surface shear and the size of the particle. The shearing stress depends on the turbulent velocity U_t.

$$k_{ij}^{PDT} = k_j^{PDT} d_i^n U_t \tag{9.23}$$

9.3.5.4 Attachment Phenomena in the Froth

It is well-known that if inadequate frother is added to a flotation cell, extensive disruption of aggregates occurs in the froth due to bubble bursting with subsequent loss of particle load. These particles may be reattached in the froth or may drain back into the pulp depending on the operating conditions. The bursting of bubbles not only occurs if inadequate frother is used, but other variables may contribute to the instability of the froth, such as inadequate mechanical removal of the froth, or the presence of particles which form a very small contact angle with the air phase.

To derive an expression for the attachment in the froth, a flow-line interception mechanism of particle collection has been assumed. In this regime the efficiency of collection is given by the ratio, $\frac{d_P}{d_{BF}}$ where d_P is the diameter of a particle and d_{BF} is the diameter of a bubble in the froth. This collection mechanism has been described by Gaudin (1957) and has been supported experimentally by Spedden and Hannan (1948) in a highly idealized flotation study which resembles the process occurring in the froth. Under these conditions the rate of attachment of particles in the froth will be given by the number of bubbles per unit of time that rise throughout the froth of height H_F, multiplied by the efficiency of collection by streamline interception. The equation for the rate constant of attachment in the froth can be written as

$$k_{ij}^{FAT} = k_j^{FAT} Q_A \left(\frac{d_P}{d_{BF}} \right) \cdot \left(\frac{H_F}{d_{BF}} \right) \tag{9.24}$$

in which the first term in brackets on the right-hand side represents the collision efficiency and the second the number of "bubbles heights" in the froth.

9.3.5.5 Detachment Phenomena in the Froth

The argument used to postulate a detachment rate constant equation for the froth is based on the fact that the particles, which are attached to a bubble, experience a shear force exerted by the fluid draining past a surface in the downward direction; the larger the particle, the larger the force for disruption. The other effect considered is that due to gravity. The resulting equation has the following form (Bascur 1982):

$$k_{ij}^{FDT} = k_j^{FDT} \rho_j d_p^n d_{pU_\infty} \tag{9.25}$$

where ρ_j is the specific gravity of mineralogical species j, U_∞ bulk fluid velocity occurring as a result of drainage. This equation is derived for flow in the Stoke's regime. In the case of flow in the Newton's regime, "n" the exponent of d_P is two. A similar expression can be developed by considering the torque required to dislodge a single particle from the surface of a bubble (Bascur 1982). In both derivations the water drainage velocity U_∞ can be approximated by

$$U_{\infty} = \frac{Q_R}{A_3\left(1-\phi_F\right)} \tag{9.26}$$

Expression for the distribution of liquid between the pulp and froth is required. This expression will be obtained by postulating a hydraulic model for the flotation cell in the section that follows. This section is the most important for dynamic modeling, a flotation cell since by manipulating the tailings pulp flowrate is the main variable for control the hydrodynamics of two-phase systems. This also applies to thickeners models as we will show in Chapter 10.

9.3.6 Flotation Pulp and Froth Hydrodynamic Model

The population balance of the species in flotation is only a part of the flotation model. The most important contribution in this work is the modeling of hydrodynamics for use in industrial operations. The tailing flow is what controls the continuity equation of two-phase flow systems as we will describe in the following paragraphs. The flotation process is a hydrodynamic environment mostly neglected by many mineral processors which focus only on the physiochemistry of this process. These are great research to identify the best chemical environment. However, the flotation process requires adjustments due to the viscosity and mineral processing disturbances in an operational plant. Hydrodynamics plays a fundamental role in the behavior of the pulp and bubbles in a two-phase flow.

As stated earlier, one of the most important factors that determine the separation of minerals and water distribution in a flotation cell/column is the froth. The approach used for modeling the water transport through the froth is essential to solving the population model approach. As such we need to use the multiphase flow. A model for a multiphase suspension has been developed to represent the general particles in sedimentation (Concha and Bascur 1977) extending the traditional Kynch theory of sedimentation (Kynch 1952) to all mineral processing suspension processes. The general conservation equation is satisfied using a sinusoidal volumetric balance between the bubbles of air and the pulp. This model is also applicable to two and three phases' systems as flotation. In summary, we have two phases with the pulp being the continuous phase in this case and the bubbles the discontinuous phase.

The bubble generation in a flotation vessel produces spherical bubbly foam with a changing bubble volume fraction in the pulp and froth phases. Bascur (1982) conducted flotation experiments under rigorous controlled conditions that measure the power, air, frother, reagents, and using industrial mineral ores (Illinois # 6 and KUCC Copper Ore) to develop a froth flotation dynamic model for process optimization. These experiments enabled the characterization of the hydrodynamic behavior of the flotation process and the development of a two-phase flow model based on the continuity equation derived by Bascur (1976).

The continuity equation for two phase flows is

$$\frac{\partial \varphi}{\partial t} + (q + \frac{\partial}{\partial z}\left\{\varphi\left(1-\varphi\right) f_{Bk}\left(\varphi\right)\right\} = 0 \tag{9.27}$$

$$\frac{\partial q}{\partial z} = 0, \text{with } q = v_B(1-\varphi)(v_B - v_w)$$

Where φ is the fraction of bubbles in the pulp, $\cdot f_{bk}(\varphi)$ is the flux density of bubbles, and q is the total flux density (bubbles and water), v_B is the bubble velocity, and v_w is the water velocity.

Based on this model, Bascur and Junge (2011) and Bascur (2019) developed Dynaflote, a dynamic simulator with application for training.

The two-phase flow analysis was defined by Wallis (1969) and the Kynch Sedimentation Theory (Kynch 1952). Recently, it has been revisited by Galvin and Dickinson (2014) as shown in Figure 9.12. It brings new light to the importance of the hydrodynamics in flotation devices and the importance of the air holdup distribution required to flote particles by the buoyancy effect as described by Jameson (2023).

Readers are referred to the sedimentation theory of Kynch and the state-of-the-art paper by Fitch (1979). They equate the individual solid particles that "settle" upwards in a system of rising foam. A small disturbance can easily lead to a rise or fall in the bubble froth, and the control of the flotation tailing rate is critical. This can be tested using a Dynamic Flotation Simulation described in Bascur and Junge (2011).

Unstable operating points exist in the system. It is well understood that operators dislike controlling the level of a bank of flotation cells. It is an important realization that for a particular set of operating conditions there is only one critical area (A_3) that satisfies the balance equations. This is one of the fundamental causes of froth height changes in many flotation vessels. It is common practice to use froth crowders in

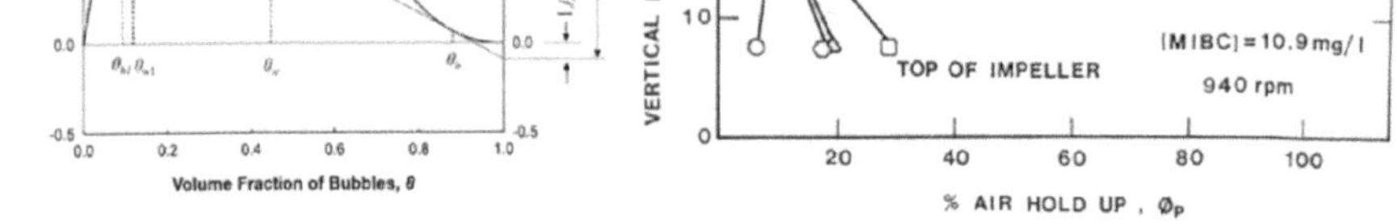

FIGURE 9.12 Drift flux curve vs volume fraction of bubbles and vertical air distribution in a flotation vessel.

Sources: Galvin, K.P., and Dickinson, J.E. 2014. Fluidized bed desliming in fine particle flotation—Part II. Flotation of a model feed. *Chemical Engineering Science* 108:299–309; Bascur, O.A. 2013. A dynamic flotation framework for performance management. In *Separation Technologies for Minerals, Coal and Earth Resources*. Edited by C. Young and G. Luttrel. Englewood, CO: SME. pp. 40–45.

flotation cells and columns to adjust the two-phase flow to reasonable conditions. For online process calculations, a simplification of equation with regard to Figure 9.9. These equations represent a discretized form of the flux equation for modeling two phase flows. It is a numerical approximation to resolve the flux equation (Bascur 1991).

$$\frac{d}{dt}(V_{LP}) = Q_{Feed} - Q_T - Q_E + Q_R \tag{9.28}$$

$$\frac{d}{dt}(V_{LF}) = Q_E - Q_R - Q_C \tag{9.29}$$

The entrainment water flow is based on the water shed around bubbles based on hydration effects at the air-water interface and by the turbulence levels in the pulp volume. The volumetric flowrate of entrained water can be calculated from the number of bubbles rising per unit of time $6Q_A / \pi d_{BP}^3$ 6, where Q_A is the volumetric air flow rate, and d_{BP} is the mean size bubble diameter in the pulp region. The volume of liquid held by a film surrounding a spherical bubble is $\pi d_{BP}^2 \delta$, so that the volume flow rate of water into the froth is

$$Q_E = \frac{6Q_A\delta}{d_{BP}} \tag{9.30}$$

In order to apply Equation 9.33, the average value of d_{BP} must be known. Extensive experimentation has been reported on the estimation of bubble size in aerated agitation vessels by Calderbank. Many experiments were performed to characterize the hydraulic behavior of an agitated flotation device by Bascur (1982). In the current study, a modified expression for the Sauter mean diameter of bubbles in a flotation cell under controlled turbulent measurement is used:

$$d_{BP} = K2 \frac{\phi_P^{.4}}{\rho_P^{.2} \left(\frac{P_g}{V_{LP}} \right)^{.4} C_F^{.344}} \tag{9.31}$$

This equation arises from the application of Kolmogorov's theory of local isotropy to aerated agitation vessels, so the form of the equation is universal for systems having similar turbulent conditions. The air holdup can be estimated by Bascur Equation 9.21.

Extensive experiments were conducted in the original research to obtain scalable hydrodynamic equations appliable to a variety of flotation cell and columns cell (Bascur 1982, 2013). Estimation of the film thickness δ around the bubbles is obtained from Levich (1962) and Clift et al. (1980):

$$\delta \, \alpha \left(\mathrm{d_{BP}} v_{\mathrm{eff}} / \mathrm{U_B} \right)^{1/2} \tag{9.32}$$

Where v_{eff} is an effective kinematic viscosity and U_B is the slip velocity of the bubble through the fluid. Since the estimation of v_{eff} is very difficult in the system under consideration, it will be lumped into an overall constant. For estimation of the bubble slip velocity, Levich's approximation for Reynold numbers above 200 was used, that is, $U_B \alpha (g\ d_{BP})1/2$, which is in agreement with the data presented by Clift et al. (1980). Unfortunately, no good correlation between U_B and frother concentration is available for the size range of bubbles under consideration (0.5 to 1 mm).

9.3.6.1 Water Drainage from the Froth

Mika and Fuerstenau (1968) have emphasized that the drainage from the froth is essential to describing the froth concentration process. Gaudin (1957) presents excellent diagrams explaining this phenomenon. The water drainage equation and the air holdup in the froth were derived following Barber and Hartland (1975) studies of foams and validated with the experimental results by Bascur (1982).

$$Q_R = KRv^{.53} V_{LF}^{.56} A_3^{.4} / \left(\sigma^{.24} d_{BF}^{1.92}\right) \tag{9.33}$$

Where v is the superficial gas velocity $((Q_A/A_3)$, A3 represents the cross-sectional area of the froth volume and KR an adjustable parameter obtained from industrial experiments. V_{LF} is the volume of liquid in the froth, σ is the frother concentration, and d_{BF} the mean bubble size in the froth. The air holdup in the froth is estimated with

$$\varnothing_{F^*} = C6 * (QA / A3) \wedge 1.2^* \left(A_3^{0.4}\right) / \left(\sigma^{0.4} d_{BF}^{1.2} * V_{LF}^{0.4}\right) \tag{9.34}$$

A_3 is the cross-sectional area of the discontinuity of the two-phase flow, σ is the frother concentration based on the frother addition to the flotation cells, and C6 a parameter obtained from industrial data.

The water flow rate to the concentrate is modeled from the studies of flow over a weir. The correlation developed by Francis was adapted to a flotation cell.

$$Q_C = KC \cdot L \cdot \left(H_F - H_C\right)^{1.5} \left(1 - \phi_F\right) \tag{9.35}$$

L is the width of the flotation cell lip, HF the total froth height referred from the bottom of the cell, HC is the height of the flotation cell, (1-ϕF) is the average liquid holdup in the froth, and KC is an adjustable parameter that depends on the type of removal technique used.

Q_{Feed} is the process feed rate to the flotation bank. Using Equations 9.30, 9.33, and 9.35 into dynamic Equations 9.28 and 9.29 represents the dynamic behavior of a two-phase flow systems, which estimate the froth/pulp interface based on the turbulence, chemistry, and geometric characteristics of the cell.

The pulp density is calculated using the following equations in real time to include the ore density disturbance as the model. This has been found of great value since the ore density of low-grade ores has a large variation which also affects the % solids measurements in the plant. % Solids by Volume is used to estimate the viscosity and the rheology of the pulps in real time.

$$\varrho_P = \phi s^* \varrho_s + (1 - \phi s)\varrho_f \tag{9.36}$$

The flotation framework can predict the pulp/froth interface bases on the hydrodynamic balance in the two-phase flow system. These sub processes are the attachment of particles to bubbles in the pup phase and in the froth phase and the particles entrainment in the pulp phase or drainage of particle from the froth. A detailed description of the model framework can be found in Bascur (2013).

This flotation framework considers all particles in several species by size entering a flotation vessel into a pulp phase where the bubbles are generated by hydrodynamics forces. The power intensity in the pulp is the most important mechanism used in the model to estimate the number of particles that will be attached to bubbles. These bubbles also carry water and entrain other gangue mineral into the froth phase. In the froth phase the bubbles burst. This burst releases the water which will help in cleaning the participles which are not wanted in the concentrate. Only the species which are tightly attached to the bubbles will report to the concentrate stream. Of course, the water in the froth will carry all fine participles, which will also report to the concentrate. As such, the flotation framework allows us to understand the behavior of the aerophilic and hydrophilic particles. The air holdup in the pulp and air holdup in the froth play an important role in the flotation process. These two metrics can be estimated based on the aeration rate, power intensity, frother concentration, and the pulp and froth volumes.

The manipulation of the tailing flow rate, Q_T, the volume flux of solids and bubbles at the bottom of the cell controls the chock wave created in a tow phase flow (Pulp = continuous phase and Bubbles = dispersed phase) turbulent environment. For this reason, the dynamic flotation model was made available as dynamic model since it does not have a steady state solution. It is a dynamic system that needs to be operated dynamically to balance continuously the flow of bubble rising with particles and a shed of water towards the froth and the drainage of the water from the bubble bursting on the top of the froth. Today, with the froth cameras the bubbles can be analyzed using machine learning techniques together with a mathematical model that provides a complete assessment of the flotation bank performance metal recovery and air holdup profiles in the pulp and froth to maximize the recovery of the metal bearing minerals for all minerals.

The evolution of computers and networks capture operational data, validate, classifies operational modes, and provides the non-line modeling capabilities for advanced data analytics. These new technologies enable us to create hybrid models using first principles flotation models (Bascur 2019), with the powerful computing power using learning algorithms to parameterize the models. These models detect the operational constraints and assist in providing best estimates of the state variables in the models.

Dynaflote is based on three fundamental laws:

1. The model of two-phase flows suspension was described by Bascur (1976) and an update of a unified separation theory of phase flows by Bascur (1991).

2. The Kolmogoroff turbulence theory in flotation cell as described by Schubert and Bischofberger (1978, 1981).
3. The Mika and Fuerstenau (1968) population balance model adapted and parameterized by Bascur and Herbst (1982).

9.3.6.2 Industrial Application of the Dynamic Model

A key application is for training using Dynaflote as dynamic simulator (Bascur and Junge 2011). Figure 9.13 shows a couple of screenshots of the simulator Dynaflote running on an iPad. It shows the flotation cell with all the manipulated variables and access to the different parameters while touching the selected areas on the displays. The passing of the finger on the displays changes to the display showing the dynamic flotation framework showing the pulp phase with the pulp area and the bubbles areas interacting via the attachment and detachment of the species. It also shows the froth phases showing the interaction of species being attached and detached from the bubbles in the froth. The most interesting aspect of this display is the characterization of the air holdups in the pulp and in the froth as well as showing the amount of water going into the froth and the water drainage from the froth back to the pulp phase. The total tailings are the summation of the species exiting via the pulp and the bubble phases. The total concentrate flow is the sum of the species for all sizes leaving the froth and with the pulp (water entrainment). Any change in the manipulated variables (aeration rate, tailing froth rate, frother addition, impeller speed, feed process feed rate, and species concentrations) are reflected in the changes in the pulp and froth levels and the concentrate and tailing shows in real time. A change in air flowrate produces a change in concentrate flow and an increase in the air holdup in both the pulp and froth. The magnitude depends on the tailings flow to absorb the change, and the hydrophobic species will increase reporting the concentrate, but the water entrainment will also increase, decreasing the grade of the concentrate. The choke wave defines the pulp and froth levels as the tailings flowrate in controlling the pulp level in the cell.

The flotation bank process variables are very important for maximizing metal production (Dahlke et al. 2004). For applications in the industry, we use the hydrodynamic equations to fine-tune the air holdup in a flotation bank while calculating the Metal Recovery from the plant using a special algorithm to validate, classify the operating modes and ore mineral characteristics (Chawda 2019; Bascur and Soudek 2019).

Figure 9.15 shows a typical flotation bank with the associated measurements required to create a predictive flotation model for optimization of the grinding and the rougher flotation circuits. The key is to get the best stable particle size distribution within the hydrodynamic capabilities of the flotation bank to maximize metal recovery at all times. This strategy focusses on estimating metal losses in the tailings stream in real time. Customers have achieved historical records using this operational strategy (Doucet 2017; Chawda 2019; Copper 2018; Bascur and Retamal 2021, 2023; Bascur 2023b).

Figure 9.15 shows the improved results when the grinding product is optimized for the best flotation according to the limitations of the classical flotation equipment

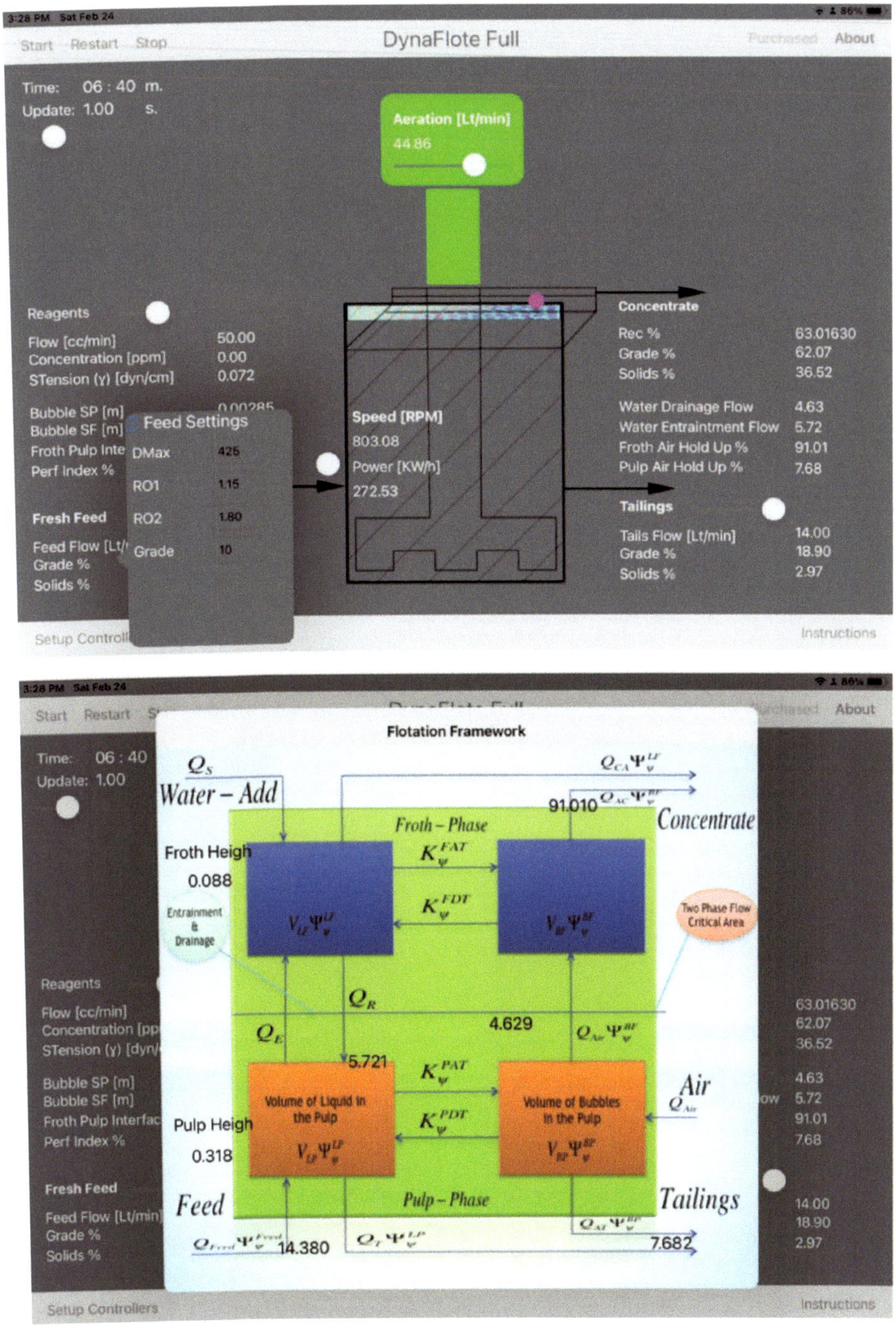

FIGURE 9.13 Dynamic Flotation Simulation Model, Dynaflote™.

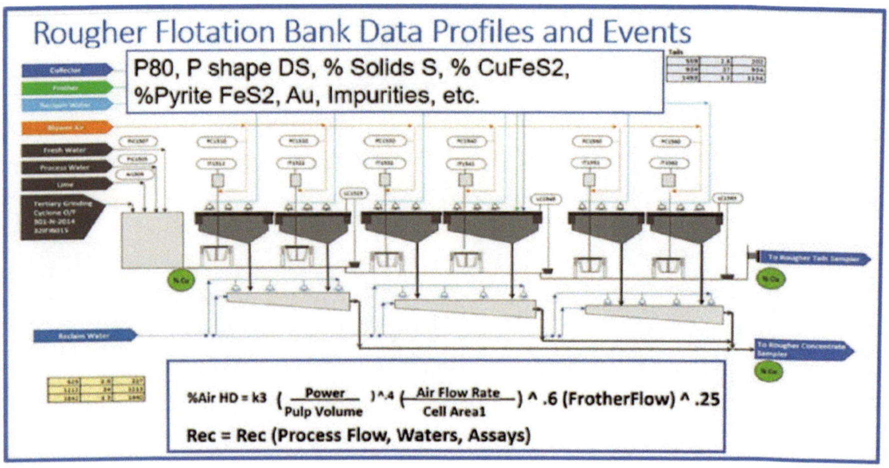

FIGURE 9.14 Measurements required to obtain an online machine learning flotation bank based on a series of unit flotation cell model.

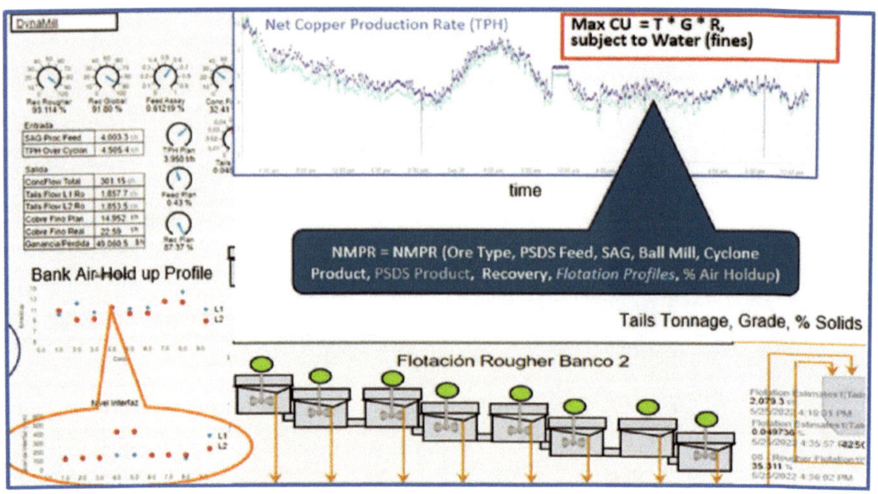

FIGURE 9.15 Industrial rougher flotation turbulence and air holdup profiles.

and using our hydrodynamic knowledge to create three types of hydrodynamic conditions in a flotation bank. Using the optimal bank hydrodynamic conditions shows a radical improvement on the copper production rate. The key is to recover the maximum amount of water from the thickeners to keep the right rheology in the grinding and classification. We will review the industrial applications later in Chapter 12.

The flotation kinetics is determined by the mineral composition of the copper and not the copper grade, which is being measured. Measuring the mineral composition of the ore being treated will provide information, which can be used to infer expected float cell behaviors and recoveries. These online analyzers are now available using laser-induced breakdown spectroscopy or x-ray fluorescence. Their information could improve the response time of the flotation circuit to a change in mineralogy of the ore being treated. The information could be used to accurately determine the ideal mass pull based on a known grade recovery curve for that mineral (Baas 2016). An online model using the mineral composition provided by the flotation model and the x-ray fluorescence measurement combination was designed using a Kalman filter, which can estimate the mineral composition in the flotation system. This strategy was presented by Bascur and Herbst (1985).

New advances in flotation are being developed to improve the fine and coarse particles problems using classical flotation cells and aerated jet flotation cells. A discussion on improving the efficiency of galena flotation using an aerated jet flotation cell is provided by Li et al. (2020). A summary of the challenges and opportunities with fine particle flotation was reviewed by Ayedzi et al. (2022).

9.3.7 Huang Model

Huang (2015) presented a first principles flotation model incorporating in each of the subprocess the hydrodynamic and physiochemistry parameters. In a form similar to Bascur's model, the flotation cell was divided into two volumes, the pulp volume and the froth volume, that separate and remove the valuable minerals. Figure 9.16 shows the two cell volumes with information to calculate the overall flotation recovery. It shows only one mechanism to take the species to the froth and one to take care of the drainage.

The subprocess considered was flotation kinetics: bubbles generation, turbulence, collision, attachment, detachment, pulp recovery, and contact angle of composite particles. If during flotation the number of bubbles is much greater than the number of particles, the flotation model can be simplified as a *pseudo first-order* process $dN_1/dt = -k_pN_1$.

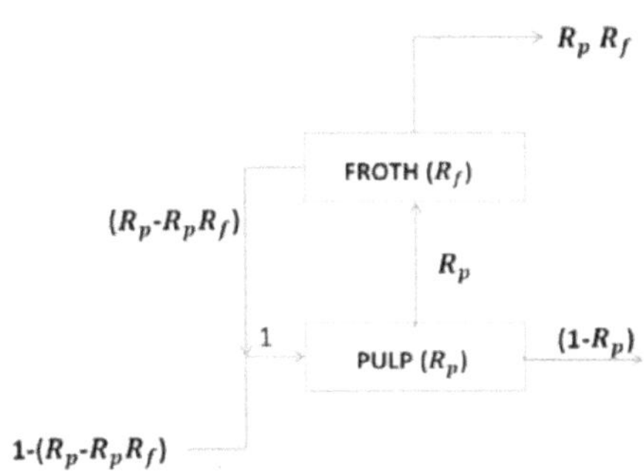

FIGURE 9.16 Schematic showing the two volumes in the flotation cell, pulp and froth.

Source: Huang, K. 2015. Validation and application of a first principle flotation model. MSc thesis in Mining and Mineral Engineering, Virginia Polytechnic Institute and State University.

TABLE 9.1
Flotation Micro Processes with Their Physicochemical Hydrodynamic Equations

Volume Phases	Physicochemical parameter	Mathematical expression	References
Recovery in the Pulp	Rate constant in the pulp phase k_p Number of bubbles N_B Particle velocity u_p Bubble velocity u_B Probability of flotation P Collision diameter d_C Energy dissipation of bubble $\in_B$	$k_p = 2^{3/2}\pi^{1/2}N_B d_C^2\sqrt{\bar{u}_p^2+\bar{u}_B^2}P$ $u_B = \left(2(\in d_B)^{2/3}\right)^{1/2}$ $d_2 = \left(\frac{2.11\gamma_{GL}}{\rho_3 \in_B^{0.66}}\right)^{0.6}$	Abrahamson (1975) Lee et al. (1987) Schulze (1977, 1983)
	Energy dissipation rate $\in$ RMS particle velocity $\bar{u}_1$ Particle diameter d_p Bubble diameter d_B Particle density ρ_p Liquid density ρ_L Water kinematic viscosity υ	$u_B = \left(2(\in d_B)^{2/3}\right)^{1/2}$ $\bar{u}_1 = 0.4\frac{\in^{4/9} d_S^{7/9}}{\upsilon^{1/3}}\left(\frac{\rho_p-\rho_L}{\rho_L}\right)^{2/3}$	Schubert (1999)
	Probability of flotation P Probability of collision P_c Probability of adhesion Pa Probability of detachment P_d Probability of transfer particle-bubble from pulp to froth P_t Reynolds Number Re Energy barrier E_1 Electrostatic, van der Waals, and hydrophobic forces E_E, E_W, E_H Interfacial liquid-gas tension γ_{LG} Contact angle θ	$P = P_c P_a(1-P_d)P_t$ $P_c = \tanh^2\left(\sqrt{\frac{3}{2}}\left[1+\frac{3}{16}\left(\frac{\text{Re}}{1+0.249R^{0.56}}\right)\right]\left(\frac{d_p}{d_B}\right)\right)$	Xu and Yoon (1989); Luttrell and Yoon (1992) Yoon and Mao (1996) Yoon et al. (1997); Do (2010)

		$P_a = \exp\left(-\frac{\gamma_{LV}\pi r_p^2 (1-\cos\theta) E_1}{0.5m_p\left((d_S + d_B)\sqrt{\frac{\in}{\upsilon}}\right)^2}\right)$	
		$E_1 = E_E + E_W + E_H$	
		$P_d = \exp\left(-\frac{\gamma_{LG}\pi r_p^2 (1-\cos\theta)^2 + E_1}{0.5m_1\left((d_p + d_B)\sqrt{\in/\upsilon}\right)^2}\right)$	
Recovery in the Froth	Recovery in the froth R_f Overall recovery R	$R_f = R_{\max}\exp(-\alpha\tau_f)$ $R = \frac{R_p R_f}{R_p R_f + (1-R_p)}$	Huang (2015)
Behavior of Composite Particles	Contact angle of composite particle $\bar{\theta}$ Contact angle of the particle i, θ_i Fractional surface liberation a_i Fitting parameter b_i	$\bar{\theta} = \exp(a_1 b_1 \ln\theta_1 + a_2 b_2 \ln\theta_2)$	Huang (2015)

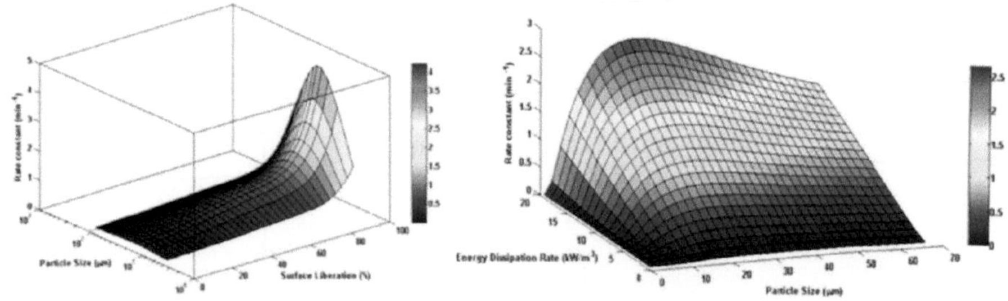

FIGURE 9.17 Effect of particle size, surface liberation, and energy dissipation rate on overall rate constant (k).

Source: Huang, K. 2015. Validation and application of a first principle flotation model. MSc thesis in Mining and Mineral Engineering, Virginia Polytechnic Institute and State University.

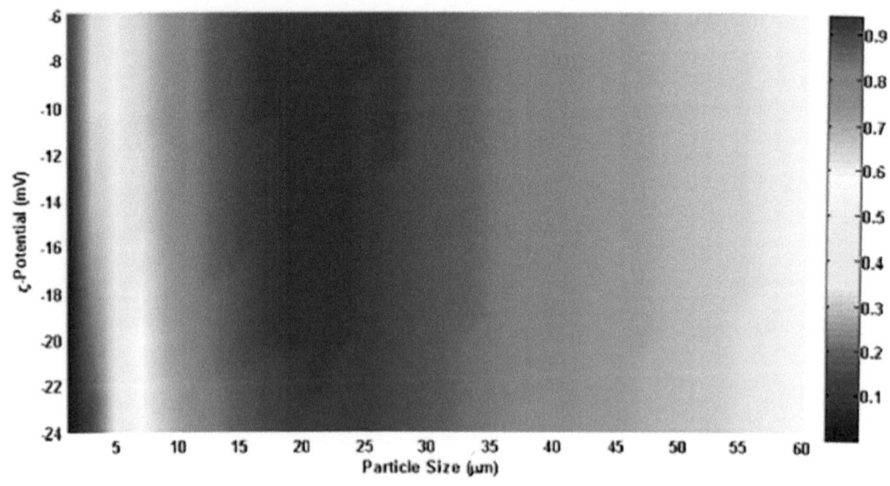

FIGURE 9.18 Effect of particle size and ζ-potential on overall rate constant (k).

Source: Huang, K. 2015. Validation and application of a first principle flotation model. MSc thesis in Mining and Mineral Engineering, Virginia Polytechnic Institute and State University.

Table 9.1 summarizes the flotation behavior equations. Huang's first-principle flotation model could predict the performance of a single flotation unit or industrial flotation circuits. Data for each simulation are expressed in Huang (2015). Figures 9.17, 9.18, 9.19, and 9.20 shows the effect of the selected variables of the flotation modeling on a single flotation cell.

For a batch flotation of galena with the following properties (Welsby et al. 2010). Table 9.2 shows the data.

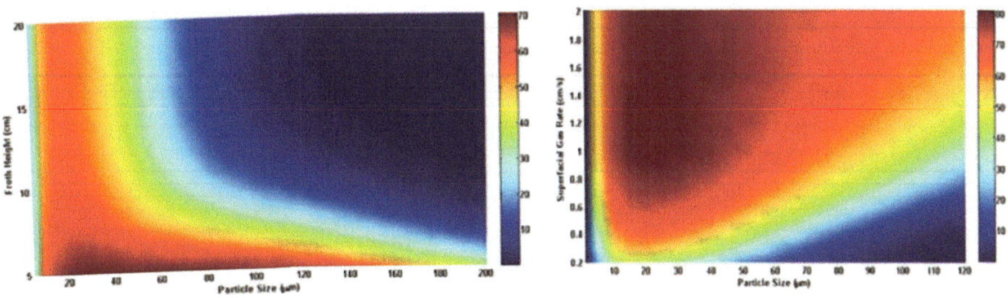

FIGURE 9.19 Effect of particle size and froth height and superficial gas rate on flotation recovery (*R*).

Source: Huang, K. 2015. Validation and application of a first principle flotation model. MSc thesis in Mining and Mineral Engineering, Virginia Polytechnic Institute and State University.

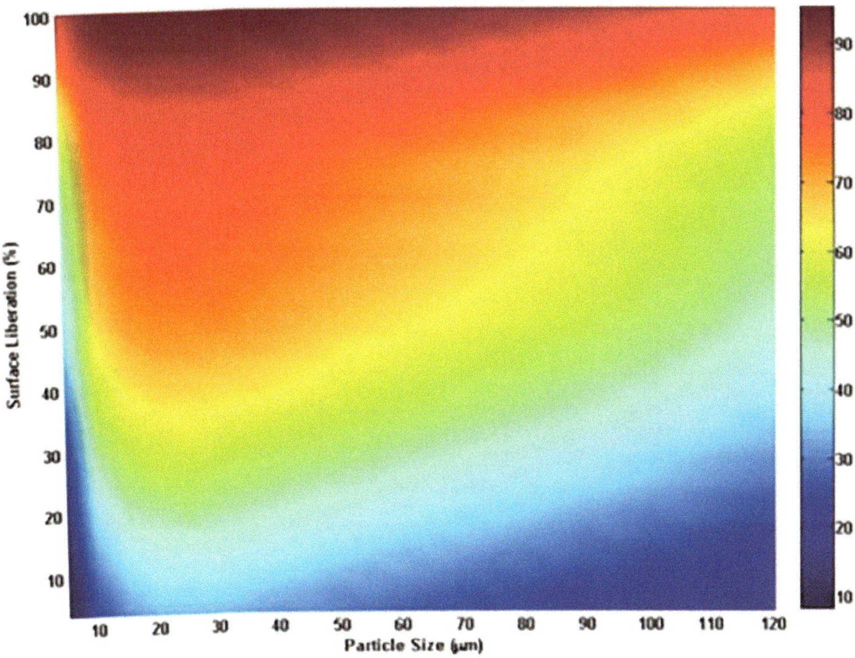

FIGURE 9.20 Effect of particle size and surface liberation on flotation recovery (*R*).

Source: Huang, K. 2015. Validation and application of a first principle flotation model. MSc thesis in Mining and Mineral Engineering, Virginia Polytechnic Institute and State University.

TABLE 9.2
Liberation Classes by Size

m_{ij} (%)	Liberation Classes (%)											
Particle Size Classes (μm)	0	0~10	10~20	20~30	30~40	40~50	50~60	60~70	70~80	80~90	90~100	100	Total Mass
+106	22.46	4.12	0.64	0.27	0.15	0.05	0.03	0.02	0.01	0.01	0.01	0.05	27.83
+75/−106	9.28	1.05	0.17	0.08	0.06	0.04	0.03	0.01	0.01	0.02	0.03	0.16	10.95
+38/−75	15.02	1.73	0.31	0.16	0.11	0.08	0.06	0.05	0.04	0.05	0.13	0.96	18.70
+28/−38	4.70	0.30	0.07	0.04	0.03	0.02	0.02	0.02	0.01	0.02	0.05	1.57	6.85
+19/−28	7.15	0.30	0.07	0.04	0.03	0.02	0.02	0.02	0.01	0.01	0.05	1.44	9.15
+10/−19	8.87	0.27	0.09	0.05	0.03	0.04	0.03	0.02	0.02	0.02	0.07	2.05	11.55
−10	11.09	0.14	0.11	0.06	0.04	0.04	0.03	0.04	0.03	0.04	0.05	3.29	14.96
Total	78.57	7.90	1.46	0.69	0 45	0.29	0.22	0.17	0.15	0.16	0.40	9.51	100.00

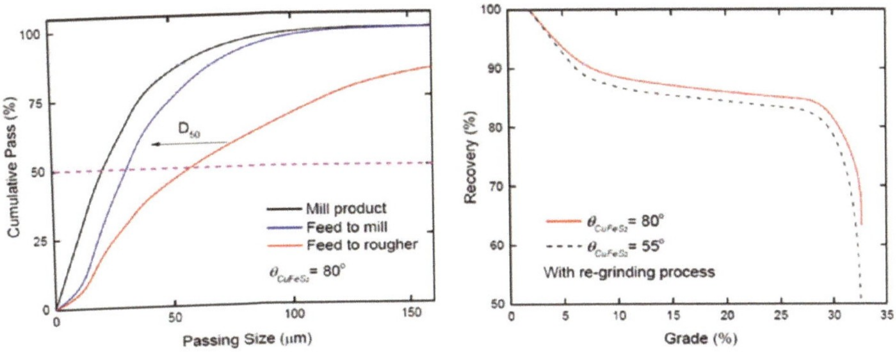

(a) Particle sizes un the circuit

(b) Recovery after regrinding

FIGURE 9.21 Effect of re-grinding to the particle size and contact angle on the grade-recovery curve of chalcopyrite with two different collectors.

9.3.7.1 Continuous Flotation Circuit

Huang (2015) simulated the performance of a flotation circuit similar to that of the Escondida Copper Flotation Plant of Chile (110.000 tpd). He studied the effects of particle size and particle hydrophobicity using different collectors in the re-grinding circuit. The results show that the re-grinding of rougher scavenger concentrate greatly increased the overall copper recovery, which can be attributed to the increased flotation rate with increasing surface liberation.

The simulation results show (see Figure 9.21) also that an increase in contact angle by way of using a stronger collector greatly increased the copper recovery at the rougher flotation circuit. These results are consistent with the plant practice, demonstrating the benefits of using a first-principle flotation model/simulator to improve the performance of the real-world flotation plants.

REFERENCES

Abrahamson, J. 1975. Collision rates of small particles in vigorously turbulent fluis. *Chemical Engineering Science* 30:1371–1379.

Ayedzi, L.D., Abaka-Wood, G.B., Zanin, M., and Skinner, W. 2022. A brief review of fine particles flotation, Procemin-Geomet. In *18th International Conference on Mineral Processing and Geometallurgy*, Gecamin, Santiago. https://ww.researchgate.net/publication/3370100865

Baas, D. 2016. *Automated Rougher Process Control Project, ENG8308, Master Engineering Practice*. Australia: University of Southern Queensland, New South Whales.

Barber, A.D., and Hartland, S. 1975. The collapse of cellular foams. *Transactions of the Institution of Chemical Engineers* 53:106–111.

Bascur, O.A. 1976. Phenomenological model of suspensions in sedimentation. Engineering thesis, University of Concepción [in Spanish].

Bascur, O.A. 1991. A unified solid liquid separation framework. *Fluid Particle Separation Journal* 4:117–122.

Bascur, O.A. 2010. A flotation framework for dynamic performance monitoring. In *Proceedings of the 2nd International Congress on Automation in the Mining Industry: Automining*, November 10–12.

Bascur, O.A. 2013. A dynamic flotation framework for performance management. In *Separation Technologies for Minerals, Coal and Earth Resources*. Edited by C. Young and G. Luttrel. Englewood, CO: SME. pp. 40–45.

Bascur, O.A. 2019. Chapter 2.6: Process control and operational intelligence. In *SME Mineral Processing and Extractive Metallurgy Handbook*. Edited by R.C. Dunne, S.K. Kawatra, and C.A. Young. Englewood, CO: SME.

Bascur, O.A. 2023a. Digital twins in mining and mineral processing. In *26th World Mining Conference Proceedings, Resourcing Tomorrow, Creating Value for Society, Artificial Intelligence Invited Lecture*, Brisbane. www.wmc2023.org

Bascur, O.A. 2023b. Dynamic water management to maximize metal production. In *International Flotation Symposium, Celebrating Science and Engineering Innovations, ARC Centre for Excellence for Enabling Beneficiation of Minerals*, University of Newcastle, Newcastle, July 14.

Bascur, O.A., and Herbst, J.A. 1982. Dynamic modelling of a flotation cell with a view toward automatic control. Presented at XIV IMPC, Toronto. www.researchgate.net/publication/284604831_Dynamic_Modelling_of_a_Flotation_Cell_with_a_view_towards_Automatic_Control.

Bascur, O.A., and Herbst, J.A. 1985. On the development of a model-based control strategy for copper ore flotation. In *Proceedings of Flotation of Sulphides Minerals*. Edited by K.S. Eric Forssberg. New York: Elsevier. pp. 409, 431.

Bascur, O.A., and Junge, P.A. 2011. Dynamill, DynaFlote and Optimill [iPad software apps]. www.apple.com/app-store/.

Bascur, O.A., and Retamal, F. 2021. Integrated rock processing strategies in the digital age. Presented at Minería Digital, Gecamin, Santiago, October 17–19.

Bascur, O.A., and Retamal, S. 2023. Increasing water recovery to maximize copper production with low-grade ores. In *Proceedings of the 62nd Conference of Metallurgists, COM 2023*. Springer Nature, Toronto, Canada.

Bascur, O.A., and Soudek, A. 2019. Grinding and flotation optimization using operational intelligence. *Mining, Metallurgy and Exploration* 36:139–149. https://doi.org/10.1007/s42461-018-0036-4.

Bu, X., Xie, G.-Y., Peng, Y.-L., Ge, L., and Ni, C. 2017. Kinetics of flotation: Order process, rate constant distribution and ultimate recovery. *Physicochemical Problems of Mineral Processing* 53(1):342–365.

Calderbank, P.H. 1958. Physical rate processes in industrial fermentation, part 1. *Transactions of Institution Chemical Engineers* 36:443–463.

Chawda, N. 2019. IoT opportunities for Industry 4.0. at Vedanta HZL: Empower your analytics with operational data. Presented at OSIsoft Mumbai Seminar, Mumbai.

Clift, R., Grace, J.R., and Weber, M.E. 1980. *Bubbles, Drops and Particles*. New York: Academic Press.

Concha, F., and Bascur, O.A. 1977. A phenomenological model of sedimentation. *Proceedings of IMPC XII: Special Publication* 1:29–46.

Concha, F., and Delgado, O. 1977. Modelo de flotación con velocidad específica distribuída. *III Seminario de Química de Superficie y Flotación, Avances en Flotación, Vol. 3*. Edited by S. Castro and J. Alvarez. Concepción: Universidad de Concepción.

Copper, A. 2018. Personal Communication Letter.

Dahlke, R., Finch., J.A., Gomez, C.O., Cooper, M., and Scott, D. 2004. Impact of air distribution profile on banks in a Zn cleaning circuit. Presented at the 36th Annual Meeting of the Canadian Mineral Processors, Ottawa, ON.

Dickinson, J.E., and Galvin, K.P. 2014. Fluidized bed desliming in fine particle flotation—Part I. *Chemical Engineering Science* 108:283–298.
Do, H. 2010. Development of a turbulent flotation model from first principles. PhD thesis, Engineering Mechanics, Virginia Polytechnic Institute and State University.
Doucet, N. 2017. Process improvement using PI system at Vale Voisey Bay. In *OSIsoft Users Conference*, San Francisco, CA.
Fitch, B. 1979. Sedimentation of flocculent suspensions: State of the art. *AIChE Journal* 25(6):913–930.
Fuerstenau, D.W. 1999. The froth flotation century. In *Advances in Flotation Technology*. Edited by B.K. Parek and J.D. Miller. Littleton, CO: SME. pp. 3–21.
Galvin, K.P., and Dickinson, J.E. 2014. Fluidized bed desliming in fine particle flotation—Part II. Flotation of a model feed. *Chemical Engineering Science* 108:299–309.
Gaudin, A.M. 1957. *Flotation*. New York: McGraw-Hill. pp. 333, 351, 359, 362.
Harris, C.C. 1976. Flotation machines. In *Flotation: A.M. Gaudin Memorial Volume*, vol. 2. Edited by M.C. Fuerstenau. New York: American Institute of Mining, Metallurgical and Petroleum Engineers. pp. 753–815.
Hay, M., and Rule, C. 2003. SUPASIM: A flotation plant design and analysis methodology. *Minerals Engineering* 16(11):1103–1109.
Herbst, J.A., and Bascur, O.A. 1985. *Alternative Control Strategies for Coal Flotation, Minerals and Metallurgical Processing*. Englewood, CO: SME. pp 1–9.
Huang, K. 2015. Validation and application of a first principle flotation model. MSc thesis in Mining and Mineral Engineering, Virginia Polytechnic Institute and State University.
Imaizumi, T., and Inoue, T. 1963. Kinetic consideration of froth flotation. In *Proceedings of the Sixth International Mineral Processing Congress*, Cannes. p. 581.
Jameson, G.J. 2023. New ways of extending the flotation of fines and large particle sizes. In *International Flotation Symposium in Honor of Distinguished Laureate Professor Graeme Jameson*. New Castle University, Australia. https://smi.uq.edu.au/blog/2023/07/international-flotation-symposium-honour-distinguished-laureate-professor-graeme-jameson
Jameson, G.J., Nam, S., and Moo Young, M. 1977. Physical factors affecting recovery rates in flotation. *Minerals Engineering* 3(9):103–118.
Jovanovic, I., and Miljanović, I. 2015. Modelling of flotation processes by classical mathematical methods: A review. *Archives of Mining Sciences* 60(4):905–919.
Kelley, K.R. 2011. Model based computer simulation of froth flotation. MSc thesis, Virginia Polytechnic Institute and State University.
Kynch, G.J. 1952. A theory of sedimentation. *Transactions of the Faraday Society* 48:166–182.
Laskowski, J. 1974. Particle-Bubble attachment in flotation. *Mineral Science and Engineering* 60(4):223–235.
Lee, C.-H., Erickson, L., and Glasgow, L. 1987. Bubble breakup and coalescence in turbulent gas-liquid dispersions. *Chemical Engineering Communications* 59(1–6):65–84.
Levich, V.G. 1962. *Physicochemical Hydrodynamics*. Englewood Cliffs, NJ: Prentice Hall.
Li, S., Wang, Y., Lu, D., Zheng, X., and Li, X. 2020. Improving separation efficiency of galena flotation using the aerated jet flotation cell. *Physicochemical Problems of Mineral Processing* 56(3):513–527.
López-Saucedo, F., Uribe-Salas, A., Perez-Garibay, R., and Magallanes-Hernandez, L. 2012. Gas dispersion in column flotation and its effect on gas recovery grade. *Canadian Metallurgical Quarterly* 51(2):111–117. http://dx.doi.org/10.1179/1879139511Y.0000000033.
Luttrell, G., and Yoon, R-H. 1992. A hydrodynamic model for bubble—particle attachment. *Journal of Colloid and Interface Science* 154(1):129–137.
Lynch, A.J., Johnson, N.W., Manlapig, E.V., and Thorne, C.G. 1981. *Mineral and Coal Flotation Circuits, Their Simulation and Control*. New York: Elsevier Scientific.

Mesa, D., Morrison, A.J., and Brito-Parada, P.R. 2020. The effect of impeller-stator design on bubble size: Implications for froth stability and flotation performance. *Minerals Engineering* 157:106533. https://doi.org/10.1016/j.mineng.2020.106533.

Mika, T.S., and Fuerstenau, M. 1968. A microscopic model of the flotation process. In *Proceedings of the Eight International Mineral Processing Congress*, vol. 2. Leningrad: Mechanobr. pp. 246–269.

Randolph, A.D., and Larson, M.A. 1971. *Theory of Particulate Processes*. New York: Academic Press.

Ruylov, N. 2020. Chapter 5: Kinetics of flotation process. In *Flotation: Physical Theory and Applications*. www.researchgate.net/publication/343255785

Schubert, H. 1999. On the turbulence-controlled microprocesses in flotation machines. *International Journal of Mineral Processing* 56(1): 257–276. Schubert, H., and Bischofberger, C. 1978. On the hydrodynamics of flotation machines. *International Journal of Mineral Processing* 5:131–142.

Schubert, H., and Bischofberger, C. 1981. On the optimization of hydrodynamics in flotation processes. In *Proceedings of the 13th International Mineral Processing Congress*, vol. 22. Edited by J. Laskowski. London: Elsevier. pp. 1261–1284.

Schubert, H., Heidenreich, E., Liepe, F., and Neese, T. 1979. *Mechanische Verfarenstechnick I*. Leipzig: Deutscher Verlag. p. 75.

Schulze, H.J. 1977. New theoretical and experimental investigations of stability on bubble/particle aggregates in flotation: A theory of the upper particle size floatability. *International Journal of Mineral Processing* 4:241–259.

Schulze, H.J. 1983. *Physicochemical Elementary Processes in Flotation*. Amsterdam: Elsevier Science Publishers. p. 348.

Smidth, F.L. 2010. *Flotation Technology*. Salt Lake City, UT: FLSmidth Minerals Processing Technology Center.

Spedden, H.R., and Hannan, W.S. Jr. 1948. Attachment of mineral particles size suspensions. *American Institute of Chemical Engineers Journal* 10(4):517–523.

Trahar, W.J. 1981. A rational interpretation of the role of particle size in flotation. *International Journal of Mineral Processing* 8:289–327.

Villeneuve, J.J., Guillaneau, C., and Durance, M-V. 1995. Flotation modelling: A wide range of solutions for solving industrial problems. *Minerals Engineering* 8(4):409–420.

Wallis, G.B. 1969. *One-Dimensional Two-Phase Flow*. New York: McGraw Hill.

Welsby, S., Vianna, S., and Franzidis, J.P. 2010. Assigning physical significance to floatability components. *International Journal of Mineral Processing* 97(1):59–67.

Woodburn, E.T., King, R.P., and Colburn, R.P. 1971. The effect of particle size distribution and the performance of a phosphate flotation process. *Metallurgical Transactions* 2:3163–3174.

Xu, Z., and Yoon, R.H. 1989. The role of hydrophobia interactions in coagulation. *Journal of Colloid and Interface Science* 132(2):532–541.

Yoon, R.-H., Flinn, D.H., and Rabinovich, Y.I. 1997. Hydrophobic interactions between dissimilar surfaces. *Journal of Colloid and Interface Science* 185(2):363–370.

Yoon, R.-H., and Mao, L. 1996. Application of extended DLVO theory, IV: Derivation of flotation rate equation from first principles. *Journal of Colloid and Interface Science* 181(2):613–626.

FURTHER READING

Bascur, O.A. 2005. Example of a dynamic flotation framework. In *Proceedings of the Centenary of Flotation Symposium at Brisbane*. Edited by M.C. Fuerstenau, G. Jameson, and R.H. Yoon. CO: SME. pp. 85–88.

BHP Billiton. 2002. Minera Escondida: Merrill Lynch Investor Visit, November 18. www.slideshare.net/robertoiq20/escondida-2002.

Bu, X., Xie, G., Chen, Y., and Ni, C. 2016. The order of kinetic models in coal fines flotation. *International Journal of Coal Preparation and Utilization* 37(3):113–123. https://doi.org/10.1080/19392699.2016.1140150.

Savassi, O. 1999. Direct estimation of the degree of entrainment and the froth recovery of attached particles in industrial flotation cells. PhD thesis, University of Queensland.

Schubert, H. 1974. Zur proze bestimmenden Rolle der Turbulenz bei Aufberei tungsprozessen-2. *Teil, Aufbereitungs Tecknik* 12:680–685.

Schubert, H. 2008. On the optimization of hydrodynamics in fine particle flotation. *Mineral Engineering* 21:980–936.

10 Solid-Liquid Separation

10.1 INTRODUCTION

Solid-liquid separation is a process that has a double function in mineral processing plants. First, the product of a mineral processing plant usually contains around 70% to 80% of water that must be lowered to a humidity of less than 10% to be suitable for transport by sea to their final destination. On the other hand, the effect of the ever-increasing world drought in mining sector requires maximizing the recovery of water from the tailings to no more than 30% before depositing them in tailing ponds. Thickening, filtration, rheology, and suspension transport, the main processes of solid-liquid separation, have many common elements, especially when it comes to the relative flow between solids and fluids. In some cases, the flow consists of particles settling through a suspension, while in others the suspension moves freely through tubes or through a consolidated particle structure. Despite their common mechanism, the various solid-liquid separation processes have been independently developed by individuals with different interests.

In recent years there has been an important advance in understanding the phenomena that occur in the solid-liquid separation process. Empirical relationships for the design of thickeners have given way to methods based on phenomenological models, which have been validated in practice. Better methods of quantifying flocculant dosing and new instruments have been designed for simultaneous and automatic measurement of sedimentation and compression parameters, along with simulation software, which allows metallurgists to predict the behavior of thickeners and establish simulation-based operation and control strategies.

10.2 THICKENING

Thickening and *clarification* involve the settling of a suspension of solid particles through a liquid form. While clarification deals with very dilute suspensions, thickening is performed to produce more concentrated pulps. Thickening is performed in a cylindrical tank with diameter much larger than its height. It is supplied with a feedwell at the top center where the dilute pulp is fed to the equipment. As soon as the suspension leaves the feedwell, it is diluted to a lower concentration and start settling. When the particles reach the bottom of the tank, it accumulates forming a bed called sediment. The sediment discharge from the lower central thickener by one or more tubes as a more concentrated pulp. The water, which is displaced from the suspension as solid particles settle, rises to the top of the tank and overflows through a ring located on the top outer area of the thickener. In order to facilitate the sediment discharge from the bottom of the thickener, its bottom has conical shape over which a mechanism that drags the sediment towards the discharge circulates. See Figure 10.1.

DOI: 10.1201/9781032614212-10

FIGURE 10.1 (Top) Industrial thickener, (bottom left) pulp discharge from the bottom, and (bottom right) water overflow.

10.2.1 Ideal Thickeners

To be able to model an industrial thickener with all its component, several models of which were shown in the previous section, we will define an *ideal thickener*, which is an idealized version consisting in a cylindrical tube with a diameter several times greater than the particle sizes of the suspension. This assumption allows one to neglect the friction of fluid and particle at the walls and consider the flow one-dimensional. A laboratory settling column is a *batch ideal thickener*. If we add to this batch ideal thickener a feed, discharge, and overflow, we have an *ideal continuous thickener*. See Figure 10.2.

10.2.2 Sedimentation

Several properties of the solid and the fluid in a suspension fed to a thickener affect the performance of the thickening process (see Figure 10.3); properties of the mineral particles such as minerology, density and particles size and size distribution; the density, viscosity, surface tension, and pH of the fluid are also important. Although it is difficult to measure that many properties, we are still able to detect their effect on some *measured variables*. The two most important are the suspension particle concentration and its settling velocity.

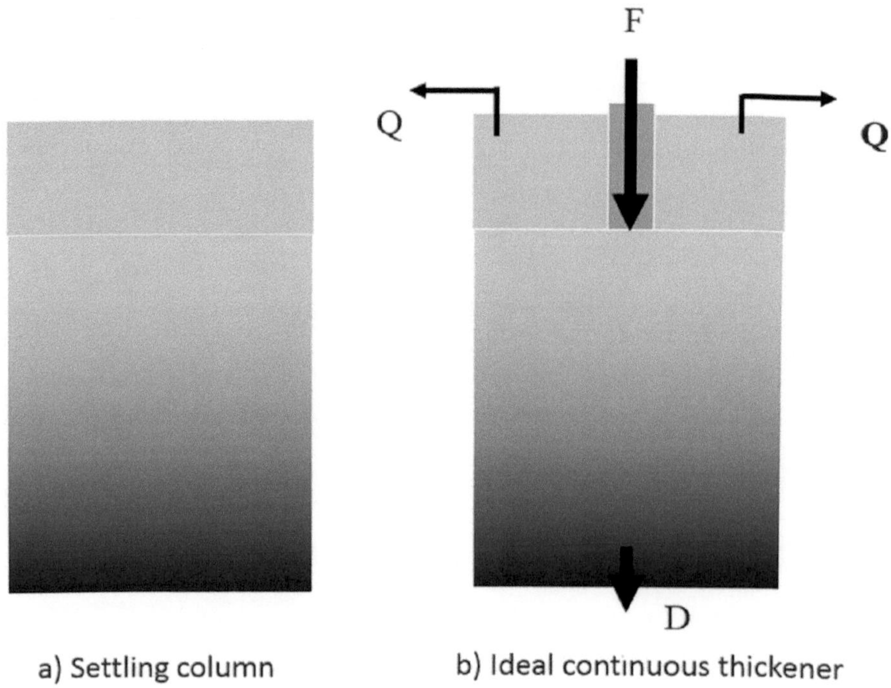

FIGURE 10.2 Ideal thickeners.

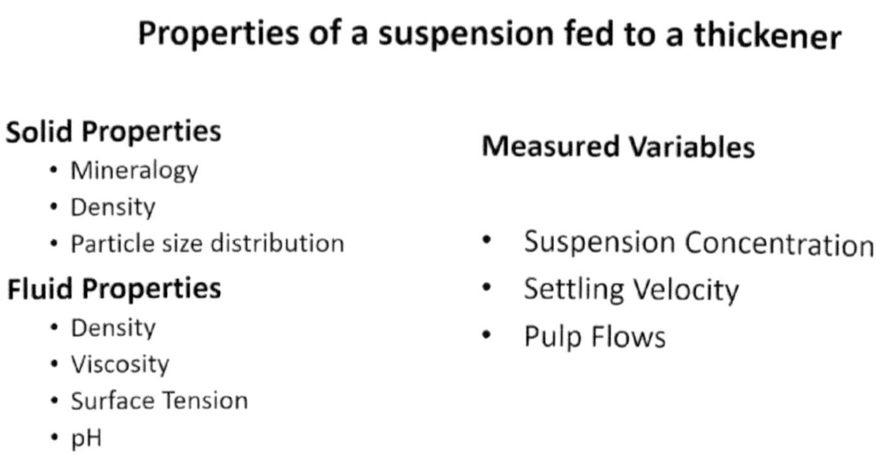

FIGURE 10.3 Material properties and measured variables.

10.2.2.1 Settling Velocity

Suspension mineral particles are quantified by their solid concentration and its settling velocity. The concentration can be measured in several ways, as fraction of solid by volume φ, percent solids, pulp density ρ, and dilution D. If the solid and fluid densities are chosen by ρ_s and ρ_f, the conversion between the different designation of the concentrations are given here.

$$\begin{aligned}
\text{Volume fraction } \varphi &= \frac{\rho-\rho_f}{\rho_s-\rho_f} = \frac{\rho_f X}{\rho_s(1-X)+\rho_f X} = \frac{\rho_f}{\rho_f+\rho_s D} \\
\text{Percent solid } X &= 100\frac{\rho_s\varphi}{\rho_f+\varphi\Delta\rho} = 100\frac{\rho_s\left(\rho-\rho_f\right)}{\rho\Delta\rho} = 100\frac{1}{1+D} \\
\text{Density } \rho &= \frac{\rho_s\rho_f}{\rho_s(1-X)+\rho_f X} = \rho_f+\Delta\rho\varphi = \frac{\rho_s\rho_f(1-D)}{\rho_s D+\rho_f} \\
\text{Dilution } D &= \frac{\rho_f(1-\varphi)}{\rho_s\varphi} = \frac{1-X}{X} = \frac{\rho_f\left(\rho_s-\rho\right)}{\rho_s\left(\rho-\rho_f\right)}
\end{aligned} \tag{10.1}$$

10.2.2.2 Settling Curves

The sedimentation of solid particles in a suspension consist of the gravity settling of particles or flocs in water. As the concentration of the suspension increases, the interaction with the other particles of the suspension will retard its sedimentation; therefore, there is an intimate relationship between the two variables sedimentation and concentration. The settling velocity of a suspension at a given solid concentration may be expressed as a curve, such as that shown in Figure 10.4. The initial slope (strait line) corresponds to the initial velocity at the concentration of $\varphi = 0.100$, and the line after 4.0 [s] represents the decreasing velocity thereafter (Garrido et al. 2004).

10.2.2.3 Solid Flux Density

Neither by themselves the settling velocity and the concentration completely define sedimentation; therefore, it is convenient to combine both variables in just one called *solid flux density* defined by

$$\begin{aligned}
f_{bk}(\varphi) &= -\varphi v_s \qquad \left[\frac{\mathrm{m_s^3}}{\mathrm{m^3}}\times\frac{\mathrm{m}}{\mathrm{h}}\right]=\left[\frac{\mathrm{m}}{h}\right] \\
F_{bk}(\varphi) &= -\rho_s f_{bk}(\varphi)\left[\frac{\mathrm{ton}}{\mathrm{m_s^3}}\times\frac{\mathrm{m}}{\mathrm{h}}\right]=\left[\frac{\mathrm{tph}}{\mathrm{m^2}}\right]
\end{aligned} \tag{10.2}$$

It is convenient to use the second version of the solid flux density, which incorporates the solid density, because it gives from a batch settling test the solid tonnage per square meter of a thickener. If the solid flux density refers to a determined thickener from which its surface area S is known, it is convenient to use the following version that gives the capacity of the thickener.

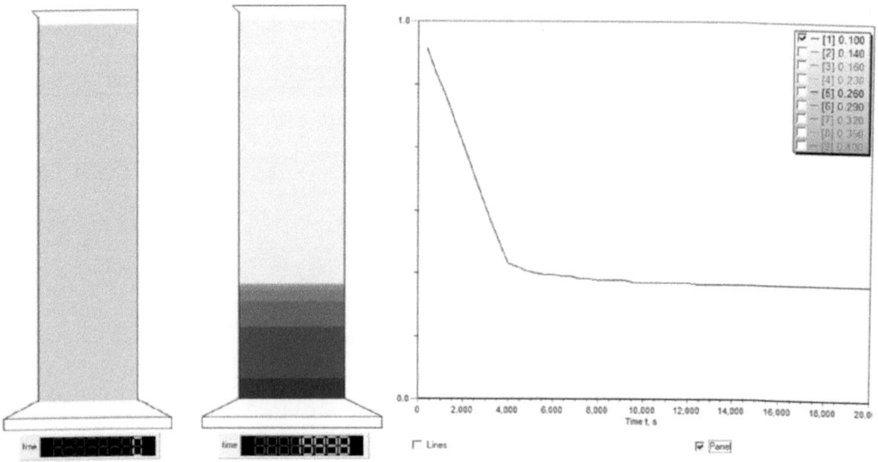

FIGURE 10.4 Sedimentation of a suspension showing the initial settling velocity at volume fraction $\varphi = 0.100$ (straight line) and concentration from $\varphi = 0.100$ to $\varphi = 0.230$.

$$F_b(t) = -\rho_s S f_{bk}(\varphi)\left[\frac{\text{tph}}{\text{m}^2} \times \text{m}^2\right] = [\text{tph}] \tag{10.3}$$

The relationship between the sedimentation velocity and concentration is usually represented by the following equation

$$v_s(\varphi) = u_\infty \varphi (1-\varphi)^c \tag{10.4}$$

Where $v_s(\varphi)$ is the settling velocity at concentration φ, u_∞ is the first parameter, the settling velocity of one particle in the fluid, and c is the second parameter. The solid flux density function corresponding to this function is given in Figure 10.5.

The importance of the solid flux density function is that, for any suspension concentration of a given material, it indicates the velocity that the particles of that suspension will acquire when it is introduced in a settling column.

If we add to the batch ideal thickener of Figure 10.2b a feed, we have a *continuous ideal thickener*, shown in Figure 10.6. For example, if $F = -0.237\left[\text{tph}/\text{m}^2\right]$ of a material of solid density of $\rho_s = 2.534\left[\text{ton}/\text{m}^3\right]$ and a pulp of volume flow rate discharge $Q_D = -4.5 \times 10^3\left[\text{m}^3/\text{h}\right]$ traveling in the thickener, there will be particles sedimenting in this pulp with a batch flux density function $F_{bk}(\varphi)\left[\text{tph}/\text{m}^2\right]$, then the total flow in continuous sedimentation will be the sum of these two flows $F_k(\varphi) = \rho_s Q_D \varphi + F_{bk}(\varphi), \left[\text{tph}/\text{m}^2\right]$. See Figure 10.6.

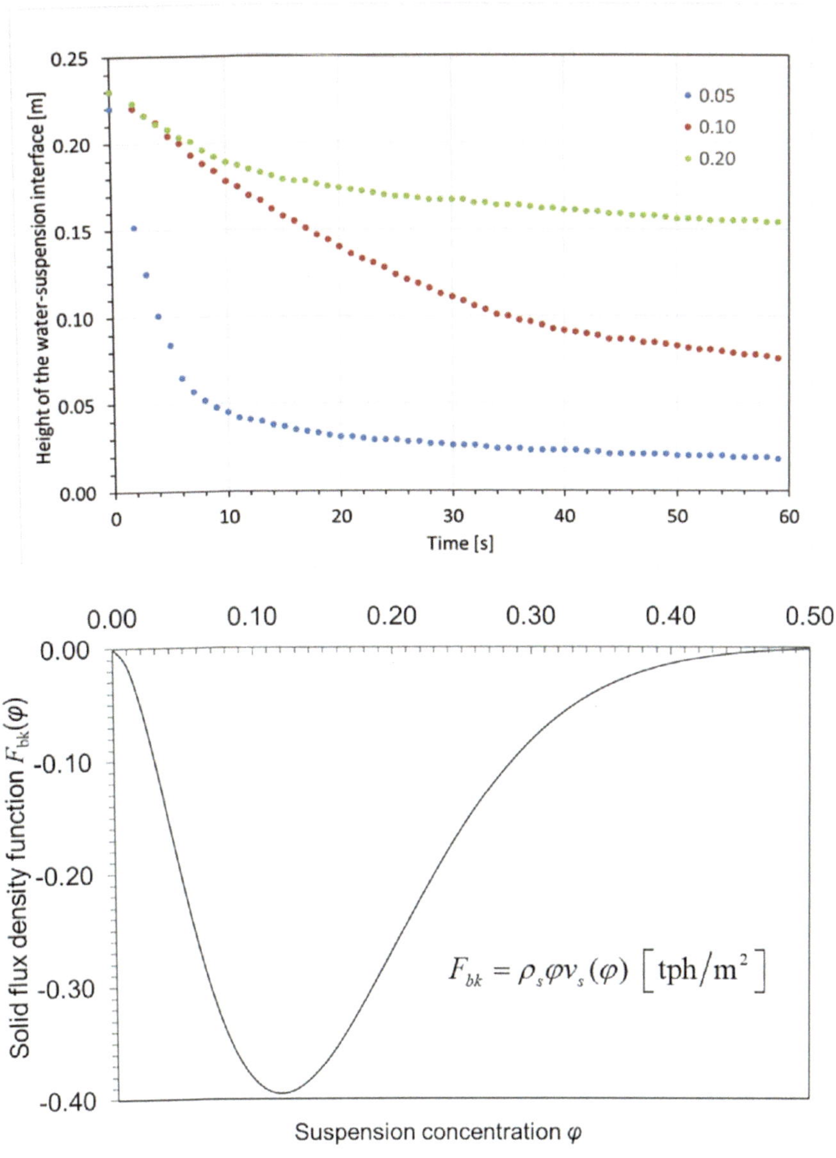

FIGURE 10.5 Settling plot and solid flux density function.

When incompressible particles, such as glass beads, settle in a batch column and reach the bottom of the vessel, they will lay one above the other, and the whole process ends (Quispe et al. 2000), as shown in Figure 10.7.

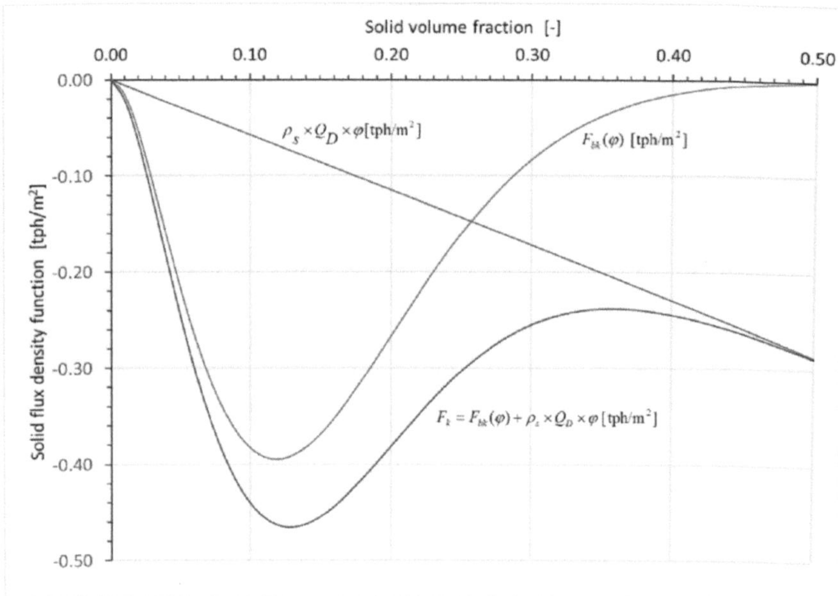

FIGURE 10.6 Batch and continuous flux density function.

But if the particles are compressible, as is the case of flocculated copper flotation tails, the weight of the bed of flocs in the sediment compresses the flocs lying underneath, expressing the water out from the pores. This phenomenon of extracting water by compression is called *consolidation.*

10.2.2.4 Consolidation

Consolidation is an important discipline in several fields besides thickening, for example in geotechnique, which studies the behavior of the ground when buildings are constructed. In this case, the applied force, called *solid effective stress*, compressing the soil causing consolidating, is external. In thickening, the solid effective stress force compressing the sediment is its own weight, which makes a big difference in the theory of consolidation due to the scale of the forces involved, internal and external.

For flocculated copper tailings, the sediment is compressible so that a concentration profile is established with porosity increasing in the direction to the bottom of the thickener. Therefore, one way to estimate the compressibility of the sediment is to measure a property, such that the differential pressure is along the sediment height. See Figure 10.8 (Concha 2014).

From this concentration profile, the solid effective stress σ_e can be obtained as a function of the sediment concentration.

$$\sigma_e(\varphi) = \alpha \exp(\beta\varphi) \tag{10.5}$$

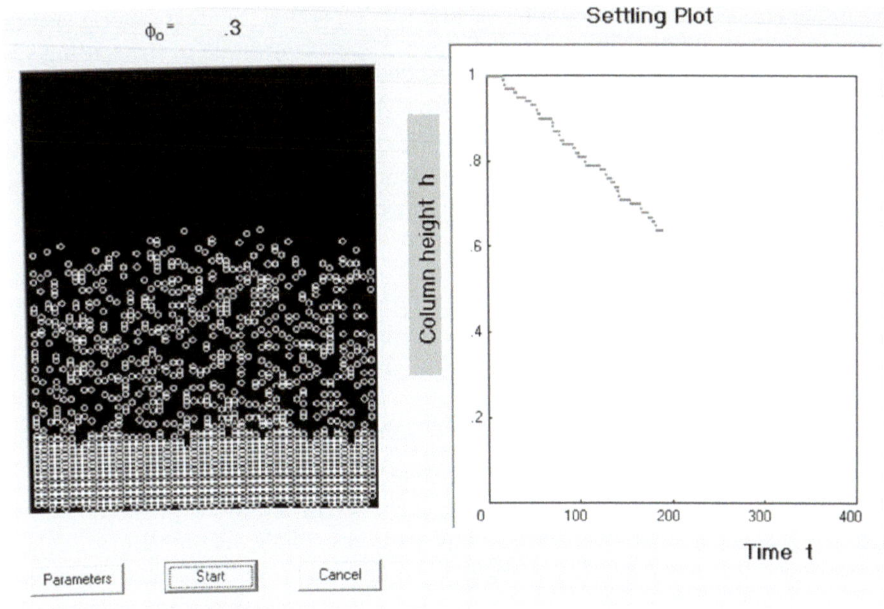

FIGURE 10.7 Settling of glass beads in water.

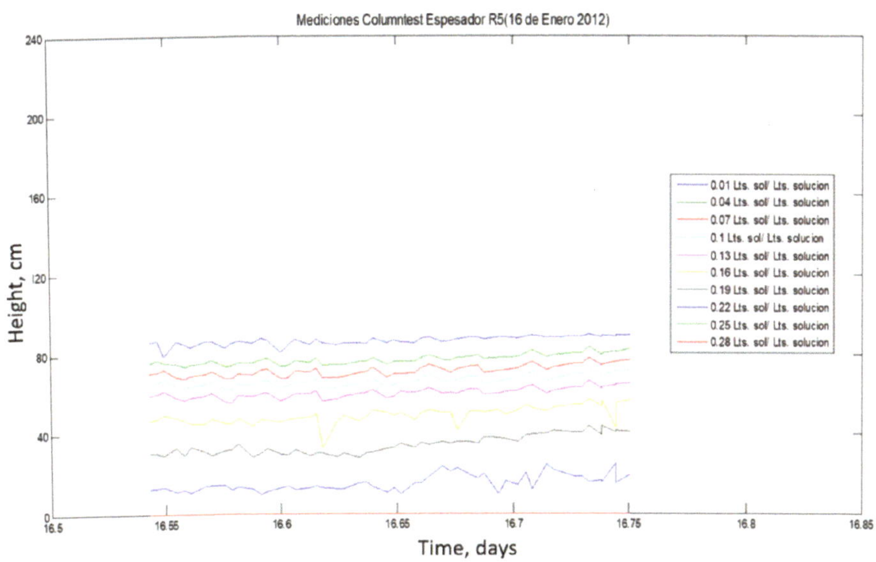

FIGURE 10.8 Concentration profile in a copper tailing thickener.

Sedimentation and consolidation are usually separated by two interfaces, the water-suspension and the suspension sediment. The concentration of the settling suspension at the first interface is called *conjugate concentration* at which the feed pulp is diluted in the feedwell. The second interface is the level of the sediment, also known as bed level, which concentration is called *critical concentration* φ_c, also known as *yield point* by authors coming from colloidal science. Figure 10.9 shows an example of the function. Figure 10.10 shows both interfaces in a copper tailing thickener.

10.2.2.5 Rheology of the Sediment

The rheology of copper sediment is a non-Newtonian fluid (see Chapter 6). Figure 10.11 presents, for a copper flotation tailing, the amount of stress to produce a certain rate of shear. Take note that the shear viscosity depends on the shear rate.

10.2.3 Phenomenological Theory of Sedimentation-Consolidation

In the previous section we described the sedimentation of compressible particles of a suspension when they are subjected to the force of gravity. We explained the relationship between the velocity acquired by the particles and the relationship with their concentration. We also explained that when the particles acquire critical concentration, they begin to be subjected to the weight of the particles resting on them. To quantify this process, the *phenomenological theory of sedimentation-consolidation* was postulated by Bascur (1976), Bascur and Barrientos (1977), and Barrientos (1978) and elaborated by Barrientos and Concha (1993), Bascur, (1991), Concha et al. (1995), Bürger and Concha (1998), and Bürger et al. (2000a).

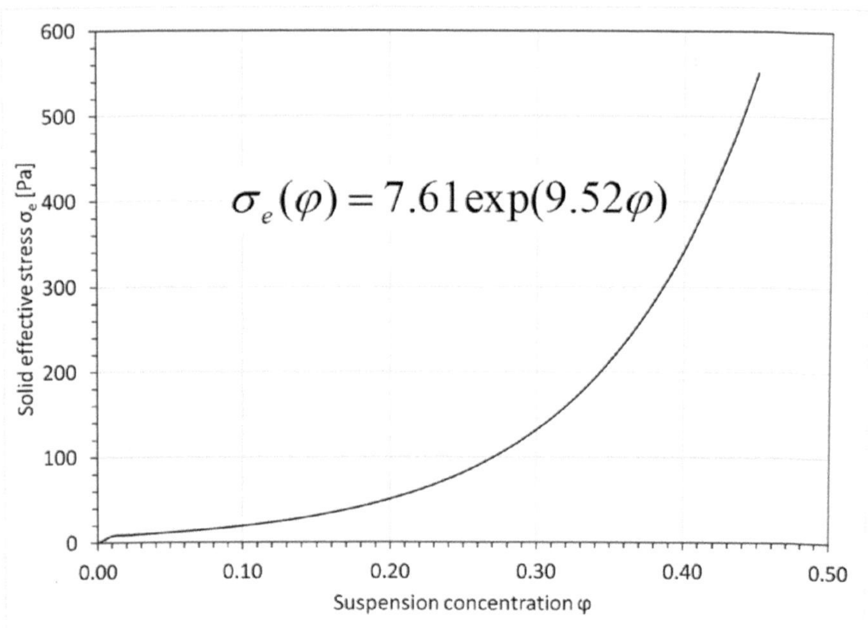

FIGURE 10.9 Solid effective stress.

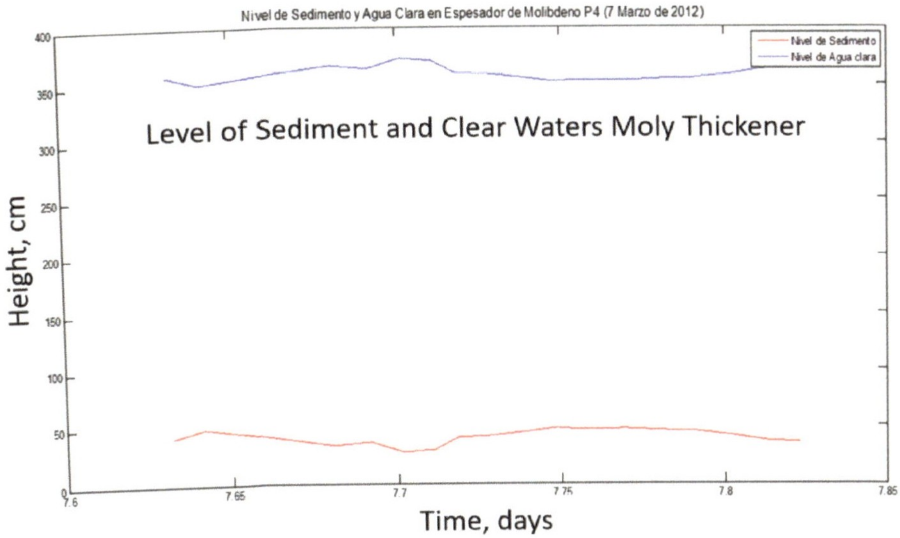

FIGURE 10.10 Water-suspension and suspension-sediment interface levels in a Molybdenum thickener.

Source: Concha 2014

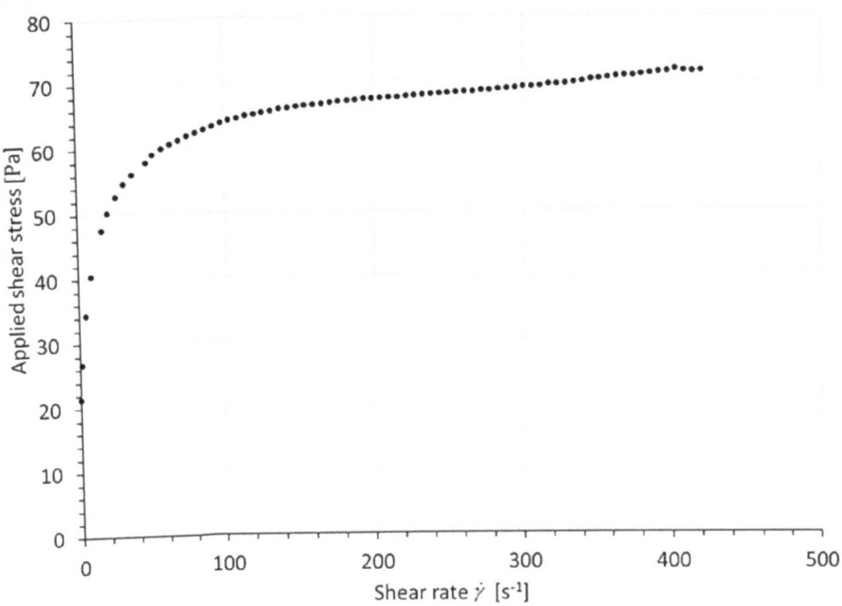

FIGURE 10.11 Rheogram of a copper flotation tailing.

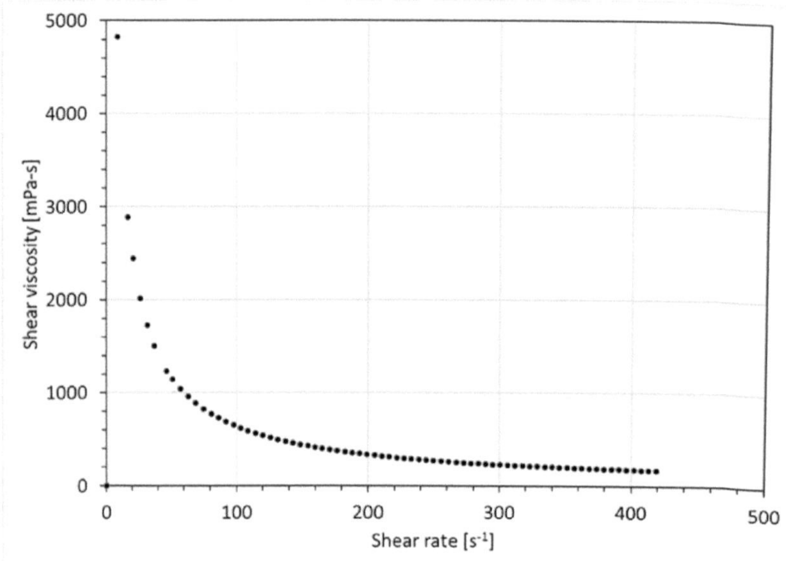

FIGURE 10.11 Continued

The phenomenological theory of sedimentation-consolidation assumes that a particulate system consists in a two superimposed continuous media that obey local mass and momentum balances and constitutive equations for the stresses and forces. The result is a nonlinear degenerate parabolic differential equation (Bürger and Concha 1998; Bürger et al. 1999) describing the sedimentation and consolidation of flocculated suspensions.

10.2.4 Dynamic Thickening Process

The properties involved in thickening are those describing sedimentation and consolidation; that is, (1) the suspension concentration, (2) the solid phase velocity, (3) the pulp velocity, (4) the solid-fluid interaction force, (5) the excess pore pressure, and (6) the compressibility of the sediment. Properties (1) to (6) can be represented by the following *field variables* (Concha 2014):

$$\begin{array}{ll} \text{Sediment concentration} & \varphi(z,t) \\ \text{Solid flux - density function} & f(z,t) \\ \text{Volume average velocity} & (z,t) \\ \text{Solid fluid interaction force} & m_d(z,t) \\ \text{Exess por pressure} & p_e(z,t) \\ \text{Solid effective stress} & \sigma_e(z,t) \end{array} \tag{10.6}$$

These variables and constitutive equation represent a *dynamic sedimentation-consolidation process* if they obey the following equations in those regions where the variables are continuous:

$$\frac{\partial \varphi}{\partial t}+\frac{\partial f}{\partial z}=0 \tag{10.7}$$

$$\frac{\partial q}{\partial z}=0, \text{ with } q=v_s-(1-\varphi)v_l \tag{10.8}$$

$$\frac{\partial \sigma_e}{\partial z}=-\Delta\rho\varphi g_z+\frac{m_d}{1-\varphi}v_r \tag{10.9}$$

$$\frac{\partial p_e}{\partial z}=-\frac{m_d}{1-\varphi}v_r \tag{10.10}$$

And in the discontinuities they need to obey the jump conditions.

$$\text{mass } \sigma[\varphi]=[f];[q]=0 \tag{10.11}$$

Since we have six field variables (Equation 10.6) and only four field equations (Equations 10.7–10.10), two experimental constitutive equations should be established to describe the properties of the material relating the kinematical variables v_s and the dynamical variables σ_e.

$$m_d=\begin{cases}-\mu K(\varphi)v_r & \text{for } \varphi<\varphi \\ -\dfrac{\mu}{k(\varphi)}(1-\varphi)^2 & \text{for } \varphi\geq\varphi\end{cases} \tag{10.12}$$

$$\sigma_e=\begin{cases}\text{constant} & \text{for } \varphi<\varphi_c \\ \alpha\exp(\beta\varphi) & \text{for } \varphi\geq\varphi_c\end{cases} \tag{10.13}$$

From Equation 10.9

$$m_d=\Delta\rho\varphi(1-\varphi)g+\frac{\partial\sigma_e}{\partial z}(1-\varphi)$$

For the settling regime $\varphi<\varphi_c$, $m_d=-\mu K(\varphi)v_r$, and

$$v_r=-\frac{\Delta\rho\varphi(1-\varphi)g}{\mu K(\varphi)} \tag{10.14}$$

and for consolidation $\varphi\geq\varphi_c$, $m_d=-\mu/k(\varphi)(1-\varphi)^2v_r$, and

$$v_r = \frac{k(\varphi)}{\mu} \times \frac{\Delta\rho\varphi g}{1-\varphi}\left(1 + \frac{\sigma_e'(\varphi)}{\Delta\rho\varphi g}\frac{\partial\varphi}{\partial z}\right) \tag{10.15}$$

Then,

$$f(\varphi) = q\varphi + f_{bk}(\varphi) \tag{10.16}$$

since $f = \varphi v_s$, obtaining v_r in terms of v_s and q from Equation 10.8 and defining $f_{bk}(\varphi)$ by

$$f_{bk}(\varphi) = -\frac{\Delta\rho\varphi^2(1-\varphi)^2 g}{\mu K(\varphi)} \text{ for } \varphi < \varphi_c \tag{10.17}$$

yields the balance equation (Equation 10.7) (Bürger et al. 1999)

$$\underbrace{\frac{\partial\varphi}{\partial t}}_{\substack{\text{Rate of change}\\ \text{of the concentration}}} + \underbrace{\frac{\partial}{\partial z}\left(q\varphi + f_{bk}(\varphi)\right)}_{\text{Flux for}\varphi<\varphi} = \underbrace{\frac{\partial}{\partial z}\left(-\frac{f_{bk}(\varphi)\sigma_e'(\varphi)}{\Delta\rho\varphi g}\right)\frac{\partial\varphi}{\partial z}}_{\text{Flux for}\varphi\geq\varphi,\text{(diffusion)}}, \text{ for } 0 \leq \varphi < 1 \tag{10.18}$$

with the two parameters $f_{bk}(\varphi)$ and $\sigma_e(\varphi)$ to be determined experimentally. Take note that for $\varphi < \varphi,\quad f = q\varphi + f_{bk}(\varphi)$.

Equation 10.18 is a *degenerate parabolic partial differential equation*. The name comes from the fact that, for values $\varphi < \varphi_c, \sigma_e'(\varphi) = 0$, so that the equation becomes *hyperbolic* (Kynch 1952):

$$\frac{\partial\varphi}{\partial t} + \frac{\partial}{\partial z}\left(q\varphi + f_{bk}(\varphi)\right) = 0, \text{ for } \varphi < \varphi_c \tag{10.19}$$

$f_k(\varphi)$ is known as the *continuous Kynch flux density function,* defined for $\varphi < \varphi_c$, and Equation 10.19 is Kynch's equation for continuous sedimentation of ideal suspensions. This result shows that *for flocculated suspensions, Kynch's equation is still valid in those regions where the concentration is less than the critical.*

10.2.5 Batch Thickening

Before sedimentation starts during batch thickening, the suspension is flocculated and homogenized by agitation so that its concentration is constant (Concha 2014).

1. At the beginning of sedimentation, all flocs settle at the same speed forming a well-defined water-suspension interface in the upper part of the column. This interface descends at the same speed as the flocs. This stage is called *hindered settling*. A diffuse interface indicates incomplete flocculation, especially of the fine particles.

2. When the flocs reach the bottom of the column, they occupy rapidly the entire available surface. From then on, the flocs accumulate one over another, compressing those lying underneath. Each layer with constant concentration φ moves upwards as more particles of the column fall over it. This upward motion receives the name of wave of the constant concentration φ. We say the sediment is under compression or consolidation.
3. The interface between the sediment and the settling suspension has no floc laying over it, therefore it suffers no compression and has the critical concentration φ_c.
4. At a certain instant, the water-suspension and the suspension-sediment interfaces meet. The coordinates at this time are called critical height and critical time, and they define the critical point where sedimentation ends and only consolidation prevails.
5. In time, consolidation ends, and a characteristic concentration profile is established in the column. Water in the upper part is followed by a concentration gradient from the critical concentration at the top to the maximum concentration at the bottom of the column.

For batch thickening, the thickening equation (Equation 10.18) becomes

$$\frac{\partial \varphi}{\partial t} + \frac{\partial f_{bk}(\varphi)}{\partial z} = 0, \text{ for } \varphi < \varphi_c \tag{10.20}$$

With boundary conditions

$$\begin{aligned} \varphi(z,0) &= \varphi_0 \quad \text{for } 0 \le z \le L \\ \varphi(L,t) &= 0 \quad \text{for } t > 0 \\ \left.\frac{\partial \varphi}{\partial z}\right|_{z=0} &= \frac{\Delta\rho\varphi g}{\sigma_e'(\varphi)} \text{ for } t \ge 0 \end{aligned} \tag{10.21}$$

Experimentally it has been found that the batch flow density $f_{bk}(\varphi)$ can be expressed as

$$f_{bk}(\varphi) = -u_\infty \varphi(1-\varphi)^c \tag{10.22}$$

So the parameter $K(\varphi)$ from Equation 10.12

$$\frac{\Delta\rho\varphi(1-\varphi)^{2-c} g}{u_\infty} = \mu K(\varphi) \tag{10.23}$$

The solution of this boundary problems is shown in Figure 10.12, which is a typical batch sedimentation curve with the interfaces and the constant concentration lines and Figure 10.13 with height versus concentration for the data of the settling plot of Figure 10.12.

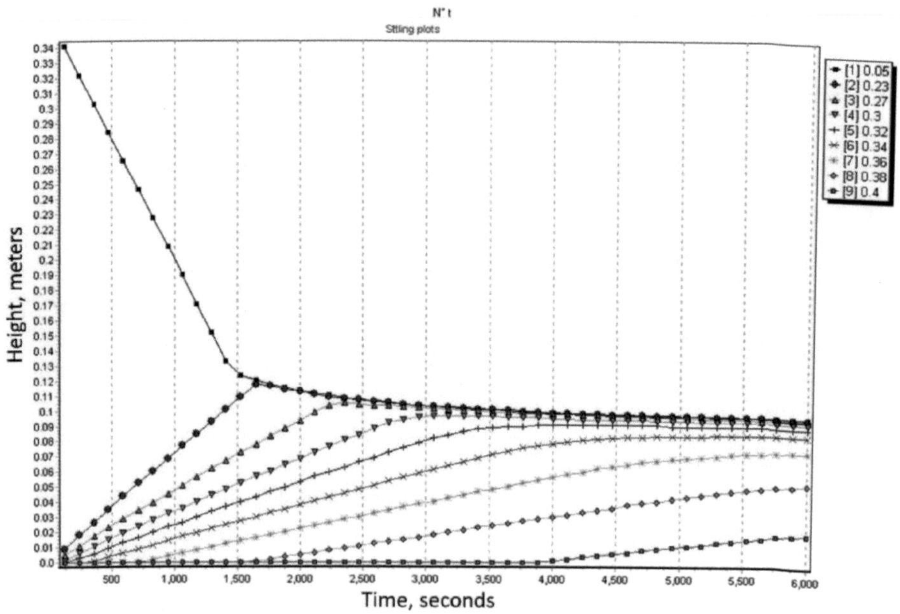

FIGURE 10.12 Sedimentation curve for a copper flotation tailing showing the water-suspension and the suspension solid interfaces and the curves of constant concentration in the sediment.

Source: Data from Becker (1982)

10.2.6 Continuous Thickening at Steady State

Equations 10.7 and 10.8 show that at steady state the solid flux density $f(\varphi)$ and que volume average velocity q are constant in the thickener and are determined by boundary conditions.

Consider a solid flux density surface source of strength $f(\varphi)$ and a volume flow of pulp $Q_F(\varphi_F)$ fed to a thickener of cross-sectional area S at level $z = L$ (see Figure 10.14).

$$f_F = \frac{-Q_F\varphi_F}{S} < 0 \tag{10.24}$$

At the underflow, $z = 0$, a well-mixed pulp is discharged by gravity at a volume flow rate of $Q_D > 0$ and concentration φ_D, through an orifice at $z = 0$, the bottom and center of the thickener.

$$f_D = \frac{-Q_D\varphi_D}{S}, \; \varphi(0) = \varphi_D \tag{10.25}$$

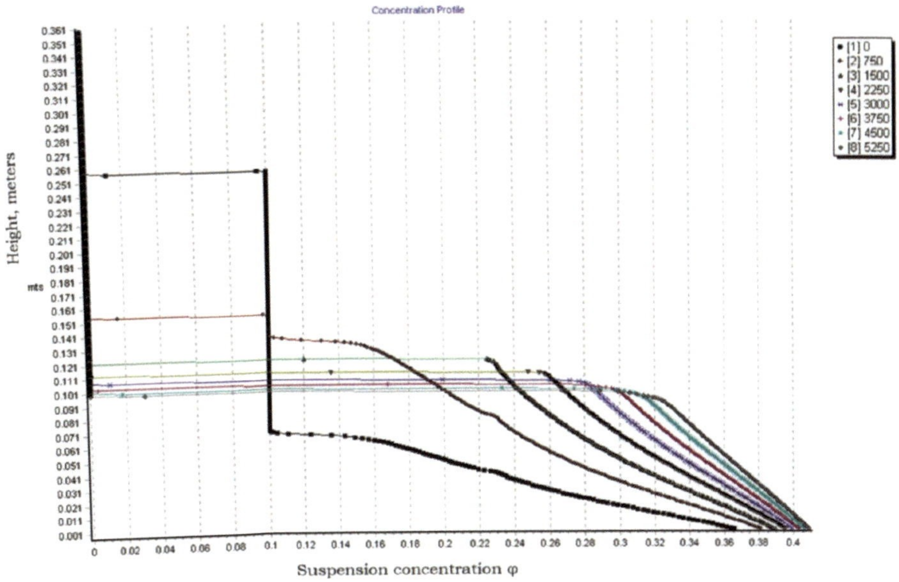

FIGURE 10.13 Height of the interfaces versus the suspension concentration.

Calling $q = -Q_D / S$ the pulp volume average velocity, we can write

$$f = q_D \varphi_D, \quad q(0) = q_D \tag{10.26}$$

No solid passes to the overflow, then the water to the overflow is

$$Q_R = Q_F(1-\varphi_F) - Q_D(1-\varphi_D) \tag{10.27}$$

Since from Equation 10.18 $f(\varphi) = q\varphi + f_{bk}(\varphi)$ then, the solid flux density function for φ_L

$$f_F = q\varphi_L + f_{bk}(\varphi_L) \tag{10.28}$$

then, the concentration at $z = L$ is

$$\varphi_L = \frac{f_F - f_{bk}(\varphi_L)}{q} \tag{10.29}$$

$$\text{And at } z = 0 \ \varphi_D = \frac{f_F}{q} \tag{10.30}$$

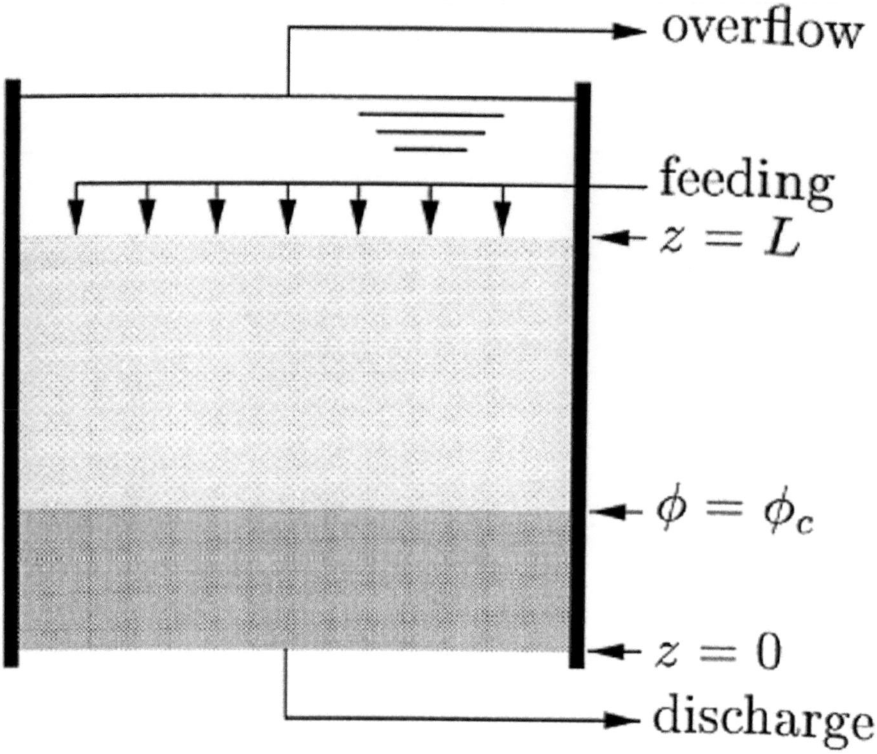

FIGURE 10.14 Ideal continuous thickener.

Source: Concha and Bürger 2003

These relationships justify calculating the conjugate and the discharge concentrations for any values of the feed flux density function from plots of the continuous flux density function and que average pulp velocity.

Figure 10.15 shows that the discharge concentration, blue dot, does not lay on the continuous flux density, red curve. The reason for this is that for the concentration to reach the discharge concentration, there is the need for compression. From Equation 10.16,

$$f_F = q\varphi + f_{bk}(\varphi)\left(1 + \frac{\sigma_e'(\varphi)}{\Delta\rho\varphi g}\frac{\partial\varphi}{\partial z}\right) \tag{10.31}$$

Obtaining the profile of concentration from this equation, $z = z(\varphi)$ write

$$\frac{\partial z}{\partial\varphi} = \frac{f_{bk}(\varphi)\sigma_e'(\varphi)}{\Delta\rho\varphi g}\frac{1}{\left(f_F - f_{bk}(\varphi) - q\varphi\right)} \tag{10.32}$$

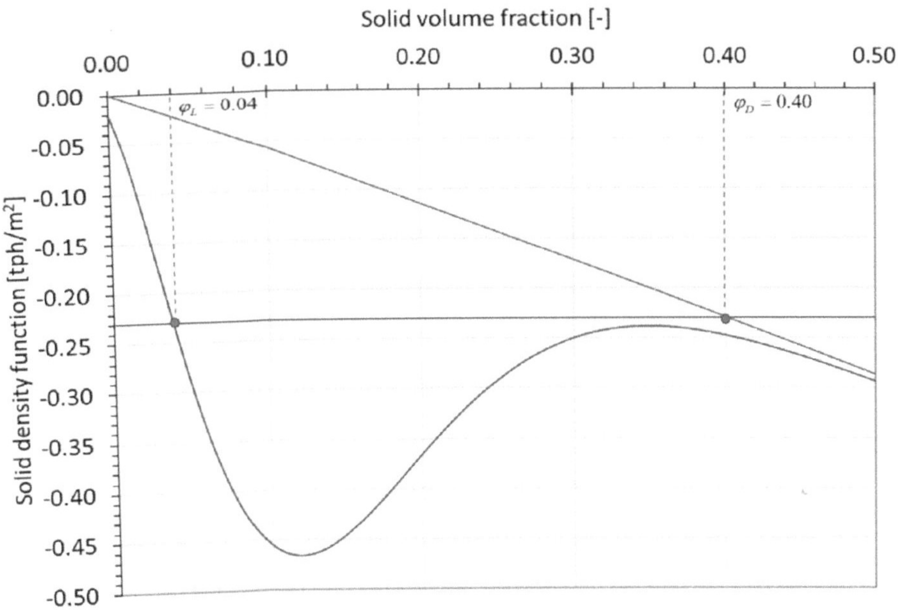

FIGURE 10.15 Determination of the conjugate and discharge concentration from the flux density plots.

and integrating with boundary condition at the bottom of the thickener, $z = 0$, $\varphi(0) = \varphi_D$, and since the volume average velocity of the pulp is $q = f_F / \varphi_D$,

$$z(\varphi) = \int_{\varphi_D}^{\varphi} \frac{f_{bk}(\xi)\sigma_e'(\xi)}{\Delta\rho\xi g\left(f_F\left(1-\xi/\varphi_D\right)-f_{bk}(\xi)\right)} d\xi \tag{10.33}$$

From this equation we can calculate the concentration profile of solid in the thickener. See Figure 10.16.

To elucidate when a steady state is possible with concentrations φ_F, φ_L, and φ_D, consider that, for a thickener to be properly operating, the gradient of the concentration should be negative or zero $\partial\varphi/\partial z \le 0$, that is, the solid concentration should increase downwards or be constant.

Equation 10.32 may be written in the form

$$\frac{\partial\varphi}{\partial z} = \underbrace{\frac{-\Delta\rho\varphi g}{\sigma_e'(\varphi) f_{bk}(\varphi)}}_{>0} \underbrace{\left(q\varphi + f_{bk}(\varphi) - f_F\right)}_{\le 0} \le 0$$

$$\text{then } f_k(\varphi) \le f_F, \left(\text{or } F_k(\varphi) \le F_F\right), \text{for } \varphi_L \le \varphi \le \varphi_D, \tag{10.34}$$

where $f_k(\varphi) = q\varphi + f_{bk}(\varphi)$, (or $F_k(\varphi) = \rho_s Q_D \varphi + F_{bk}(\varphi)$), that is

$$f_F \ge q\varphi + f_{bk}(\varphi) \tag{10.35}$$

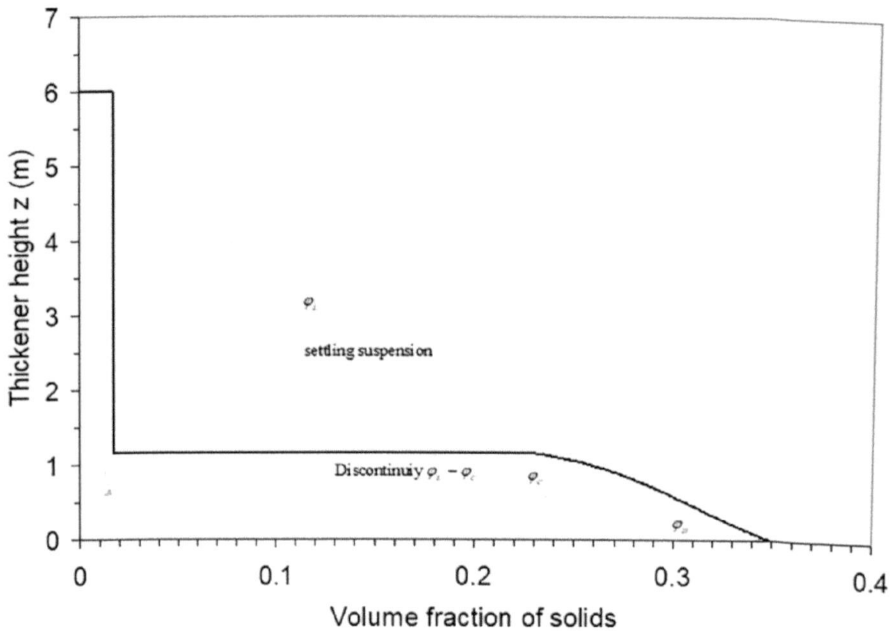

FIGURE 10.16 Typical solid concentration profile in a thickener.

which indicates that the feed line in the flux density plot must always be above the continuous flux density (all terms are negative). The maximum discharge concentration is obtained when the line of the feed is tangent to the continuous flux density function as shown in Figure 10.17.

An application of continuous thickening was done by Mehdi, R.; Majid, U.; Bahram, R., et al., (2017) in a pilot thickener treating samples from a copper plant in Iran and in a coal washing plant (2019), as shown in Figure 10.18, Figure 10.19, and Figure 10.20.

10.2.7 Determination of the Thickening Parameters

Sedimentation is characterized through the solid flux density function, which is calculated from the *initial settling velocity* and the *suspension concentration* and compression characterized by the *solid effective stress*.

10.2.7.1 Initial Settling Velocity

Experiments with different solid concentration are performed in five one liter settling columns, and the initial settling velocity is determined from the slope of the settling curve. These experiments may be performed automatically in a novel instrument *SediRack* (Cettem Instrument), as shown in Figure 10.21 and Figure 10.22.

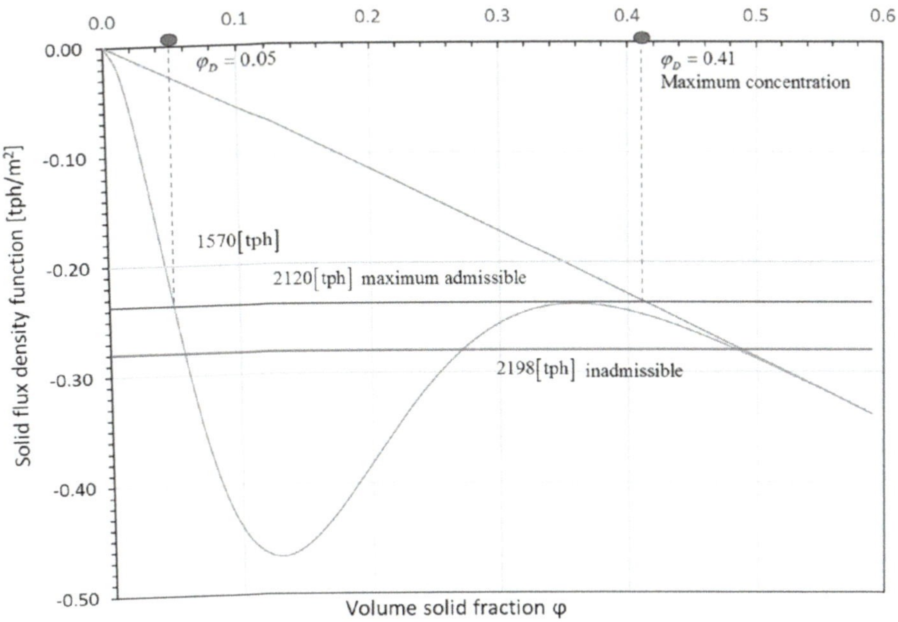

FIGURE 10.17 Admissible and not admissible feed solid flux density and maximum discharge concentration achievable, obtained when the line of the feed is tangent to the continuous flux density function.

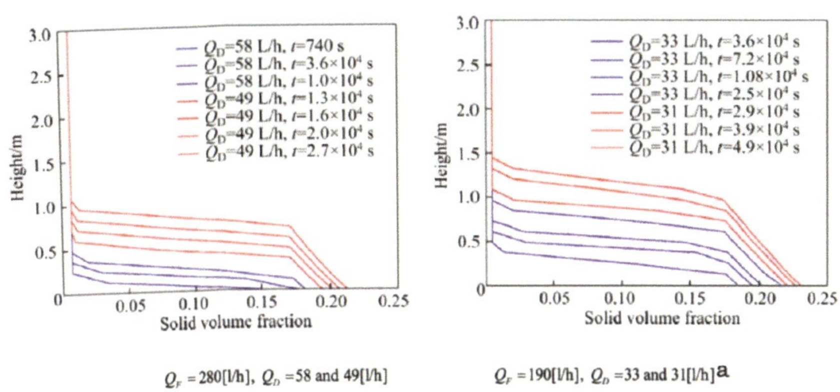

FIGURE 10.18 Dynamic simulation of the bed height for a pilot thickener treating a copper tailing.

Source: Mehdi, R., Majid, U., Bahram, R., and Ali, A. 2017. Dynamic simulation of pilot thickener operation using phenomenological model with results validation for continuous and discontinuous tests. *Journal of Central South University* 24:1207–1216. https://doi.org/10.1007/s11771-017-3524-7

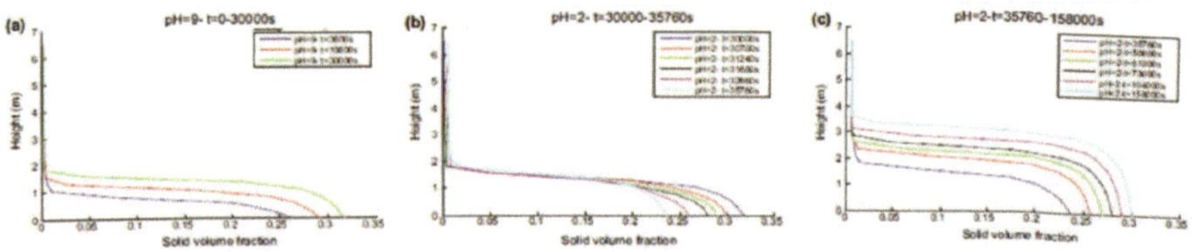

FIGURE 10.19 Dynamic simulation of the bed height in a coal washing plant for a step change from a pH of 9 to 2.

Source: Mehdi, R., Majid, U., Bahram, R., and Ali, A. 2017. Dynamic simulation of pilot thickener operation using phenomenological model with results validation for continuous and discontinuous tests. *Journal of Central South University* 24:1207–1216. https://doi.org/10.1007/s11771-017-3524-7

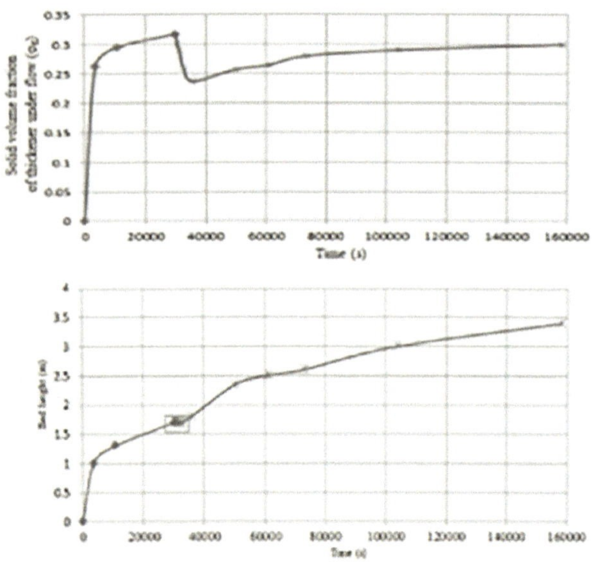

FIGURE 10.20 Dynamic simulation of the underflow concentration and the bed height in a coal washing plant, with an addition of 40 g/l of a flocculant and a step change of feed for the step change of pH from 9.0 to 2.0.

Source: Mehdi, R., Majid, U., Bahram, R., and Ali, A. 2017. Dynamic simulation of pilot thickener operation using phenomenological model with results validation for continuous and discontinuous tests. *Journal of Central South University* 24:1207–1216. https://doi.org/10.1007/s11771-017-3524-7

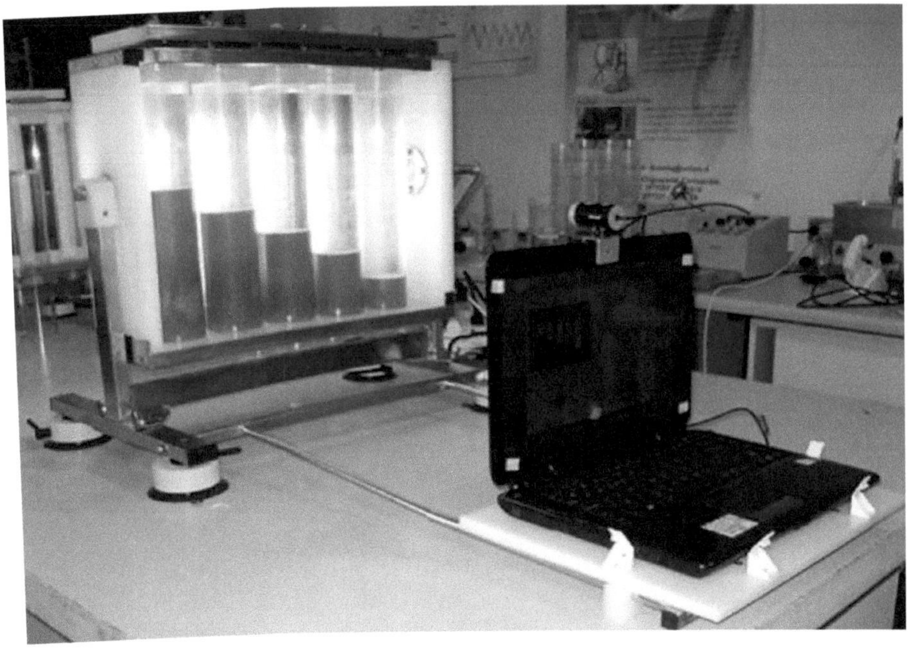

FIGURE 10.21 SediRack for the automatic determination of the settling plot of a suspension at five different concentrations.

10.2.7.2 Suspension Concentration

The concentration of a suspension may be determined in two new pieces of equipment: *DensiTest* and *Industrial Pycnometer* (see Figure 10.23). A certain amount of a suspension is placed in vessel and placed in any of the two instruments. In the DensiTest the vessel with suspension is weight. Data of the suspension volume, density of the solid and fluid is entered by keyboard on the instrument with results given as volume fraction, percent solid, density, or dilution.

Both instruments are based on determining the concentration of a suspension by alternatively measuring the weight of an empty vase, the vase with water, the vase with the liquid, if it is different from that of water, and the vase with the suspension. The density of the solid and the liquid are entered by keyboard. The result is obtained, according to interest, as volumetric fraction, percent solid, pulp density, or dilution. The difference between them is that the DensiTest uses a strain gage while the Pycnometer is a scale.

10.2.7.3 Solid Effective Stress

The solid effective stress is the parameter quantifying the compressibility of the sediment. Since this stress is due to the self-weight of the sediment, a concentration

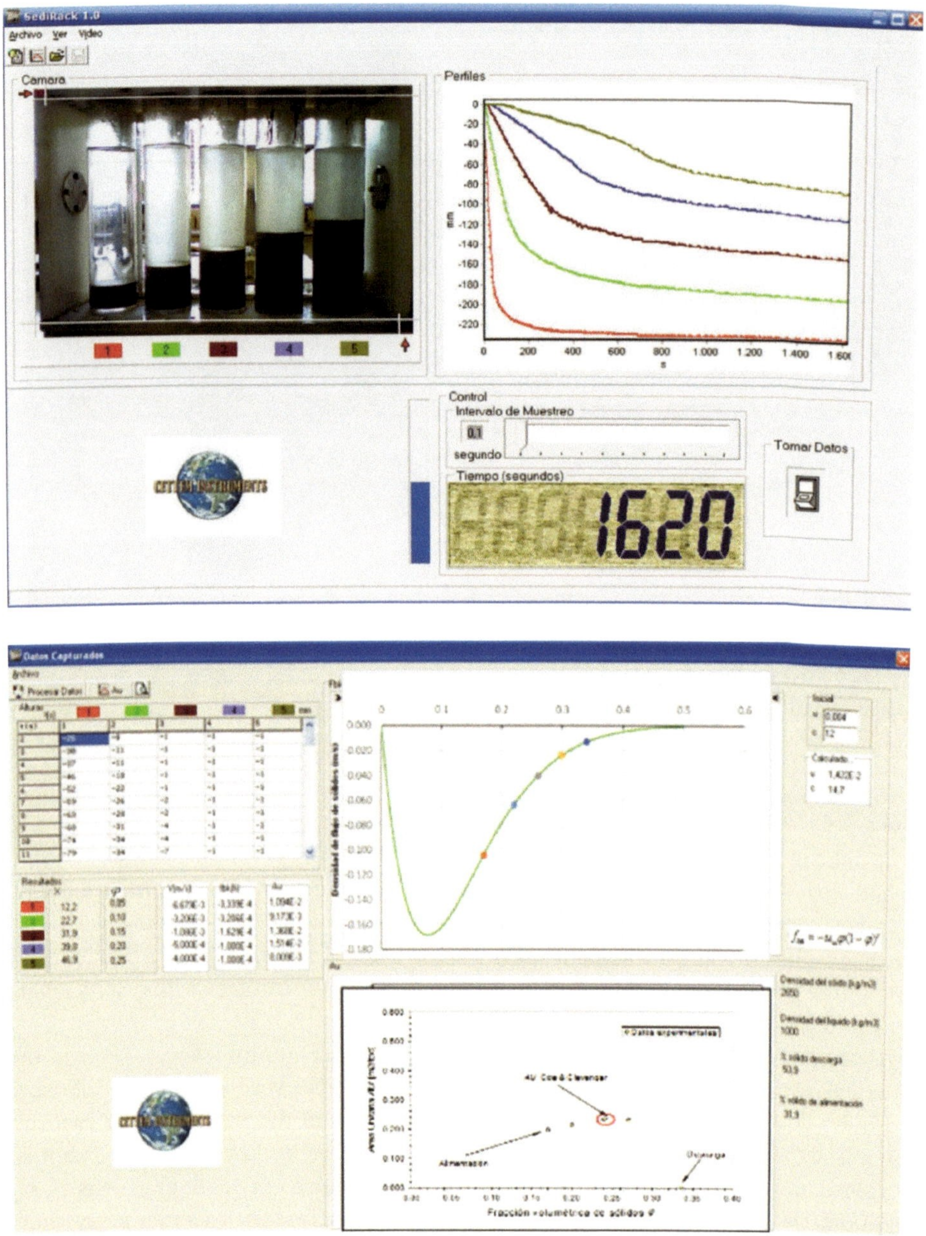

FIGURE 10.22 Results of the measurements, settling velocities, solid flux density function, and unit area.

gradient is established with the concentration increasing to the bottom. Measuring this concentration profile, the solid effective stress may be calculated from Equation 10.21.

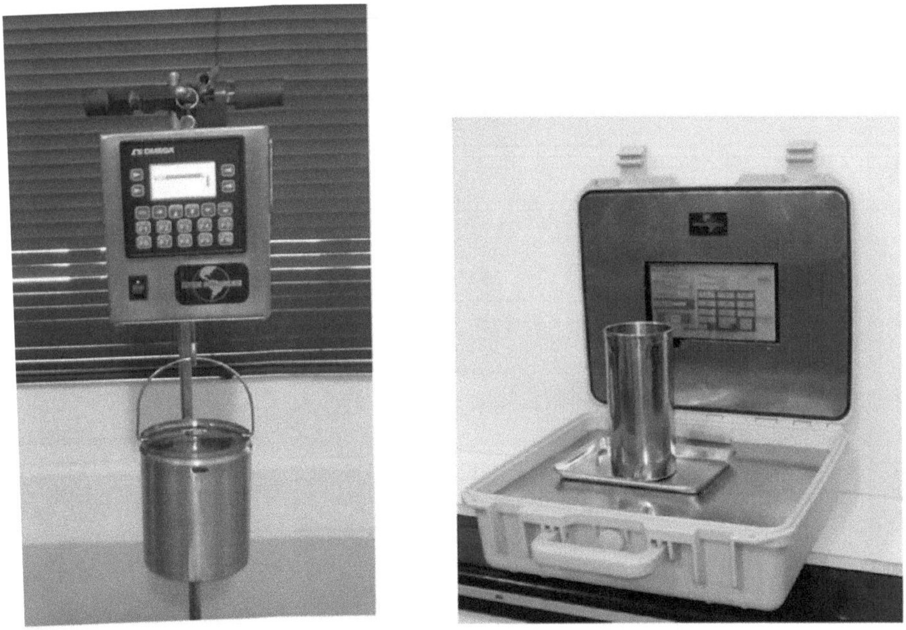

DensiTest Industrial Pycnometer

FIGURE 10.23 Instrument to measure the concentration of a suspension in the laboratory and in an industrial environmental.

$$\sigma_e(\varphi) = \int_0^{\varphi_c} \frac{\Delta\rho\xi g}{d\xi / dz} d\xi \tag{10.36}$$

The conversion between conductivity and concentration is done with an empirical equation shown here

$$\varphi = \frac{(k_s + 2k_w)(1-K)}{(k_w - k_s)(K+2)}$$

Where φ and k_s are the concentration and the conductivity of the suspension, k_f is the conductivity of the fluid, k_s is the conductivity of the solid and $K = k_s / k_f$. The equipment software chooses the conductivities at long times and convert them into concentration. These are plotted as concentration profiles such in Figure 10.24 and Figure 10.25.

10.2.7.4 Solid Inventory in the Thickener

Once the concentration profile in the thickener is known, the amount of solid accumulated inside can be calculated with the equation, as shown in Figure 10.26.

FIGURE 10.24 SediTest and conductivity values at different heights in the settling column as a function of time.

$$M_I = \rho_s S\left\{(L - z_c)\varphi_L + \int_0^{z_c} \varphi(z)dz\right\} \tag{10.37}$$

10.2.8 Thickening Design

From the various methods of thickener design proposed, the most widely used today is that of Coe and Clevenger (1916). This method is based on a mass balance in an ideal thickener resulting in the equation, which, in the nomenclature used in this book, is (Concha and Barrientos 1993):

$$UA(\varphi_F, \varphi_D) = \underbrace{\max}_{\varphi_F \le \varphi \le \varphi_D} \frac{1}{\rho_{sk} f_{bk}(\varphi)}\left(\frac{\varphi}{\varphi_D} - 1\right) \tag{10.38}$$

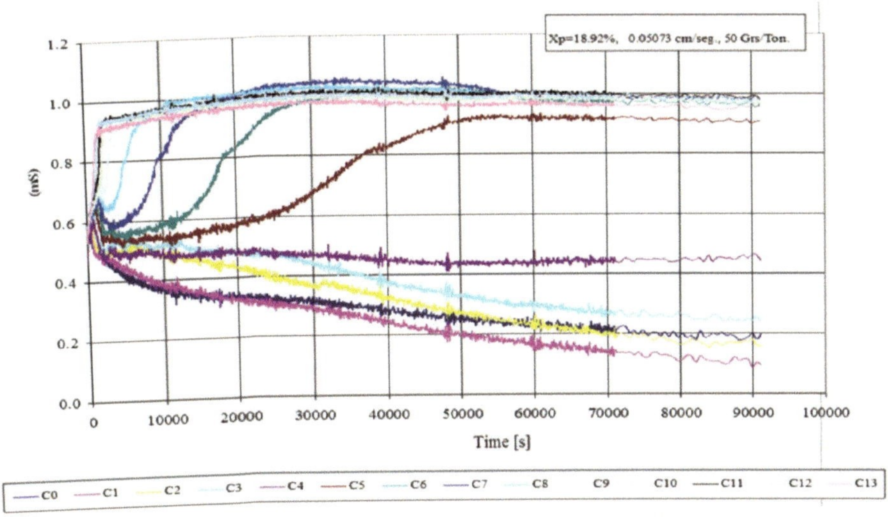

FIGURE 10.24 Continued

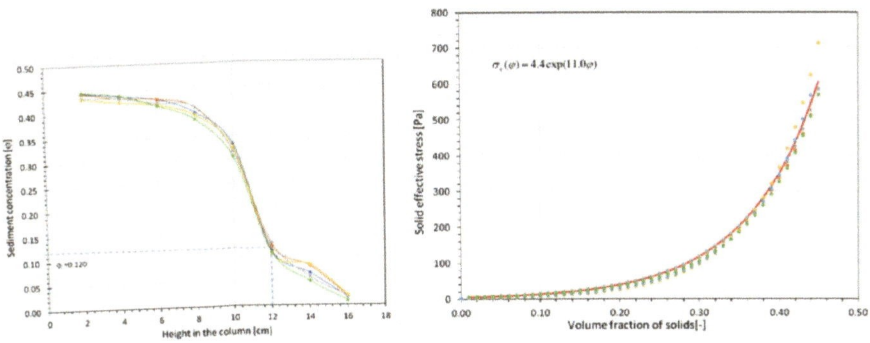

FIGURE 10.25 Concentration profile $\varphi = \varphi(z)$ at the end of sedimentation and the solid effective stress $\sigma_e(\varphi)$.

Where UA is the Unit Area defined as the maximum area S per feed flow F_F of a thickener obtained from settling velocity experiments for a given pulp.

10.2.8.1 Design Method Based on the Phenomenological Theory

We demonstrate that for a steady state to exist in a continuous thickener, it is necessary that the following inequality is obeyed (Equation 10.28) $f_F \geq q\varphi + f_{bk}(\varphi)$, for $\varphi_L \leq \varphi \leq \varphi_D$. Since $q = f_F / \varphi_D$, the following expression is valid for the steady state: $f_F \geq (f_F / \varphi_D)\varphi + f_{bk}(\varphi)$; dividing by φ results in $f_F / \varphi \geq f_F / \varphi_D + f_{bk}(\varphi) / \varphi$ and since $f_F = -\rho_s F_F / S$, $\dfrac{F_F}{S} \leq \dfrac{\rho_s f_{bk}(\varphi)}{1 - \varphi / \varphi_D}$ for $\varphi_L \leq \varphi \leq \varphi_D$. With $UA = S / F_f$,

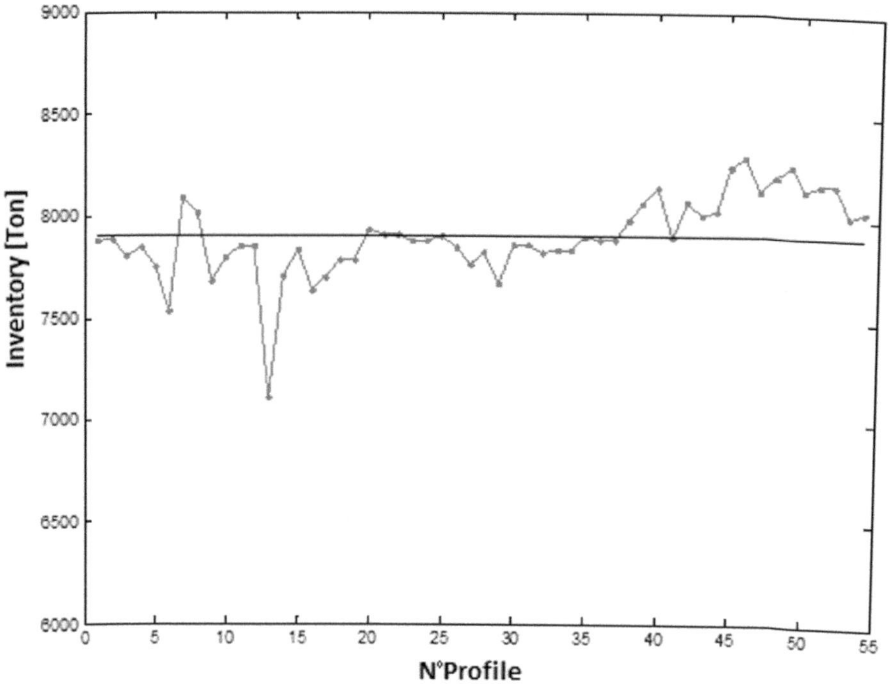

FIGURE 10.26 Solid inventory in the thickener.

$$\frac{S}{F_F} \geq \frac{1}{\rho_s f_{bk}(\varphi)}\left(\frac{\varphi}{\varphi_D} - 1\right), \text{for } \varphi_L \leq \varphi \leq \varphi_D$$

$$UA = \underbrace{\max}_{\varphi_L \leq \varphi \leq \varphi_D} \frac{1}{\rho_s f_{bk}(\varphi)}\left(\frac{\varphi}{\varphi_D} - 1\right), \text{for } \varphi_L \leq \varphi \leq \varphi_D \tag{10.39}$$

It is interesting to note that Equations 10.2.38 from Coe and Clevenger's method and (Equation 10.39) of the phenomenological method have the same from, except for the small but important difference that Coe and Clevenger assumed that the range of concentration where the value of the *UA* should be searched is the *feed* and the underflow concentrations, while phenomenological method emphasizes that, when the feed enters the thickener it will dilute to the conjugate concentration and therefore the range of concentration should be $\varphi_L \leq \varphi \leq \varphi_D$ instead of $\varphi_F \leq \varphi \leq \varphi_D$. Figure 10.27 shows the difference.

In designing a thickener, the desired feed flow rate F_F and underflow concentration φ_D, together with the thickening parameters $f_{bk}(\varphi)$. $\sigma_e(\varphi)$ and φ_c, the solid and liquid densities ρ_s and ρ_f must be known. With these values, the function $UA(\varphi_L, \varphi_D)$ can be calculated.

The problem to calculate $UA(\varphi_L, \varphi_D)$ is that the conjugate concentration φ_L is unknown until the value of F_F is calculated, and to do so, we need to know the

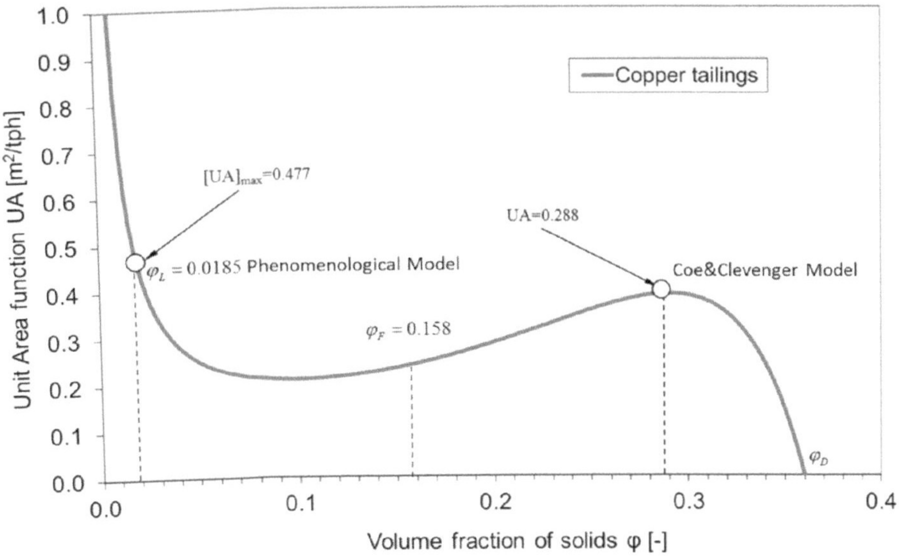

FIGURE 10.27 Comparison of the phenomenological and Coe and Clevenger's method to determine the unit area for a copper flotation tailing.

thickener's cross-sectional area S, which is the value we are seeking for. Therefore, the problem is undetermined, and to solve it, a value of φ_L must be assumed, for example $\varphi_L = 0.01$. Once φ_L is chosen, the value of $AU(\varphi_L, \varphi_D)$, the area S and diameter D of the thickener and the flowrate F_F and underflow volume rate Q_D can be established, and therefore, the value of φ_L be recalculated.

We still have one problem. There is no clue of the height the sediment will reach in the designed thickener, which could be very small or very large. The solution is to consider the compression of the sediment and set a target for the value of the sediment height z_c (e.g., 20% of the available height L of the thickener) and iterate the design changing the value of D to obtain the desired bed height z_c. This thickener diameter is then approximated to the next integer (or to the next commercially available thickener diameter). The simulation is repeated to obtain the final unit area $UA(\varphi_L, \varphi_D)$, the thickeners cross-sectional area S, the solid inventory M_I, the sediment height, and the correct value of the conjugate concentration φ_L. Figure 10.28 gives the algorithm of calculations. If the size of the thickener obtained is too large, outside of the range of the available equipment in the market, we choose a reasonable value of diameter and iterate to obtain the number of thickeners necessary to handle the desired feed.

10.2.8.1.1 Example

Determine the size and number of thickeners to process 120.000 tpd of a flotation copper tailing with the following properties:

Solid density	$\rho_s = 2650\ \text{kg/m}^3$
Liquid density	$\rho_f = 1000\ \text{kg/m}^3$
Feed concentration	$w_F = 30.0\%$ solid by weight $(\varphi_F = 0.139)$
Critical concentration	$w_c = 40.0\%$ solid by weight $(\varphi_c = 0.200)$
Conjugate concentration	$\varphi_L = 0.01$
Underflow concentration	$w_D = 60.0\%$ solid by weight $(\varphi_D = 0.361)$
Settling parameter	$f_{bk}(\varphi) = -1.72\times10^{-4}\times\varphi\times(1-\varphi)^{15.6}$ m/s
Compression parameter	$\sigma_e(\varphi) = 20.0\exp(22\varphi)$ Pa

The results of the example are shown in Figures 10.31, 10.32, and 10.33.

$$\text{Design module}: UA = \underbrace{\max}_{\varphi_L \le \varphi \le \varphi_D} \frac{1}{\rho_s f_{bk}(\varphi)}\left(\frac{\varphi}{\varphi_D} - 1\right), \text{for } \varphi_L \le \varphi \le \varphi_D \tag{10.40}$$

The results indicate a 178 m thickener diameter. Since it is out of the range of existing thickener, we will choose a 100 m thickener.

10.2.8.2 Ideal Clarifier-Thickener Model

An ideal clarifier-thickener (ICT) was defined by Bürger et al. (2000b) and distinguishes four zones, as shown in Figure 10.34, the sediment zone $0 < z < z_c$, clear liquid zone $z_c \le z \le H$, the overflow zone $z > H$, and the underflow zone $z < 0$. The unit is continuously fed by a singular source at $z = L$ by a volume rate of suspension $Q_F \ge 0$ and concentration as volume fraction φ_F. The volume underflow rate is Q_D with volume concentration φ_D.

The macroscopic volume balance yields as follows:

$$Q_F = Q_D + Q_R \tag{10.41}$$

Where Q_R is the overflow volume rate.

In a clarifier-thickener the solid flux density function has different values for each zone:

$$f(z,\varphi,t) = \begin{cases} Q_R / S(\varphi - \varphi_F) & \text{for} \quad z > H \\ Q_R / S(\varphi - \varphi_F) + f_{bk}(\varphi) & \text{for} \quad L < z < H \\ Q_D / S(\varphi - \varphi_F) + f_{bk}(\varphi) & \text{for} \quad 0 < z < L \\ Q_D / S(\varphi - \varphi_F) & \text{for} \quad z < 0 \end{cases} \tag{10.42}$$

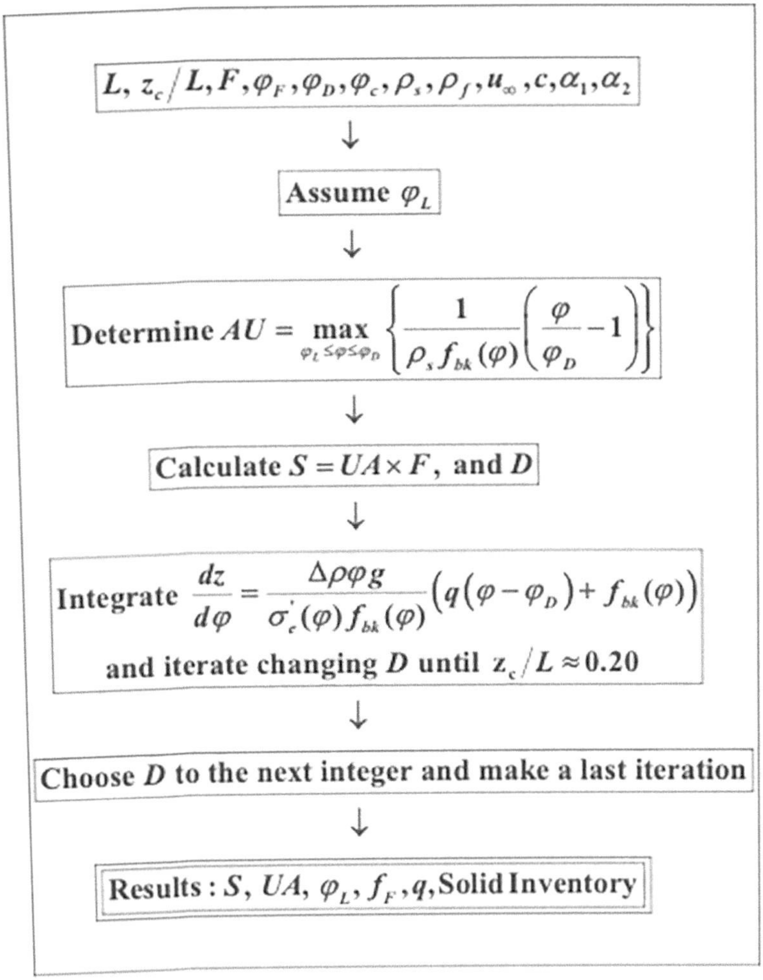

FIGURE 10.28 Algorithm for the design of continuous thickeners with Cettem's method.

Source: Concha, F.A. 2014. *Solid Liquid Separation in the Mining Industry.* Cham, Switzerland: Springer. pp. 298–299; 303–306.

10.2.8.3 High-Capacity Thickener at Steady State

In the sediment $0 \le z \le z_c$, where z_c is the top of the sediment level, we can write

$$\frac{d\varphi}{dz} = \frac{\Delta\rho\varphi g}{f_{bk}(\varphi)\sigma_e^{'}(\varphi)S}\left(-Q_D\left(\varphi(z) - \varphi_D\right) + Sf_{bk}\left(\varphi(z)\right)\right), 0 \le z \le z_c \tag{10.43}$$

This equation can be solved with the boundary condition $\varphi(0) = \varphi_D$ giving the concentration profile in the sediment, as shown in Figure 10.33.

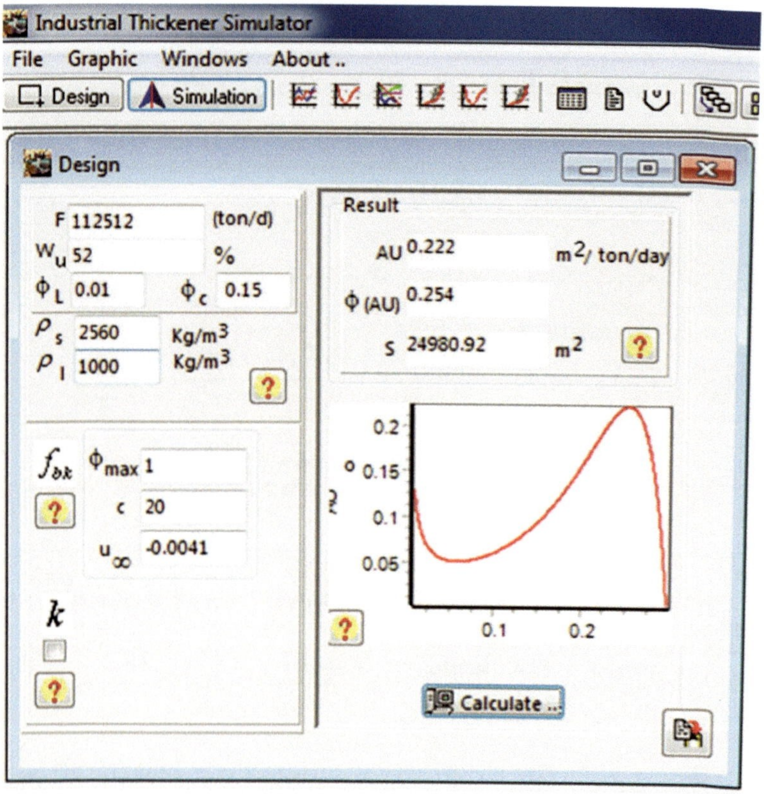

FIGURE 10.29 Design module.

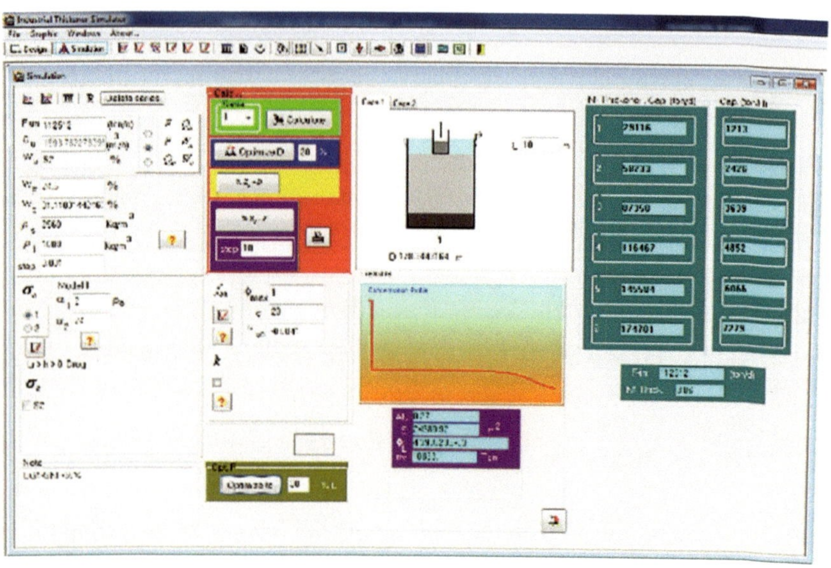

FIGURE 10.30 Simulation module.

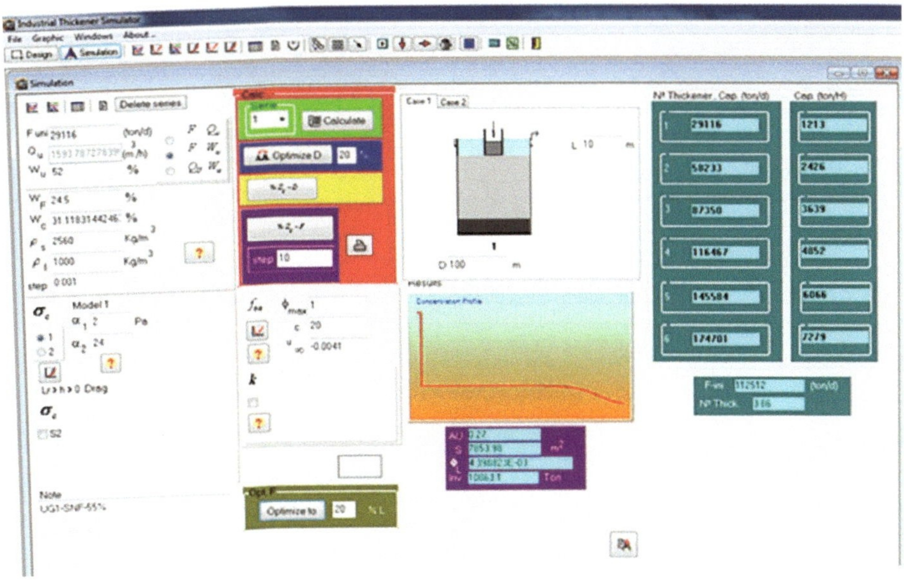

FIGURE 10.31 Final simulation module: each of the 100 m thickener treats 29.116 tpd, so we need four thickeners.

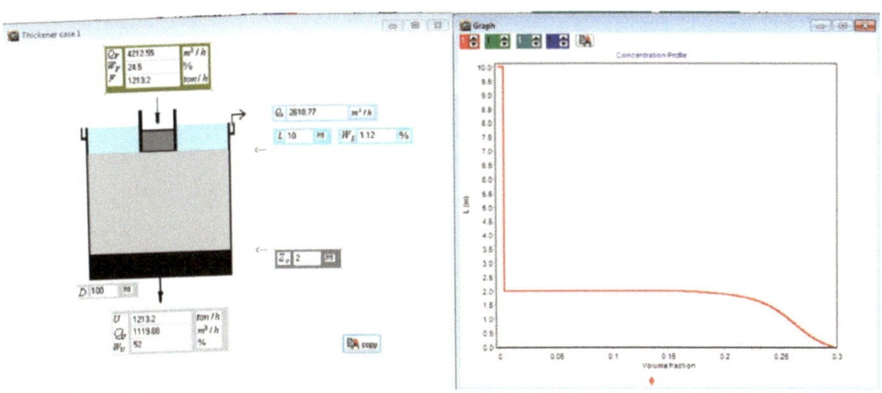

FIGURE 10.32 Mass balance and concentration profile in the set of four thickeners.

10.3 ONLINE THICKENER ESTIMATION OF DERIVED VARIABLES TO AUGMENT THE WATER RECOVERY IN MINERAL PROCESSING PLANT

Based on current sedimentation theory, an online thickener model was developed and coupled with an extended Kalman filter to obtain additional information such as the compression zone level and concentration, a corrected underflow concentration, the inventory, and some rheological information. The online model was installed and tested for robustness for periods up to three days in an industrial environment giving

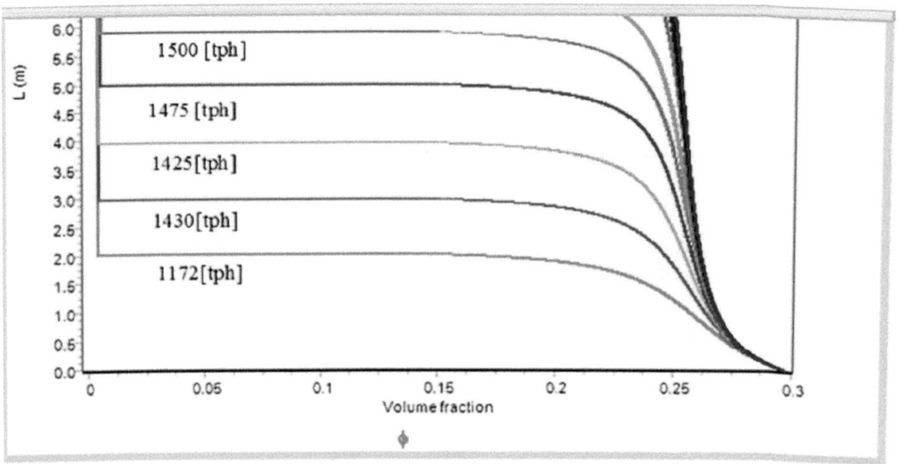

FIGURE 10.33 Simulation: capacity of the designed thickener for different sediment height.

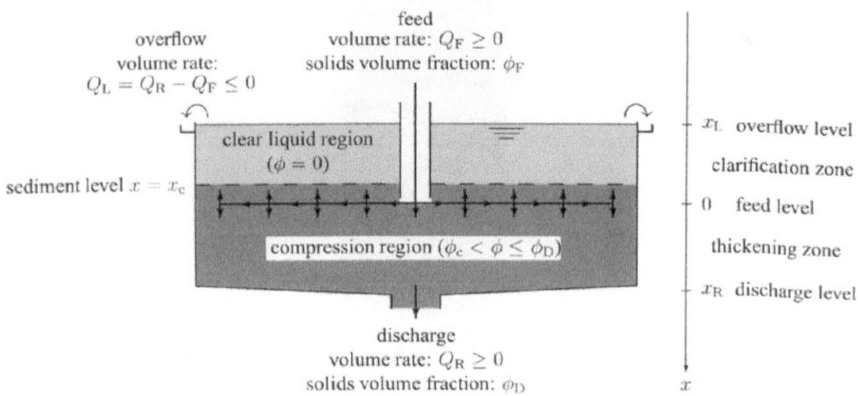

FIGURE 10.34 Clarifier-thickener.

good results. Emphasis was given to signal processing and selecting the appropriate input variables for filtering.

The analysis of the thickener from a phenomenological point of view provides for better operation, establishes the limiting operating conditions, and assists in the design of control strategies.

Online state prediction estimation techniques have been successfully applied to several critical chemical processes (Fisher and Seborg 1976; Ray 1980) and mineral-processing operations (Herbst and Bascur 1984). The work to be described here was motivated by the need to achieve maximum usage of the thickeners during drastic change in the rheological characteristics of the ore causing a slowdown in pumping. Accurate predictions of the inventory, concentrations, and zone levels are necessary to avoid high-torque conditions. These requirements led to the use of the Kalman filtering technique both to do estimation of unmeasured variables and thickening parameters and to infer rheological pulp characteristics.

The work described here was performed in conjunction with a 180 ft diameter thickener in a copper concentrator. The concentrator performance is monitored using a distributed control system (DCS) and a real-time computer system.

10.3.1 Thickener Online Model Development

There have been several attempts to describe the dynamic behavior of clarifiers for water treatment plants (Alkema 1971; Mijares-Lopez 1973; Attir and Denn 1978; Paterson and Denn 1980). In contrast, only a few attempts to describe the dynamic behavior of a thickener have been made (Adorjan 1975; Bascur 1979).

Several reviews of thickener theory were utilized in the development of the candidate models. Some of the relevant literature includes the work of Wakeman (1981), Schubert (1981), Fitch (1979), Selim et al. (1983), and Smith (1983). Experimental verification of the models has been missing to provide the necessary feedback to the theory. The procedure described uses the plant as an experimental system so direct comparison of the predictive equations is performed. The final model is semi empirical with adaptive coefficients based on a phenomenological sedimentation concept. A drastic change in the adaptive coefficients can be used for diagnosis of plant performance and for control purposes.

The description of the methodology to obtain a suitable online model is discussed first (Bascur and Herbst 1986). Initially, a description of suitable model candidates for online modeling of thickener behavior will be presented. Parallel to the model development, an assessment of the necessary instrumentation for major disturbance detection and observation measurement will be given. Then the selected algorithm for estimation and the problems in its implementation will be described. Finally, the benefits of such a study and future recommendations are given.

In the development of the model which follows, an attempt has been made to include in a simple manner all possible physical variables of importance in the control of thickening, that is, manipulated, controlled, and measured variables.

For purposes of model development, an idealization of the process is made as depicted in Figure 10.35. Figure 10.35 also shows the major instrumentation that was used during this study. In the sedimentation of heterogeneous particles with a wide size particle distribution, several subprocesses may be distinguished for a continuous thickener. The feed pulp is centrally fed. Because it is denser than the clarified overflow liquid, it passes through this layer and produces the feed layer, which spreads out at a level of hydrostatic equilibrium. This condition establishes the position of the sedimentation zone. In the clarification zone, the solids concentration is very low, provided that the original sizing of the equipment is appropriate and that the operating conditions are normal. The position of the mudline is determined by the particles or flocs, which—due to their low settling rate—cannot escape the rising fluid stream or which have been "suspended" in the settling zone by eddy formation created by the inlet flow. In this development, the concentration of solids in the clarification zone will be assumed to be equal to zero.

At the top of the sedimentation zone, the solids concentration changes markedly, and typically its value closely corresponds to that of the feed pulp. The particles or the flocs settle in this zone against the flow of the rising fluid.

In the compression zone, the interaction forces between the solids and liquid within the flocs must be overcome to achieve further thickening. In passing downward through it, the flocs are subjected to an ever-increasing solids stress or "squeeze" because of the increasing weight of the compressing solids previously. The moderate stirring by the rotating rake mechanism improves the effectiveness of the compression.

Based on this description, the thickener is divided into four zones: the clear liquid zone, settling zone, compaction zone, and removal zone.

The settling zone height is modeled by considering the movement of two boundaries, one between the clear zone and the settling zone and the other between the settling zone and the compaction zone.

The equation describing the volume of the settling zone is given by the following:

$$Ah_S = A\left(z_1 - z_2\right) \tag{10.44}$$

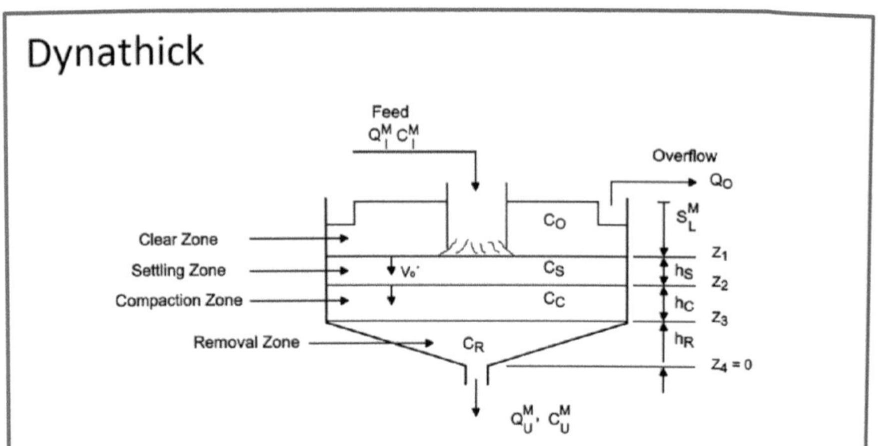

FIGURE 10.35 Schematic representation of the thickener. Measured variables are Q_I^M, C_I^M, S_L^M, Q_U^M, and C_U^M.

Source: Bascur, O.A., and Herbst, J.A. 1986. Improved thickener performance through the use of an extended Kalman filter. In *Design and Installation of Concentration and Dewatering Circuits*. Edited by A.L. Mular and M.A. Anderson. Littleton, CO: SME. Pp. 835–845.

Which in dynamic form can be expressed as follows:

$$\frac{dh_S}{dt} = \frac{dz_1}{dt} - \frac{dz_2}{dt} = v_1 - v_2 \tag{10.45}$$

The compaction zone height can be obtained in an analogous manner. In the case of the removal zone, we define a constant height, h_R, which coincides with the starting of the conical bottom of the thickener and ending at the underflow solids removal point. Thus, for the compaction zone we have the following:

$$\frac{dh_C}{dt} = v_2 - v_3; v_3 = 0.0(\text{fixed}) \tag{10.46}$$

In order to model the movement of these boundaries, the theory of "continuity shock waves" is used. The mathematical representation of this velocity according to Wallis (1969) is as follows:

$$v_i = j + \frac{\partial j_{ws}}{\partial c} \tag{10.47}$$

Where ν_i is the velocity of the shock wave, j is the pulp flux (m/min), j_{ws} is the solid drift flux, and C is the volumetric fraction of solids. This mode of particle solids settling is the basic assumption of the Kynch theory of thickening.

For the clear liquid zone and settling zone boundary, this equation can be written in discrete form as follows:

$$v_1 = Q_0 / A - \left[v_o^S (1 - C_o)^{m_1} C_o - V_o^S (1 - C_S)^{m_1} C_S\right] / (C_o - C_S) \tag{10.48}$$

And since the concentration in the clear zone is set to zero ($C_o = 0$), we obtain the following:

$$v_1 = Q_0 / A - v_o^S (1 - C_S)^{m_1} \tag{10.49}$$

Which says that the settling zone height is defined by an equilibrium between the terminal settling of the smallest (and lightest) particles and the volume flow rate of the overflow. In this model, a modified form of the Richarson-Zaki equation is used for estimation of the settling velocity (Schubert 1981), that is,

$$v_S = v_o^S (1 - C_S)^{m_1} \tag{10.50}$$

$$v_o^S = k_f k_S \frac{(\rho_S - \rho_f) d_T^2 g}{18\mu} \tag{10.51}$$

Where k_f is a volume shape factor for the particle (or floc), k_S is a hindered settling factor, and d_T is the diameter of the smallest flocs.

Equation 10.49 can be rearranged to obtain one limiting and often neglected characteristic for the design of the thickener, the clarifying area, when dealing with broad size distributions, that is,

$$A = \frac{18\mu Q_0}{k_f k_S \left(\rho_S - \rho_f\right) d_T 2_g} \tag{10.51}$$

In order to evaluate the change in position of the upper boundary of the compaction zone, the general Equation 10.47 for shock waves is applied, resulting in

$$v_2 = -\frac{Q_U}{A} - v_o^C \left(1 - C_C\right)^{m_2 - 1} \left[1 - \left(m_2 + 1\right) C_C\right] \tag{10.52}$$

This equation represents the limiting flux condition in a thickener and corresponds to the typical design equation based on Kynch theory.

Since the total volume of the thickener is fixed, the total volumetric flow of pulp must be constant, so that

$$\mathrm{Q_I} = \mathrm{Q_0} + \mathrm{Q_U} \quad \mathrm{Q_0} = \mathrm{Q_I} - \mathrm{Q_U} \tag{10.53}$$

Equation 10.45 with Equation 10.43 and Equation 10.46 with Equation 10.52 define the position of the settling zone and compaction zone, respectively. Both $\mathrm{Q_I}$ and $\mathrm{Q_U}$ can be measured, and therefore $\mathrm{Q_0}$ can be calculated.

The coupling of these equations with the mass balance equations for the solids in each zone constitutes the overall online thickening model. To complete the proper filter setup, the measurement model equations are also required.

For the settling zone we have the following:

$$\frac{dh_S C_S}{dt} = \frac{Q_I C_I}{A} - \left[v_o^S \left(1 - C_S\right)^{m1} + \frac{Q_U}{A}\right] C_S \tag{10.54}$$

For the compaction zone we have the following:

$$\frac{dh_C C_C}{dt} = \left[v_o^S \left(1 - C_S\right)^{m1} + \frac{Q_U}{A}\right] C_S - \left(\frac{Q_U}{A}\right) C_C \tag{10.55}$$

For the underflow removal zone, we have the following:

$$\frac{dC_U}{dt} = \frac{Q_U \left(C_C - C_U\right)}{V_R}$$

$$V_R = 7.500\, m^3 \tag{10.56}$$

The resulting measurement and possible inferred variables from the states obtained are shown in Table 10.1.

In summary, Equation 10.45 with Equation 10.43, Equation 10.46 with 10.46, and Equations 10.54 through 10.56 can be used to predict the state variables of the thickener.

The application of these equations requires the specification of a cut size d_{50} which should be as small as possible for good clarification performance. In order to apply this model, one needs information on specific gravity, size distribution, and so forth, of the flocs. Another difficulty results from the fact that the dynamics of the flocculation are not usually complete when settling begins, and therefore flocculation and sedimentation are occurring simultaneously. These effects contributed to a classification problem encountered in the thickener understudy. The analysis suggests that the lightest particles (flocs) require a longer flocculation time so they settle toward the walls of the thickener, which then, due to their different rheological properties, do not compress adequately and do not move efficiently with the available rakes, causing accumulation. In this case the rakes need to be lifted to continue operation; reduction in feed flowrate or dilution of the feed is then necessary to remove the accumulated fines. The feedwell design is critical when such ore types are present.

The difficulties associated with the evaluation of these physical properties has been partially overcome using the extended Kalman filtering technique. By comparing selected measured values with model predictions, the model parameters are adapted constantly. The parameters m_1 and m_2 were kept constant (m_1 = 8.5 and m_2 = 11.0). This concept can be used to diagnose improper flocculation dosages and overloading/underloading conditions. In order for the thickener to operate properly, it should be loaded to the point which produces the maximum throughput at a desired underflow concentration without carrying particles to the overflow, from which the reference parameters are obtained.

10.3.2 Experimental

The measurement of the basic disturbances to the thickener was found to be crucial in the performance of the filter. During the model-development phase a suggestion was made to install a feed volumetric flowrate sensor. Initially, a volume flow estimate

TABLE 10.1
Measurement and Inferred Variables Equations

Underflow solids density (measurement)	$\rho_u = C_U(\rho_S - 1) + 1$
Settling zone level (measurement)	$S_L = 3.3 + h_S + H_C$
Torque (measurement)	$T = T_K(T_L C_U + (58 - T_L)C_C h_C)$
Where $T_L = (2.41 - \text{RAKPOS})10.0417$	
Thickener inventory (inferred)	$T_{INV} = V_R C_U + h_C C_C A + h_S C_S A)\rho_S$
Thickener inventory rate (inferred)	$R_{INV} = (Q_I C_I - Q_U C_U)\rho_S$
Underflow apparent viscosity number (inferred)	$\text{VISCA} = 1/(1 - C_C)^{(m_{2-}2)}$

obtained from the total concentrator feed was used. An estimate of the fraction of solids by weight was obtained from two sources: (1) an x-ray analysis of the total tails sample and (2) a global plant water and solids mass balance.

The upper level of the settling zone was measured very accurately with a sonic detector. The underflow flowrate and pulp specific gravity measurements were of very high quality. Unfortunately, unmeasured dilution water was added periodically when fines accumulation occurred. A correlation of valve position was used to correct the measurement to obtain an adequate total mass balance. An attempt was made to use rake torque measurement and rake position to predict the states of the thickener when operated in critically overloaded or underloaded conditions. In both instances the thickener was not operated as a thickener so that the basic sedimentation equations did not apply.

It was found that the online thickener model worked without problems when the thickener operated close to typical operating conditions. These conditions are shown in Table 10.2.

The filter was made to operate in situations outside the regular operating conditions by reinitializing it to the new conditions. The thickener was sometimes operated at half capacity for periods up to three days. For these operating conditions, the model which was tuned for full capacity would diverge on some occasions.

Several experimental verifications of the filter estimates of unmeasured variables were made by making direct experimental measurements with the thickener in operation. In order to obtain the thickener concentration profile, a buoyancy probe was built. The calibration of the buoyancy probe was made using accurate densitometers and controlled pulp consistency. The probe was immersed in the thickener using a rope. Stainless steel washers were sent through the rope to the buoyancy probe. The position and number of washers was recorded. The operating conditions of the thickener were maintained as constant as possible during the profile determination.

The model predicted accurately and the filter performed without problems when the thickener operated as a true thickener and was able to predict the behavior up to the point of highly overloaded conditions. It did not function acceptably when the tank was under operated slurry removal conditions (underloaded conditions) due to the drastic change in mechanisms prevailing. The inference of the states could, however, be tracked with the level probe measurement, torque, and rake position, so the inventory could be calculated at all times.

TABLE 10.2
Operating Conditions

Feed pulp volume flowrate	$Q_I = 40.00\ m^3/min$
Feed volume fraction of solids	$C_I = 0.11$ (27% solids)
Underflow volume flowrate	$Q_U = 17.00\ m^3/min$

It is interesting to note that the settler zone concentration decreased due to the transfer of solids to the compression zone. The tracking of the two observations (slime level from top of thickener and removal zone concentration) was very good.

With the values of the zone height and the predicted concentrations, a good estimate of the total slurry in the thickener was obtained. This inventory estimate was used to plan the concentrator operations to minimize the overall energy consumptions based on the different energy costs during the weekends. It was found the pipeline was the limiting factor of the concentrator, and it had to be redesigned to handle the large variations on the ore type. A critical change in ore density was detected and incorporated in the density measurements due to high levels of magnetite in the feed to the plant. A Cu/Fe ratio measurement in the grinding circuits was used to adapt the mass balance calculation to improve the density corrections necessary to the high variability of ore type characteristics.

Table 10.3 shows the thickener online model operator display with the measured and estimated variables and soft sensors implemented to extend the operating time of the concentrator due to the pipeline underdesigned characteristics. The thickener was pushed to an inventory limit to extend the availability of the plant and to look

TABLE 10.3
Thickener Online Model Operator Display

Display No.		Thickener Online Model	Time
Covariances		**Measured Variables**	
P(1,1)	2.47	Underflow density	1.525 SGU
P(2,2)	0.1162	Underflow velocity	7.89 FPS
P(3,3)	2.455	X-ray solids	20.1%
P(4,4)	0.1031	Calculated solids	21.5%
P(5,5)	0.0010	Total feed	1206.0 tph
P(6,6)	0.1890	Slime level	0.7879 ft
P(7,7)	0.0356	Torque	71.93%
		Rake position	54.94%
		Valve position	54.91%
Calculated Variables		**Estimated Variables**	
Sludge level	0.7870 ft	Clay settling	0.0064 ft/min
Compaction level	5.746 ft	Compress zone	0.00359 ft/min
Solids density	3.319 SGU	Settling zone level	4.959 ft
Corrected density	1.588 SGU	Settling zone concentration	9.427% S
Settling rate	0.0492 ft/min	Compression zone level	5.074 ft
Inventory	18398 ton	Compression zone concentration	45.51% S
Volume flowrate of feed	1057 cu ft/min	Removal concentration	53.01% S
Volume flowrate of underflow	492 cu ft/min	Viscosity number (T_Q)	86.3 CP
Volume flowrate of overflow	533.9 cu ft/min	Viscosity number (% S)	44.1 CP

on ways to run the plant during the weekends to lower the overall energy costs of running the plant.

The thickener has been provided to the operator with the estimated concentrations, height, and operating indexes such as compaction and settling rates. This information has been found to be very useful for identifying problems in thickener operation.

10.3.3 Conclusions

The analysis of the thickener using the proposed method was useful in identifying the sedimentation characteristics of the ore and in the identification of the critical mechanisms responsible for the overloading of the system. This dynamic analysis method provided an understanding of the process which was translated into better operator control. The systematic approach of identifying the important measurements and assessing their reliability contributed to the quality of the online calculations performed in the plant.

10.4 FILTRATION

Filtration is the process of solid-liquid separation of passing the suspension through a porous bed, known as *filter medium*, where the bed retains the particles while the fluid passes through the filter medium and receives the name of *filtrate*. The mechanism of filtration is the application of a pressure difference across the filter medium. There are several forms to do this depending on the driving force, for example: gravity, vacuum, applied pressure and vacuum and applied pressure combined, centrifugal force, and the establishment of a saturation gradient. Usually, the different driving forces give rise to different filtration types of filters.

Pressure filters are presently the most reliable equipment for the filtration of concentrates in the mineral industry delivering cakes with 9% humidity or less if required. Filtration is always performed in cycles beginning with the admission of pulp into the filter and ending with the discharge of the filter cake. During each cycle two stages are essential, cake formation and cake dewatering. The first stage is accomplished by applying pressures to the suspension, making the liquid pass through the filter medium, and the second, by blowing air directly through the saturated cake formed in the previous stage. In special cases, other stages may be added such as expression and washing the filter cake followed by a new dewatering stage. In most mineral processing plants, filter cakes of metallic concentrate are approximately incompressible; therefore, the theory of flow through rigid porous medium becomes valid as the basis to develop a filtration theory.

Understanding filtration fundamentals including cake formation and dewatering is a fundamental requisite for proper selection, sizing and design of filtration plants.

10.4.1 Cake Formation Cycle

Cake formation is the first stage of the process. It starts when a suspension is pumped into the filter and ends with the formation of a saturated filter cake. Cake formation is described by Darcy's law:

$$q(t) = \frac{k(\varepsilon_0)}{\mu_f} \frac{\partial p_e}{\partial z} \tag{10.57}$$

Where p_e is the constant applied pressure, q is the percolation velocity, and ε_0 and $k(\varepsilon_0)$ are the porosity and permeability of the filter cake, both considered constants.

In Figure 10.36, p_0 is the atmospheric pressure, l and l_m are respectively the thickness of the cake and the filter medium, and $\Delta p = \Delta p_1 + \Delta p_2$ is the sum of the pressure drop through the cake and the filter medium, with $p_\ell - p_0 > 0$. The total pressure drop across the filter cloth and the filter cake is

$$\Delta p_1 = \int_0^{l(t)} \frac{\mu}{k(\varepsilon_0)} q dz = \frac{\mu l(t)}{k(\varepsilon_0)} q$$

$$\Delta p_2 = \int_{-l_m}^{0} \frac{\mu}{k_m} q dz = \frac{\mu l_m}{k_m} q$$

$$\Delta p = \mu \left(R_m + \frac{l(t)}{k(\varepsilon_0)} \right) \tag{10.58}$$

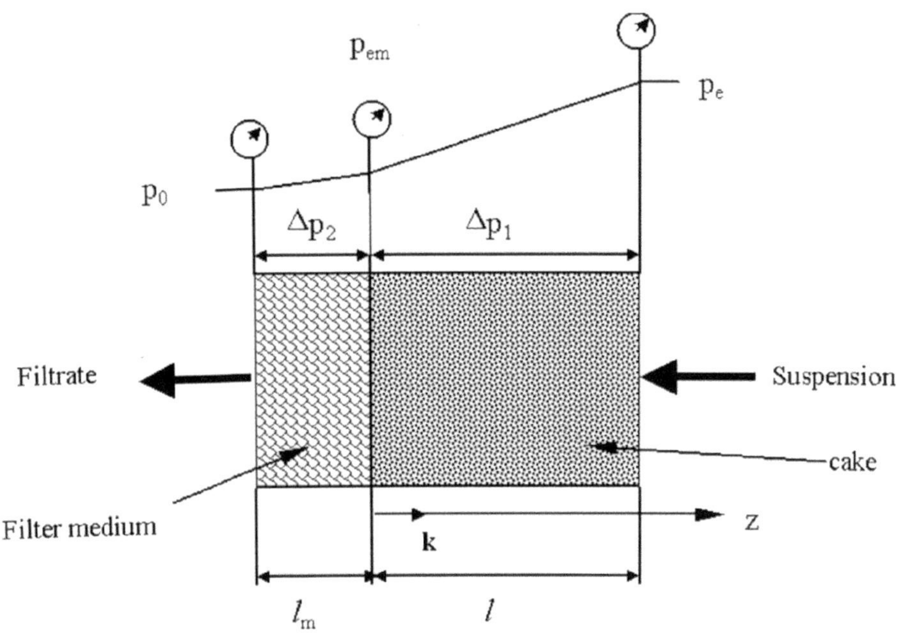

FIGURE 10.36 Filtration with incompressible cake.

Since the volume flow rate of filtrate of filtrate is $Q_f(t) = q_0(t)S = dV_f / d_t$, Equation 10.58 may be expressed in the form

$$\Delta pS = \mu\left(R_m + \frac{l(t)}{k(\varepsilon_0)}\right)\frac{dV_f}{dt} \tag{10.59}$$

$$\frac{\Delta pS}{\mu}t = R_m V_f + \frac{1}{k(\varepsilon_0)}\int_0^{V_f} l(t)d\xi \tag{10.60}$$

To obtain the relationship of the thickness of the filter mass $m(t)$, filtrate volume $V_f(t)$, and the thickness of the filter cake $l(t)$ and the concentration of the pulp fed to the filter, consider

$$\varphi_0 = \frac{m(\varepsilon_0,t)/\rho_s}{m(\varepsilon_0,t)/\rho_s + \left(V_f(t) + Sl(t)\varepsilon_0\right)} \tag{10.61}$$

Substituting the value of the mass of the cake $m(\varepsilon_0,t) = \rho_s(1-\varepsilon_0)Sl(t)$ in Equations 10.61 yields

$$l(t) = \frac{\varphi_0}{1-\varphi_0-\varepsilon_0}\frac{V_f(t)}{S} \rightarrow m(\varepsilon_0,t) = \frac{\varphi_0(1-\varepsilon_0)}{1-\varphi_0-\varepsilon_0}\rho_s V_f(t) \tag{10.62}$$

Substituting these terms in Equation 10.62 gives

$$\text{Cake thickness: } l(t) = k(\varepsilon)\left\{\left[R_m^2 + \frac{2\Delta p}{\mu k(\varepsilon_0)}\frac{\varphi_o}{1-\varphi_o-\varepsilon_0}t\right]^{1/2} - R_m\right\} \tag{10.63}$$

$$\text{Cake mass: } m(\varepsilon_0,t) = \rho_s S(1-\varepsilon_0)k(\varepsilon_0)$$
$$\left\{\left[R_m^2 + \frac{2\Delta p}{\mu k(\varepsilon_0)}\frac{\varphi_o}{1-\varphi_o-\varepsilon_0}t\right]^{1/2} - R_m\right\} \tag{10.64}$$

$$\text{Formation time: } t = \left\{\left(\frac{m_s}{\rho_s S(1-\varepsilon_0))k(\varepsilon_0)} + R_m\right)^2 - R_m^2\right\}$$
$$\left\{\frac{2\Delta p}{\mu k(\varepsilon_0)}\frac{\varphi_o}{1-\varphi_o-\varepsilon_0}\right\}^{-1} \tag{10.65}$$

$$\text{Filtrate volume: } V_f(t) = S\frac{1-\varphi_0-\varepsilon_0)}{\varphi_0}$$

$$\left\{\left[R_m^2 + \frac{2\Delta p}{\mu k(\varepsilon_0)}\frac{\varphi_o}{1-\varphi_o-\varepsilon_0}t\right]^{1/2} - R_m\right\} \tag{10.66}$$

10.4.2 Blowing Cycle

When the main part of the liquid is removed from the suspension during the step of cake formation, it leaves a saturated cake, where saturation s is the fraction of the cake pore volume filled with liquid. The only way to eliminate this retained liquid is by blowing air to displace it from the cake. When air is blown through the filter cake, liquid, air, and solid particles of the cake become in contact constituting a three-phase flow in which part of the liquid is retained by capillary forces and other is displaced from the pores, leaving simultaneously with the air during the process.

Darcy's law also applies to the two-phase flow, liquid-air, through the filter cake with two permeabilities, a liquid *effective permeability* and an *air effective permeability*, which are functions of the saturation in addition to the porosity of the cake.

$$q_a(s) = -\frac{k_a\left[(\varepsilon_0,s)\right]}{\mu_a}\frac{\partial p_e}{\partial z} \text{ and } q_l(s) = -\frac{k_l\left[(\varepsilon_0,s)\right]}{\mu_l}\frac{\partial p_e}{\partial z} \tag{10.67}$$

Where $q_a(s)$ and $q_l(s)$ are the flow rate of air and liquid, $\partial p_e / \partial z$ is the pressure drop, k_a and k_l are the effective permeabilities, and μ_a and μ_l the viscosity of the air and the liquid, respectively. Since the effective permeabilities are functions of the cake permeability and the saturation, we can define *relative permeabilities* k_{ra} and k_{rl} that, multiplied by the *cake permeability*, $k(\varepsilon)$ reproduce the *effective permeabilities.*

$$k_a(s) = k_{ra}(s)k(\varepsilon_0) \text{ and } k_l(s) = k_{rl}(s)k(\varepsilon_0) \tag{10.68}$$

then

$$q_a(s) = -\frac{k_{ra}(s)k(\varepsilon_0)}{\mu_a}\frac{\partial p_e}{\partial z} \text{ and } q_l(s) = -\frac{k_{rl}(s)k(\varepsilon_0)}{\mu_l}\frac{\partial p_e}{\partial z} \tag{10.69}$$

Considering that the capacity of filtration depends on the length of the filtration cycles, it is important to know the associated parameters. Integrating Equation 10.69 yields the cake formation time t_{CF} and the blowing time t_B of the corresponding cycle for a given temperature $T\left[{}^\circ\text{C}\right]$.

$$t_{CF}/V_f(t) = \underbrace{\frac{R_m \mu_l(T)}{S\Delta p}}_{a} + \underbrace{\frac{\mu_l}{2S^2\Delta p} \frac{\varphi_0}{\left(1-\varphi_0-\varepsilon_0\right)k\left(\varepsilon_0\right)}}_{b} V_f(t) \tag{10.70}$$

$$t_B = \frac{\mu_l(T) l \varepsilon_0}{\Delta p_e}\left(R_m(1-s) + \frac{1}{k\left(\varepsilon_0\right)}\int_s^1 \frac{d\eta}{k_{rl}(\eta)}\right) \tag{10.71}$$

The temperature relation of the viscosity of the water and air are (Kinnas n.d.)

$$\mu_w = 3\times10^{-8}t^4 - 9\times10^{-6}t^3 + 0.001t^2 - 0.0558t + 1.788 \;\; ; \;\; \mu_a = 0.0046t + 1.7209$$

The specific resistance of the filter medium R_m and the permeability $k\left(\varepsilon_0\right)$ of the filter cake can be determined from the values of the terms "a" and "b" of the plot in Equation 10.70.

To obtain the blowing time from Equation 10.71, functions defining the relative permeability $k_{rl}(s)$ have to be specified, such as

$$k_{rl} = \alpha \exp(\beta s), \text{ with } \alpha\text{=9.86 and } \beta\text{=4.97}\times10^{-4}\left[\text{m}^{-1}\right] \tag{10.72}$$

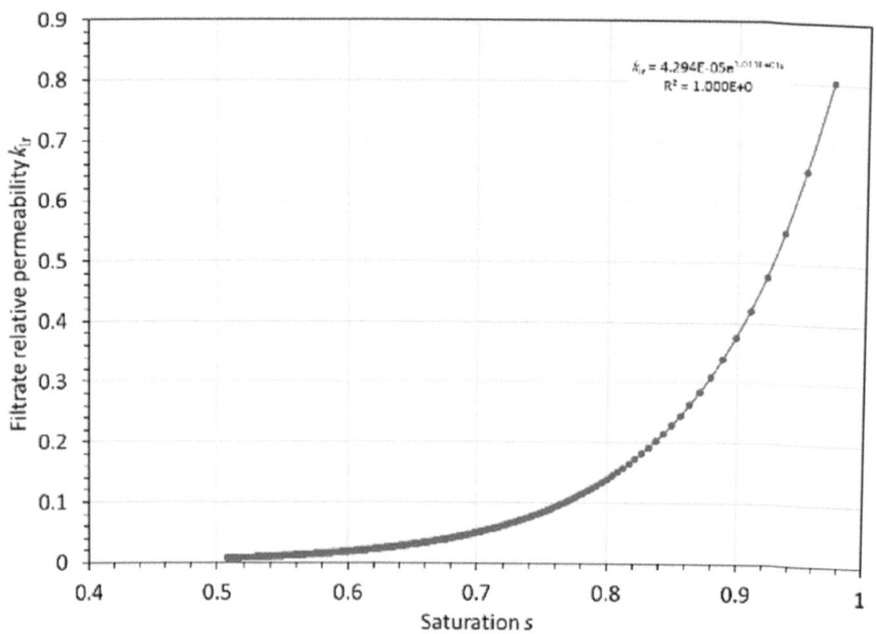

FIGURE 10.37 Relative permeability of the filtrate.

It is convenient to express Equation 10.71 in terms of the humidity h instead of the saturation, where $s = \frac{\rho_s}{\rho_f}\frac{1-\varepsilon}{\varepsilon}\frac{h}{100-h}$, then

$$t_B = \frac{\mu_l l \varepsilon_0}{\Delta p_e}\left\{\begin{array}{l} R_m\left(1-\frac{\rho_s}{\rho_l}\frac{1-\varepsilon_0}{\varepsilon_0}\frac{h}{100-h}\right)- \\ \frac{l}{\beta \times k(\varepsilon_0)}\left(\exp(\alpha)-\exp\left(-\alpha \times \frac{\rho_s}{\rho_l}\frac{1-\varepsilon_0}{\varepsilon_0}\frac{h}{100-h}\right)\right) \end{array}\right\} \tag{10.73}$$

Figure 10.38 shows the plot of Equation 10.73, showing that to reach a humidity of 10%, 344 [s] of blowing time is needed.

Another important parameter is the capillary curve that shows the saturation of the cake accomplished by the applied pressure. In this curve two parameters are important; the *entry pressure* p_b is the minimum pressure that must be applied to begin dewatering the saturated filter cake and the *residual saturation* S_∞ is the value of the minimum saturation, which does not diminish, with an increase in the applied pressure. These to parameters establish the range for the blowing air pressure. Figure 10.39 shows an example of such curve.

10.4.3 Optimization

To improve the efficiency of a filtration cycle, properties such as particle size, insoluble content of the concentrate, cake thickness, and filtration pressure have been

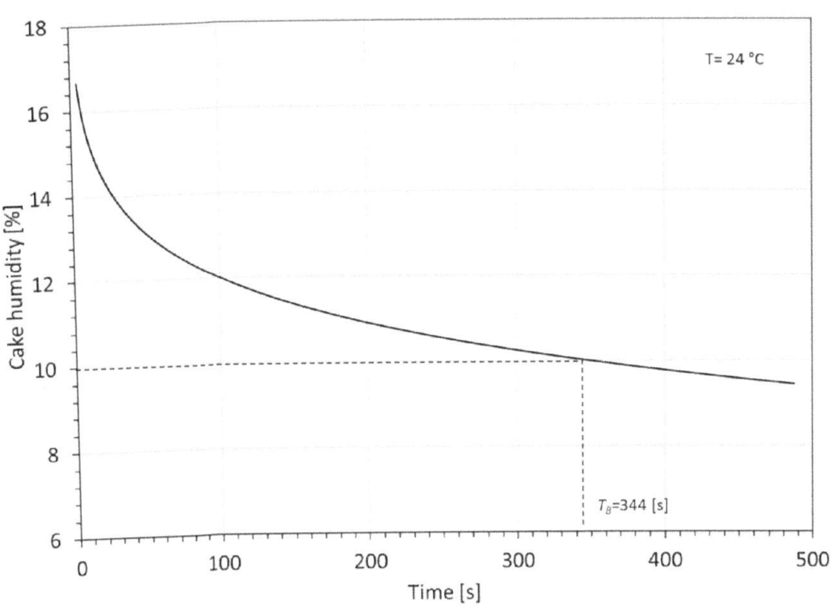

FIGURE 10.38 Cake humidity versus blowing time.

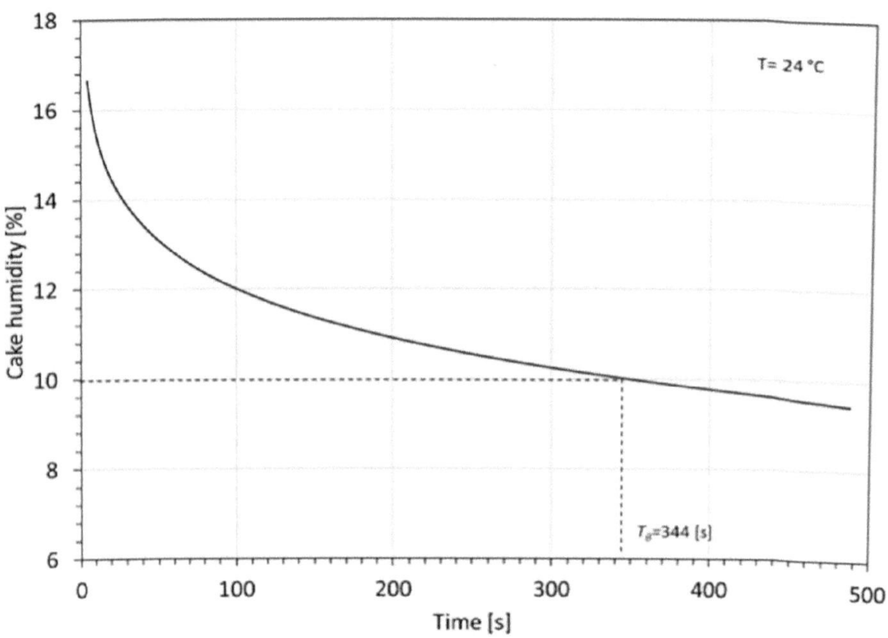

FIGURE 10.39 Capillary curve: pressure drop versus saturation.

studied to reach an optimum performance. We can take advantage of variables such as surface tension and blowing air temperature.

10.4.4 Surfactant

If due to several causes, such as variation of insoluble in a copper concentrate, filtration plants cannot achieve the specified capacity or cake humidity, we can take advantage of the properties of surfactants that diminish the water surface tension. See this from Equation 10.73.

Work for a Chilean copper mining plant (Segovia et al. 2011) showed that surface-active reagents made it feasible of reaching humidity of 8.4%, decreasing the blowing cycle around 40 to 58%. This work was performed using the instrument FiltraTest from Bokela company that can reproduce the operation of an industrial filter by means of laboratory tests. Figure 10.40 shows the case of a copper concentrator lowering its capacity while its insoluble content increased for several months, and Figure 10.41 shows how the use of surface-reagent allowed to solve the problem. In these figures the saturation was replaced by the humidity with the relationship (Concha 2014).

10.4.5 Blowing Air Temperature

Steam filtration used to remove water from a filtration cake helps by reducing viscosity and capillary pressure as the temperature of the liquid increased. A steam

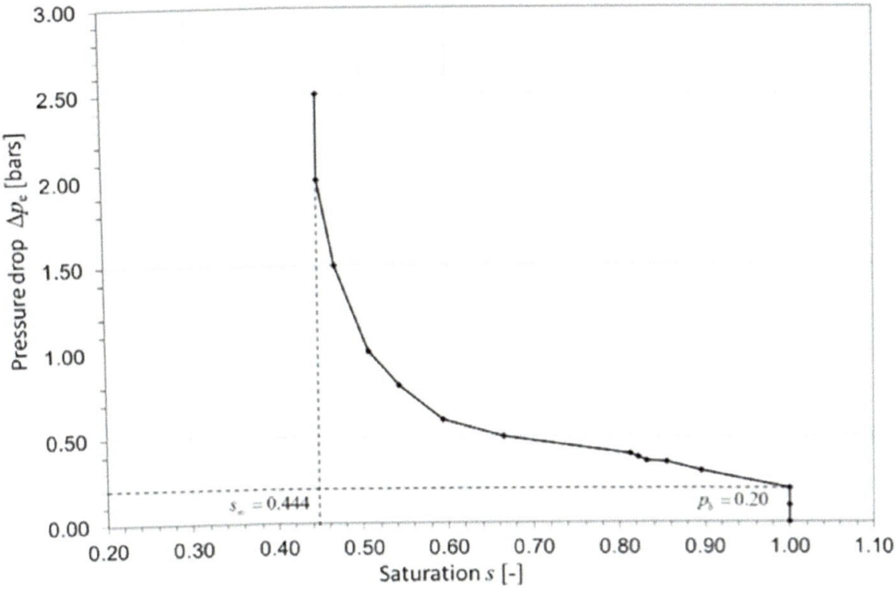

FIGURE 10.40 Capillary curves for various reagent dosage.

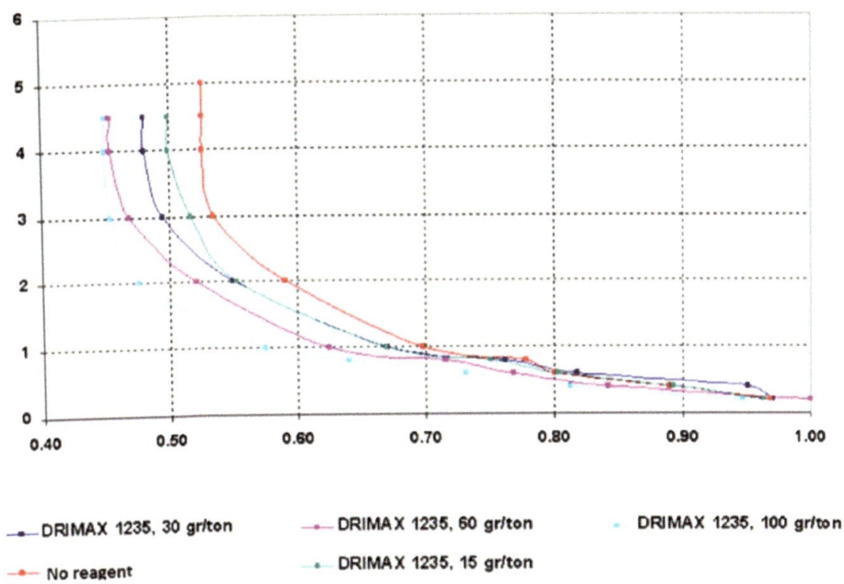

FIGURE 10.41 Humidity versus blowing time.

condensate front moves through the cake, replacing the water in the pores. This mechanism combines mass and heat transfer between steam and the saturated cake (Bott and Langeloh 1996; Gerl and Stahl 1996). Heating the dry (or moist) air before blowing is another way of diminishing the viscosity of the liquid without introducing the complication of using steam and its condensate.

Using the instrument FiltraSet (MR), delivering air at temperatures at several temperatures, a sample of copper concentrate was filtered with the air heated at 20, 30, 40, 50, and 60°C. The results indicated that heating the air increases the time of cake formation by 10%, while the reduction in blowing time was 75%.

Important variable is water viscosity represented by the following expression $\mu_l = 3\times10^{-8}t^4 - 9\times10^{-6}t^3 + 0.001t^2 - 0.0558t + 1.788$. The air viscosity $\mu_a = 0.0046t + 1.7209$ is 100 times smaller than that of water could be neglected. With Equation 10.17, the humidity of the filter cake for a given time was calculated for all five temperatures, as shown in Figure 10.43.

Figure 42 shows that temperature has a significant effect on the blowing time. As the air temperature increases, the blowing time needed to reach 10% humidity decreases dramatically. At 24°C the required blowing time is 365 [s], which at 60°C was reduced to 90 [s].

To evaluate the overall improvements in filtration due to the change in air temperature in cake formation time and blowing time, we have to take into account that, in addition to the useful cycles, in this case cake formation and blowing, there are idle cycles that do not contribute to filtration. Table 10.4 shows cake formation and air

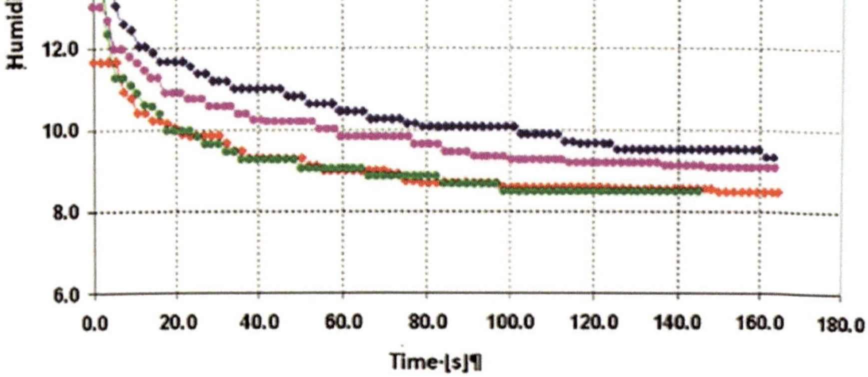

FIGURE 10.42 Humidity of the filter cake as a function of blowing time, circles are experimental data and lines simulation from the equation.

TABLE 10.4
Typical Filtration Cycles in a Horizontal Press Filter

Typical cycle time	Filtration Cycle's duration [s]				
Action	**24**	**30**	**40**	**50**	**60**
Close filter	50	50	50	50	50
Open feed valves	1	1	1	1	1
Star feeding pump	0	0	0	0	0
Feeding and cake formation	163	174	171	180	184
Close valve	0	0	0	0	0
Open drains	0	0	0	0	0
Wash feed pipes	13	13	13	13	13
Close drain	0	0	0	0	0
Close wash water supply	0	0	0	0	0
Press filter cake	0	0	0	0	0
Release compressed air	5	5	5	5	5
Open/close drains	30	30	30	30	30
Blowing air and dewatering	365	348	169	115	90
Open drain valves	10	10	10	10	10
Open filter	50	50	50	50	50
Discharge filter cake	18	18	18	18	18
Wash filter cloth	72	72	72	72	72
Total time [s]	777	771	589	544	523
Total filtration cycle time decrease [%]	0	0,8	24,2	30,0	32,7

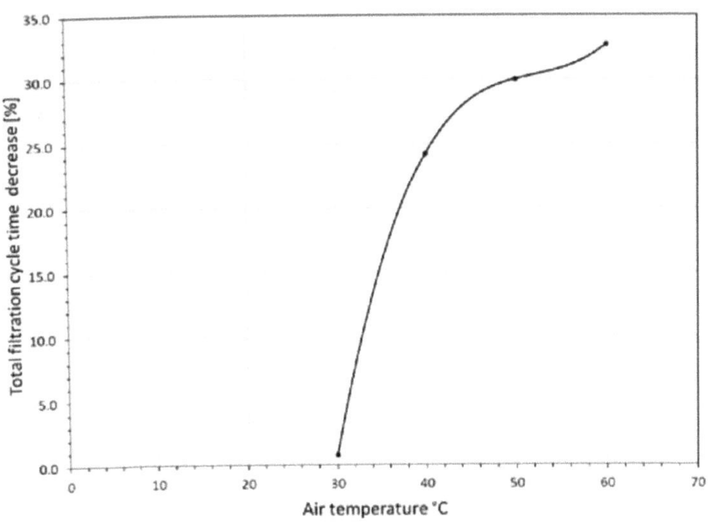

FIGURE 10.43 Percent total filtration cycle decrease by heating the air.

blowing time together with typical idle cycles for a horizontal filter press (Droguett 2000).

Figure 10.43 shows the decrease in total filtration cycle time due to air heating.

REFERENCES

Alkema, K.L. 1971. *The Effect of Settler Dynamics on the Activated Sludge Process*, M.S. Thesis, University of Colorado, Boulder, Colorado.

Adorjan, L.A. 1975. A theory of sediment compression. In *Proceedings of the Eleventh International Mineral Processing Congress*. Melbourne: Australasian Institute of Mining and Metallurgy. pp. 297–318.

Attir, U., and Denn, M.M. 1978. Dynamics and control of the activated sludge wastewater process. *AIChE Journal* 24(4): 693–698.

Barrientos, A. 1978. Parameters determination in the phenomenological model of sedimentation. Engineering thesis, University of Concepción [in Spanish].

Barrientos, A., and Concha, F.A. 1993. A critical review of thickener design methods. *Kona Powder and Particles* 11:79–104.

Bascur, O.A. 1976. Phenomenological model of suspensions in sedimentation. Engineering thesis, University of Concepción [in Spanish].

Bascur, O.A. 1979. Dynamic model of sedimentation. *Technological Notes in Metallurgy* 3(3).

Bascur, O.A. 1991. A unified solid liquid separation framework. *Fluid Particle Separation Journal* 4:117–122.

Bascur, O.A., and Barrientos, A. 1977. Phenomenological model of sedimentation: Experimental parameters. Presented at the IV Chilean Chemical Engineering Congress, University de Concepcion.

Bascur, O.A., and Herbst, J.A. 1986. Improved thickener performance through the use of an extended Kalman filter. In *Design and Installation of Concentration and Dewatering Circuits*. Edited by A.L. Mular and M.A. Anderson. Littleton, CO: SME. pp. 835–845.

Becker, R. 1982. Continuous thickening: Design and simulation of thickeners. Engineering thesis, University of Concepción [in Spanish].

Bott, R., and Langeloh, T. 1996. Steam pressure filtration: Advanced version of the continuous pressure filtration (Hi-Bar filtration). *Aufbereitungs-Technik* 37(4).

Bürger, R., Bustos, M.C., and Concha, F.A. 1999. Settling velocities of particulate systems: 9. Phenomenological theory of sedimentation processes: Numerical solution of the transient behavior of flocculated suspensions in an ideal batch or continuous thickener. *International Journal of Mineral Processing* 55(4):267–282.

Bürger, R., and Concha, F.A. 1998. Mathematical model and numerical simulation of the settling of flocculated suspensions. *International Journal of Multiphase Flow* 24(6):1005–1023.

Bürger, R., Evje, S., Karlsen, K.H., and Lie, K.-A. 2000a. Numerical methods for the simulation of the settling of flocculated suspensions. *Chemical Engineering Journal* 80:91–104.

Bürger, R., Wendland, W., and Concha, F.A. 2000b. Modelling equations for gravitational sedimentation-consolidation processes. *ZAMM* 80:1–18.

Coe, H.S., and Clevenger, G.H. 1916. Methods for determining the capacity of slime settling tanks. *Transactions of the AIME* 55:663–668.

Concha, F., and Barrientos, A. 1993. Critical review of thickener methods. *KONA* 11:1–26.

Concha, F.A. 2014. *Solid Liquid Separation in the Mining Industry*. Cham: Springer. pp. 298–299; 303–306.

Concha, F.A., Barrientos, A., and Bustos, M.C. 1995. Phenomenological model of high capacity thickening. In *Proceedings 22nd International Mineral Processing Congress*, vol. 2. Littleton, CO: SME. pp. 75–79.

Concha, F.A., and Bürger, R. 2003. Thickening in the 20th century: A historical perspective. *Mining and Metallurgical Processing* 20(2).

Droguett, M.H. 2000. Optimization of the filtration system of the Coloso plant, Minera Escondida. Engineering thesis, University of Concepción [in Spanish].

Fisher, O.G., and Seborg, O.E. 1976. *Multivariable Computer Control—A Case Study*. New York: American Elsevier.

Fitch, B. 1979. Sedimentation of flocculent suspensions: State of the art. *AIChE Journal* 25(6):913–930.

Garrido, P., Burgos, R., Concha, F.A., and Bürger, R. 2004. Settling velocities of particulate systems: 13. A simulator for batch and continuous sedimentation of flocculated suspensions. *International Journal of Mineral Processing* 73:131–144.

Gerl, S., and Stahl, W. 1996. Improved dewatering of coal by steam pressure filtration. *Coal Preparation* 17.

Herbst, J.A., and Bascur, O.A. 1984. Mineral processing control: Realities and dreams. In *Control '84: Mineral Metallurgical Processing*. Edited by J.A. Herbst, D.B. George, and K.V.S. Sastry. Littleton, CO: SME. pp. 197–215.

Kinnas, S. n.d. Tables-fluids 13.pdf. www.caee.utexas.edu.

Kynch, G.J. 1952. A theory of sedimentation. *Transactions of the Faraday Society* 48:166–176.

Mehdi, R., Majid, U., Bahram, R., and Ali, A. 2017. Dynamic simulation of pilot thickener operation using phenomenological model with results validation for continuous and discontinuous tests. *Journal of Central South University* 24:1207–1216. https://doi.org/10.1007/s11771-017-3524-7

Mijares-Lopez, C.J. 1973. Simulating the effect of settler dynamics on the activated sludge system. MS thesis, University of Colorado, Boulder, CO.

Paterson, R., and Denn, M.M. 1980. *Design and Control of an Activated Sludge Process for Municipal Wastewater Treatment*. Newark, DE: Water Resource Center, University of Delaware.

Quispe, J., Toledo, P.G., and Concha, F.A. 2000. Discrete sedimentation model of ideal suspensions. *Chemical Engineering Journal* 80(1–3):135–140.

Ray, W.H. 1980. *Advanced Process Control*. New York: McGraw-Hill.

Schubert, H. 1981. On the theory and design of sedimentation. *Aufbereitungtechnik* 8:427–441.

Segovia, J.P., Concha, F., and Sbárbaro, D. 2011. On the control of sludge level and underflow concentration in industrial thickeners. In *Proceedings of the 18 IFAC World Congress*, Milan.

Selim, M.S., Kothari, A.C., and Turian, R.M. 1983. Sedimentation of multisized particles in concentrated suspensions. *AIChE Journal* 29(6):1029–1038.

Smith, T.N. 1983. Limitations to the capacity of continuous thickeners. *Chemical Engineering Research and Design* 61:45–50.

Wakeman, R.J. 1981. Thickening and filtration: A review and evaluation of recent research. *Transactions of the Institution of Chemical Engineers* 59:147–160.

Wallis, G.B. 1969. *One-Dimensional Two-Phase Flow*. New York: McGraw Hill.

FURTHER READING

Bürger, R., and Concha, F.A. 1997. Simulation of the transient behavior of flocculated suspensions in a continuous thickener. In *Proceedings of the 20th International Mineral Processing Congress*. Melbourne: Australasian Institute of Mining and Metallurgy. pp. 91–101.

Bürger, R., and Concha, F.A. 2001. Settling velocities of particulate systems: 12. Batch centrifugation of flocculated suspensions. *International Journal of Mineral Processing* 63:115–145.

Bürger, R., and Karlsen, K.H. 2001. On some upwind difference schemes for the phenomenological sedimentation-consolidation model. *Journal of Engineering Mathematics* 41:145–166.

Bürger, R., and Wendland, W. 1998a. Existence, uniqueness and stability of generalized solutions of an initial-boundary value problem for a degenerating quasilinear parabolic equation. *Journal of Mathematical Analysis and Applications* 218:207–239.

Bürger, R., and Wendland, W. 1998b. Mathematical model and viscous splitting approximation for the sedimentation-consolidation processes. In *Proceedings of the 4th International Congress on Numerical Methods and Applications*. pp. 19–23.

Concha, F.A. 2000. *Manual de Filtración y Separación*. Concepción: Cettem. pp. 428–437.

Concha, F.A., and Bürger, R. 2002. A century of research in sedimentation and thickening. *Kona Powder and Particle* 20:38–70.

Concha, F.A., and Bustos, M.C. 1991. Settling velocities of particulate systems. Part 6. Kynch sedimentation processes: Batch settling. *International Journal of Mineral Processing* 32:193–212.

Concha, F.A., and Bustos, M.C. 1992. Settling velocities of particulate systems. Part 7. Kynch sedimentation processes: Continuous thickening. *International Journal of Mineral Processing* 34:33–51.

11 Water in the Mining Industry

11.1 INTRODUCTION

Water is mainly found in the seas and oceans, which cover 71% of the earth's surface with 97% of salt water and only 3% is freshwater, from which 2% freshwater is frozen at the poles and only 1% natural liquid freshwater. Besides, there is an unequal distribution of this resource in the different continents, creating abundance zones and scarcity zones.

Several articles define the water as the new Blue Gold (Rivas and Sapag 2008). *National Geographic* published a special issue to describe the water in our thirsty world (Johns 2010). He reports that water, "this precarious molecular edge on which we survive," as Barbara Kingsolver (2010) says in this publication, will only grow more precarious. By 2025, 1.8 billion people will live where water is scarce (Johns 2010). A deep analysis of the huge challenge by the civilization to avoid the abyss of climate change is discussed by Fueyo (2022).

Most parts of the word are in some stage of water stress, some of which are desert and coincide with the largest mining companies of the world; for example, the copper ore deposits in Chile, Peru, Mexico, the United States, Canada, North and South Africa, Australia, and Russia. Water scarcity in these mining areas has become one of the most complicated issues for mining companies. Figure 11.1 shows the state of stress in 2013 and projections to 2040.

11.2 PROBLEMS IN THE MINING INDUSTRY TODAY

The copper mining industry currently faces numerous challenges, the management of which may well determine the extent of its future success. These challenges are the reduction of copper grades, the increase in the consumption and cost of energy, large production of waste and pollution of the environment. A mining company must deal with all these problems and ensure an efficient and sustainable operation. All these challenges are closely related to water, as we will see in the next section, using the case of copper mining.

11.2.1 Decrease in Copper Grades

The significant drop in copper grades by 50% in recent years has brought with it the need in this period and in the future to significantly increase the mass of ore to be treated to maintain or increase production, as shown in Figure 11.2.

DOI: 10.1201/9781032614212-11

FIGURE 11.1 Water stress by countries.

Source: Gassert, F., Luck, M., Landis, M., Reig, P., and Shiao, T. 2013. Aqueduct Global Maps 2.1. In Working Paper. Washington, DC: World Resources Institute. https://files.wri.org/d8/s3fs-public/Aqueduct_Global_Maps_2.1.pdf. Accessed August 2023.

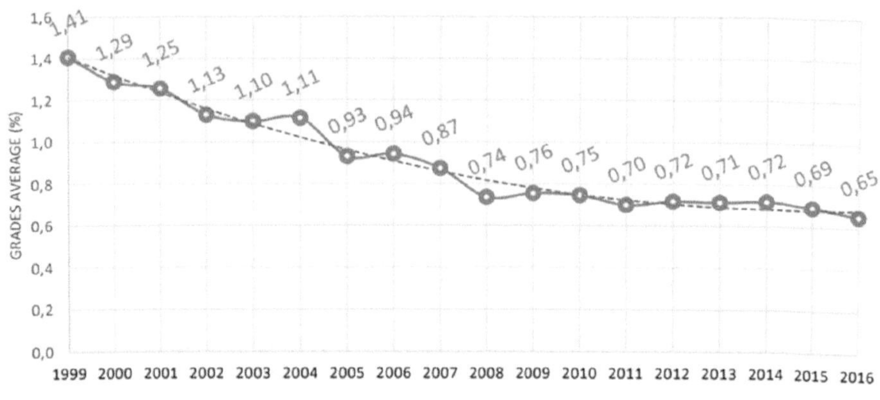

FIGURE 11.2 Decrease in copper ore grades in the last 20 years.

Source: Cantallopts, J. 2017. *Chilean Copper Mining Costs.* Santiago, Chile: Chilean Copper Commission. www.cochilco.cl/Presentaciones%20Ingls/Chilean%20Copper%20Mining%20Costs.pdf. Accessed August 2023.

11.2.2 Increase in Consumption and Variability in the Cost of Energy

The mining industry is one of Chile's largest consumers of electricity, with 21.0 TWh in 2017, an amount that is expected to increase to 30 TWh by 2027. The energy situation in Chile is more of a political problem than a technical one. In

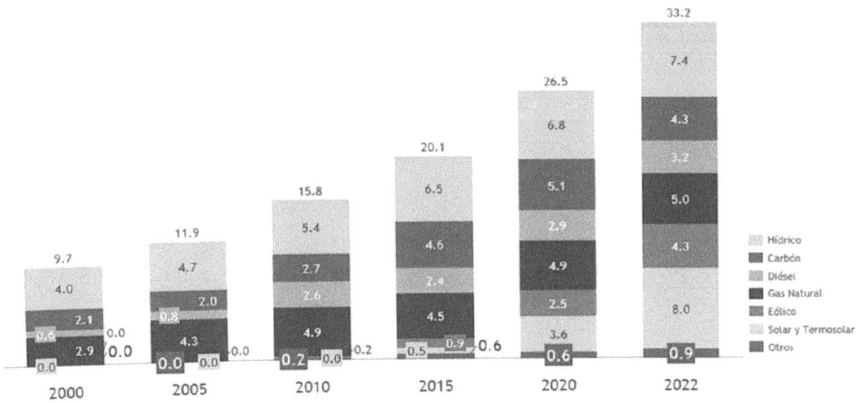

FIGURE 11.3 Chile's national electrical system installed capacity in Gigawatt.

Source: http://energiaabierta.cl/visualizaciones/capacidad-instalada

2019 about 70% of energy came from fossil fuels: coal, gas, and oil. The first step to achieve greater energy autonomy is to have the political will to achieve it, and Chile committed to walk the route of the nonconventional renewable energy (NCRE) reaching 20% in 2019, generating 60% of energy needs with NCRE by 2035 and 70% by 2050. The installed electric matrix capacity in Chile is shown in Figure 11.3 from current data from Chilean Energy Commission (energiaabierta.cl 2023)

Until recently, in 2015, the cost of energy in Chile was the highest of all copper-producing countries, at 0.15 US$/kWh, while the average value for all countries was 0.08 US$/kWh. However, with technological advances and the increase in the efficiency of renewable energies by 2020, the unit cost will be between 0.06 and 0.08 US$/kWh, corresponding to the average of copper-producing countries. Figure 11.4 shows the cost of energy in copper-producing countries.

The largest consumer of electrical energy in copper mining in Chile is the mineral concentrator with 14% of 25.0 TWh in 2020 representing 8% of the total operating costs of this industry as shown in Figure 11.5. The projection to 2031 of 20% of 33.4 TWh. This is the main reason to devote studies in the comminution systems as presented in Chapter 7.

11.2.3 Large Waste Production

The residues resulting from the concentration of copper ores are barren discard rocks and tailings. Minerals with a content of 0.60% copper producing concentrates of 35%, the volume of tailings produced is sidereal 1.4 million tons of tailings per day, and considering the production increase scheduled for the next 10 years, tailings production would increase to approximately 1.6 million tons per day in 2030. Addressing these shortcomings is a serious territorial problem.

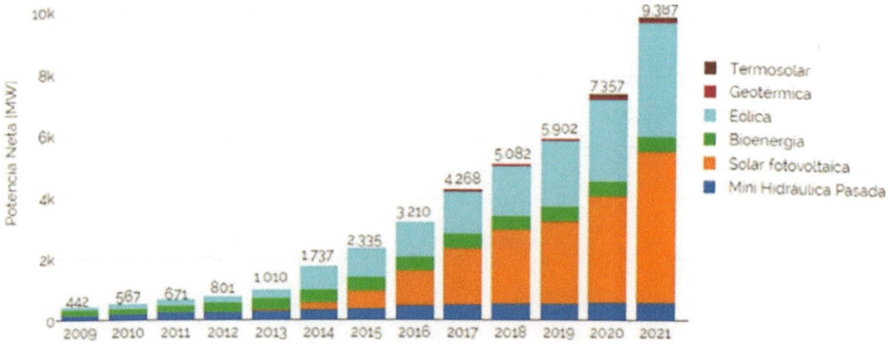

FIGURE 11.4 Evolution of the Nonconventional Renewal Energies currently installed in Chile as of 2018.

Source: Cantallopts, Pino and Brantes. 2021. www.cochilco.clDEPP06/2021

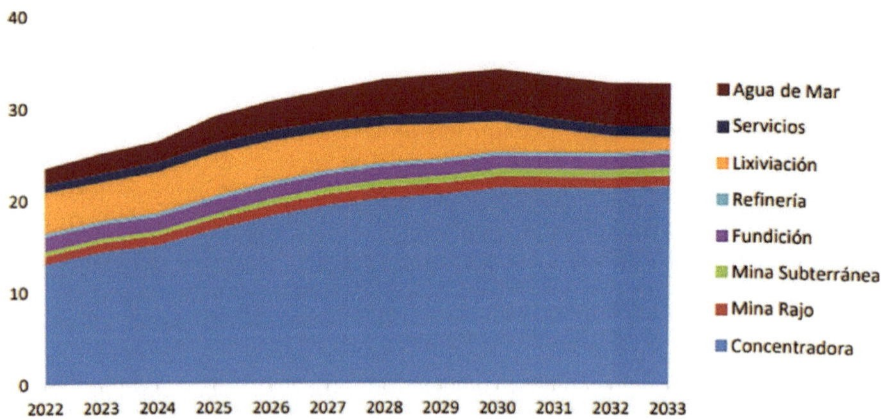

FIGURE 11.5 Expected electricity consumption (percent) of copper mining nationwide by process, 2022 and 2033.

Source: www.cochilco.cl/ COCHILCO (Chilean Copper Commission). DEP 2/2023

11.2.4 Pollution of the Environment

The main environmental contaminations in the copper industry are from the air and water. In the case of mineral concentration, the problem is acid mine drainage, which consists of the introduction of heavy metals into the water from the degradation of solid material, in addition to leaching and cyanidation in hydrometallurgy.

11.3 WATER IN THE MINING INDUSTRY

The water resources available to mining are continental water, seawater, and recycled water. For example, in 2020, Chile used a total of 64[m/s] of water in its entire

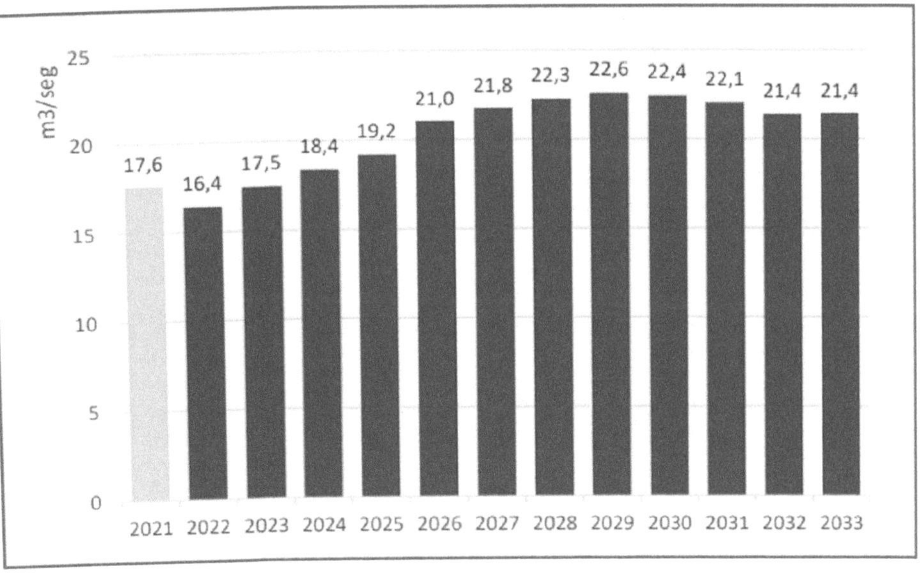

FIGURE 11.6 Water consumption in 2020 in Chile.

Source: www.cochilco.cl/ COCHILCO (Chilean Copper Commission). 2020. Water consumption in the copper mining industry in 2019. DEPP14/2020. (In Spanish).

operation with a contribution of 12[m^3/s] for continental, 5[m/s] for sea water, and 47[m/s] of recovered and recycled water. The projected water demand in Chilean copper mining is shown in Figure 11.6. Low grade ore mines require a range from 2.1 to 3.1 m3/ton of water.

11.4 WATER BALANCE IN A CONCENTRATOR

Mineral processing of sulfide mineral involves, as we have described in Chapter 1, the crushing, grinding, and classification of the ore, followed by flotation, thickening, filtration, and transportation of concentrates and tailings to the filtration plant or tailing dams. Continental and seawater are used in a every one of these production units, where the main water users are classification and flotation. The three units responsible to recover and recycle process water are thickening, tailing disposal, and filtration (Figure 11.7). Two types of thickeners are profusely utilized today, conventional and high-density thickeners. Paste thickeners are recently introduced in the copper industry, and filtration of tailings has yet to be considered.

Kotiranta et al. (2015) made simulations of water balance and economic calculating of four different tailings management, conventional thickening, high-density thickening, paste thickening, and tailings filtration, as shown in Figure 11.8. A comparison of the freshwater makeup in cubic meters per ton of concentrate produced [m/ton] is shown in Figure 11. 9.

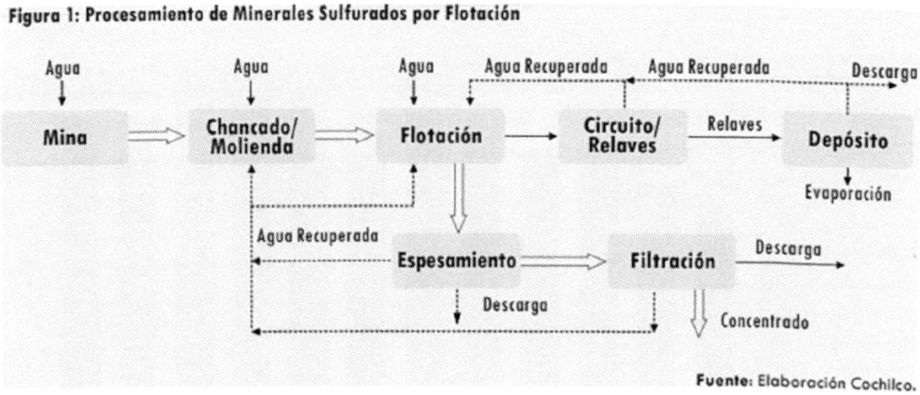

FIGURE 11.7 Simplified flowsheet of a sulfide mineral concentration plant.

Source: COCHILCO (Chilean Copper Commission). Brantes, R. 2010. Water Consumption in the copper industry 2010, Chapter 2.

Bernal gives similar information in cubic meters per tons of mineral processed [m/ton] in Figure 11.10.

11.4.1 Water Economics

From the total amount of process water used in a copper concentrator, one cubic meter can treat one ton of ore, and 30% of it is lost and must be replaced with makeup water coming from either three sources, continental, desalinated, or raw seawater. From the 75% of process water recovered, the thickener recovers 50%, and 23% is recovered in the tailing dams. The mining industry has made great efforts to reach this water recovery and recycling, but this amount must increase in the future to reach a number of close to 80% or more. There are several common methods used to reduce the amount of fresh makeup water used in mineral concentration (Bruce and Seaman 2014):

- Water control and recycling
  - Adequate measurement and control of water inputs to the plant
  - Use of high efficiency thickeners to decrease water losses to tailing
  - Recovery of water infiltration from tailings for use in the process
  - Recycling of tailings runoff to the process
- Water substitution
  - Use of desalination plants for water supply
  - Direct use of sea water
  - Use of wastewater/gray water in the process

While water recovered in thickeners and tailing dams have a cost of 0.3 to 0.5 [US/m], desalinated seawater, including desalination and transport to the mining sites, costs 3 to 5 [US$/m] or more. (Bernal 2015).

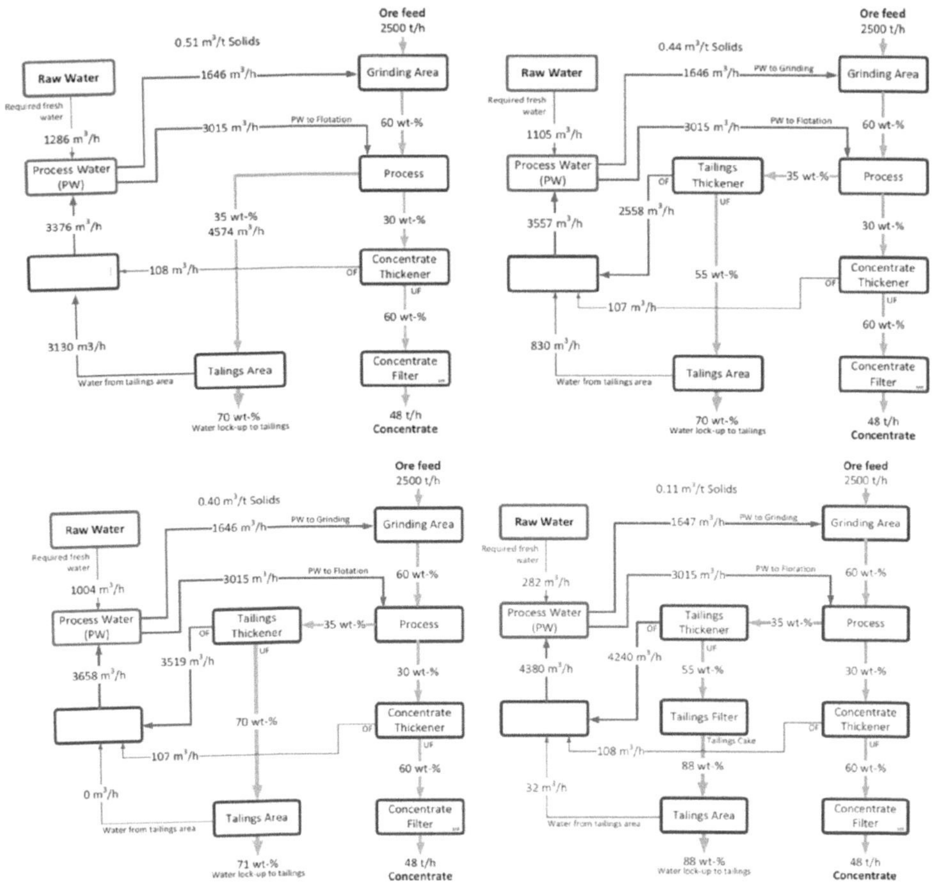

FIGURE 11.8 Water balance of tailings management, conventional thickening (a), high density thickening (b), paste thickening (c), and tailings filtration (d).

Source: Kotiranta, T., Horn, S., Jansson, K., and Reuter, M.A. 2015. Towards a "minimum impact" copper concentrator: A sustainability assessment. In *Proceedings of the 11th International Mineral Processing Conference: Procemin 2015.*

The use of raw seawater, that is, direct without treatment, is only possible for ores containing less than 5% of pyrite, and it is very convenient from an economic point of view, avoiding the cost of the desalination plant of about 6000 [US$/m]. Concha, Castro, and Vergara (2016) obtained 1.9 to 4.5 [US$/m].

11.5 DYNAMIC WATER MANAGEMENT USING A DIGITAL TWIN STRATEGY IN MINERAL PROCESSING COMPLEXES

The water resource has been a critical process constraint in mineral processing plant treating low grade ore. The tonnage required for these plants to be profitable are very large, thus requiring larger amounts of water and electricity like never

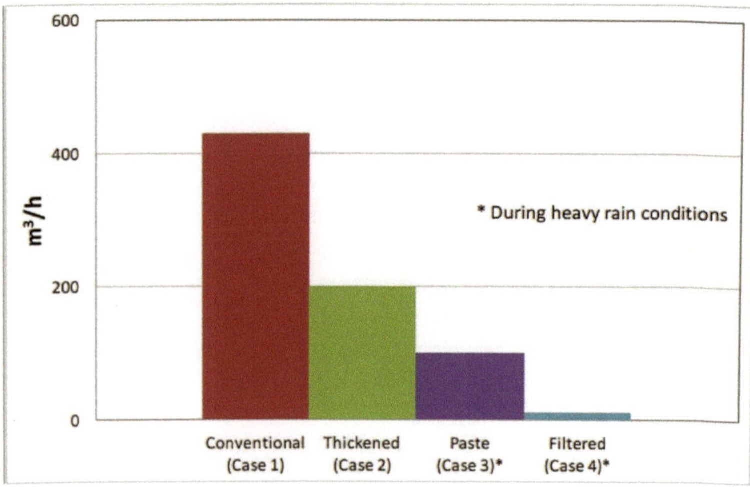

FIGURE 11.9 Comparison of the freshwater makeup in cubic meters per ton of concentrate produced [m/ton].

Source: Kotiranta, T., Horn, S., Jansson, K., and Reuter, M.A. 2015. Towards a "minimum impact" copper concentrator: A sustainability assessment. In *Proceedings of the 11th International Mineral Processing Conference: Procemin 2015*. www.gecamin.cl.

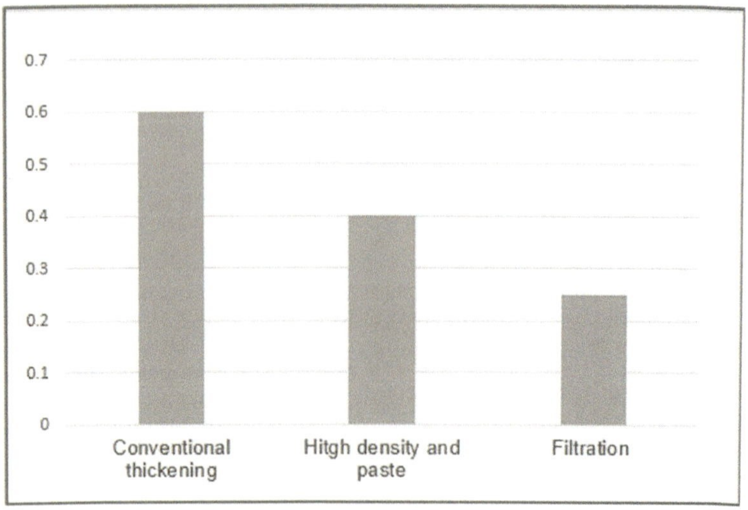

FIGURE 11.10 Makeup water in cubic meters per tons of mineral processed [m/ton].

Source: Bernal, L. 2015. Strategies to reduce water consumption in concentrator plants (part 1). *Process Analysers* [blog]. https://plapl.com.au/concentrator-plants-water-consumption-2/. Accessed August 2023.

before. As such, the hydric requirement for the grinding and flotation is sometimes what hinges on getting to the design required throughput as conceived in the industrial capital project. New mineral processing plants have started to use seawater or constructing osmosis water plants (Castro and Laskowski 2018). However, all these plants require large investments cost, and they use additional energy, making the processing of the low grade's ores less sustainable. As such, many of the operating practices for grinding have been to push the rheological constraints of the mill. It is well-known that once the hydrocyclones start roping, the grinding product exhibits a bimodal size distribution due to over-grinding. This over-grinding creates finer particles that are difficult to recover by conventional flotation cells and present difficult flocculation results. The typical P80 measures do not account for the tail of the size distribution. A detailed analysis conducted by the plant show the bimodal distribution, which causes flocculation problems and water losses in the tailing ponds.

One of the major improvements that can be achieved with today's technologies is the development of an integrated approach to manage the water consumption and water recovery from specialized hydrocyclones, thickeners, and the tailings ponds for increasing the water recovery in large industrial complexes. What is measured can be controlled and optimized. As such, conducting a production and water mass balance of an entire mineral processing enables to identify the critical steps in recovering the maximum amount of water.

The first step is to create the proper database and evaluation of the operating states of all process units in the chain supply. Figure 11.11 shows a block process flow diagram of a mineral processing plant with the major areas for analysis in terms of the production flow, energy, water flow for the rock process, water processing, and mineral processing. This strategy uses a digital twin object model to process the raw data against triggers that define the operational states of each major process unit. A digital twin is an abstraction of the physical entity using data processing and predictive models to avoid trespassing or violating costly equipment or process constraints. (Additional details are described in Chapter 12.) These operational states are as follows: Is the unit running okay, unit running in troubles, unit in startup mode, unit idle, units down, or unit in maintenance. An operational event frame algorithm is used to transform the data into information at the right degree of details for developing the predictive analytics models using the process unit operating state.

Figure 11.11 shows the integrated plant, including the water management at the thickeners and tailings. It shows that first there is a ROCK Processing step when mining low grade ores with a water management in parallel to be able to recover the necessary water for the grinding and rougher flotation processes (Bascur et al. 2022). And then to continue with the mineral processing, cleansing the rougher flotation concentrate and recovery of the (Chalcopyrite) copper and (Molybdenite) molybdenum minerals. As such, we can analyze the entire mineral processing to have the consumption of water by area and recovery of water from the thickeners and tailing ponds. An example is shown in Figure 11.12. This figure shows a screen capture

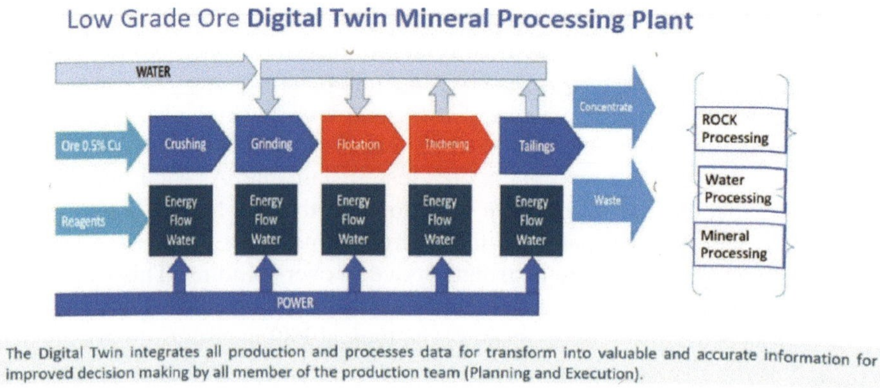

FIGURE 11.11 OSB Digital Twin integrated mineral processing plant and water management.

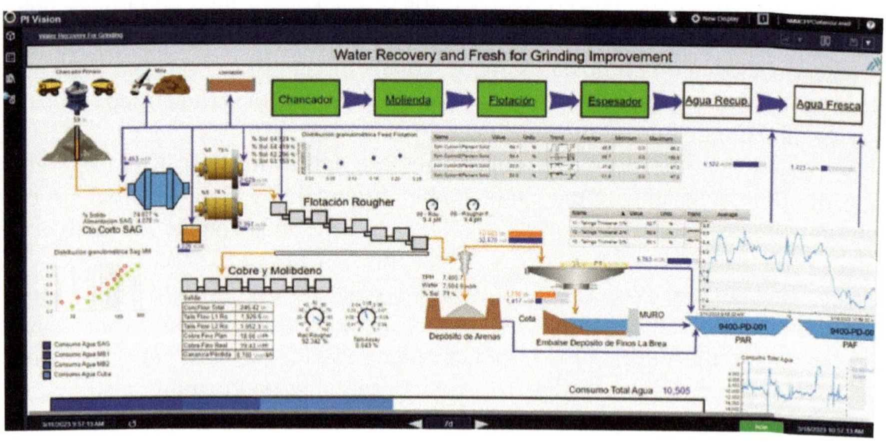

FIGURE 11.12 Digital concentrator of digital twin real-time integrated process flow diagram (plant and water management).

of simplified process block showing the state of the major areas (shown in green, meaning that they are running OK) and the key variables for the analysis. It shows the water recovery from the thickeners. The analysis of process models shows that one of most important variables in the water recovery from the thickeners is the particle size shape produced in the grinding circuit and the flocculant addition in the feed to the thickeners. As such, further analysis shows the shape of the grinding size distribution drastically affects the recovery of water due to the content of fines. The analysis shows that when the particle size distribution as a particle size distribution shape is greater than 0.37, the content of fines is bimodal, meaning that too many fines are present to have a good flocculation of all the particles. Thus, entrapping water in the coagulated flocs which report to the underflow of the thickener. By controlling the grinding circuits, adding the right amount to the sump to maintain % solids to the cyclones below 65% and the right amount of water to the ball mill feed to maintain a % solids below 75% produce a particle size distribution with an M larger than the 0.37 to get a 40% increase in water recovery and considerable improvement in the copper production rate. These results are presented in Figure 11.12.

The block diagram shown in Figure 11.12 has been implemented to improve the efficiency of grinding the ore (we use the word *rock* for grinding, rougher flotation, and water separation on purpose) to increase the water recovery from the thickeners and from the tailing's ponds. This increment of water availability enables us to improve control over the particle size distribution shape to avoid fines and to improve the flotation hydrodynamics of both the fines and coarse particles.

Figure 11.13 shows an example of using the digital twin model to build a water recovery predictor model based on the grinding product, flocculation process, and thickener operating variables. Inspection of this data shows that the most influential variables as per the statistical analysis provided by the p value are the particles size distribution shape from the grinding product M L1 and M L2 with a *p* value of 2e-314, and the percent solids of the discharge of the rougher tailing (feed to the thickener) (p = 1e-44) and the flocculant addition rate (p = 2e-21). The last two variables are the two traditional variables. The particle size analysis performed with particle size analyzer online to provide an inferred particle size distribution shape is novel contribution with brilliant and motivating results for improvement.

Legend

Variable	Name
$a	Alim Water Esp Filter
$b	Dos Floc Esp2 Filter
$c	m L2 Filter
$d	%sol Over Filter
$e	P80 L2 Filter
$f	P80 L1 Filter
$g	%Sol Tails Filter
$h	%sol Descarga Esp1 Filter
$i	%sol Descarga Esp2 Filter
$j	m L1 Filter
$k	Dos Floc Esp1 Filter
$l	Tails Flow Filter

Coefficients

Name	Coefficient	Std. Error	P-Value
$a	0.3552	0.0041	0
$b	-15.76	0.2224	0
$c	7781	73.594	0
$d	-43.18	0.388	0
$e	-4.759	0.0451	0
$f	0.5565	0.046	1e-33
$g	2.6308	0.1874	1e-44
$h	2.8666	0.3129	5e-20
$i	-6.467	0.4707	7e-43
$j	1940.4	51.045	2e-314
$k	-2.341	0.2466	2e-21
$l	1.3444	0.0133	0

FIGURE 11.13 Water recovery model and multivariable statistical advanced analysis.

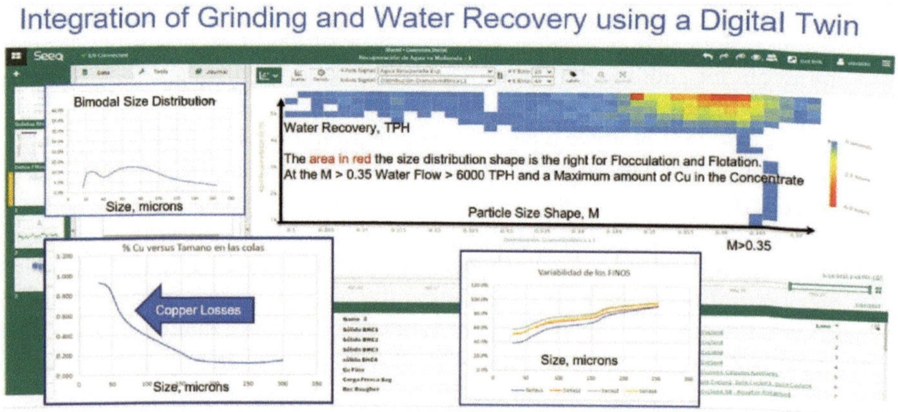

FIGURE 11.14 Effect of fines in water recovery and metal production.

The analysis can be enhanced by looking at a heat map of the water recovery from the thickener and the grinding product particle size distribution shape (M) as presented in Figure 11.14. Figure 11.14 shows that the water recovery goes up to more than 6000 TPH when the size distribution is M > 0.37 (in the dark area of the diagram). This means that less fines have been generated, and the size distribution is not bimodal. The bimodal size distribution example is shown at the left on Figure 11.14. The operational data analysis provides clear visual representations which are very difficult problems to discover unless a data classification of the raw data is implemented. Figure 11.14 indicates that maintaining a size distribution with fewer fines maximizes water recovery and an increase in copper production rate (less than 40% -325 mesh). The maximization of copper production is dependent on the water recovery from the thickening and the tailing ponds. It has been found that optimizing the size of the distribution shape of the grinding circuit clearly affects simultaneously the production of copper and water recovery.

Readers can find additional discussion of the grinding, flotation, and thickener analysis in Chapter 12.

The effect of solids concentration, flocculant dosage, and particle size on dewatering behavior of tailings as Shahrebanck copper complex has been reported by Garmsiri and Hossini (2018). Results indicates that not only settling flux but also compressibility of the suspension was improved by reducing suspension solids concentration. It was found that increasing flocculant dosage from 15 to 45 g/t led to an increase in settling flux about seven times. Both settling flux and compressibility were reduced with fine particles. They show that increasing −75 microns fraction in thickener feed from 47 to 59% led to a decrease in the underflow solids concentration from 63 to 55%.

Additional research has been reported by Wang et al. (2020) on the influence of coarse tailings on flocculation statement.

REFERENCES

Bascur, O.A., Retamal, F., and Espinosa, S. 2022. Treating low grades ores using a digital in mining and mineral processing plants. In *Proceedings of the Copper 2022 International Conference, Vol. 7: Process Optimization through Innovation, Technology, Automation and Design*, IIMChile, Santiago.

Bernal, L. 2015. Strategies to reduce water consumption in concentrator plants (part 1). *Process Analysers* [blog]. https://plapl.com.au/concentrator-plants-water-consumption-2/. Accessed August 2023.

Bruce, R., and Seaman, T. 2014. Reducing fresh water use in the production of metals. www.teck.com/media/CESL-Publication-Copper-reducing-fresh-water-use-in-the-production-of-metals.pdf. Accessed August 2023.

Castro, S., and Laskowski, J. 2018. Sea water flotation practices. In *Proceedings of 29th International Mineral Processing Council (IMPC)*. Carlton: IMPC.

Concha, F., Castro, S., and Vergara, M. 2016. Economic evaluation of alternatives for using desalinated and non-desalinated seawater in Cu/Mo flotation Plants. In *Proceedings of the 5th International Congress on Water Management in Mining*, Santiago. www.Gecamin.cl. http://energiaabierta.cl/visualizaciones/capacidad-instalada/

Fueyo, J. 2022. *Blues para un Planeta Azul*. Barcelona: Penguin Random House.

Garmsiri, M.R., and Hossini N.M. 2018. The effect of solids concentration, flocculant dosage and particle size on dewatering behavior of tailings sample at Shahrebanck copper complex. *Amirkabir Journal of Civil Engineering* 49(4):201–204.

Johns, C. 2010. Water: Our thirsty world. *National Geographics* 214(4):4.

Kingsolver, B. 2010. Fresh water. *National Geographics* 217(4):44–49.

Kotiranta, T., Horn, S., Jansson, K., and Reuter, M.A. 2015. Towards a "minimum impact" copper concentrator: A sustainability assessment. In *Proceedings of the 11th International Mineral Processing Conference: Procemin 2015*.

Rivas, C., and Sapag, R. 2008. Agua caliente, Capital, Edicion No. 225, April 18th May 1st., Ediciones e Impresos S.A., Santiago.

Wang, S., Wang, X., Chen, Q., Song, X., Qin, J., and Ke, Y. 2020. Influence of coarse tailings on flocculation settlement. *International Journal of Mineral, Metallurgy and Materials* 27:1065–1074.

FURTHER READING

Cantallopts, J. 2017. *Chilean Copper Mining Costs*. Santiago: Chilean Copper Commission. www.cochilco.cl/Presentaciones%20Ingls/Chilean%20Copper%20Mining%20Costs.pdf. Accessed August 2023.

Cantallopts, J., Pino, F., and Brantes, R. 2021. Las energías renovables en la minería chilena. www.cochilco.clDEPP 06/2021.

COCHILCO (Chilean Copper Commission). 2020. Projection of water consumption in the copper mining industry in 2020–2031. DEPP/35/2020. www.cochilco.cl. [in Spanish].

COCHILCO (Chilean Copper Commission). 2020. Water consumption in the copper mining industry in 2019. DEPP/14/2020. www.cochilco.cl [in Spanish].

COCHILCO (Chilean Copper Commission). 2022. www.cochilco.cl/Listado%20Temtico/Proyecci%C3%B3n%20de%20demanda%20de%20agua%20en%20la%20miner%C3%ADa%20del%20cobre.pdf

COCHILCO (Chilean Copper Commission). 2023. DEP 2/2023. www.cochilco.cl/Mercado%20de%20Metales/Las%20energias%20renovables%20en%20la%20mineria%20chilena%20del%20cobre%202021.pdf

Gassert, F., Luck, M., Landis, M., Reig, P., and Shiao, T. 2013. *Aqueduct Global Maps 2.1*. In Working Paper. Washington, DC: World Resources Institute. https://files.wri.org/d8/s3fs-public/Aqueduct_Global_Maps_2.1.pdf. Accessed August 2023.

World Resources Institute. 2015. Aqueduct projected water stress country rankings. www.wri.org/data/aqueduct-projected-water-stress-country-rankings. Accessed August 2023.

12 The Future of Mineral Processing

Digital Transformation in Mining Plant Operations

12.1 INTRODUCTION

The world is experiencing one of the greatest and fastest change in human history caused by these new technologies which are transforming the economy, society, and the way we live. The *digital disruption* produced in industry by *automatization* of processes, *artificial intelligence, big data, and machine learning* present an opportunity for great technological improvements. Big data is a discipline of the field of information and communication technologies, which shows, by collecting and ordering data sets too large or complex to be treated by traditional data, how to present their visualization, representation, and predictions and apply them to software analysis from the observed patterns. This discipline is based on three elements, large data volume, data treatment speed, and variety of data formats. The existence of big data allows machine learning, the autonomous learning of a machine that does not need to be programmed. This is what we call Industry Revolution 4.0 (Schwab 2016).

Currently the major roadblock for the *digital transformation* is the cultural change required to adapt to new ways for business process workflows. These roadblocks are as follows:

- Too many data silos and legacy systems
- Too many versions of the truth
- Cultural adaption and new skills
- Lack of management involvement and having the right information
- Resistance to redesign the new business work processes in the digital age
- Not empowering people to use computers to analyze data

The mission of the cultural business process reengineering and the adoption of the new cloud-based tools running advanced predictive analytics are the main goal and requires the collaboration of all. This is not only an automation project or a control project but also the convergence of new tools to integrate the subject matter using them for collaboration, locally and externally, with communities worldwide. Today, the data travels at the speed of light and computers, and networks have tremendous computing power that requires us to manage this mission. The tasks of a company in the digital era are as follows:

DOI: 10.1201/9781032614212-12

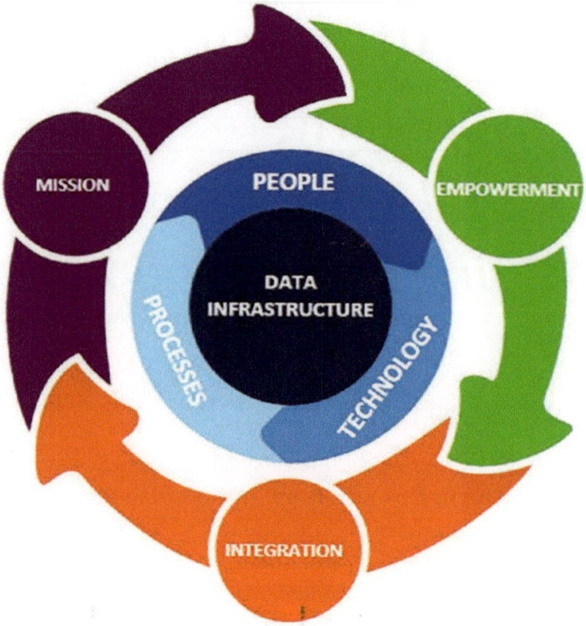

FIGURE 12.1 Digital transformation in the process industries

Source: Bascur, O.A. 2020. Digital Transformation in the Process Industries: A Road Map. Boca Raton, FL: CRC Press.

- Start with your business goal
- Reduce the hidden production (energy and water losses)
- Reduce equipment downtime
- Maximize your product recovery
- Transform the organization from firefighting to a new culture of continuous improvement and innovation
- Transform the data into knowledge
- Use the locked historical production data to achieve historical production records by learning and instructing the people, for the almost free computing power, to avoid troubles, idle times, and downtimes
- Get the costs in real time to be able to control them

12.2 MINING AND MINERAL PROCESSING FACTORS

The future of mining lies in the implementation of the Industrial Revolution 4.0 to mining. These new technologies, called Mining 4.0, should permit the integrated planning for smother production with rapid response to all types of perturbations in the composition of an ore and the process and equipment conditions. Importantly, drastic reduction of energy and waste with prevention of asset breakdown, safety, and environmental issues are achieved. Dark data are mostly unstructured data collected by real-time historic records in process plants. Among many new improvements are an increase in the productivity, safety, and well-being of workers and the use of phenomenological models in mineral processes control systems, all of which was too complex and slow to be used in the past. The amount of *dark* data collected by the current mining and mineral processing sites is huge, requiring the need of new tools to integrate the many silos, to validate, classify, and transform data into meaningful insights to become available to advisory systems and people.

Critical problems make integrating the silos (mine, mill, maintenance, engineering, environmental, and safety) difficult since each silo has its own version of the fact. The use of spreadsheet to get the operation data (and *dark* data) without *subject matter expert* validation, cleansing, noise filtering, classification, and aggregation of the real-time operation data is inconvenient. Due to the complexity of the task, people stay uninformed, working with basic data.

The decades between 1990 and 2010 saw the parallel development of three important aspects of modern mineral processing: (1) measurement devices designed especially for the particulate systems encountered in these industrial plants, (2) mathematical models constructed for the analysis and description of these systems, and (3) control strategies developed using the capabilities of digital computers. Starting in 2016 the development of intelligent sensors (advanced control systems based on object technologies and communications networks) has transformed the way plants are managed:

1. By measuring and managing unproductive times, personnel find new ways of avoiding them, improving the profitability of the plant. The *information* created by real-time analytics enables the calculation of product recovery in real time and develop predictive models to handle large variations in their ores, catching the hidden problems, energy and water losses, type of ore, support, and unmeasured disturbances, to find the best operating condition based on the type of ore currently mined. This information permits the creation of new workflows and collaboration between mining, concentrator, and the company including services and providers.
2. A *net metal production rate* (NMPR) model presents the optimal mill throughput in real time, with the right grinding particle size, solid concentration, metal flotation recovery, and water recovery. The right manipulated variables can then be known to maximize the NMPR. The development of advanced predictive capabilities such as the particle size soft sensors (PSDS) and operational derived variables, such as air holdup, energy intensity, and bank profile in flotation recovery is now possible.
3. Basic process control is an integral part of most mineral and metal processing plants. It has helped many operations to reduce costs and increase productivity and performance. New developments over the past decade in process control of the mineral-metallurgical processing industry have been greatly influenced by the existence of new measurements, advances in computer software and communications, and internet-cloud technologies. Another extremely important factor has been the selection of strategies to

link control actions to process capacities to form an overall control and plant management system. These three capacities, *measurements, process control strategies, and computer hardware*, form a triangle of integrating elements.

12.2.1 Forcing Factors

The mining industry in the world is facing several forcing factors that deal with ore falling grades, rising energy costs, large ore type variations, increase in water consumption and its decreasing availability, remote operations, stringer safety and environmental regulations, and increasing skilled operator labor costs (see Figure 12.2) (Bascur et al. 2011).

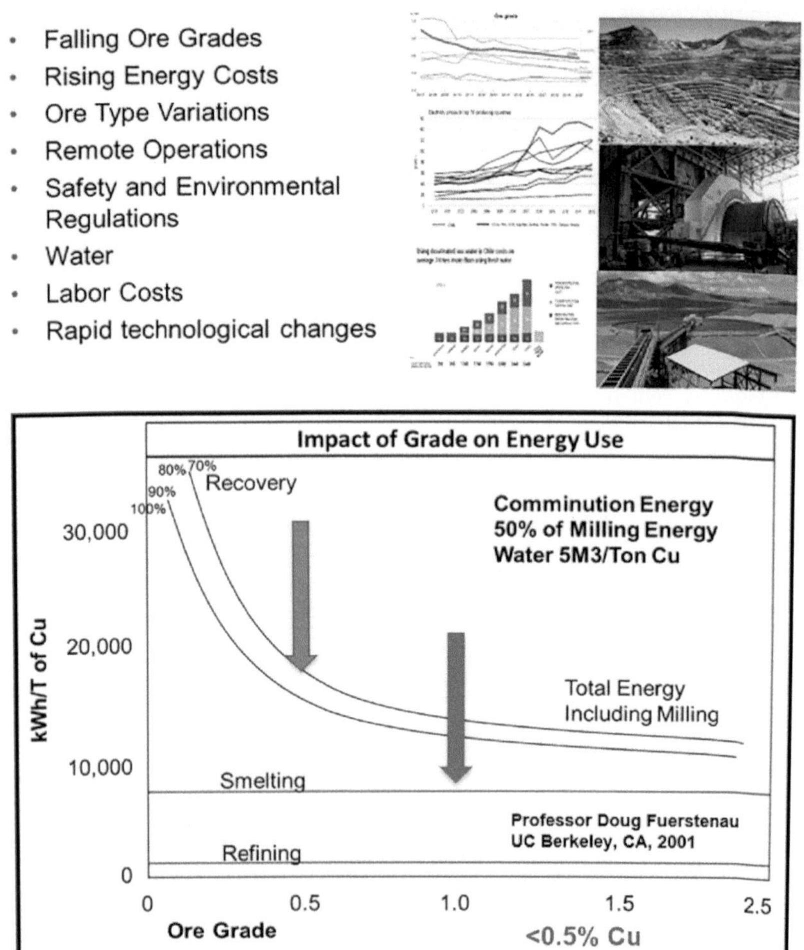

FIGURE 12.2 Key mineral processing factors, increase in water and energy resources in operating copper ores with less than 0.5% grade operations.

12.3 PLANT DATA AND CONTROL HIERARCHY

A data and control hierarchy pyramidal diagram with two sides is shown in Figure 12.3. The left side of the pyramid represents *control hierarchy* and the right-side symbolizes the *Integration hierarchy*, which provides actionable information to all functions of the plant to improve productivity. At the base of the left-hand side are the *regulatory controls* and at the right-hand side the *condition-based maintenance*.

12.3.1 Control Hierarchy

Control hierarchy shows the traditional real-time control levels fed by the sensors and data collected from the field. At the lowest level exist the instrumentation level, which consists of devices for acquiring data from sensors, field displays, and hardware safety interlocks for ensuring emergency shutdowns. The *process measurements* require many factors involved in implementing the right measurements in the very harsh environment of the mineral and metals industry. It is essential to have good, updated process flow diagrams electronically available for all people to understand the cause and effect of the process controls and manipulated variables (Bascur 2019).

The instrumentation level, which is implemented using control hardware such as *Distributed Control System* (DCS) and *Programmable Logic Controller* (PLC), provides the integration of the real-time data for the *regulatory control* level. It is one of the most important levels because it must be extremely robust for industrial process continuity and operational safety.

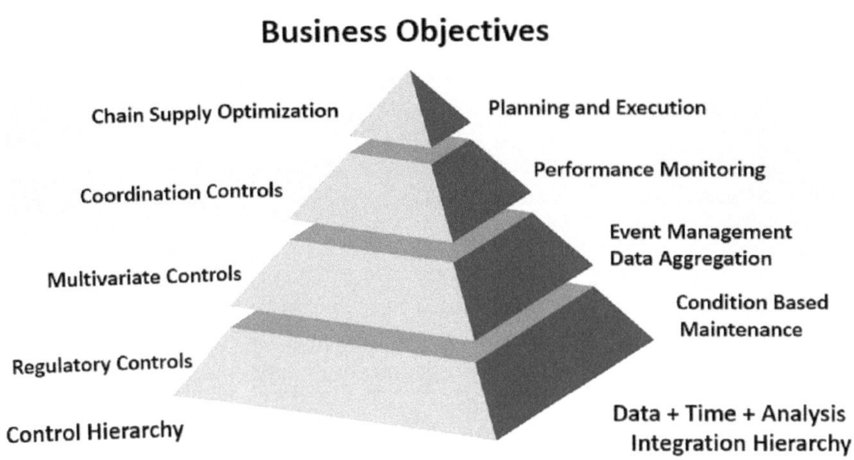

FIGURE 12.3 Plant data and control hierarchies for process control and operational intelligence.

Source: Bascur, O.A. 2019. Chapter 2.6: Process control and operational intelligence. In *SME Mineral Processing and Extractive Metallurgy Handbook*. Edited by R.C. Dunne, S.K. Kawatra, and C.A. Young. Englewood, CO: SME.

12.3.1.1 Regulatory Controls

Regulatory controls maintain the process variables at their prescribed set point, stabilizing the variations due to local disturbances occurring at a time scale of seconds to minutes. The disturbances can be caused by many factors, including weather conditions, changes in the raw material characteristics, ore hardness changes, particle size distributions, or mineralogical compositions and shutdown of other sections of the chain supply of the plant. In addition, this level allows the operator to take control manually of the plant in case of an emergency. The DCS and PLCs require software tools to configure the control strategies and require a human operator interface to monitor the behavior of the plant under control. The level sends the streaming data collected from the plant to a dedicated industrial historian (like the black box of an airplane). The stream data are used for enhanced equipment condition-based maintenance and for control tuning and process operational improvements.

12.3.1.2 Multivariate Controls

Multivariate controls, or model-based process control, have evolved in recent years due to advances in hardware and computing capabilities. In a processing plant, the problems are typically multivariable with many control interactions due to the nonlinearities of the process, process equipment constraints, and unknown process disturbances. Because of the possible interaction among the variables, all the control setting must be coordinated. Control actions are taken to accommodate longer duration disturbances at a time scale of approximately minutes. Mineral and metal process dynamics are important, and a *dynamic process model* is necessary. The output of this level is used by the digital control systems (DCS) as a set point to drive the final control actuators that operate the plant. The final control elements are the valves, variable-speed pumps, and weight feeders, among others. This level is metaphorically equivalent to the hypothalamus, which maintains brain chemistry and controls body functions, which stresses the importance of having a hierarchy in process control design. It is highly recommended that priorities are clarified when implementing and maintaining overall control strategies.

12.3.1.3 Coordination Controls

Coordination controls are implemented to balance the overall constraints and find the optimal steady-state operating conditions of the plant, based on the current production requirements and factors such as raw materials, energy, consumable costs, and production demand. These process controls require the left-side level to analyze operations and identify the current process and equipment constraints.

12.3.1.4 Chain Supply Optimization

The planning and scheduling activities are usually determined by the integration of the plant's industrial data infrastructure with the enterprise business systems, which contain the production plans and utility costs. For example, in mineral processing plants, the mine production, mill, tailing, and port production areas are integrated to optimize the overall energy and water use based on the type of ore being mined (drilling, explosives, and blasting strategies). The lowering of the ore grades requires a greater detailed analysis for classifying the operating strategies and lineups of the plant for the best and most profitable alternative. Mill data are now being used to optimize mine-blasting operations for minimum operating costs while maximizing throughput.

12.3.2 Integration Hierarchy

12.3.2.1 Condition-Based Equipment Monitoring

Operational real-time data is used to generate events to create actionable insights based on operational intelligence. It is critical to be vigilant about maintaining equipment to achieve high productivity and availability levels to remain competitive in the mining and mineral processing industry. Reusing data from the industrial data infrastructure is the most cost-effective way of achieving high productivity levels. Today, operations and maintenance teams must work with the same data to keep the instrumentation, process controls, and equipment online 24 hours seven days for optimal production levels. Using the process sensors and capabilities of a modern industrial data infrastructure, maintenance personnel can implement condition-based maintenance strategies to monitor and predict the performance of moving equipment in real time. Now there is powerful data analytics to filter data to detect problems prior to a failure. Because process and equipment data are always available, employees can use the data to generate notifications to initiate a programmed shutdown as opposed to dealing with an unscheduled shutdown.

When regulatory controls and multivariable control loops underperform can cause process variability that adversely affects profitability. Having the stream data available in an easily accessible format for advanced analysis simplifies the continuous improvement process required to support the controls at all levels for all process units in the plant.

The increasing use of real-time data is elevating the role of plant personnel to that of *knowledge workers* who are expected to make real-time decisions to improve business performance. The high variability of ore qualities, rapid changes in energy costs, reduced availability of process water (or, in some cases, too much water in equatorial regions) require adaptive real-time performance strategies.

12.3.2.2 Event Management Data Aggregation

The second level of the righthand side supports the validation and classification of data to develop actionable information which is the key ingredient for implementing a continuous improvement and innovation program in process industries. This operation intelligence program allows the sensor data and the operational events to enable online analytics to evaluate the production and operational costs on a shift-by-shift basis. These operational events are the new transactions to optimize production and reduce operating costs. Transactional systems are now integrated with this new data to proactively improve the overall production performance of mineral and metal processing plants (Bascur 2019, 2020a, 2020b, 2021a, 2021b, 2021c).

12.3.2.3 Performance Monitoring

The operational metrics can be calculated by using proper data classification and aggregation at a high level of detail. This new strategy enables faster communication and collaboration within the functional teams at the plant and within the global business via the cloud. At this level, the business and time contexts are added to the process measurement. This level is used to generate tangible benefits for improvements. The data and events are analyzed to determine the best course of action for the plant.

12.3.2.4 Planning and Execution Management

Planning and execution management, the integration of operational analysis and business intelligence is the last level, where plans are constantly being adapted by defining the best targets to optimize all process areas, beginning with the mine, process plants, and tailings management. Botin (2009) describes the latest strategies to integrate sustainability into a mining organization. Data management and automation are what allow these strategies to be successfully implemented.

12.4 DIGITAL TWINS

Today digital technologies enable us to model an object using advanced analytics to create *digital twin*. A digital twin is a digital representation of a physical object or system, a virtual replica of physical devices that can be used to run simulations before actual devices are built and deployed (Shaw and Frülinger 2019). The next figure shows a typical digital tween of a person. Building a digital twin is complex, and there is as yet no standardized platform for doing so.

This new technique, which has been implemented in several industries, requires top access to historical data base and the transformation of real-time series into information. It is paramount to understand all process involved that debottleneck the plant, the integration of the planning targets with the real-time cost which enables the estimation of the maximum profit while avoiding operation and equipment constraints in the plant. The maximization of production is integrated with an asset reliability and prescriptive maintenance. For a mineral processing plant, a digital twin would look like Figure 12.5.

12.4.1 Transforming Data and Events in Information and Knowledge

Event-framed operational data classification permits transformation of the data into operational insights. This was demonstrated, for a large iron ore, a gold and copper-gold mine, lead and zinc mining complexes by Shuen (2016), Strack (2017), Plourde et al. (2017), Steyn et al. (2018), Chawda (2019), and Bascur (2020a, 2020b, 2020c).

Figure 12.6 shows a schematic of the Digital Twin Design Concept. It integrates the operational process data with a Digital Twin that contextualizes data, provides predictive analytics with planning, process control and engineering, safety, sustainability, with the visualization tools for analysis and human interface.

The key is to provide the time-series data at the desired level of aggregation for the users and predictive algorithms for continuous improvement and innovation. The Digital Twin simplifies the access to information for people to collaborate, store the process knowledge and organize the reporting functions between the many functions in industrial complexes.

FIGURE 12.4 Digital twin of a person.

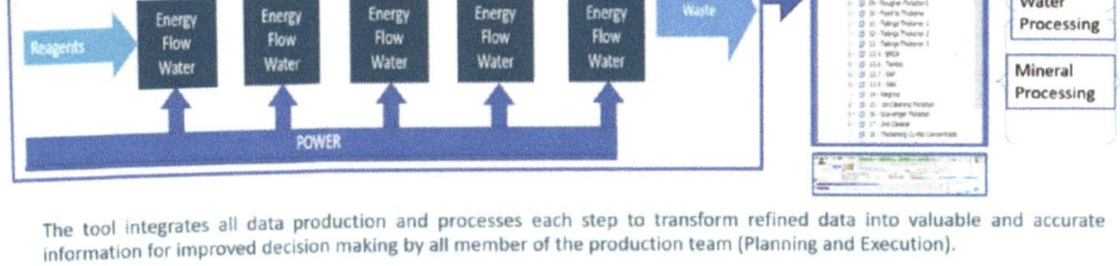

FIGURE 12.5 Digital plant twin into a digital object-oriented database.

Process data is the primary data source for predictive and prescriptive analytics in industrial environments. Process engineers tend to know the data well, and with the right tools, they can cleanse, filter, and create predictive models using their process chemical and metallurgical basic principles models. Many of them are interested in expanding their analytics efforts with access to Python code and libraries (Raschka

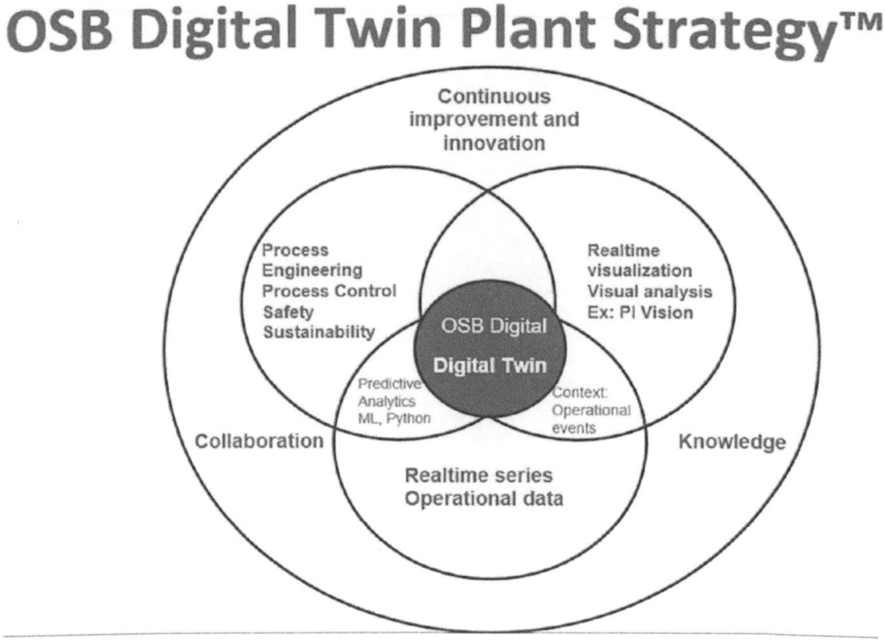

FIGURE 12.6 OSB digital twin plant strategy.

2015; Kelleher et al. 2015). An exceptional new treatment of machines learning tools for engineering and scientific validation can be found in Brunton and Kutz (2019). This approach is what we are using in this chapter to transform times series data into predictive models using regression analysis. Both Data Scientists and Process Engineers can easily collaborate seamlessly while analyzing the operational process data.

Data scientists tend to know the algorithms well. Moving data around to make it accessible for analytics can be costly and problematic. Predictive and prescriptive analytics outputs must trace back seamlessly to source data and processes to be actionable in a high-consequence environment.

The collection and contextualization of real-time data is essential to provide sophisticated offline predictive data analysis tools. Figure 12.7 shows the basic steps to transform sensor data into information by adding the correct physical and time context. It shows the recursive nature of building knowledge from real-time data. There is continuous learning loop comparing the generated insights and decisions. Another important loop that refines the data requirements and gives context to the data. These iterations should be simple to configure and maintain.

Data hierarchy infrastructure is shown in Figure 12.8. Here the first level of raw data is captured by sensors through the diverse control systems (Bascur 1988, 2019, 2020b).

In building a digital twin plant model, a layered process data knowledge is used. These are specialized data acquisitions and control devices that reside near the

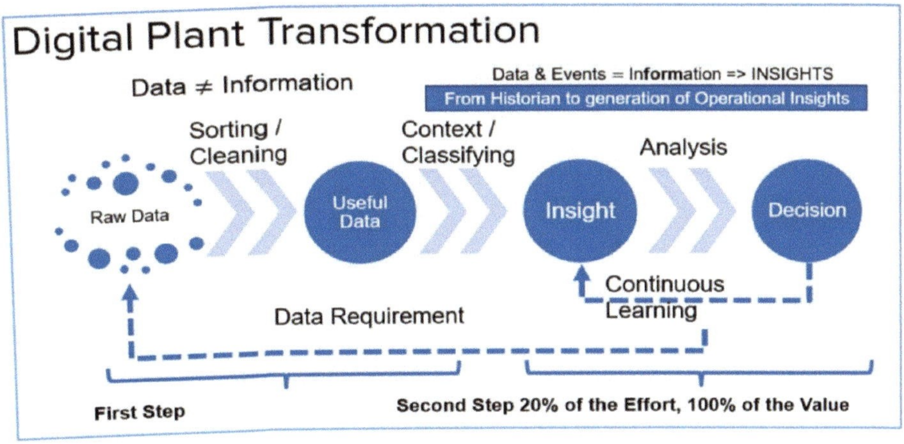

FIGURE 12.7 From dark data to knowledge decision-making.

Source: Bascur, O.A. 2020. Digital Transformation in Mining Plants, Automining 2020 Proceedings, Gecamin, Santiago, Chile.

process or in mobile equipment. Data validation is required to check the validity of the sensor signal. These data are classified to coordinate the integrated processes for overall plant optimization. A heuristic method defines the operating conditions of the process data and defines the event start and end times in order to transform the raw data into more valuable information. The resulting data can be used to develop empirical predictive analytics models and a predictive process model to be used within the correct range.

Data modeling is conducted by an operating mode, performing calculations of equipment condition alerts, process efficiencies, energy management, and estimation of inferred process variables-based real-time simulations. The first step is to increase the necessary information and system to support decision-making. It is the right side of the brain of an operating plant (Bascur 2019, 2020b). On the other side of the brain are the advanced process control where the known facts are applied to maintain a reasonable stable operation. Special tools record and process the discoveries, putting them in a dashboard, such as the Seeq Organizer (Seeq 2020), for engineers and managers to use and distribute to actionable information. See Figure 12.9.

Dashboards are real-time documents containing a detailed journal with every step done in the Workbench engineering environment, including all the tools for identification of operational patterns, encapsulation, cleaning, filter, aggregation, and creation of process predictive models which are especial, and not a typical data regression, building a robust process model of each part of a plant.

12.4.2 Digital Twin of a Mineral Processing Plant

In many mining operations, ore is crushed and wet milled to liberate the valuable mineral. This slurry is concentrated for metal by flotation and then filtered to form a dry mineral concentrate that is shipped to refineries to produce metallic products.

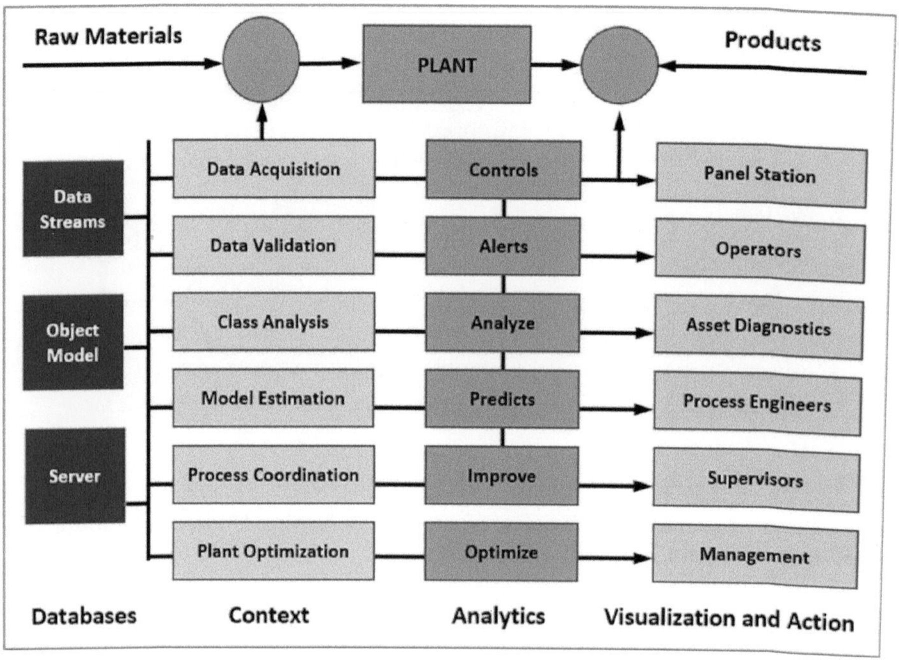

FIGURE 12.8 Layers of process analytics from validation to plant optimization.

Source: Bascur, O.A. 2020. Digital Transformation in the Process Industries, CRC Press, p. 76.

The type of filtration equipment required depends on the particle size, mineralogy, and shipping requirements. As with all mining operations, the required equipment is robust and designed to be reliable even under the toughest operating conditions.

For a given circuit in a mineral process plant, the optimal throughput is usually limited by several constraint such as the ore hardness that can alter the particle size distribution and the transport mechanisms in the mill due to the complex rheology of the pulp of finer particle size distribution and high percent solids. The strategies to improve the particle size distribution that affects the grinding efficiency, flotation, and the water recovery during thickening and filtration are very important for low grade ores. By operating closer to these constraints, considerable mill capacity is gained. Flotation control technology has matured considerably in the last decades in understanding the interaction of the grinding circuit variations and how these affect the froth flotation process performance.

The curves in Figure 12.10 are not unique and change constantly, thus information must be handled on a timely basis in a computer system, which will provide for adapting values of the moving constraints and setpoints. This is known as *online adaptive decision-making* based on control (Bascur 1991a). In this way, the recovery curves for the plant are calculated in real time using a gross error detector algorithm as part of a digital twin digital operational model.

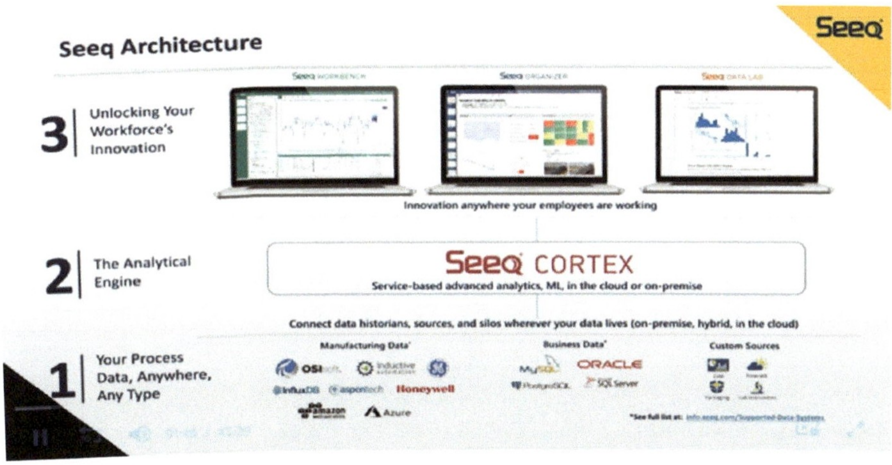

FIGURE 12.9 Latest tools dive into seas of data to rescue diamonds for collaboration.

Source: Seeq. 2020. *Seeq Workbench*. Seattle, WA: Organizer and Data Lab. www.seeq.com

Figure 12.11 shows the key objective of maximizing *metal production rate* in a copper mine, a major contributor to model this process. In the figure, *P* represents the metal production rate, which is the most important key performance Indicator (KPI) to maximize the profitability in mineral processing, *H* is the copper grade to the plant, and *R* is the metal recovery.

The optimal tonnage processed is a function of the mineral type, the mineralogical characteristics of the pulp, rheology, the asset availability, and the operating mode of all process unit interacting in an industrial complex. The objective is to find the optimal cut of grade at the mine and the optimal cut size at the comminution to achieve the best recovery grade combination to maximize profits. At mills, energy costs have increased to liberate the metal-bearing particles. The typical process control objective is lowering operating production and mine costs and maximizing the metal yield. Because grades at mines have been lowered, large volumes of ore are processed to achieve economical production targets. While obtaining these targets, to transport the ore and dispose of the waste has become more expensive. Recent research (Dunne et al. 2019; Bascur 2019; Bascur and Gorrie 2021) document new strategies of operational problems to reduce the hidden operational losses, maximize copper recovery and production, and become more proactive, using grinding, flotation, and thickening predictive models to optimize the mineral processing plant.

It is well-known that in processing low grade ore, it is very important to reduce the variability in the grinding circuit to get the right *grind cut point* and a particle size distribution in the mill that improves the water recovery rate in the tailing thickeners. The *metal production maximization strategy* is shown in Figure 12.12 with its left side exhibiting each of the process units in operation to obtain the best results in the rougher flotation recovery. On the right side, the figure shows the

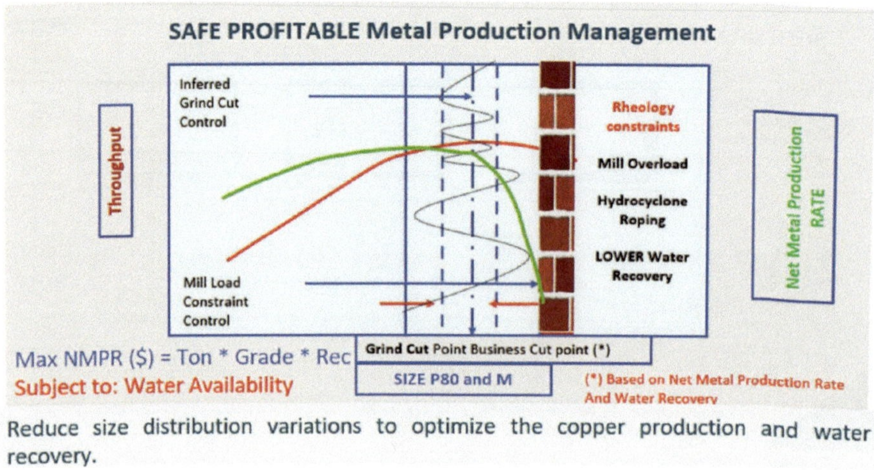

FIGURE 12.10 Opportunities for optimization of the net metal production and water recovery.

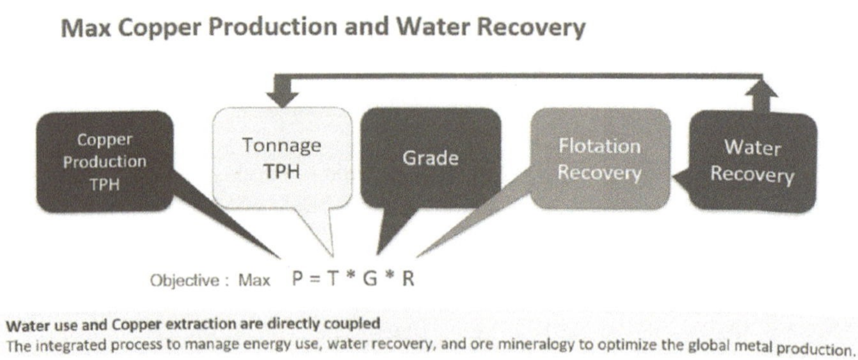

FIGURE 12.11 Net metal production rate factors.

well-known *flotation elephant curve* showing that minerals in very small and large particles sizes do not concentrate well in traditional flotation equipment (Lynch et al. 1981; Trahar 1981). It is also known that particles with a wide size distribution do not flocculate well and are carried out with the overflow water in thickeners. These examples make particle size distribution in the flotation and thickener's feed a major target.

Following the continuous improvement and innovation methodology, the operation detects the origin causing problems and evaluate the opportunities for improvement defined by the engineering and management team through the fishbone analysis, as shown in Figure 12.13.

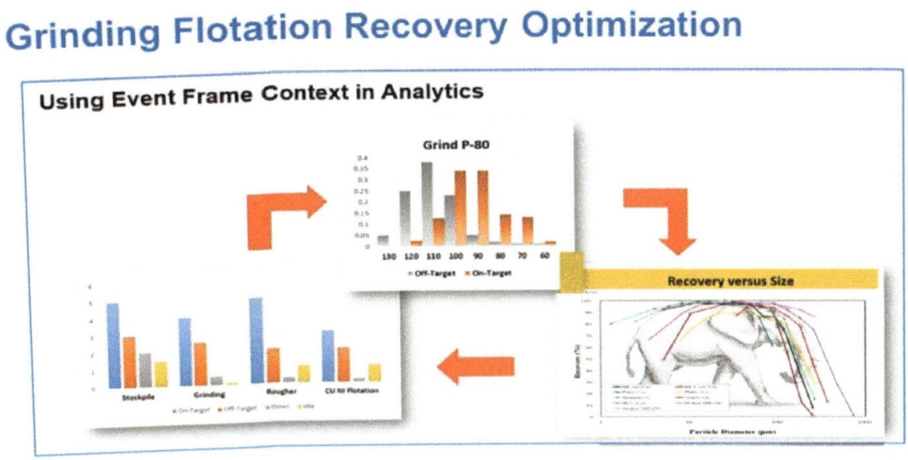

FIGURE 12.12 Net metal production rate maximization strategy.

Source: Bascur, O.A. 2020. Digital transformation in mining plants. In *Automining 2020 Proceedings*, Gecamin, Santiago.

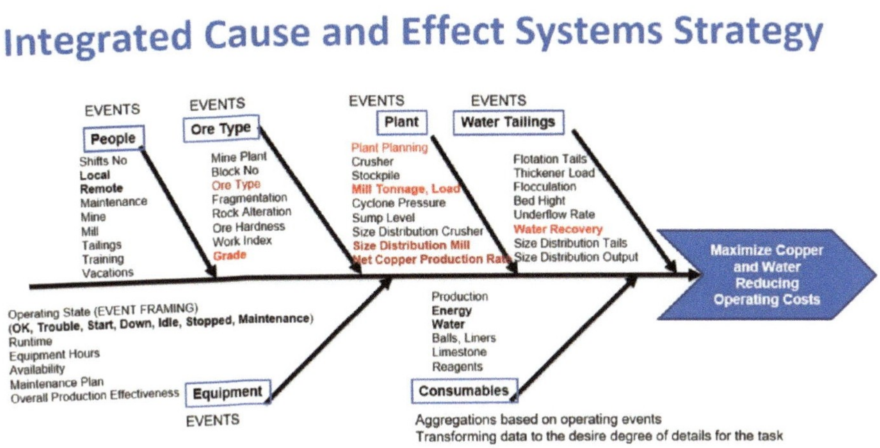

FIGURE 12.13 Fishbone diagram showing key events causing maximizing yields.

Source: Bascur, O.A. 2019. Chapter 2.6: Process control and operational intelligence. In *SME Mineral Processing and Extractive Metallurgy Handbook*. Edited by R.C. Dunne, S.K. Kawatra, and C.A. Young. Englewood, CO: SME.

The fishbone analysis shows the effect of the recovery grades and losses based on operating parameters, the equipment events, the operating shifts, the material grades and the amount of energy, water and reagents used to achieve the recovery and grades. Classification of the operating data enables operators to build this predictive model using regression analysis and other models provided by the new software tools available. Grasping in time these events it is possible to identify the

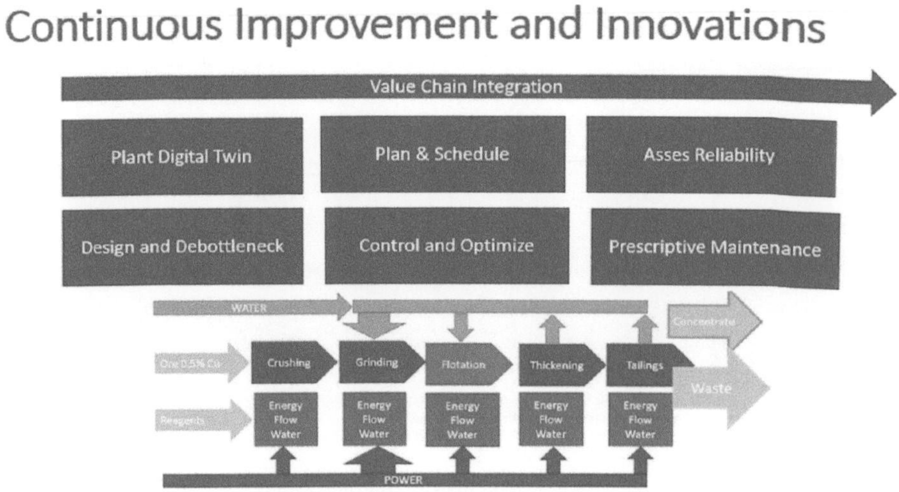

FIGURE 12.14 Continuous improvement and innovations based on advanced analytics.

real opportunities to maximize *metal recovery* in an industrial mining and mineral processing complex.

Figure 12.14 shows a summary of the new scope in building a digital twin for a mineral processing plant based on advanced analytics in the digital age.

12.5 BUILDING A DIGITAL PLANT MODEL

The creation of the digital twin plant model starts with a good up-to-date process flow diagram of the plant, the way of communicating the process knowledge (Turton et al. 2018). The process variables and key process controllers should be shown to build on these variables the necessary process mass balances and online state simulation models of the plant. In this way a digital twin is created by an online observer predicting the outcomes from the operational state and the process variables.

Figure 12.15 shows how unit operations are used to model the whole plant for production variance improvement and coordination. This data model enables us to identify the process constraints to satisfy the maximization of the metal production. All violations are reported, and the time series data is aggregated at the desired degree of details using a modern process historian and advanced analytics tool. It is impossible for people to do this task, so automating many of the traditional exercises used in a typical spreadsheet should be made. This strategy is sometimes difficult to understand. The computer is trained to do many of these process engineering calculations.

Digital twin processes the raw data, checks with the plant economic plant, and detects the variances for all units in the supply chain. Any anomality is tracked immediately using *Event Framing* methodology. The computer puts a mark on the database to be able to combine the time series raw data and integrate them with the operational events. Only the *Running OK* events can be used to build reasonable process information from predictive models. The *Down*, *Idle*, and *Trouble* are sent

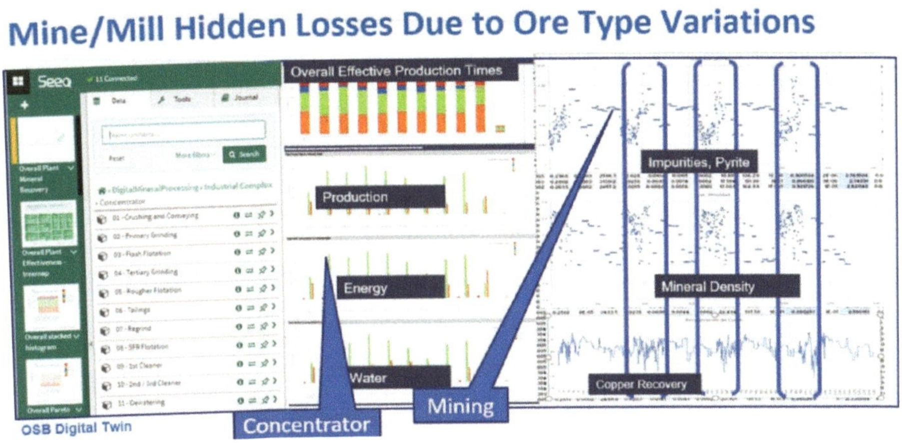

FIGURE 12.15 Digital plant model and dashboard showing Pareto plots for production, energy, and water results.

Source: Bascur, O.A., and Gorrie, T. 2021. Maximizing copper production using a Digital Twin with predictive analytics. In *Procemin GEOMET 2021*, Gecamin, Santiago.

to operational personnel immediately to get to the root of the problem and establish if they can solve them. When it is not possible to fix the problem, the alerts are escalated to the competence center for further analysis and resolution.

The continuous improvements and innovations obtained by this methodology, the data and events at the desire degree of details, are used by the process engineers to build economic cases to invest in the new strategies or equipment necessary for optimizing the economic performance of the plant. The best return on investment projects is analyzed to invest in resolving the operational problems due to equipment, process constraints, retuning process control strategies, and training the operating team.

12.5.1 Detecting the Hidden Production Losses

Online Overall Production Effectiveness is the most important result where all the process units are automatically analyzed by the system. It determines the hidden production losses in all the units, by the ore type and by shift, for the entire Digital Plant. *Seeq Workbench* is the latest tool to clean, calculate, and contextualize data and work with model results. Seeq Organizer is a way to present the results in dashboards in a clear, succinct form for collaboration and further socialization of the improvement possibilities. Process workflow led to the creation of valuable information, which is further improved using machine learning tools by starting with cleansed gridded datasets, developing models, and operationalizing the findings.

Engineers, managers, and the supporting group (local or remote) can collaborate better than ever by using the new tools available using cloud computing. Figure 12.16 shows the Seeq Workbench presenting the plant *Overall Production Effectiveness (OPE) in real time* and converting the real-time series data, and the metadata associated for each of the area of the plant, into process information for *improved decision support*. The figure shows the whole plant with the pareto diagrams for each of the units in the plant. These graphics show the losses over time and estimate the operating and consumable losses for each of the operating selected for analysis and aggregation. Now the patterns of the best conditions can by unified to create a model based on the historical cleansed and gridded data sets. Predictive analytic models are built to forecast the occurrence of insipient process problems or equipment failures on the horizon.

12.6 DYNAMIC GRINDING-CLASSIFICATION CIRCUITS

To maximize the metal production, it is a top priority to maintain the desired throughput with the right cut size and the proper particle size distribution to maximize the rougher flotation recovery with a control strategy able to absorb the disturbances caused by the variability of the ore. There are four principal indices that can assist in maximizing the throughput.

12.6.1 Ball Mill Load Index

A technique was developed to infer the ball mill loading based in the power, bearing pressure and on the pulp rheological model. The raw mill power is processed by a least squares second order filter to obtain the rate of change, acceleration and moving average, and the integral calculation of the load in the mill for its residence time. The residence time of the mill was chosen for integration of the load parameters over time.

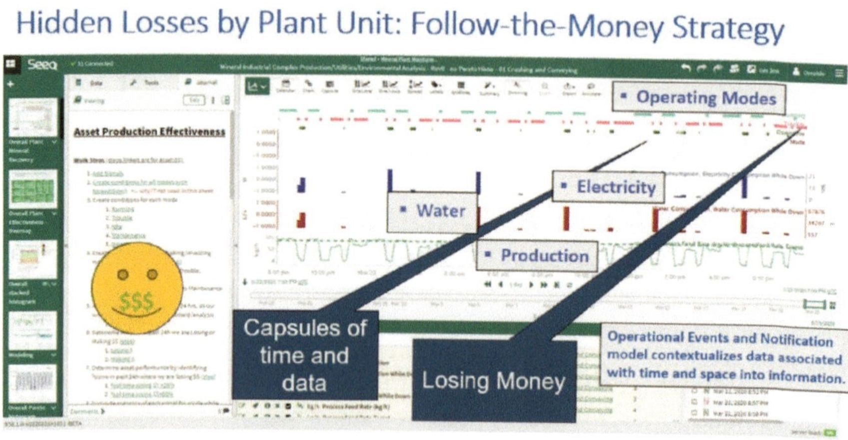

FIGURE 12.16 Digital plant model showing the operating mode via capsules showing the losses for each event.

Source: Bascur, O.A., Retamal, F., and Espinosa, S. 2022. Treating low grades ores using a digital in mining and mineral processing plants. In *Proceedings of the Copper 2022 International Conference*, IIMChile. https:/copper2022.cl.

12.6.2 Grind Index

The size distribution soft sensors, in conjunction with dual measurement sensors and the water addition to the sump, in agreement to its availability, have been used to maintain the particle size in control. A selected algorithm can be used in practice to generate the right *grind cut index* P_{80} *and particle size distribution shape (PSDS)*. The PSDS can be represented by the Gaudin-Schumann model via the M parameter (F3(d) = (D/Dmax)^ M. The target is to maintain a stable product to flotation with the right particle size distribution for the best attachment of the hydrophobic particle with the air bubbles in the flotation bank. In the past, grinding and flotation was treated independently, however today they are tied together.

12.6.3 Grinding Transport Pulp Index

The *transport pulp index* in a mill is a regression equation based on the pulp viscosity calculated from the percent solids and the particle size distribution (Hemming and Boyes 1977; Bascur 1985, 1991b). The pulp transport index assists in creating the right cataract effect for the balls to maximize the number of successful impacts breaking the large particles in the mill and protect the balls and mill liners to extend their duration over time. This saves energy and reduces the wear of the balls and the replacement of the liners

12.6.4 Sump Level Index

The hydrocyclone feed sump plays an important stabilizing action in a ball-mill circuit. The large circulating load and the sump water addition can create instabilities in the circuits. Dynamic grinding circuits simulators assist the design of the proper best combinations of hydrocyclone, pump size, and sump level for a stable operation of a grinding circuit (Bascur et al. 1986; Bascur and Junge 2011).

Figures 12.17 and 12.18 show mineral processing plant with the key measurements for particle size and the estimation of particle size distribution using, in this case, the Gaudin-Schumann equation to track the number of fines and coarse particles to include their effect in the rheological model.

12.7 DYNAMIC HYDRODYNAMIC FLOTATION MODEL

Flotation dynamic models were developed by Bascur (1982), Bascur and Herbst (1986), Bascur (2005, 2013). More recently by Bascur and Junge (2011) presented *Dynaflote*, a dynamic model that uses the hydrodynamic flotation cell of Wallis (1969), integrates Bascur's the hydrodynamic model with the particle population balance described by Mika and Fuerstenau (1968) and Fuerstenau (1999). Recently, Bernal et al. (2020) tested the importance of the manipulation of the flotation cells speed. This was proposed originally as a key manipulated variable by Bascur and Herbst (1982).

Crushing, grinding, rougher flotation, scavenger flotation, cleaning flotation, dewatering, and filtration are used to estimate the mass balance for each unit. The

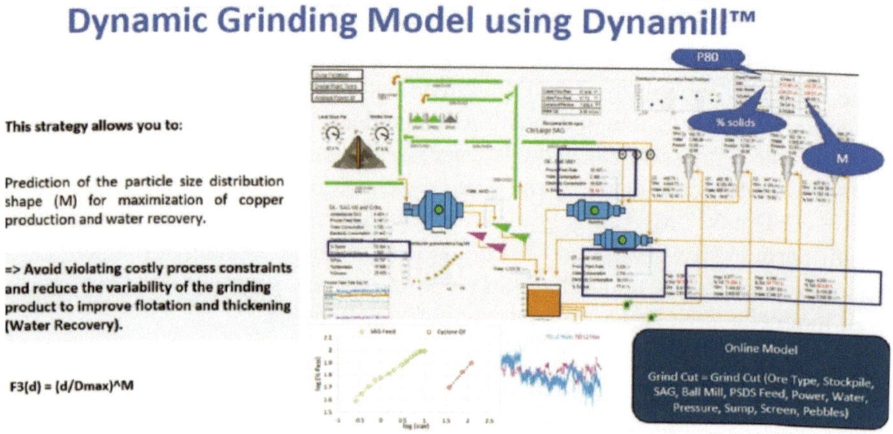

FIGURE 12.17 Mineral processing plant grinding and flotation predictive models.

Source: Bascur, O.A., Retamal, F., and Espinosa, S. 2022. Treating low grades ores using a digital in mining and mineral processing plants. In *Proceedings of the Copper 2022 International Conference*, IIMChile. https:/copper2022.cl.

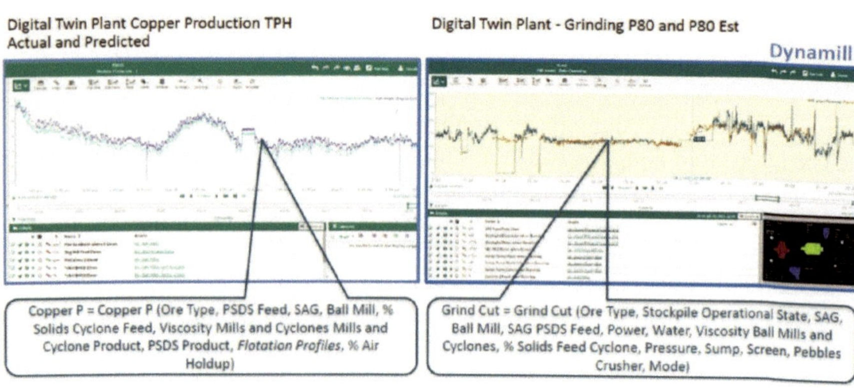

FIGURE 12.18 Digital twin—copper production rate and grinding P_{80} and P_{80} estimates.

concentrate and tailing flows are estimated using the flotation bank mass balance in real time as shown in Figure 12.19. Figure 12.20 shows the flotation dynamic, which, for a given mineral type, maximizes the metal production rate. In this particular case, the first cells operate to recover all fast and coarse particles, the second set of flotation cells are set to recover the fines, and the last cells to recover the slow floating particles.

The flotation bank air holdup has an enormous importance in managing the flotation performance. In the air holdup correlation of Bascur (1982, 2005, 2013) the air flowrate, cell pulp/froth area, cell power, pulp level provides an estimate of the cell air holdup. Having the rougher metal assay, particle size, particle size distribution,

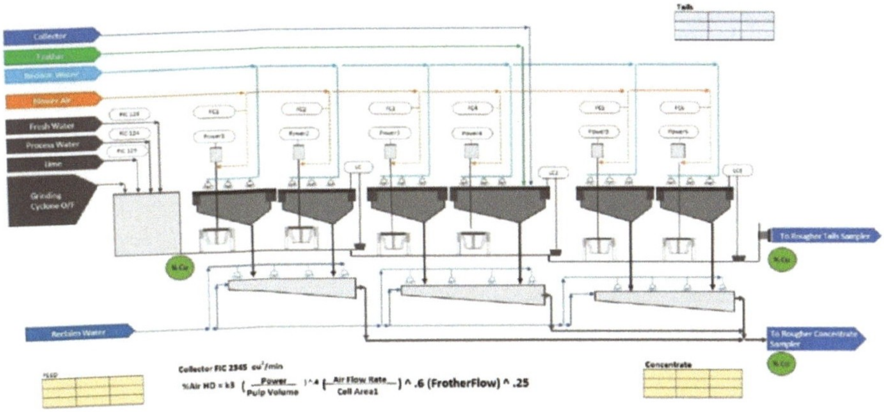

FIGURE 12.19 Typical rougher flotation bank with the powers, levels, air fluxes, frother, and metals x-ray assays.

Source: Bascur and Soudek 2019

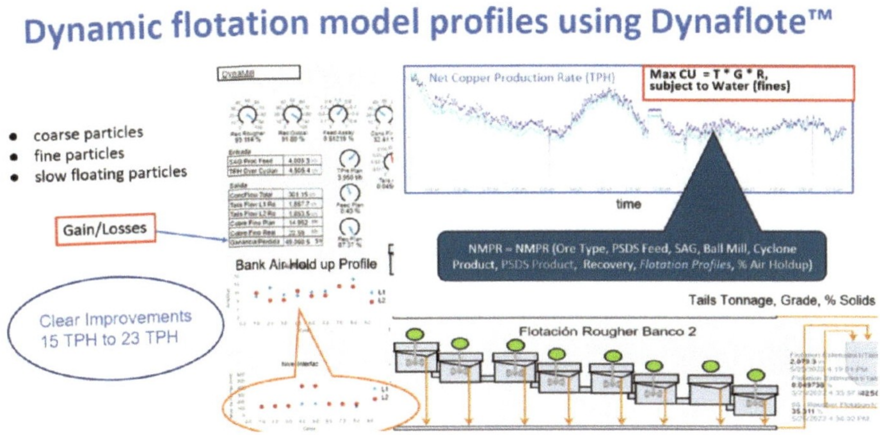

FIGURE 12.20 Rougher flotation hydrodynamic profiles to maximize the metal production based on the plant schedule values.

Source: Bascur, O.A., Retamal, F., and Espinosa, S. 2022. Treating low grades ores using a digital in mining and mineral processing plants. In *Proceedings of the Copper 2022 International Conference: IIMChile.*

air holdup, flotation cell power, froth height, feed flow, and reagent flow operating variables, a predictive model is derived using machine learning tools.

12.7.1 Online Rougher Metal Production Predictive Analytics

Having the rougher metal assay, particle size, PSDS, air holdup, flotation cell power, froth height, feed flow, and reagent flow operating variables, a predictive model is derived using machine learning tools. Figure 12.21 shows copper production rate predictive model using Seeq Workbench tools. The tools are using the real-time data cleansed data set using the Running Ok operating model to predict recovery using their multiple linear regression. The disturbances manipulated and observed variables can be used to find the best set of operating condition for a given ore type.

The grinding model is using the SAG Mill Feed PSD and PSDS with all the traditional operating variables such as power, second order power, load index, water addition flows, number of cyclones, percent solids by volumes, sump level, and pulp power to mention a few.

Figure 12.21 shows the results for the metal production rate model based on the augmented variables created using the Dynaflote created variables such power intensity, Bubble Size, PSDS, P80, P80 Squared, air holdup, tails flow, concentrate flow, frother and water additions, ore type, and Running Ok mode. As such, the operating personnel can move the plant together the best net metal production rate value knowing the key variables. From the set of variables there is one set that is always at a constraint. As such, they need to only see if they can move the variables which are available for each type of mineral processing ore type.

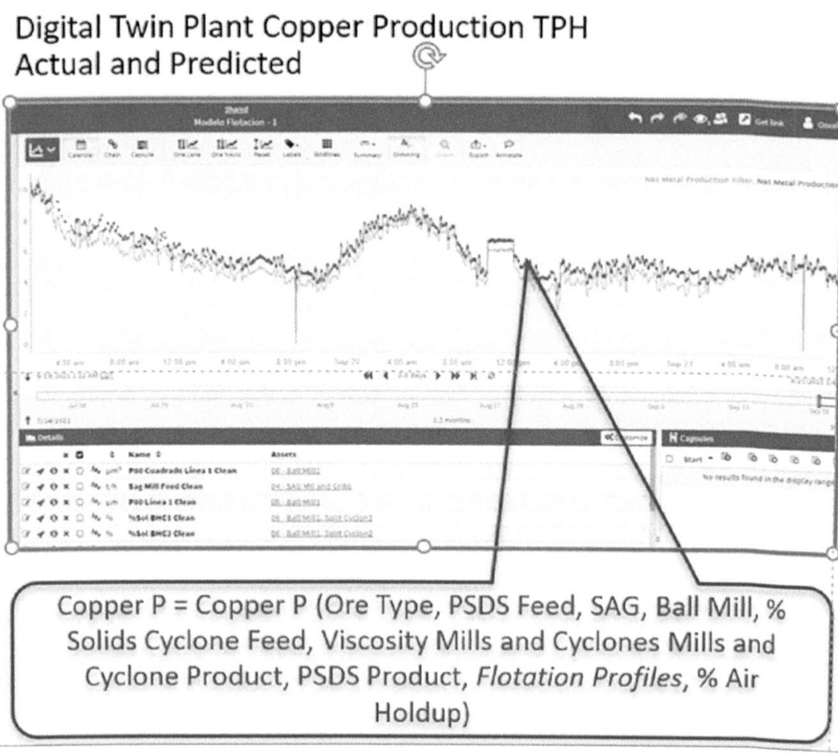

FIGURE 12.21 SEEQ workbench showing the results of the metal production model.

12.8 DYNAMIC WATER RECOVERY MANAGEMENT

We have discussed the importance of providing the best grinding product in rougher flotation to liberate the valuable mineral and minimize the metal losses in the tails. In a concentrator, for lower grade ores, 99% of the contained solids will report to the rougher flotation tailings and the thickeners feed. Therefore, having the right size distribution during grinding assist in having the right size distribution for flocculation in thickeners. The two important controls of a thickener are related to the properties of the thickener feed and underflow. For example, the presence of very fine clays particles in the feed greatly affects the thickener settling velocity and an appropriate flocculant, and automatic flocculation dosage is necessary. The principal property of the underflow is the solid concentration, which is properly controlled by maintaining defined level of the sediment in the thickener by manipulating the underflow rate.

Figure 12.22 shows the underflow manipulation to keep the right inventory in the thickener to compress the pulp obtaining the highest underflow concentration compatible with an adequate pulp transport (viscosity) to recover as much water as possible.

12.8.1 Maximizing Net Metal Production While Improving Water Management

The strategy used to move toward maximizing the NMPR is a steepest ascent called EVOP (Hessing 2022). Evolutionary operations methodology, also referred as EVOP, is a design of experiment (DOE) statistical method to improve the process or product quality. This process was first developed by George Box, 1957 (Box et al. 2016). The basic purpose of the evolutionary operations methodology (EVOP) is to improve a process through systematic changes in the operating conditions of a given

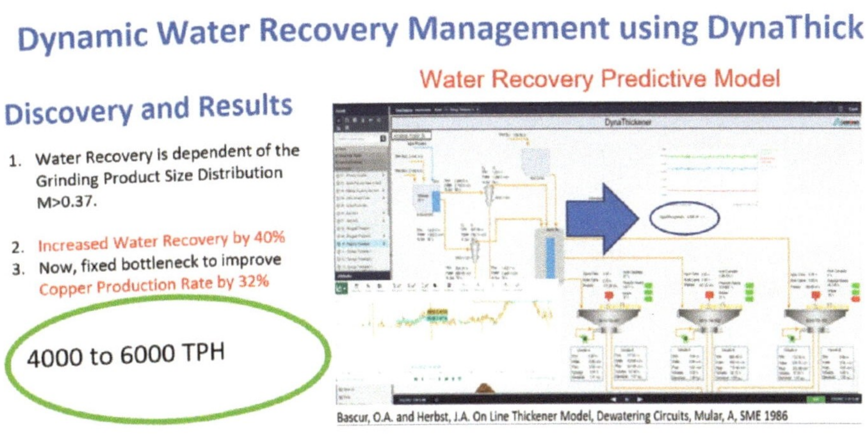

FIGURE 12.22 Digital twin showing DynaThick mass water balance and predictions.

Source: Bascur, O.A., Retamal, F., and Espinosa, S. 2022. Treating low grades ores using a digital in mining and mineral processing plants. In *Proceedings of the Copper 2022 International Conference*: *IIMChile.*

set of factors. An experimental design is established and conducted through a series of phases and cycles. The effects are tested for statistical significance against experimental error when such error can be calculated. When a factor is found to be significant, the operating conditions for that factor are reset and the experiment conducted again. This process continues until no further gain is achieved. Hence, the concept of an evolution is established.

Figure 12.23 shows the results for a given type of ore which provides the maximum copper production rate and the allowed maximum tonnage while within the process constraints at the grinding/classification. It shows that beyond a 4300 TPH tonnage the copper production rate drops due to the production of too many fines in the grinding circuits when the cyclones feed is higher than 67% solids and the ball mill % solids is higher than 76% solids. The capabilities to predict the particle size distribution shape with the "M" Distribution Module parameter is a key for maximizing the copper production rate and as we are going to discuss the capability to recover the necessary amount of water for the grinding classification. The right amount of water to the sump and the right water flow rate to the underflow of the hydrocyclone, as we have indicated in the grinding profitable grinding controls, are the key manipulated variables that can be used to drive the plant to a maximum profitable production. Adapting quickly to the ore type variability of a must to avoid large losses of copper to the tails by maintaining the right hydrodynamics in the grinding/classification, flotation, and thickening processes.

Bascur and Barrientos run several experiments with Andina Ore to identify the rheological characteristics of porphyric ores describing the effect of the pulp fluidity using a compression cell while developing their thickening operational models (Bascur and Barrientos 1977; Bascur 1991a) shown in Figure 12.24. These results have validated by many others, such as He and Forssberg (2006), Shi et al. (2000), Concha et al. (2000), and recently by Gutiérrez (2021).

Figure 12.25 shows how the historical time series data is integrated with hybrid machine learning algorithms to augment the operational raw data using the real-time digital twin, classifying, validating, and providing the best process values estimates to find the optimal tonnage while avoiding process and equipment constraints. The decision support teams become the guiding intelligence for the best economic production plan by updating the models with operating experience gathered for the type of minerals being mined. As shown, process setpoint are feedback into the operating controls as a true optimizing controller.

The major benefits of the implementation of a digital twin mineral processing are the simplicity to access data, events, and information for the following:

1. Identification of the process constraints on a daily basis and identify the hidden losses.
2. Augment the knowledge of the whole from crushing to water recovery thickeners to maximize the copper production rate.
3. Creation of a dynamic integrated SAG mill, grinding/classification model which includes disturbances such as ore type, silo level, and size distribution analysis in real time, pebble crushers, Sump, 4 hydrocyclones banks, and two large ball mills.

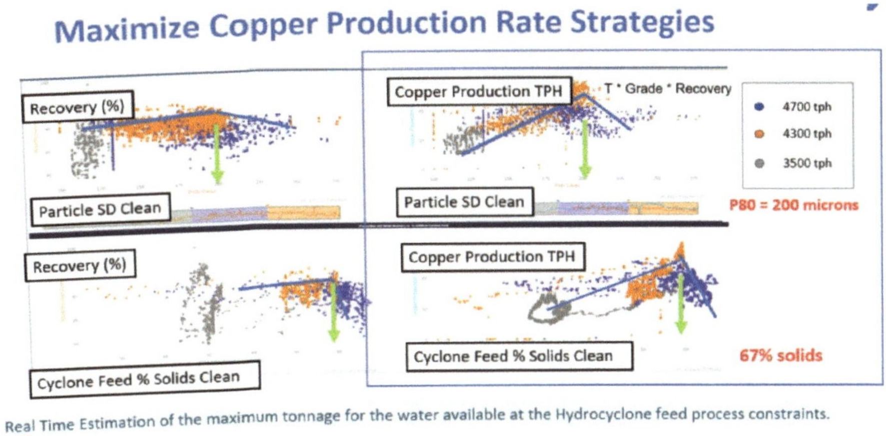

FIGURE 12.23 Advanced analysis showing the optimal throughput for the maximum Cu production rate.

Source: Bascur, O.A., Retamal, F., and Espinosa, S. 2022. Treating low grades ores using a digital in mining and mineral processing plants. In *Proceedings of the Copper 2022 International Conference*, IIMChile. https://copper2022.cl/.

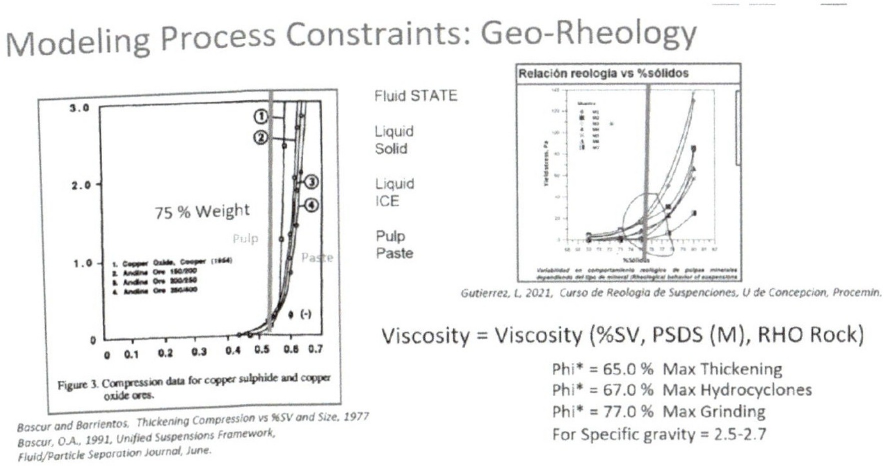

FIGURE 12.24 Suspensions in sedimentation nonlinearities to avoid from pulp and paste phases.

4. Creation of a dynamic flotation model which includes the grinding product quality information, capability to estimate the right cell level, aeration, turbulence, and air holdup estimate and to visualize the copper recovery, mass flow rates of the concentrate and tails, losses of copper in the tails and the net copper production copper rate in real time.

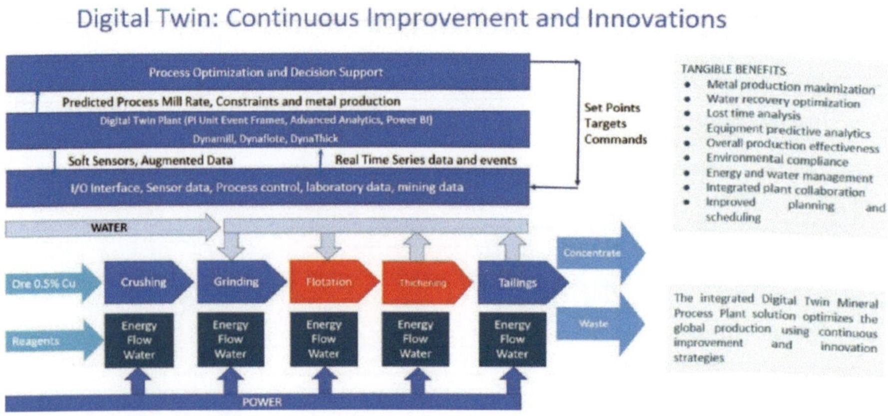

FIGURE 12.25 Digital twin for creation of value at a mineral processing plant.

5. Creation of a dynamic for water recovery from the thickeners and see the effect of the early processes effect in the results of maintaining the right flocculant additional level, solids level; underflow % solids. The integration of the mine, mineral processing plant, rougher flotation tailings thickeners, and tailing treatment management into one vision enabled to develop a dynamic recovery model which enabled to identify that the grinding product had a tremendous effect of the water recovery from the thickeners due to excess fines created by not adding the right amount of water. The increase of 40% water recovery enabled the plant improved the Copper Recovery from 86% to 92% which represent an average of 25–32% copper production rate.
6. Creation of a process unit analysis template to analyze all the process units at any time and to trigger alerts to solve problems and to learn how to fix them when they arise.
7. Most importantly, creation of digital twin to analyze the whole plant and to develop predictive models. The digital twin serves to coordinate, ore silo, SAG mill, peebles crushers, and vibrating screens, sump, cyclones, flotation, and water recovery thickener all at any time simultaneously.

Most importantly, creation of digital twin to analyze the whole plant and to develop predictive models. The digital twin serves as to coordinate ore silo, SAG mill, peebles crushers and vibrating screens, sump, cyclones, flotation, and water recovery thickener all at any time simultaneously.

As such the key results of implementing such a strategy using a digital twin plan connected to real time historical and using an advanced analytics capabilities in the cloud are as follows:

- Reduce variances between production schedule and real executed results
- Cleanse data and add missing information for management
- Augment data into information using engineering knowledge
- Manage new KPIs for the digital age (real time metrics)

- Effective communications and improve decision support
- Avoid hidden production losses and equipment failures
- Innovate with new business processes using the CLOUD
- Digital disruption promotes dynamic performance management

12.8.2 Managing Water Recovery for Maximizing Copper Production

Figure 12.26 shows a typical integrated mineral processing plant, including the water and tailings water recovery system, to enable to maximize copper production while increasing the water availability. These low-grade ores mineral processing plants require large investment costs, and they need additional energy and water resources, making the processing of the low grade's ores less sustainable. As such, many of the operating practices for grinding have been to push the rheological constraints of the mill. It is well-known that once the hydrocyclones start roping the grinding product, it exhibits a bimodal size distribution due to over-grinding. This over-grinding creates finer particles that are difficult to recover by conventional flotation cells and present difficult flocculation results. The typical P80 measures do not account for the tail of the size distribution. A detailed analysis conducted by the plant shows a bimodal particle size distribution, which causes flocculation problems and water losses in the tailing ponds.

In this case, the thickener output was getting 55% solids in the underflow. When the mill operated with the right size distribution, the detailed analysis of the water recovery showed an increase in the three thickeners to 60% solids in the underflow, augmenting the water recovery by a factor of 42%. This increase in the water recovery allows to increase the tonnage to maximize the copper production rate within the rheological process constraints in the grinding/classification circuits. In Chapter 11, we review the highlights of a detailed analysis, which enables to recover more copper with the increase of the water recovery from the thickeners.

A reasonable conclusion is that adding up the right amount of water to the sump for the right, the % solids in the feed to the cyclones is crucial as well as putting the right amount of water to the ball mills feed to get the right apparent viscosity of the mill.

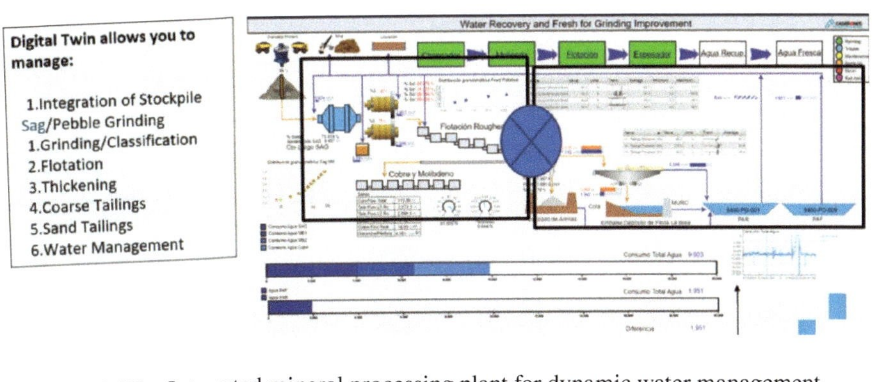

FIGURE 12.26 Integrated mineral processing plant for dynamic water management

Several earlier mine-to-mill detailed engineering projects never analyzed the tailings product and the effects of the grinding on the water recovery. The current digital twin for low grades takes into account an integrated analysis of the mine, mill, and tailings water recovery for the plant to improve the net metal production rate.

12.9 THICKENER MULTIVARIABLE PROCESS CONTROL

To date, most of the process control applications on a mineral processing site have predominantly focused on SAG mill, general grinding circuit and various aspects of flotation control. Bascur and Herbst (1986) implemented an online industrial version. This online model was useful in identifying one of the two thickening parameter that describe the thickening operation, the *flux density function* $f(x)$ (Concha 2014). Two instruments have been reported to perform this task (Phathom turbidity sensors [Phathom 2023] and SediRack-Unitube [Cettem n.d.]). This parameter represents all the perturbations of the suspension to the thickener's feed, such as mineralogy, density and particle size, and distribution for the solid and density, viscosity, surface tension, and pH for the liquid, by two variables: the settling velocity $v_s(\varphi)$ and the settling zone concentration φ, $f(x) = \varphi v_s(\varphi)$. The importance of this parameter is that, without knowing the perturbations occurring in the feed, the flowrate $F_F = \rho_s S\, f(\varphi)[\text{tph/m}^2]$, where S is the thickener surface area and ρ_s the solid density, permits the automatic control the flocculant addition. The second thickening parameter, the solid effective stress $\sigma_e(\varphi)$, representing the compressibility of the sediment, was represented by the solid inventory in the thickener whose prediction permitted controlling the thickener underflow concentration constrained by the capacity of the pipeline. An instrument to measure the sediment-suspension interfaces is the *SmartDiver* (Precision Light & Air Process Analyzers, n.d.) and another *SediRack-Unitube*, which measures the solid concentration profile and the solid inventory in the thickener.

The robust control strategy for a thickener must consider the three main objectives:

- Particle settling control
- Solid inventory knowledge
- Thickener rake protection

Once the inventory is determined correctly, the particle settling is controlled correctly and the thickener protection control is implemented, the thickener will operate in a more stable and reliable manner.

Weidenbach and Lombardi (2012), by following these considerations, presented as a result an increase in throughput and recovered water with a reduction in flocculant of approximately 20% to 30% shown in Figure 12.26. Additional information describing control strategies for mineral thickening are described by Bernal (2015). The current trend in improving the operation and control in thickeners is applying advanced expert control techniques. To implement these systems, measurements are critical. Thus, it is required to populate all points of the circuit with the necessary instrumentation to characterize input and output flows.

Also, to complement the control tasks is necessarily to use modern instrumentation that can monitor the internal layers of the thickener and even the in-line measurement of the yield stress or viscosity of the pulp.

Additional thickener controls are described by Schönbrunn et al. (2022). The authors use an expert system strategy and present a good overview of the process instrumentation and the rules to consider meeting the control objectives.

12.10 MINE-TO-MILL BIG DATA ANALYSIS

One of the most valuable tools today is the availability to provide context to real-time data for big data analysis using self-service tools such as Microsoft Power BI and machine learning tools.

Consumables such as power and water are monitored in real time, as well as the average, maximum, and minimum targets for each operational mode by shift, month, or any interval of time. Monitoring process events enables operators to aggregate the data by ore type, shift operator, and process modes to evaluate the copper grade, recovery energy, and water consumption based on the ore type (Bascur and Soudek 2019; Bascur 2019). In this example, the assets are organized by production line with the monitoring of consumables. Real-time statistics are used to calculate the minimum, maximum, and standard deviation of each consumable by asset, production line, and the mill. Users can modify the time range they are interested in from

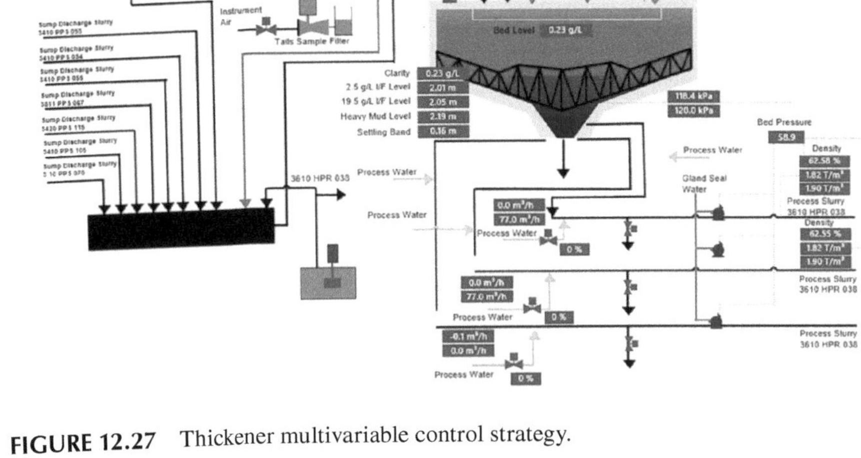

FIGURE 12.27 Thickener multivariable control strategy.

Source: Bascur, O.A. 2019. Chapter 2.6: Process control and operational intelligence. In *SME Mineral Processing and Extractive Metallurgy Handbook*. Edited by R.C. Dunne, S.K. Kawatra, and C.A. Young. Englewood, CO: SME.

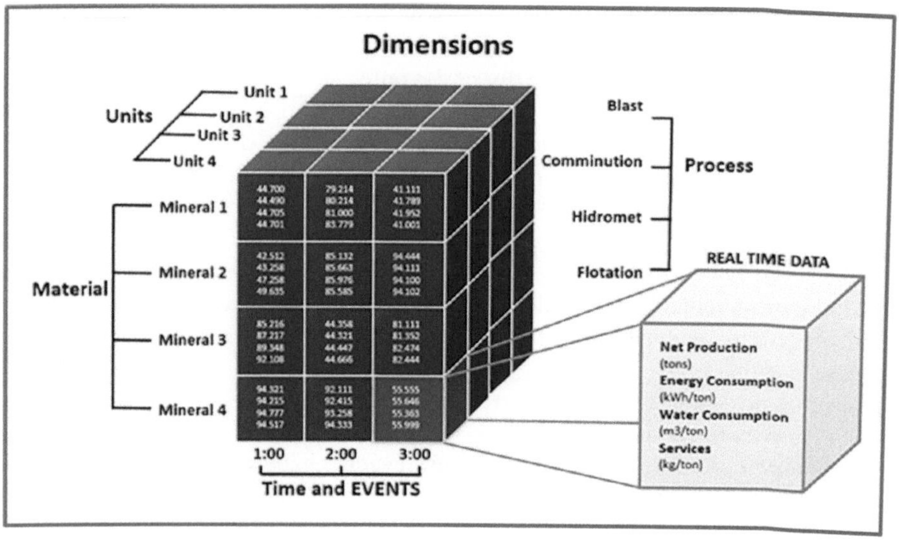

FIGURE 12.28 Multidimensional cube of real-time data with context and data services.

Source: Bascur, O.A. 2020. Digital Transformation in the Process Industries: A Road Map. Boca Raton, FL: CRC Press.

one hour to a shift or a day. The statistics are recalculated in real time based on this time range. This way, operations and the plant floor are using the same assets as the business systems, but these departments have different views of the same data for their own purposes and needs (Bascur 2019). Figure 12.28 shows the handling of the process data, process events, equipment status, process areas in consideration, material type (ore/blend), and process time. This figure shows the analysis of the results shown using Power BI in Figure 12.28.

Figure 12.29 shows a Microsoft Power BI dashboard, where the selected event is sliced into data segments. These filters can be based on many selected criteria, such as block number, process area, ore type, and shift team for all the variables selected, such as total production, reagent consumption, water consumption, and electricity consumption for the process lines in the mill. The variables are time interval, ore type, ore type block number, number of production lines (bin, mill, flotation), and production line availability (running, in trouble, down, stopped, in maintenance). Bascur et al. (2012), Kanchibotla (2014), and Bennett et al. (2014) discuss advances in the integration with blasting and mine-to-mill implementations.

12.11 DYNAMIC WATER MANAGEMENT IN MINERAL PROCESSING COMPLEXES

Today, the capabilities of integrating the mine, plant, water recovery, tailings plants, and ports is a reality (Plourde 2016, 2019; Plourde et al. 2017).

Extending the fishbone analysis and identifying the coordination of grinding, flotation, and thickening for optimal metal production is a reality (Bascur et al. 2022).

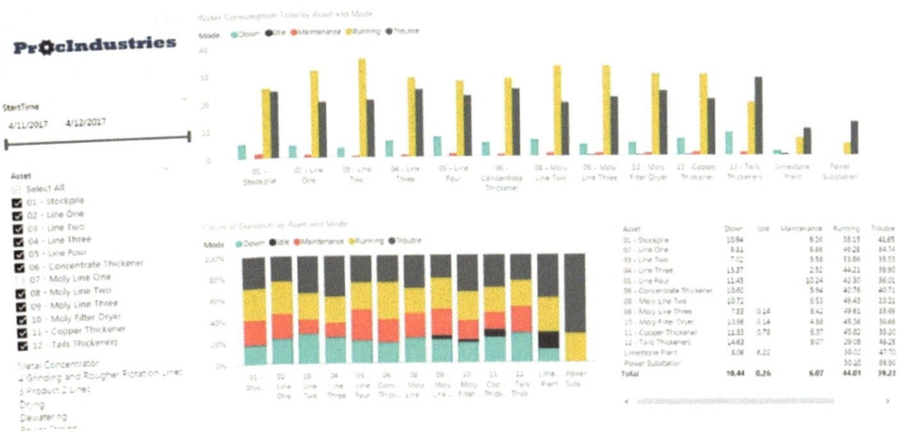

FIGURE 12.29 Example of a multidimensional analysis report to perform dynamic operation analysis using Microsoft Power BI,

Source: Bascur, O.A. 2019. Chapter 2.6: Process control and operational intelligence. In *SME Mineral Processing and Extractive Metallurgy Handbook*. Edited by R.C. Dunne, S.K. Kawatra, and C.A. Young. Englewood, CO: SME.

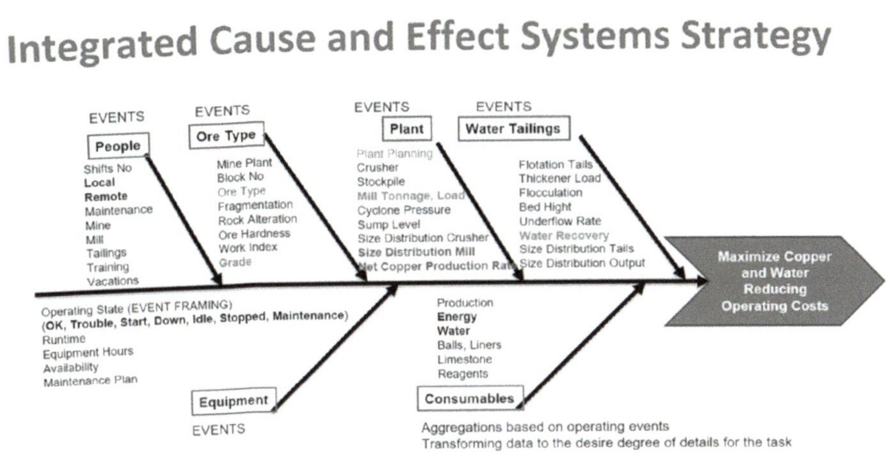

FIGURE 12.30 Integrated cause and effect fish bone for water recovery.

Figure 12.30 shows the key variables which can be used to maximize the metal production while increasing the required amount of water to optimize the product size distribution in the grinding circuits for both improved flocculation of suspension of the rougher tails. Details of this strategy were discussed in Chapter 11, devoted to the water in the mining industry.

12.12 DATA RECONCILIATION AND YIELD ACCOUNTING

There are several requirements for developing a methodology to implement a data reconciliation system. First, the algorithm able to balance and reconcile the plant data must be robust and must perform correctly against any process topology or configuration. This mathematical algorithm must have an information infrastructure, and the right infrastructure connecting the object-oriented model to the real-time process data must be in place so that a database allows storage and manipulation of elements. The system must adapt to changes in the process topology because a piece like an instrument going out of service is enough to change the mass balance of a process network. Additionally, this object-oriented database should be able to communicate with the real-time information system. Third, the real-time system acts as a repository of process data and inventory data and as a metallurgical laboratory of both the raw and reconciled data. The results are distributed to all staff, including the operators and the plant manager.

The typical problems with process data in industrial plants are as follows:

- An overwhelming amount of data
- Low confidence in the available data
- Lack of consistency; the data do not make sense
- The data violate known constraints (mass and energy balances)

A concentrator plant mass balance model includes the operational data, metal assays, metal inventories, and receipts of mineral from the mine, stockpiles, bins, thickeners, crushers, grinding, flotation circuits, and the overall flow and composition balance of the chemical components. Once the mass flow, stream, and inventory compositions are available from the system, many calculations and reports can be performed. In the process of solving the network problem, the measurements are validated, and gross errors are detected prior to providing a solution (Bascur and Kennedy 2002, 2004).

Poor data quality creates a decision-making "fog" at all levels of an organization and results in a financial penalty (fine). These problems have been addressed in the past using traditional methods to capture the required information. The main problem with these traditional methods is that human errors can occur during manual data entry, and operational events, which are typical in metallurgical complexes, may be ignored.

Figure 12.30 describes the daily procedure to reconcile process data. The unbalanced data are collected from the real-time information system. Once the data are in the system, a set of analysis rules is executed to detect gross errors that can negatively affect results. These gross errors are eliminated before the final data reconciliation is run. The results provide a unified balance for the whole complex. The assay errors are identified and reported.

The reconciled data can then be used to improve yield performance, balancing the optimal recovery and grades while minimizing metal losses in the tails. The data can be reused to improve process planning and to determine the optimal set points for steady-state optimization of the plant. Bascur and Linares (2005), Bascur and Soudek (2009a, 2009b), and Hodouin (2011) described additional uses for this type of methodology.

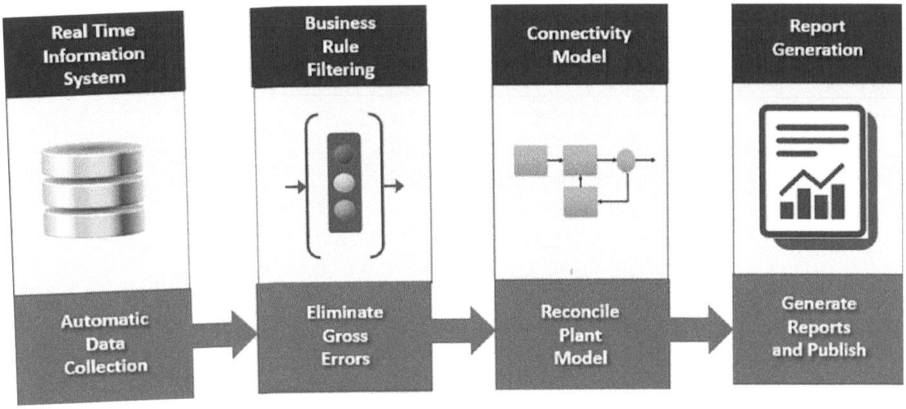

FIGURE 12.31 Daily procedure for data reconciliation.

Source: Bascur, O.A. 2019. Chapter 2.6: Process control and operational intelligence. In *SME Mineral Processing and Extractive Metallurgy Handbook*. Edited by R.C. Dunne, S.K. Kawatra, and C.A. Young. Englewood, CO: SME.

12.13 DISRUPTION IN OPERATIONAL DECISION SUPPORT

Today's technologies are enabling mineral and metallurgical processors to develop competency centers where they can manage their mine and mineral processing plants remotely to coordinate mining operations with downstream operations. The centers can be located near the mill or based remotely to enable a virtual, cloud-based operation. Integration of the operational data and events transforms the organization into a continuous improvement and innovation culture. An operational excellence program should be an ongoing process of improvement to achieve environmentally safe, profitable operations.

The availability of a central data infrastructure database allows operational information to be shared with external collaborators. As shown in Figure 12.32, Seeq Advanced Analytics can be used to share and expand the use of the operational data. Cloud-based strategies simplify access to real-time information through simple registration to Seeq running in AWS Cloud Services or running on Microsoft Azure. A plant asset framework data model is selectively shared using Seeq between the enterprise and an equipment vendor.

The data published by the mineral processing company pertain to a set of equipment used for material separation such as filters. The operational data of the filters are shared by the customer and the service provider in real time. This new capability provides a secure, auditable, and low-maintenance exchange of real-time processes and equipment. It drastically reduces the time spent accessing data and enhances the use of data for equipment vendors, which can make recommendations about the optimal use of energy. Many exciting new possibilities are currently being explored by service providers using the data at the original resolution. This allows for the monitoring of sophisticated equipment by experts, integration with utilities, automatic restocking, and integration with joint ventures (Bascur et al. 2016).

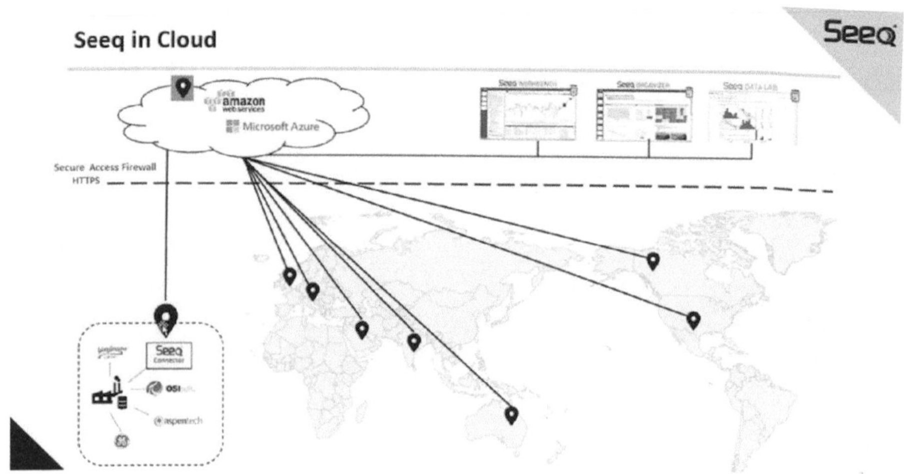

FIGURE 12.32 Seeq in cloud computer architecture example connecting an enterprise to its service partners.

Source: Seeq. 2020. *Seeq Workbench*. Seattle, WA: Seeq. www.seeq.com/product/workbench.

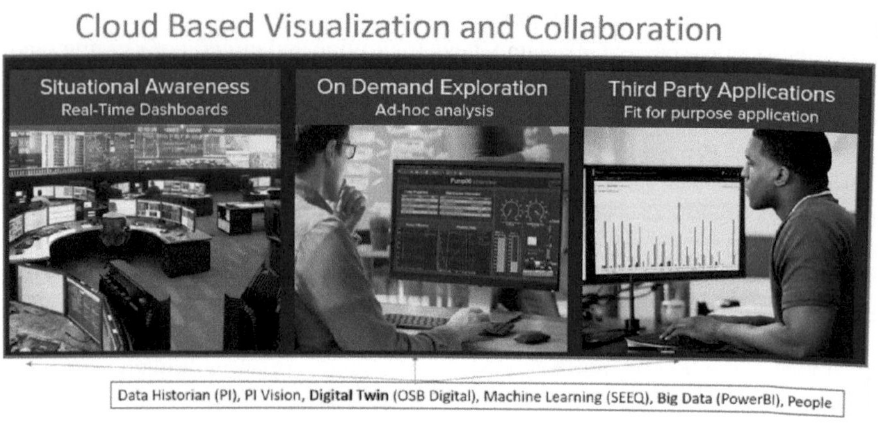

FIGURE 12.33 Cloud-based visualization, analysis, and collaboration.

Source: Bascur, O.A. 2020. Digital Transformation in the Process Industries: A Road Map. Boca Raton, FL: CRC Press.

Figure 12.33 shows some of the tools used to transform the operational real-time data into practical information to be used by people at the plant floor and remotely to collaborate for continuous improvement and innovation. A change in culture has been necessary, and by following basic guidelines, it has been a success. The key strategy has been to use an object oriented creating of templates to provide the right attributes for people consumption with the right operational context with the digital twin unit strategy as described earlier. As such, the process variables are easily coupled for any type of analysis by the latest data analysis tools. Now, the teamwork and use of the tools by operators, engineers, maintenance, and management is a reality. Awareness is improved by having the system provide real-time visualization, operational alerts, and event frame in real time. The ease of the real-time information, with right data in the Digital Twin Model, simplifies the use of the historical facts by any person anywhere. The historical events and data aggregation at the right degree of the detail is available in the CLOUD to be analyzed using tools such as Microsoft, Power BI for overall production effectiveness for improved decision support at the control room, plant, and remotely.

12.14 PLANT OPERATIONAL INTELLIGENCE STRATEGY CONCLUSIONS

Despite the advances in automatic data collection and archiving, business decision-makers face the problem of exploiting information that is relevant for plant operations and the sustainability of the business enterprise. An operational plant intelligence, using machine learning and business intelligence strategy tools, add operative context to the data, transforming them into operational insights for further analysis. This novel strategy improves collaboration and provides real-time feedback on decisions made at all levels of the organization. This strategy allows the operating and support team to proactively detect shifts in process performance. It can provide structure and a means of determining the root cause of the shifts and, if needed, propose corrective action. The competence team is in charge of identifying process improvement opportunities and providing sufficient information and root-cause diagnostics to identify areas of improvement and identify new performance indicators. For example, the efficient use of water, energy, and resources is critical and requires the infrastructure in place to be able to conduct small, focused projects, collaborate among different teams in the organization, and understand that this is a continuous improvement process.

REFERENCES

Bascur, O.A. 1985. *A Device for Mill Load and Slurry Transport Measurements* [internal report]. Tucson, AZ: Duval.

Bascur, O.A. 1988. A control data framework with distributed intelligence. *Advances in Instrumentation (ISA)* 88:1553–1169.

Bascur, O.A. 1991a. A unified solid liquid separation framework. *Fluid Particle Separation Journal* 4:117–122.

Bascur, O.A. 1991b. Integrated grinding/flotation controls and management. In *Proceedings of the Copper 91—Cobre 91 International Symposium—Volume II*, August 18–21. Edited by G.S. Dobby, S.A. Argyropoulos, and S.R. Rao. Ottawa, Canada: Pergamon Press.

Bascur, O.A. 2005. Example of a dynamic flotation framework. Presented at the Centenary of Flotation Symposium, Brisbane.

Bascur, O.A. 2013. A dynamic flotation framework for performance management. In *Separation Technologies for Minerals, Coal and Earth Resources*. Edited by C. Young and G. Luttrel. Englewood, CO: SME. pp. 40–45.

Bascur, O.A. 2019. Chapter 2.6: Process control and operational intelligence. In *SME Mineral Processing and Extractive Metallurgy Handbook*. Edited by R.C. Dunne, S.K. Kawatra, and C.A. Young. Englewood, CO: SME.

Bascur, O.A. 2020a. Digital Transformation in Mining Plants. In *Automining 2020 Proceedings*, Gecamin, Santiago.

Bascur, O.A. 2020b. *Digital Transformation in the Process Industries: A Road Map*. Boca Raton, FL: CRC Press.

Bascur, O.A. 2020c. *Transformacion Digital en la Industria Minera: Conceptos y Mejores Prácticas, Automining 2020*. Santiago, Chile: Gecamin.

Bascur, O.A. 2021a. Digital transformation in the mining industry [online course]. www.osb-digit.com.

Bascur, O.A. 2021b. Digital transformation in mining and metallurgical complexes. Technical Paper and Plenary Lecture. In *Proceedings of the 60th Annual Conference of Metallurgists (COM 2021)*. Montreal: Canadian Institute of Mining, Metallurgy and Petroleum.

Bascur, O.A. 2021c. Digital transformation in mining and metallurgical complexes. In *SME Annual Meeting Proceedings*, Englewood, CO.

Bascur, O.A., and Barrientos, A. 1977. Phenomenological model of sedimentation: Experimental parameters. Presented at the IV Chilean Chemical Engineering Congress, University de Concepcion.

Bascur, O.A., Freeh, E.J., and Herbst, J.A. 1986. Dynamic simulation and estimation of industrial grinding circuit control characteristics. *ISA Transactions* 15(2):77–84.

Bascur, O.A., and Gorrie, T. 2021. Maximizing copper recovery using digital twins plant. In *Proceedings of Procemin Geomet 2021: 17th International Conference on Mineral Processing and Geometallurgy*, Gecamin, Santiago.

Bascur, O.A., Halhead, M., Garrigues, L., and Jarvis, M. 2016. Mineral processing plant asset and energy optimization: The calming cloud over operations. In *Proceedings of the XXVIII International Mineral Processing Congress*. Westmount, QC: Canadian Institute of Mining, Metallurgy and Petroleum.

Bascur, O.A., and Herbst, J.A. 1982. Dynamic of a flotation cell for automatic control. Presented at IMPC XIV, Toronto.

Bascur, O.A., and Herbst, J.A. 1986. Improved thickener performance through the use of an extended Kalman filter. In *Design and Installation of Concentration and Dewatering Circuits*. Edited by A.L. Mular and M.A. Anderson. Littleton, CO: SME. pp. 835–845.

Bascur, O.A., Hertler, C., and Benavides, N. 2012. Drastic specific energy and water reduction in mine to mill operations. In *Proceedings of the 3rd International Congress in the Mining Industry: Automining*. Santiago: Gecamin.

Bascur, O.A., Hertler, C., and Kolz, R. 2011. Improving energy and water specific consumptions strategies: Remote supervision and diagnostics. In *Proceedings of the 8th International Mineral Processing Seminar (Procemin)*. Santiago: Gecamin.

Bascur, O.A., and Junge, P.A. 2011. Dynamill, Dynaflote and Optimill [iPad software apps]. www.apple.com/app-store/.

Bascur, O.A., and Kennedy, J.P. 2002. Reducing maintenance costs using process and equipment event management. In *Mineral Processing Plant Design, Practice, and Control: Proceedings*. Edited by A.L. Mular, D.N. Halbe, and D.J. Barratt. Littleton, CO: SME. pp. 507–527.

Bascur, O.A., and Kennedy, J.P. 2004. Are you really using your information to increase the effectiveness of assets and people? In *Plant Operators' Forum*. Edited by E.C. Dowling and J.I. Marsden. Littleton, CO: SME. pp. 47–62.

Bascur, O.A., and Linares, R. 2005. Grade recovery optimization using data unification and real time gross error detection. *Minerals Engineering* 19(6–8):696–702.

Bascur, O.A., Retamal, F., and Espinosa, S. 2022. Treating low grades ores using a digital in mining and mineral processing plants. In *Proceedings of the Copper 2022 International Conference: IIMChile. Vol. 7. Process Optimization through Innovation, Technology, Automation and Design*, Santiago.

Bascur, O.A., and Soudek, A. 2009a. Implementation strategies for energy effectiveness and sustainability—Example of Anglo-American platinum. In *Proceedings of the 12th AusIMM Mill Operators' Conference*, Australasian Institute of Mining and Metallurgy, Carlton.

Bascur, O.A., and Soudek, A. 2009b. *Real-Time Integration of Mining and Metallurgical Information for Efficient Use of Energy and Water*. SME Preprint No. 09-103. Littleton, CO: SME.

Bascur, O.A., and Soudek, A. 2019. Grinding and flotation optimization using operational intelligence. *Mining, Metallurgy and Exploration* 36:139–149. https://doi.org/10.1007/s42461-018-0036-4.

Bennett, D., Tordoir, A., Walker, P., La Rosa, Valery, W., and Duffy, K. 2014. Throughput forecasting and optimization at the Phu Kham copper-gold operations. In *Proceedings of the 12th AusIMM Mill Operator's Conference*, Australasian Institute of Mining and Metallurgy, Carlton.

Bernal, F., Ennis, S., and Blanco, E. 2020. Optimization of the rougher flotation cell mechanical energy at Morenci Mill. In *Proceedings of the Annual Meeting*. Englewood, CO: SME.

Bernal, L. 2015. Control strategies for minerals thickening (part 2). https://plapl.com.au/control-strategies-for-minerals-thickening-part-2/

Botin, J.A. 2009. Integrating sustainability into the organization. In *Sustainable Management of Mining Operations*. Edited by J.A. Botin. Littleton, CO: SME. pp. 71–132.

Box, G., Jenkins, G.M., Reinsel, G.C., and Ljung, G.M. 2016. *Time Series Analysis: Forecasting and Control*, 5th ed. New York: Wiley.

Brunton, S.L., and Kutz, J.N. 2019. *Data-Driven Science and Engineering, Machine Learning, Dynamical Systems, and Control*, 2nd ed. Cambridge: Cambridge University Press.

Cettem. n.d. SediRack. https://en.cettem.com/sedirack.

Chawda, N. 2019. IoT opportunities for Industry 4.0. at Vedanta HZL: Empower your analytics with operational data. Presented at OSIsoft Mumbai Seminar, Mumbai.

Concha, F., Castro, O., and Munoz, L. 2000. Shear viscosity of a copper ore pulp in a grinding—classification circuit. *Mineral Processing and Extractive Metallurgy Review* 20(1):155–165.

Concha, F.A. 2014. *Solid–Liquid Separation in the Mining Industry*. Cham: Springer.

Dunne, R.C., Kawatra, S.K., and Young, C.A., eds. 2019. *SME Mineral Processing and Extractive Metallurgy Handbook*. Englewood, CO: SME.

Fuerstenau, D.W. 1999. The froth flotation century. In *Advances in Flotation Technology*. Edited by B.K. Parek and J.D. Miller. Littleton, CO: SME. pp. 3–21.

Gutiérrez, L. 2021. *Rheology Course, University of Concepcion, Procemin*. Santiago: Gecamin.

He, M., and Forssberg, K.S.E. 2006. Influence of slurry rheology on stirred media milling of limestone. In *Advances in Comminution*. Edited by S.K. Kawatra. Littleton, CO: SME. pp. 243–260.

Hemming, C.E., and Boyes, J.M. 1977. An on-line viscometer technique for improved operations and controls of wet grinding circuits. Presented at IMPC XXII, Sao Pablo.

Hessing, T. 2022. Evolutionary Operations Methodology (EVOP). https://sixsigmastudyguide.com/evolutionary-operations-methodology-evop/.

Hodouin, D. 2011. Methods for automatic control, observation, and optimization in mineral processing plants. *Journal of Process Control* 21(2):211–225.

Kanchibotla, S. 2014. Mine mill value chain optimization: Role of blasting. In *Mineral Processing and Extractive Metallurgy: 100 Years of Innovation*. Edited by C.G. Anderson, R.C. Dunne, and J.L. Uhrie. Englewood, CO: SME.

Kelleher, J.D., Namee, B.M., and D'Arcy, A. 2015. *Fundamentals of Machine Learning for Predictive Analytics*. Cambridge, MA: MIT Press.

Lynch, A.J., Johnson, N.W., Manlapig, E.V., and Thorne, C.G. 1981. Mineral and coal flotation circuits: Their simulation and control. In *Developments in Mineral Processing*, vol. 3. New York: Elsevier Scientific.

Mika, T.S., and Fuerstenau, M. 1968. A microscopic model of the flotation process. In *Proceedings of the Eight International Mineral Processing Congress*, vol. 2. Leningrad: Mechanobr. pp. 246–269.

Phathom. 2023. Thickener control using turbidity measurement. www.phathom.co.nz

Plourde, M. 2016. Heavy haul equipment logistic strategies in iron ore mining. *OSIsoft.com*. http://www.osisoft.com/Presentations/Heavy-Haul-Equipment-Logistic-Strategies-in-Iron-Ore-Mining.

Plourde, M. 2019. PI system at the heart of a mining integrated operations center. https://www.osisoft.com/Presentations/The-PI-system-at-the-heart-of-a-Mining-Integrated-Operations-Center--ArcelorMittalx/

Plourde, M., Bascur, O.A., Paquet, S., and Gervais, D. 2017. Digital innovation in modern engineering and operational excellence. Presented at the 2017 SME Annual Conference and Expo, Denver, February 19–22.

Precision Light & Air Process Analysers. n.d. SmartDiver—bed level measurement. https://plapl.com.au/mud-diver/.

Raschka, S. 2015. *Python Machine Learning*. Birmingham: Packt.

Schönbrunn, F., Hales, L., and Bedell, D. 2022. Strategies for instrumentation and control of thickeners and other solid-liquid separations circuits. In *Mineral Processing Plant Design, Practice, and Control: Proceedings*. Edited by A.L. Mular, D.N. Halbe, and D.J. Barratt. Littleton, CO: SME. pp. 2164–2172.

Schwab, K. 2016. The forth industrial revolution. *World Economic Forum*. www.weforum.org

Seeq. 2020. *Seeq Workbench*. Seattle, WA: Seeq. www.seeq.com/product/workbench.

Shaw, K., and Frülinger, J. 2019.What is a digital twin and why it's important to IoT. *Network World*, January 31.

Shi, F.N., Napier-Munn, T.J., and Asomah, I.K. 2000. Rheological effect in grinding and classification. *Mineral Processing and Extractive Metallurgy Review* 20(1):123–131.

Shuen, D. 2016. *Real Time Data Critical to Mining, Improving Metal Recoveries*. San Francisco, CA: PI World. www.osisoft.com/presentations/real-time-data-critical-to-mining/.

Steyn, J., Bascur, O.A., and Gorain, B. 2018. Metallurgy analytics: Transforming plant data into actionable insights. *Mining Engineering* 70(9):18–29.

Strack, G. 2017. How data was used to improve gold recovery, Aurelia mining. Presented at the PI Users Conference, Mining, Metals and Material Session, San Francisco, CA.

Trahar, W.J. 1981. A rational interpretation of the role of particle size in flotation. *International Journal of Mineral Processing* 8:289–327.

Weidenbach, M., and Lombardi, J. 2012. Tailing thickener operability and control improvements at Prominent Hill. Presented at the 11th AusIMM Operators' Conference, Hobart.

FURTHER READING

Bascur, O.A. 1976. Modelo Fenomenológico de Suspensiones en Sedimentación, Tesis de Habilitación Profesional para recibir los títulos de Ingeniero Civil Químico e Ingeniero Civil Metalúrgico, Universidad de Concepción.

Bascur, O.A. 1990a. Expert process operator advisor. In *Control 90: Mineral and Metallurgical Processing*. Edited by R. Rajamani and J.A. Herbst. Littleton, CO: SME. pp. 67–77.

Bascur, O.A. 1990b. Profit-based grinding controls. *Minerals and Metallurgical Processing* (February): 9–15.

Bascur, O.A. 1991. Integrated grinding/flotation controls and management. Presented at the Copper 91/Cobre 91 International Symposium, Ottawa, ON.

Bascur, O.A. 2011. A flotation model framework for dynamic performance management. In *Proceedings of Procemin 2011*. Edited by R. Kuyvenhoven. Santiago: Gecamin.

Bascur, O.A., and Barrientos, A. 1977. Phenomenological model of sedimentation: Experimental parameters. Presented at the IV Chilean Chemical Engineering Congress, University de Concepcion.

Bascur, O.A., and Benavides, N. 2013. Improving flotation grade recovery using operational data at Southern Peru Copper Cuajone. Presented at Flotation 2013: Mineral Engineering Conference, Cape Town.

Bascur, O.A., Freeh, E.J., and Herbst, J.A. 1986. Dynamic simulation and estimation of industrial grinding circuit control characteristics. *ISA Transactions* 15(2):77–84.

Bascur, O.A., and Gorrie, T. 2021. Maximizing copper production using a Digital Twin with predictive analytics. Procemin GEOMET 2021, Gecamin, Santiago.

Bascur, O.A., and Halhead, M. 2013. Energy effectiveness and sustainability management at Anglo American Platinum. In *Proceedings of Copper/Cobre 2013*, vol. 1. Edited by D. Hernandez. Santiago, Chile: IMMCh.

Bascur, O.A., and Herbst, J.A. 1985a. Dynamic simulators for training personnel in the control of grinding/flotation systems. Presented at IFAC MMM: Automation for Mineral Resource Development, Brisbane.

Bascur, O.A., and Herbst, J.A. 1985b. On the development of a model-based control strategy for copper-ore flotation. In *Flotation of Sulphide Minerals. Developments in Mineral Processing, 6*. Edited by K.S.E. Forssberg. London: Elsevier. pp. 409–431.

Bascur, O.A., and Kennedy, J.P. 1995. Measuring, managing and maximizing performance in industrial plants. In *XIX IMPC Proceedings*. Littleton, CO: SME.

Bascur, O.A., and Kennedy, J.P. 1996. Measuring, managing and maximizing refinery performance. *Hydrocarbon Processing* 75(1):111–116.

Bascur, O.A., and O'Rourke, J. 2020. Measuring, managing and transforming data into operational insights. In *Smart Manufacturing: Concepts and Methods*. Edited by M. Soroush, M. Baldea, and T. Edgar. London: Elsevier.

Bascur, O.A., and Retamal, F. 2021. Integrated rock processing strategies in the digital age. Presented at Minería Digital, Gecamin, Santiago, October 17–19.

Castro, S., and Laskowski, J. 2018. Sea water flotation practices. Presented at XXIX IMPC, Moscow.

Concha, F., and Bascur, O.A. 1977. A phenomenological model of sedimentation. *Proceedings of IMPC XII, Special Publication* 1:29–46.

Cooper, A. 2008. Considering the use of in-pit crushing and conveying (IPCC) as an alternative to truck. International Report. Perth: Snowden.

Ericson, G., Martens, J., and Rohm, W.A. 2016. What is azure machine learning studio? https://docs.microsoft.com/en-us/azure/machine-learning/studio/what-is-ml-studio.

Fuerstenau, D.W. 2001. Processing challenges facing the mineral industry. In *Proceedings of VI Southern Hemisphere Symposium on Mineral Technology*, Rio de Janeiro, May 2001.

Fuerstenau, D.W. 2005. *The Froth Flotation Century*. Littleton, CO: SME. p. 19.

Goldratt, E.M. 2014. *The Goal: A Process of Ongoing Improvement*, 30th anniversary ed. Great Barrington, MA: North River Press.

Hemming, C.E., and Boyes, J.M. 1977. An on-line viscometer technique for improved operations and controls of wet grinding circuits. In *IMPC XXII*, Sao Pablo.

Herbst, J.A., and Bascur, O.A. 1984. Mineral processing control: Realities and dreams. In *Control '84: Mineral Metallurgical Processing*. Edited by J.A. Herbst, D.B. George, and K.V.S. Sastry. Littleton, CO: SME. pp. 197–215.

Herbst, J.A., and Pate, W. 1999. Developments in flotation control: One part evolution, two parts revolution. In *Advances in Flotation Technology*. Edited by B.K. Parekh and J.D. Miller. Littleton, CO: SME. pp. 399–412.

Herbst, J.A., and Pate, W.T. 1998. Dynamic simulation of size reduction operations from mine to mill. *Australasian Institute of Mining and Metallurgy Publication Series* 43(4):243–248.

Karageorgos, J., Davies, S., Broers, E., and Goh, J. 2009. Current trends in countercurrent decantation and thickener circuit operability and control. Presented at the 10th AusIMM Mill Operators' Conference, Adelaide.

Knoster, T., Villa, R., and Thousand, L. 2000. A framework for thinking about systems change. In *Restructuring for Caring Effective Education*. Edited by R. Villa and J. Thousand. Baltimore, MD: Paul H. Brookes. pp. 93–128.

Kynch, G.J. 1952. A theory of sedimentation. *Transactions of the Faraday Society* 48:166–182.

Lynch, A.J. 1977. Mineral crushing and grinding circuits, their simulation, optimization, design and control. In *Developments on Mineral Processing*, vol. 1. New York: Elsevier Scientific.

Mular, A. 2003. Size separation. In *Principles of Mineral Processing*. Edited by M.C. Fuerstenau and K.H. Han. Littleton, CO: SME. pp. 119–172.

Mular, A., and Jull, N.A. 1980. The selection of cyclone classifiers, pumps and pump boxes for grinding circuits. In *Mineral Process Plant Design*, 2nd ed. New York: AIME. pp. 376–403.

Napier-Munn, T.J., Morrell, S., Morrison, R.D., and Kojovic, T. 1996. *Mineral Comminution Circuits: Their Operation and Optimization*. Queensland, Australia: Julius Kruttschnitt Mineral Research Centre, University of Queensland.

Paredes, R., and Riquelme, C. 2013. Endesa CMD: Aplicaciones Centro Monitoreo y Diagnostico. PI System Seminar, Santiago.

PowerBI. 2020. Microsoft, Bellevue, WA. www.microsoft.com

Sbarbaro, D., and del Villar, R. 2010. *Advanced Control and Supervision of Mineral Processing Plants*. London: Springer-Verlag.

Seeq. 2020. *Seeq Workbench*. Seattle, WA: Organizer and Data Lab. www.seeq.com

Turton, R., Shaeiwitz, J.A., Bhattacharyya, D., and Whiting, W.B. 2018. *Analysis, Synthesis, and Design of Chemical Processes*, 5th ed. Boston, MA: Prentice Hall.

Wallis, G.B. 1969. *One-Dimensional Two-Phase Flow*. New York: McGraw-Hill.

Weichert, R., and Herbst, J.A. 1986. An ultrafast load cell device for measuring particle breakage. In *Proceedings, 1st World Congress on Particle Technology, 2*, Elsevier, Nuremberg, Germany. pp. 3–14.

Wills, B.A., and Finch, J.A. 2016. *Wills' Mineral Processing Technology: An Introduction to the Practical Aspects of Ore Treatment and Mineral Recovery*, 8th ed. London: Butterworth-Heinemann.

Index

Note: Page numbers in *italic* indicate a figure and page numbers in **bold** indicate a table on the corresponding page.

E

F

G

H

U

V

W

Y

Z